28th Winter Workshop on Nuclear Dynamics 2012

(WWND-28)

Journal of Physics: Conference Series Volume 389

Dorado Del Mar, Puerto Rico
7-14 April 2012

ISBN: 978-1-62276-852-3
ISSN: 1742-6588

Printed from e-media with permission by:

Curran Associates, Inc.
57 Morehouse Lane
Red Hook, NY 12571

Some format issues inherent in the e-media version may also appear in this print version.

Printed by Curran Associates, Inc. (2013)

For permission requests, please contact the Institute of Physics
at the address below.

Institute of Physics
Dirac House, Temple Back
Bristol BS1 6BE UK

Phone: 44 1 17 929 7481
Fax: 44 1 17 920 0979

techtracking@iop.org

Additional copies of this publication are available from:

Curran Associates, Inc.
57 Morehouse Lane
Red Hook, NY 12571 USA
Phone: 845-758-0400
Fax: 845-758-2634
Email: curran@proceedings.com
Web: www.proceedings.com

TABLE OF CONTENTS

28th Winter Workshop on Nuclear Dynamics (WWND-28)

Rene Bellwied and Claude A Pruneau

PREFACE

These are the proceedings of the 28th Winter Workshop on Nuclear Dynamics, which was held in Dorado Del Mar, Puerto Rico. As in previous years the unique character of this conference series has allowed us to bring together nuclear scientists from very different fields to discuss recent progress and scientific achievements.

At the high-energy frontier very exciting results from heavy-ion collisions at the LHC were shown. At the lower energies the RHIC beam energy scan is underway to span the gap between the SPS and RHIC and search for the QCD phase transition and critical phenomena in the nuclear matter phase diagram. The nuclear structure and astrophysics communities have started to construct dedicated facilities at Michigan State University (F-RIB) and the GSI in Germany (FAIR).

Theory progress is made across all these different energy regimes, and the new data, in particular from the LHC and RHIC, are motivating more detailed modeling and a deeper understanding of the underlying physics.

These proceedings of the 28th Winter Workshop on Nuclear Dynamics again provide a snapshot of the status of the field. The articles, many of which are written by some of the most promising young scientists in the field, are documenting the excitement and achievements that are characteristic for all subfields of nuclear science.

Rene Bellwied (University of Houston)
Claude Pruneau (Wayne State University)

Peer review statement

All papers published in this volume of *Journal of Physics: Conference Series* have been peer reviewed through processes administered by the proceedings Editors. Reviews were conducted by expert referees to the professional and scientific standards expected of a proceedings journal published by IOP Publishing.

Standard Model Higgs boson searches in pp collisions at $\sqrt{s} = 7$ TeV with the ATLAS detector

Réda Tafirout on behalf of the ATLAS Collaboration
TRIUMF, 4004 Wesbrook Mall, Vancouver BC V6T 2A3, Canada

Abstract. We report on searches for the Standard Model Higgs boson with the ATLAS detector at the LHC. Several production mechanisms and decay channels are considered and a wide range of hypothetical Higgs boson masses explored. The datasets correspond to integrated luminosities ranging from 4.6 fb^{-1} to 4.9 fb^{-1} of proton-proton collisions collected at $\sqrt{s} = 7$ TeV in 2011. At the 95% confidence level the excluded Higgs boson mass ranges are 111.4 GeV to 116.6 GeV, 119.4 GeV to 122.1 GeV, and 129.2 GeV to 541 GeV, while the mass range from 120 GeV to 560 GeV is expected to be excluded in the absence of a signal. An excess of events is observed at a Higgs boson mass around 126 GeV with a local significance of 2.9σ. The global probability for the background to produce an excess at least as significant anywhere in the mass range explored corresponds to a significance of approximately 1σ.

1. Introduction

The Lagrangian density of electroweak interactions in the Standard Model (SM) [1]-[3] is based on $SU(2)_L \times U(1)_Y$ symmetry. To preserve gauge invariance and renormalization, fermions and gauge bosons need to have zero mass. The Higgs mechanism[4]-[9] consists of introducing a complex doublet of scalar fields which breaks spontaneously the symmetry. Mass is then generated for the weak bosons, leaving the photon massless. Fermions acquire their mass via Yukawa interactions with the Higgs field. A massive scalar particle is then predicted, namely the Higgs boson (H). The mass of the Higgs is a free parameter in the SM, which means that Higgs mass hypotheses over a wide range must be considered. However, for a given mass, production cross sections and decay modes branching fractions are well predicted.

Extensive searches have been conducted at e^+e^- and $p\bar{p}$ colliders for many years. From the LEP experiments a Higgs with a mass below 114.4 GeV is excluded at 95% confidence level (CL) [10], while from the Tevatron experiments the mass range excluded is between 147 GeV and 179 GeV [11].

One of the primary goals of the Large Hadron Collider (LHC) is to search for the SM Higgs boson and probe other mechanisms of electroweak symmetry breaking (EWSB). The expected SM Higgs properties and observables at the LHC are compiled in Ref. [12]. In the following sections we are presenting searches for the Standard Model Higgs boson made with the ATLAS detector [13] on the full pp collisions datasets collected at $\sqrt{s} = 7$ TeV in 2011.

2. Search channels and combination

The Higgs boson can decay into several channels. The ones considered are $H \to \gamma\gamma$, $H \to ZZ^{(*)}$, $H \to WW^{(*)}$, $H \to \tau^+\tau^-$ and $H \to b\bar{b}$. The $ZZ^{(*)}$ and $WW^{(*)}$ channels can be used to explore the entire mass range while the other channels are more sensitive to the lower mass region due to their higher branching fraction. Moreover, a different analysis strategy is adopted depending on the decay modes of the W, Z, and τ. The $H \rightarrow \gamma\gamma$ and $H \rightarrow ZZ^{(*)}$ provide the highest mass resolution. Table 1 summarizes the individual channels and the Higgs mass range considered for each. In order to enhance the signal sensitivity, each channel is treated and optimized separately. Some channels are also broken down into sub-categories as they have different signal and background compositions and are subject to different systematic uncertainties. The observables used in the combination are either the reconstructed invariant mass or the transverse mass depending on the channel considered.

For each Higgs mass hypothesis, the signal strength, μ, is defined as the ratio of a given Higgs boson production cross section (σ) to its SM value (σ_{SM})), $\mu = \sigma/\sigma_{SM}$. A value of $\mu = 0$ corresponds to a background-only model and $\mu = 1$ to a SM Higgs. The combination procedure described in details in [16] is based on the profile likelihood ratio test statistic $\lambda(\mu)$ [15], which extracts the information on the signal strength from the full likelihood including all the parameters describing the systematic uncertainties and their correlations. Exclusion limits are based on the modified frequentist approach (CLs prescription) [14]; a value of μ is regarded as excluded at the 95% (99%) CL when CLs takes on the corresponding value. The analysis corresponding to each channel is summarized below:

Table 1. Summary of the individual Higgs channels considered. The integrated luminosity and targeted mass range for each decay channel is also listed. ℓ is either an electron or a muon.

Higgs Channel	m_H Range [GeV]	L[fb^{-1}]	Ref.
$H \rightarrow \gamma\gamma$	110-150	4.9	[17]
$H \rightarrow ZZ^{(*)} \rightarrow \ell^+\ell^-\ell^+\ell^-$	110-600	4.8	[18]
$H \rightarrow ZZ^{(*)} \rightarrow \ell^+\ell^-\nu\bar{\nu}$	200-280-600	4.7	[20]
$H \rightarrow ZZ^{(*)} \rightarrow \ell^+\ell^-q\bar{q}$	200-300-600	4.7	[21]
$H \rightarrow WW^{(*)} \rightarrow \ell^+\nu\ell^-\bar{\nu}$	100-200-300-600	4.7	[22]
$H \rightarrow WW^{(*)} \rightarrow \ell\nu q\bar{q}'$	300-600	4.7	[23]
$H \rightarrow \tau^+\tau^- \rightarrow \tau_{lep}\tau_{lep}$	110-150	4.7	[24]
$H \rightarrow \tau^+\tau^- \rightarrow \tau_{lep}\tau_{had}$	110-150	4.7	[24]
$H \rightarrow \tau^+\tau^- \rightarrow \tau_{had}\tau_{had}$	110-150	4.7	[24]
$WH \rightarrow \ell\nu b\bar{b}$	110-130	4.7	[25]
$ZH \rightarrow \ell^+\ell^-b\bar{b}$	110-130	4.6	[25]
$ZH \rightarrow \nu\bar{\nu}b\bar{b}$	110-130	4.6	[25]

- $H \rightarrow \gamma\gamma$: the search consists in selecting two isolated photons with high transverse energy. The background contributions come for the irreducible diphoton continuum, and the reducible photon+jet and jet+jet, where a jet passes the photon selection. The analysis is conducted using nine independent event classifications of varying signal and background compositions and mass resolution. The classification is based on the pseudo-rapidity of each photon, whether reconstructed as a converted or unconverted photon, and the kinematics of the diphoton system. The diphoton invariant mass distribution is fitted using a Crystal Ball and Gaussian functions to estimate the background and is used as a discriminating variable to distinguish signal and background. The mass resolution is approximately 1.7% for a Higgs mass of 120 GeV. In Figure 1 (left) the invariant mass distribution of the diphoton system is shown for the selected candidate events, the fitted background shape and the

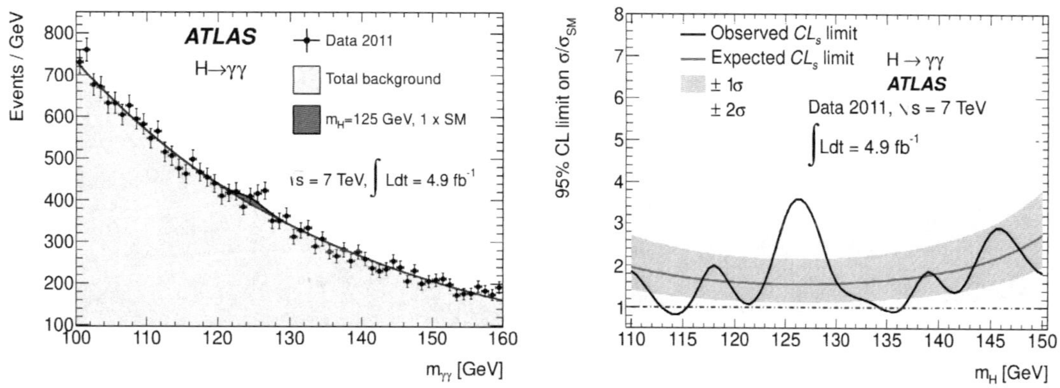

Figure 1. Left: the invariant mass distributions for the selected candidate events, the fitted background shape and the expected Higgs signal for $m_H = 125$ GeV in the $H \to \gamma\gamma$ channel [16]. Right: the observed (full line) and expected (dashed line) 95% CL limits on the SM Higgs boson production normalized to the predicted cross section as a function of m_H is shown [17]. The dotted curves show the median expected limit in the absence of a signal and the green and yellow bands indicate the corresponding $\pm 1\sigma$ and $\pm 2\sigma$ intervals.

expected Higgs signal for $m_H = 125$ GeV. The observed and expected 95% CL limits on the SM Higgs boson production normalized to the predicted cross section as a function of m_H is shown on Figure 1 (right). The largest excess is observed at 126.5 GeV with a local significance of 2.8σ. The excluded mass ranges at 95% CL are from 113 GeV to 115 GeV and from 134.5 GeV to 136 GeV.

- $H \to ZZ^{(*)} \to \ell^+\ell^-\ell^+\ell^-$: this channel provides a clean signature with a very small background. The analysis consists in selecting two pairs of oppositely-charged same-flavor leptons with high transverse energy for two of the leading leptons. The invariant mass of one lepton-pair is required to be consistent with the Z mass, while for the second pair the constraint applied is m_H dependent. The main irreducible $ZZ^{(*)}$ background is estimated using Monte Carlo simulation. The reducible Z+jets background, which has an impact mostly for low four-lepton invariant masses, is estimated from control regions in the data. The top quark ($t\bar{t}$) background normalization is validated using a dedicated control sample. The events are categorized according to the lepton flavour combinations. The mass resolutions are approximately 1.5% in the four-muon channel and 2% in the four-electron channel for m_H 120 GeV. The invariant mass of the four-lepton system is used as a discriminating variable and the distributions are shown in Figure 2 (left) for the lower mass region. The observed and expected 95% CL limits on the SM Higgs boson production normalized to the predicted cross section as a function of m_H is shown on Figure 2 (right). In this channel, the largest excess is observed at 125 GeV, 244 GeV and 500 GeV, with a local significance of 2.1σ, 2.2σ and 2.1σ repsectively. This excess is comparable to the $H \to \gamma\gamma$ channel. The range of Higgs masses excluded at 95% CL are from 134 GeV to 156 GeV, 182 GeV to 233 GeV, 256 GeV to 265 GeV and 268 GeV to 415 GeV.
- $H \to ZZ^{(*)} \to \ell^+\ell^-\nu\bar{\nu}$: the search consists in selecting a pair of oppositely-charged electrons or muons with a high transverse energy and a large missing transverse energy due to the neutrinos. The invariant mass of the electrons or muons should be consistent with the Z mass. To increase the sensitivity, the analysis is further optimized for separate Higgs mass regions below and above 280 GeV, lepton flavour categories, and different data

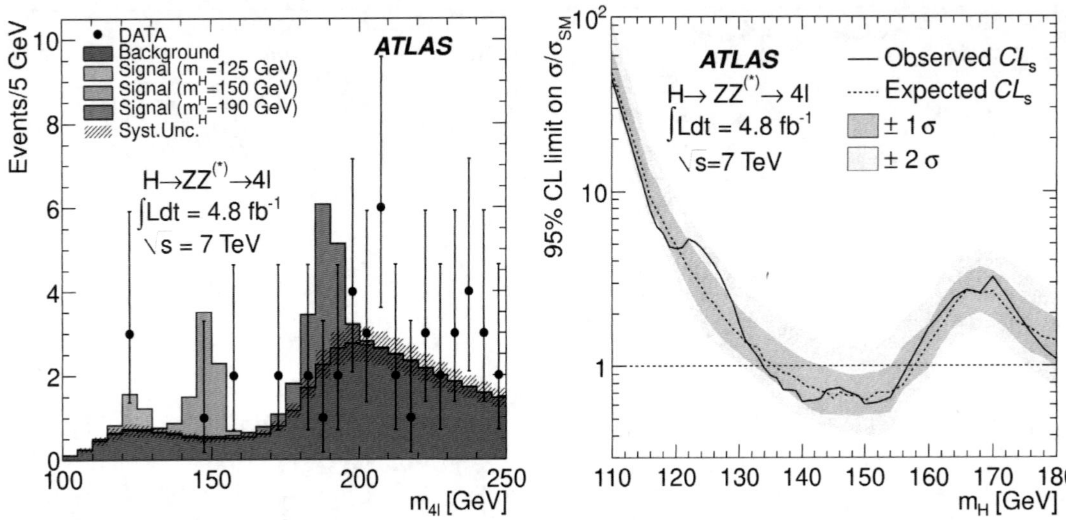

Figure 2. Left: invariant mass distributions for the selected candidate events, the total background and the signal expected in the $H \to ZZ^{(*)} \to \ell^+\ell^-\ell^+\ell^-$ channel in the mass range between 100 GeV and 240 GeV [19]. Right: the observed (full line) and expected (dashed line) 95% CL limits on the SM Higgs boson production normalized to the predicted cross section as a function of m_H [19]. The dotted curves show the median expected limit in the absence of a signal and the green and yellow bands indicate the corresponding $\pm1\sigma$ and $\pm2\sigma$ intervals.

taking conditions. For instance the average number of interactions per bunch crossing during collisions in the latter part of 2011 was 12. The transverse mass of the dilepton and missing transverse energy system is used as a discriminating variable. No significant excess is observed in this channel and a Higgs mass range from 320 GeV to 560 GeV is excluded at 95% CL. The 95% CL limits are shown in Figure 4.

- $H \to ZZ^{(*)} \to \ell^+\ell^- q\bar{q}$: the candidate events are selected by requiring a pair of oppositely-charged electrons or muons and a pair of jets with high transverse energy, and each pair invariant mass should be consistent with a Z boson. The analysis is separated and optimized into Higgs mass regions above and below 300 GeV. The dominant background arises from Z+jets production. A sizable fraction of Z's decay into a pair of b-quarks. The analysis is divided into two categories, the first containing events where the two jets are b-tagged and the second using events with less than two b-tags. Using the Z boson mass constraint improves the mass resolution of the $\ell^+\ell^- q\bar{q}$ system by approximately 10% which is used as a discriminating variable. No significant excess is observed in this channel and Higgs mass ranges from 300 GeV to 310 GeV and from 360 GeV to 400 GeV are excluded at 95% CL. The 95% CL limits are shown in Figure 4.

- $H \to WW^{(*)} \to \ell^+\nu\ell^-\bar{\nu}$: the search consists in selecting a pair of electrons, muons, or electron-muon with opposite charges and high transverse energy. The lepton pair must not be consistent with a Z. Missing transverse energy is also required to be present. The search is optimized for $m_H < 200$ GeV, 200 GeV $\leq m_H \leq 300$ GeV and $m_H > 300$ GeV due to spin correlations in the $WW^{(*)}$ and kinematic features of the two leptons. The events are classified into jet multiplicities of 0, 1, or 2. The 2-jet category selection targets vector boson fusion (VBF) production process by requiring events to have two jets with a large difference in pseudo-rapidity, a large reconstructed invariant mass and no additional energetic jet in

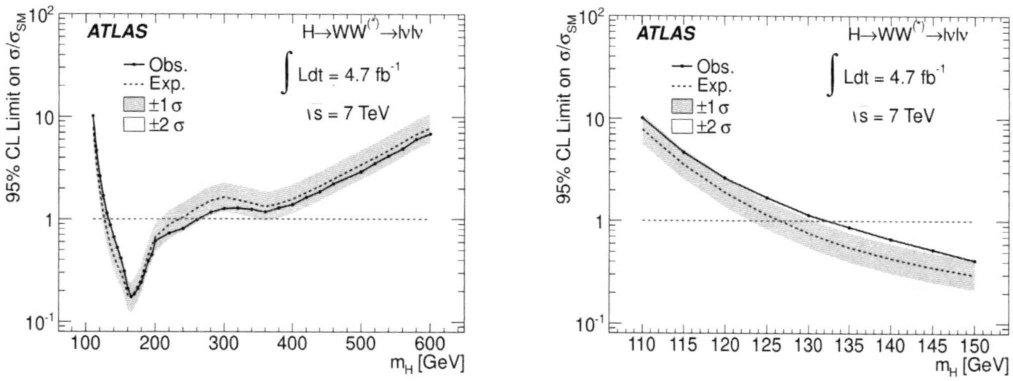

Figure 3. The observed (full line) and expected (dashed line) 95% CL limits on the SM Higgs boson production normalized to the predicted cross section as a function of m_H are shown for the entire mass range (left) and the low mass region (right) [22]. The dotted curves show the median expected limit in the absence of a signal and the green and yellow bands indicate the corresponding $\pm1\sigma$ and $\pm2\sigma$ intervals.

between. The transverse mass of the WW system distribution is used as a discriminating variable. No significant excess is found in this channel and a Higgs mass range from 130 GeV to 260 GeV is excluded at 95% CL. The 95% CL limits are shown in Figure 3.

- $H \to WW^{(*)} \to \ell\nu q\bar{q}'$: the analysis requires an isolated electron or muon, two jets and missing transverse energy. Additional jets are used to further classify the events as in the previous channel. The invariant mass of the two jets must be consistent with a W. The reconstructed $\ell\nu q\bar{q}'$ mass distribution is used as a discriminating variable. It is estimated using the W boson mass constraint to reconstruct the mass of the $\ell\nu$ system. Events with at least one b-tagged jet are rejected to reduce backgrounds from $t\bar{t}$ production. No significant excess is found in this channel and no Higgs mass range excluded from this channel alone. The 95% CL limits are shown in Figure 4.
- $H \to b\bar{b}$: this is an important channel to explore in the low mass region because of the high branching fraction, however the multijet background is very large. One has to leverage on the associated Higgs production with a W or Z boson. All three channels below require the events to have exactly two b-tagged jets with high transverse energy. The invariant mass of the b jets m_{bb} is used as a discriminating variable. Additional requirements are the following depending on the associated vector boson and its decay:
  - $WH \to \ell\nu b\bar{b}$: an isolated electron or muon and missing transverse energy.
  - $ZH \to \ell^+\ell^- b\bar{b}$: a pair of electrons or muons with opposite charges consistent with a Z and no significant missing transverse energy present to reduce background from $t\bar{t}$.
  - $ZH \to \nu\bar{\nu}b\bar{b}$: a large missing transverse energy and missing transverse momentum using tracks associated with the primary vertex.

The Higgs will recoil against the W or Z. This feature is exploited to increase the sensitivity further by looking at four separate regions of W, or Z transverse momentum when decaying into charged leptons. For $Z \to \nu\bar{\nu}$ process, the channel is separated into three missing transverse energy regions. These regions have different signal and background compositions. No significant excess is observed in this channel and no Higgs mass region is excluded from this channel alone. The 95% CL limits are shown in Figure 4.

- $H \to \tau^+\tau^-$: although less sensitive due to a small branching fraction and presence of a dominant background from Z $\to \tau^+\tau^-$, it is nevertheless an important channel to look at. The search is split into three different channels depending on the τ decay mode either leptonic (τ_{lep}) or hadronic (τ_{had}). The invariant mass of the $\tau^+\tau^-$ system $m_{\tau\tau}$ is used as a discriminating variable and is calculated using different techniques depending on the event topology. A hadronic tau decay candidates is identified as a narrow jet with a low track multiplicity. The requirements for each channel considered are listed below:
  - $H \to \tau^+\tau^- \to \tau_{lep}\tau_{lep}$: two isolated and oppositely-charged leptons (either electrons or muons) are required along with a large missing transverse energy present in the event. The analysis is separated into different jet multiplicities. For the 0-jet category only an electron-muon pair is considered, while for the 1-jet and 2-jet categories all flavour combinations are used. The 2-jet category aims for the VBF contribution as described previously and for the associated production with process with a Z, or W decaying hadronically with specific requirements on the two jets selected. For the 0-jet category the invariant mass between the missing transverse energy and four-momenta of the two selected leptons is used as a discriminating variable. For the other categories the mass reconstructed in the collinear approximation is used.
  - $H \to \tau^+\tau^- \to \tau_{lep}\tau_{had}$: an isolated electron or muon along with a τ jet having opposite charges are required, and missing transverse energy. No additional electron or muon should be present to suppress background from Z and top quark production. The new Missing Mass Calculator technique is used [26] for the discriminating variable.
  - $H \to \tau^+\tau^- \to \tau_{had}\tau_{had}$: missing transverse energy, two oppositely-charged τ jets are required and in order to reduce the large multijet background an additional jet with high transverse energy is required. No electrons or muons should be present in the event. The collinear approximation is used when estimating $m_{\tau\tau}$.

  No significant excess is observed in this channel and no Higgs mass region is excluded from this channel alone. The 95% CL limits are shown in Figure 4.

3. Combination and Results

All the channels described in the previous section have been combined into a global likelihood fit. A detailed description of the combination of all the channels can found in Ref. [16], including the treatment of statistical, systematic and theoretical uncertainties. The expected and observed 95% CL limits are shown in Figure 5 for the entire Higgs mass range considered in this search. At the 95% confidence level the excluded Higgs boson mass ranges are 111.4 GeV to 116.6 GeV, 119.4 GeV to 122.1 GeV, and 129.2 GeV to 541 GeV, while the mass range from 120 GeV to 560 GeV is expected to be excluded in the absence of a SM Higgs signal. An excess of events is observed at a Higgs boson mass around 126 GeV with a local significance of 2.9σ. This excess is observed in the $H \to \gamma\gamma$ and $H \to ZZ^{(*)} \to \ell^+\ell^-\ell^+\ell^-$ channels. The global probability for the background to produce an excess at least as significant anywhere in the mass range explored corresponds to a significance of approximately 1σ. The local p_0 probabilities are shown in Figure 6 (left). The combined best-fit of the signal strength as a function of the Higgs boson mass hypothesis in the low mass range is shown in Figure 6 (right). The signal strength $\hat{\mu}$ at 126 GeV is 1.1 ± 0.4, which is compatible with a SM Higgs boson at that mass ($\mu = 1$).

4. Conclusion

We reported on searches for the Standard Model Higgs boson with the ATLAS detector at the LHC. The datasets correspond to integrated luminosities ranging from 4.6 fb^{-1} to 4.9 fb^{-1} of

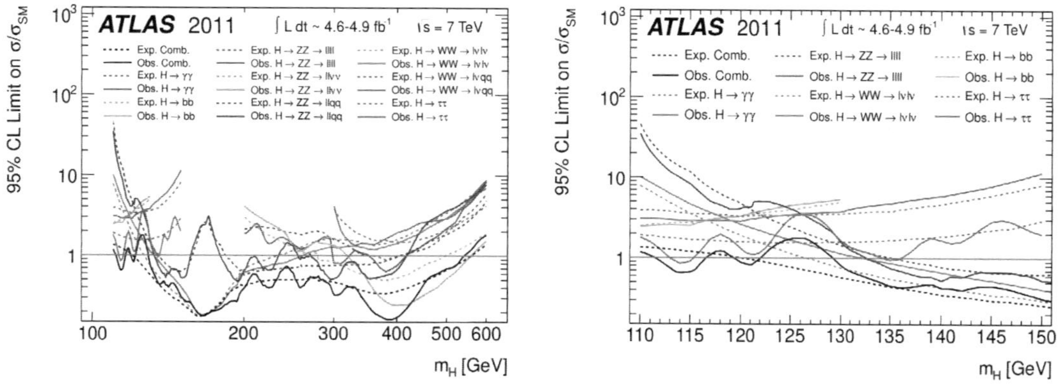

Figure 4. The observed (solid) and expected (dashed) 95% CL cross section upper limits for the individual search channels and the combination, normalized to the SM Higgs boson production cross section, as a function of the Higgs boson mass for the full Higgs boson mass (left) and low mass hypotheses range (right) [16]. The expected limits are those for the background-only hypothesis i.e. in the absence of a Higgs boson signal.

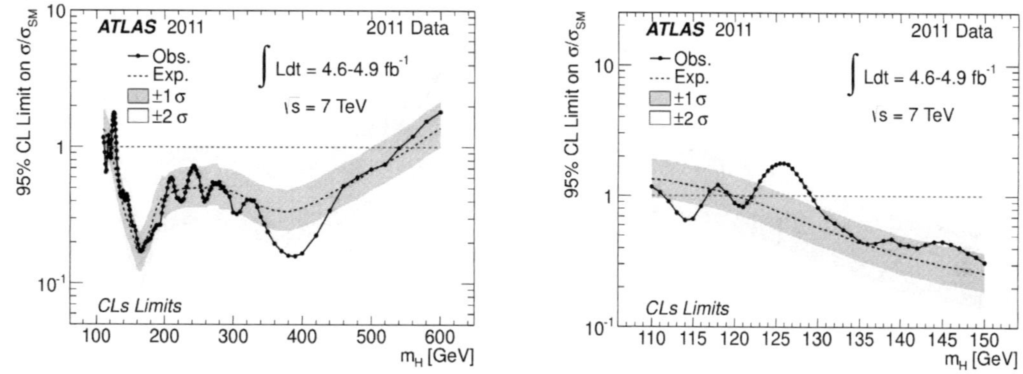

Figure 5. The observed (full line) and expected (dashed line) 95% CL combined upper limits on the SM Higgs boson production cross section divided by the SM expectation as a function of m_H in the full mass range (left) and low mass region (right) considered in this analysis [16]. The dotted curves show the median expected limit in the absence of a signal and the green and yellow bands indicate the corresponding $\pm 1\sigma$ and $\pm 2\sigma$.

proton-proton collisions collected at $\sqrt{s} = 7$ TeV in 2011. Several channels were considered and a comprehensive statistical combination performed over the mass range from 110 GeV to 600 GeV. An excess is observed and it is compatible with a SM Higgs boson in the low mass region at around 126 GeV. The datasets collected in 2012 at $\sqrt{s} = 8$ TeV will be essential in understanding this excess further. The Higgs production cross section in the low mass region is enhanced by about 20% at that energy and therefore looks promising for subsequent searches.

5. Acknowledgements

We gratefully acknowledge the support from the Natural Science and Engineering Research Council of Canada (NSERC), the National Research Council (NRC) and Canada Foundation for Innovation (CFI).

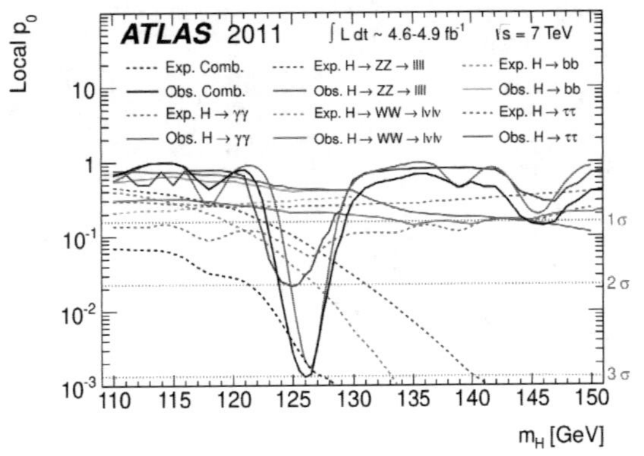

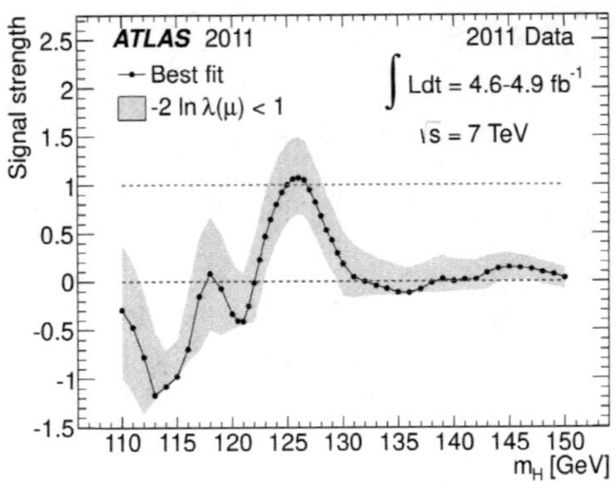

Figure 6. On the left, the local probability p_0 for a background-only experiment to be more signal-like than the observation, for individual channels and the combination in the low mass range of 110 150 GeV [16]. The full curves give the observed individual and combined p_0. The dashed curves show the median expected value under the hypothesis of a SM Higgs boson signal at that mass. The horizontal dashed lines indicate the p_0 corresponding to significances of 1σ, 2σ, and 3σ. On the right, the combined best-fit signal strength as a function of the Higgs boson mass hypothesis in the low mass range is shown [16]. The interval around the best fit value of corresponds to a variation of -2 $\ln \lambda(\mu) < 1$.

References

[1] S. L. Glashow, Nucl. Phys. **22**, 579 (1961)
[2] S. Weinberg, Phys. Rev. Lett. **19**, 1264 (1967)
[3] A. Salam, in Elementary Particle Theory, p. 367 (Almqvist and Wiksell, Stockholm, 1968)
[4] F. Englert and R. Brout, Phys. Rev. Lett. **13**, 321 (1964)
[5] P. W. Higgs, Phys. Lett. **12**, 132 (1964)
[6] P. W. Higgs, Phys. Rev. Lett. **13**, 508 (1964)
[7] G. Guralnik, C. Hagen, and T. Kibble, Phys. Rev. Lett. **13**, 585 (1964)
[8] P. W. Higgs, Phys. Rev. **145**, 1156 (1966)
[9] T. Kibble, Phys. Rev. **155**, 1554 (1967)
[10] ALEPH collaboration, DELPHI collaboration, L3 collaboration, OPAL collaboration and the LEP Working Group for Higgs boson searches, Phys. Lett. **B 565**, 61 (2003)
[11] TEVNPH (Tevatron New Phenomena and Higgs Working Group)(2011), arXiv:1107.5518 [hep-ex]
[12] LHC Higgs Cross Section Working Group, S. Dittmaier, C. Mariotti, G. Passarino, and R. Tanaka (Eds.), CERN-2011-002(2011), arXiv:1101.0593 [hep-ph]
[13] ATLAS Collaboration, JINST **3**, S08003 (2008)
[14] A. L. Read, Presentation of search results: The CL(s) technique, J. Phys. **G28** (2002) 26932704.
[15] G. Cowan, K. Cranmer, E. Gross, O. Vitells, Asymptotic formulae for likelihood-based tests of new physics, Eur. Phys. J. **C71** (2011) 1554.
[16] ATLAS Collaboration, submitted to Phys. Rev. D (2012), arXiv:1207.0319 [hep-ex]
[17] ATLAS Collaboration, Phys. Rev. Lett. **108**, 111803 (2012)
[18] ATLAS Collaboration, Phys. Lett. **B 710**, 383 (2012)
[19] ATLAS Collaboration, ATLAS-CONF-2011-162, http://cdsweb.cern.ch/record/1406357
[20] ATLAS Collaboration, submitted to Phys. Lett. B (2012), arXiv:1205.6744 [hep-ex]
[21] ATLAS Collaboration, submitted to Phys. Lett. B (2012), arXiv:1206.2443 [hep-ex]
[22] ATLAS Collaboration, submitted to Phys. Lett. B(2012), arXiv:1206.0756 [hep-ex]
[23] ATLAS Collaboration, submitted to Phys. Lett. B(2012),arXiv:1206.6074 [hep-ex]
[24] ATLAS Collaboration, submitted to JHEP(2012), arXiv:1206.5971 [hep-ex]
[25] ATLAS Collaboration, CERN-PH-EP-2012-138, submitted to Phys. Lett. B(2012)
[26] A. Elagin, P. Murat, A. Pranko, and A. Safonov, Nucl. Instrum. Meth. **A654**, 481 (2011)

Journal of Physics: Conference Series **389** (2012) 012002 doi:10.1088/1742-6596/389/1/012002

Higgs searches in CMS

Jónatan Piedra for the CMS Collaboration

E-mail: piedra@cern.ch

Abstract. These proceedings report the results on the Higgs searches at the CMS experiment with data collected during the 2011 LHC run at 7 TeV, corresponding to an integrated luminosity of about 5 fb^{-1}. The search for the standard model Higgs boson has been narrowed to a mass window of just 16 GeV. In the searches beyond the standard model no significant deviation is found, and limits on the Higgs boson mass are set for several physics scenarios.

1. Introduction

The discovery of the mechanism for electroweak symmetry breaking is one of the goals of the physics programme at the Large Hadron Collider (LHC). Here we report on several Higgs boson searches carried out in proton-proton collisions at $\sqrt{s} = 7$ TeV using the Compact Muon Solenoid (CMS) detector [1] at the LHC. The analysed data recorded in 2011 correspond to an integrated luminosity up to 5 fb^{-1}, depending on the search channel. The standard model (SM) of Particle Physics describes very precisely the experimental measurements up to now, but one of its key ingredients has not yet been observed: the Higgs boson, which is at the source of the electroweak symmetry breaking and provides a mechanism to assign mass to particles. It is clear, however, that the SM theory breaks at larger scales, and some major open points are the unification of couplings, hierarchy problem, dark matter issue and neutrino masses. Theories have been proposed that attempt to answer some of these open questions such as supersymmetry (SUSY) or other beyond the standard model (BSM) scenarios, and are currently under experimental test. The latest Higgs searches results in the following scenarios are reported: standard model, an extension to the standard model that includes a fourth generation of fermions, a fermiophobic Higgs boson, and beyond the standard model Higgs searches.

2. Standard model Higgs

In the $\mathrm{H} \to \gamma\gamma$ analysis [2] a search is made for a narrow peak in the diphoton mass distribution in the range 110-150 GeV, on a large irreducible background from QCD production of two photons. To enhance the sensitivity of the analysis, candidate diphoton events are separated into mutually exclusive categories of different expected signal-to-background ratios, based on the properties of the reconstructed photons and by the presence of two jets satisfying criteria aimed at selecting events in which a Higgs boson is produced through the vector boson fusion (VBF) process. The analysis uses multivariate techniques for the selection and classification of the events. The limit set on the cross section of a Higgs boson decaying to two photons relative to the SM expectation is shown in Figure 1 (left). The median expected limit on the cross section obtained in the previously reported analysis [3] of the same dataset is also shown. The use of multivariate techniques has increased the sensitivity by about 20% across the entire mass range. The analysis of the data excludes at 95% confidence level (CL) a standard model Higgs boson decaying into two photons in the mass ranges 110-111, 117.5-120.5, 128.5-132.0, 139-140 and 146-147 GeV. The largest excess of events above the expected standard model background is observed for a Higgs boson mass hypothesis of 125 GeV with a local significance of 2.9 standard deviations.

In the H $\rightarrow$ ZZ $\rightarrow 4\ell$ decay mode [4] a search is made for a narrow four-lepton mass peak in the presence of a small continuum background. Since there are differences in the reducible background rates and mass resolutions between the subchannels 4e, 4μ, and $2e2\mu$, they are analysed separately. The main background source is an irreducible four-lepton contribution from direct ZZ production via $q\bar{q}$ and gluon-gluon processes. The low mass range is shown in Figure 1 (right) together with the mass of each candidate and its uncertainty. Upper limits at 95% CL exclude the Higgs boson mass ranges 134-158 GeV, 180-305 GeV, and 340-465 GeV. Excesses of events are observed at the low end of the explored mass range, around masses of 119 and 126 GeV.

The decay channel H $\rightarrow$ ZZ $\rightarrow 2\ell 2\nu$ [5] is especially sensitive in the high-mass range 250-600 GeV. Events with a lepton pair (ee or $\mu\mu$ are selected, with mass consistent with that of an on-shell Z boson, and large missing transverse energy is asked for, assuming that it is due to the undetected neutrinos from a Z $\rightarrow \nu\nu$ decay. No significant excess is found above the background expectation, and the Higgs boson mass range 270-440 GeV is excluded at 95% CL.

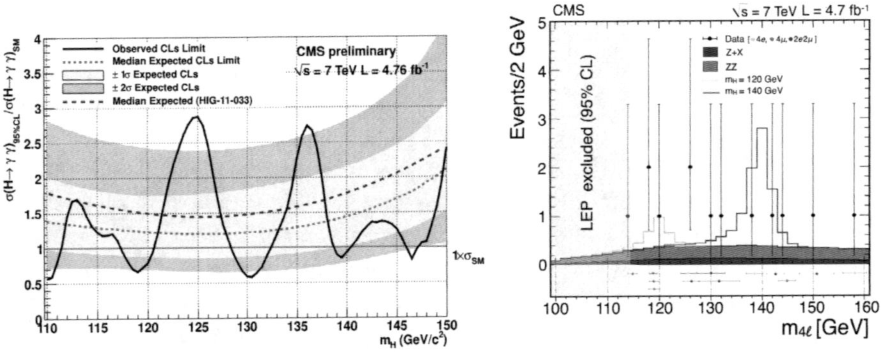

Figure 1. Left: The 95% CL limit on $\sigma/\sigma_{\text{SM}}$ for a Higgs boson decaying to two photons, relative to the SM expectation. The symbol σ denotes the production cross section times the relevant branching fractions. The background only expectations are represented by their median (red line) and by the 68% and 95% CL bands. The expected limit obtained in the earlier analysis of the same dataset is shown for comparison. Right: Distribution of the four-lepton mass for the 4ℓ channels.

The decay mode H $\rightarrow$ WW is highly sensitive to a SM Higgs boson in the mass range around the WW threshold of 160 GeV. With the development of tools for lepton identification and missing transverse energy reconstruction optimised for pileup conditions it is possible to extend the sensitivity down to 120 GeV. The most sensitive analysis in the low mass region is based on the final state in which both W bosons decay leptonically, resulting in a signature with two isolated, oppositely charged leptons (electrons or muons) and large missing transverse energy due to the undetected neutrinos [6]. To further improve the signal sensitivity, the events are separated into three mutually exclusive categories according to the jet multiplicity. Its main backgrounds are irreducible non-resonant WW production and reducible W+jets processes, where a jet has been misidentified as a lepton. Some categories are also contaminated by top-quark decays and Drell-Yan production. The event selection in the two-jet category is optimised for the VBF production mechanism. After applying the analysis selection, the resultant sample is dominated by non-resonant WW events, as it can be seen in Figure 2 (left), where $\Delta\phi_{\ell\ell}$ is shown, for the 0-jet category. To enhance the sensitivity to a Higgs boson signal, a multivariate technique is used after the analysis selection. A boosted decision tree (BDT) is trained for each Higgs boson mass hypothesis and jet category to discriminate signal from background. The expected and observed 95% CL upper limits are shown in Figure 2 (right). The presence of the SM Higgs boson with a mass in the range 129-270 GeV is excluded at 95% CL.

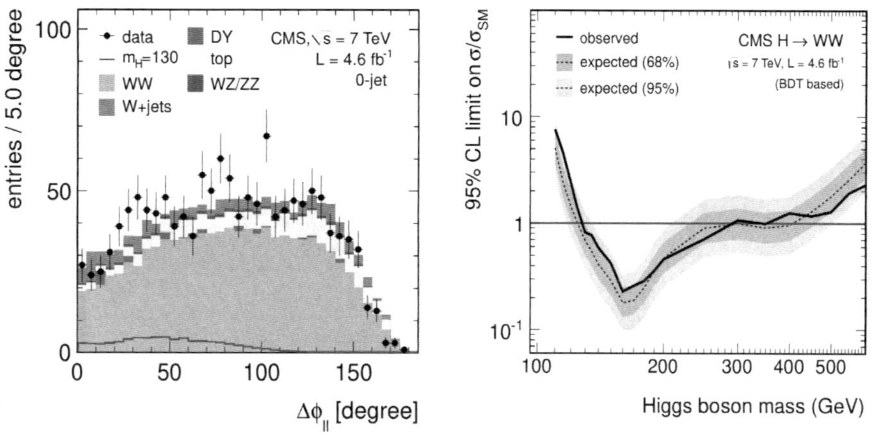

Figure 2. Left: Distribution of the azimuthal angle difference between the two selected leptons ($\Delta\phi_{\ell\ell}$) for the zero-jet category in the H $\rightarrow$ WW search. The signal expected from a Higgs boson with a mass $m_{\mathrm{H}} = 130$ GeV is shown superimposed. Right: The 95% CL limit on $\sigma/\sigma_{\mathrm{SM}}$ for a Higgs boson decaying, via a W boson pair, to two leptons and two neutrinos, relative to the SM expectation.

The H $\rightarrow \tau\tau$ search is performed using the final-state signatures eμ, $\mu\mu$, eτ_{h}, and $\mu\tau_{\mathrm{h}}$ [7, 8], where electrons and muons arise from leptonic τ-decays and τ_{h} denotes hadronic τ-decays. Each of these categories is further divided into three exclusive sub-categories according to the nature of the associated jets, and a broad excess in the reconstruced $\tau\tau$ mass distribution is searched for. In addition, a search for H $\rightarrow \tau\tau$ is performed for a Higgs boson produced in association with a W boson [9]. No significant excess of events over background expectations is observed, and 95% CL upper limits are set.

For $m_{\mathrm{H}} < 135$ GeV, the decay H $\rightarrow$ bb has one of the largest branching fractions, but the signal is overwhelmed by QCD production of bottom quarks. The analysis [10] is therefore designed to search for a dijet resonance in events where a Higgs boson is produced at high transverse momentum and in association with a W or Z boson, which largely suppresses the QCD background. Five independent search channels are explored, depending on the decay of the vector boson. An upper limit of 6 to 13 times greater than the predicted SM value is set at 95% CL in the mass range $100 < m_{\mathrm{H}} < 140$ GeV.

The combination [11] of the SM Higgs boson searches requires simultaneous analysis of the data from all individual search channels, accounting for all statistical and systematic uncertainties and their correlations. The combination is repeated for 183 Higgs boson mass hypotheses in the range 110-600 GeV. The overall statistical methodology used in the combination was developed by the CMS and ATLAS collaborations in the context of the LHC Higgs Combination Group. Figure 3 (left) shows the 95% CL upper limits on the signal strength

σ/σ_{SM}. The best fit values are shown in Figure 7 (right). The expected excluded mass range in the absence of the standard model Higgs boson is 114.5-543 GeV at 95% CL. The observed results exclude the standard model Higgs boson in the mass range 127.5-600 GeV at 95% CL. An excess of events above the expected standard model background is observed at the low end of the explored mass range. The largest excess, with a local significance of 2.8 standard deviations, is observed for a Higgs boson mass hypothesis of 125 GeV. More data are required to ascertain the origin of the observed excess.

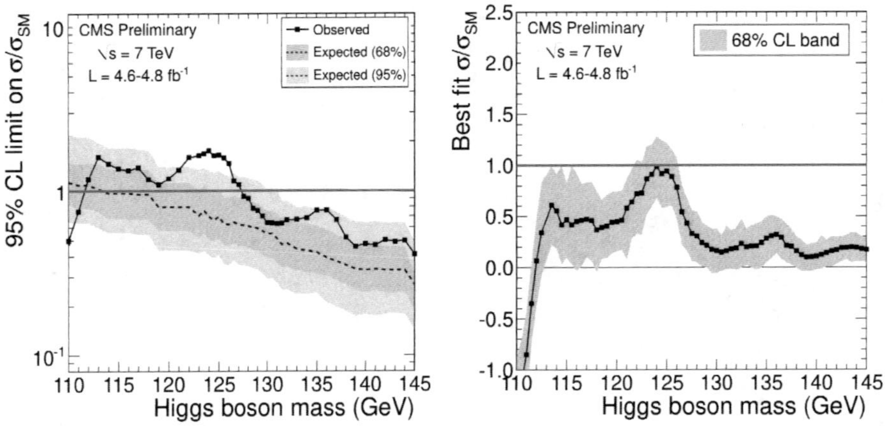

Figure 3. Left: The observed and expected 95% CL upper limits on the signal strength σ/σ_{SM} for the SM Higgs boson hypothesis as a function of the Higgs boson mass in the range 110-145 GeV. Right: The observed best-fit σ/σ_{SM} as a function of the SM Higgs boson mass in the range 110-145 GeV. The band corresponds to the ± 1 standard deviation uncertainty on the σ/σ_{SM} value, and the symbol σ denotes the production cross section times the relevant branching fractions.

For an extension of the standard model including a fourth generation of fermions (SM4), the SM4 Higgs boson is excluded in the mass range 120-600 GeV at 95% CL.

3. Minimal supersymmetric standard model Higgs

In the minimal supersymmetric standard model (MSSM), the standard scalar Higgs boson is substituted by three neutral ϕ = (h, H, A) and two charged ($H^{\pm}$) Higgs particles, and all decays to down-type fermions are enhanced by a factor of tan β. For relatively high tan β the BR($\phi \rightarrow \tau\tau$) is about 10% which is much lower than the corresponding branching ratio of the b-decay mode. Neutral MSSM Higgs bosons are searched in the $\tau\tau$ final state, which is preferred for its clear signature in the two leptons final states (electrons or muons) and in the lepton plus a hadronic decaying τ final state. The dilepton channel was searched in dimuons and in electron-muon final states. The hadronic τ is reconstructed in 1 and 3 prongs. To extract the $\tau\tau$ mass a kinematic fit is applied to the measured components, including the missing energy, with an improvement of about 20% on the measured mass [12]. In Figure 4 (left) is shown the exclusion plot for the neutral MSSM Higgs mass versus tan β. For tan β = 20 Higgs masses up to 300 GeV are excluded. The charged MSSM Higgs bosons are searched in the top decays $t \rightarrow bH^{\pm}$ with the tau final states $H^{+} \rightarrow \tau\nu$. The $t\bar{t}$ production yields with tau final states are modified by Higgs diagrams if the Higgs mass is lower than the top mass. The Higgs particle is searched in isolated τ decays plus b-jets and possibly an isolated lepton in the final state, depending on the second top decay chain in the $t\bar{t}$ events. Results from this analysis [13] on

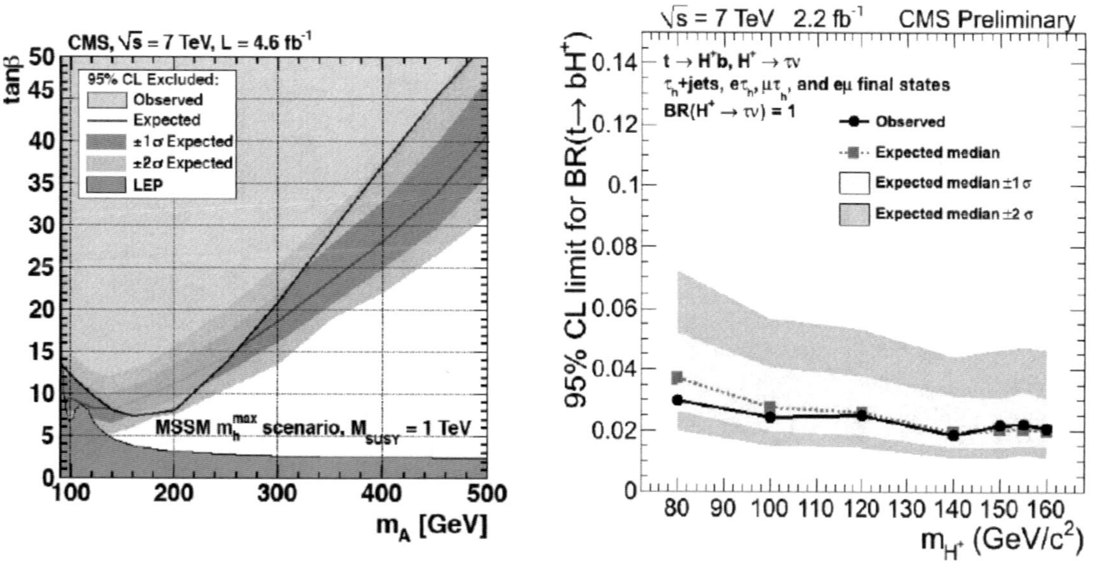

Figure 4. Left: Region in the parameter space of tan β versus $m_{\rm A}$ excluded at 95% CL in the context of the MSSM scenario. The 95% CL upper limit and the expected one- and two-standard-deviation ranges are shown together with the observed excluded region. Right: Upper limit on BR(t $\rightarrow$ bH$^+$) assuming BR(H$^+$ $\rightarrow \tau\nu$) = 1, as a function of $m_{\rm H^+}$.

BR(t $\rightarrow$ bH$^+$) are shown in Figure 4 (right). Values of BR(t $\rightarrow$ bH$^+$) > 4% are excluded for all the possible Higgs mass values.

4. Light pseudoscalar Higgs

The presence of a light pseudoscalar CP-odd Higgs is predicted within the next to minimal supersymmetric extension to the standard model. This search [14] has been performed in the sidebands of the $\Upsilon \rightarrow \mu\mu$ decays, namely $5.5 < m_{\mu\mu} < 9$ GeV and $11.5 < m_{\mu\mu} < 14$ GeV. A special high level trigger conceived for charmonium states studies was set up and this analysis has been performed with a data sample corresponding to a luminosity of 1.3 fb^{-1}. Results are shown in Figure 5 with no excess found in the dimuon spectrum. An upper limit on the cross section σ(pp $\rightarrow$ a $\rightarrow \mu\mu$) below 5 pb is set for all the masses in the two search intervals.

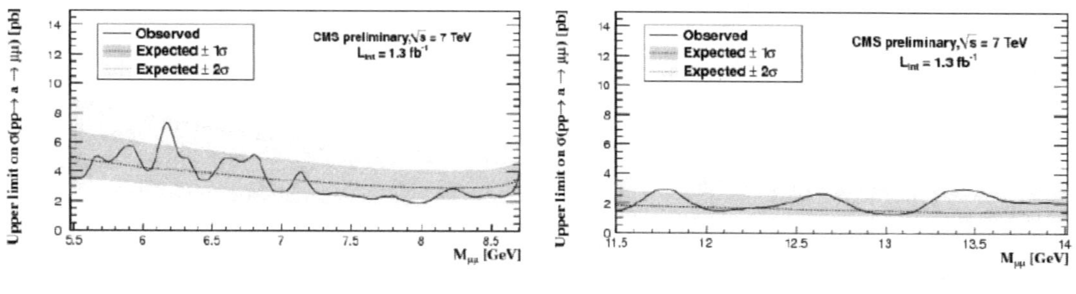

Figure 5. Limits on the cross section σ(pp $\rightarrow$ a $\rightarrow \mu\mu$) for the two mass intervals in the Υ sidebands.

Journal of Physics: Conference Series **389** (2012) 012002 doi:10.1088/1742-6596/389/1/012002

5. Doubly charged Higgs

These exotic Higgs bosons are predicted within the type II see-saw model and are related to the presence of a light neutrino mass. H^{++} decay to two same charged leptons and obviously do not have any physical background in the SM. They are produced in pairs or together with a single charged Higgs through the processes $Z/\gamma \rightarrow H^{++}H^{--}$ and $W^{+} \rightarrow H^{++}H^{-}$ (charge conjugates included), giving final states with four or three leptons, same charge resonant. No excess is observed in the CMS data [15]. In Figure 6 (left) are shown the mass limits for the different leptonic final states and four benchmark points of the see-saw mechanism.

6. Fermiophobic Higgs

In the Fermiophobic model the gluon-gluon process of Higgs production is forbidden and the production cross section is suppressed by an order of magnitude, with the VBF and the Higgs-strahlung (VH) becoming the two most important contributions to Higgs production. On the other hand, the diphoton decay $H \rightarrow \gamma\gamma$ is enhanced by another order of magnitude. This analysis [16] is based on the selection of two high transverse momentum photons and three tag classes with electron, muon or dijets in the final states, corresponding to different decays in the associate production. The results are presented in Figure 6 (right) and show a small excess at 126 GeV, which is diluted when the diphoton channel is combined [11] with $H \rightarrow WW$ and $H \rightarrow ZZ$. Two intervals of Higgs mass are excluded at 95% CL, $110 < m_H < 124$ GeV and $128 < m_H < 136$ GeV.

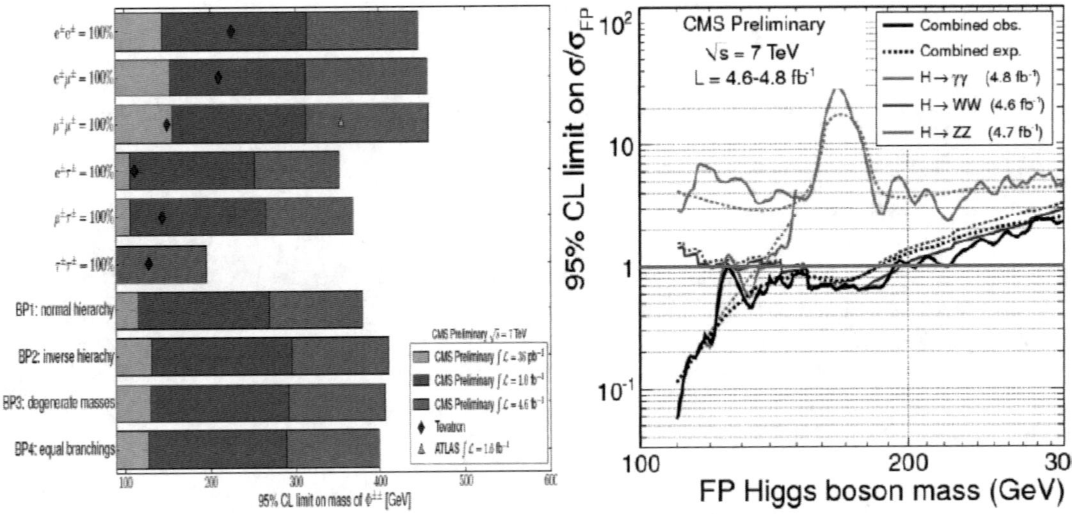

Figure 6. Left: Limits on the mass of the doubly charged Higgs bosons for different final states. Right: The 95% CL upper limits on the signal strength parameter σ/σ_{FP} for the fermiophobic Higgs boson hypothesis, as a function of the Higgs boson mass. It includes the three explored Higgs boson decay modes.

7. Conclusions

The observed results exclude the standard model Higgs boson in the mass range 127.5-600 GeV at 95% CL, and an excess of events above the expected standard model background is observed at the low end of the explored mass range, making the observed limits weaker than expected in absence of a signal. The largest excess, with a local significance of 2.8 standard deviations, is observed for a Higgs boson mass hypothesis of 125 GeV. In addition, a broad program of BSM

Higgs bosons searches has been presented. Model independent inclusive searches together with well defined new physics scenarios have been probed during 2011 with a luminosity of about 5 fb^{-1}. A large fraction of the MSSM Higgs parameters are constrained by the $H \rightarrow \tau\tau$ analysis. A small excess on the $H \rightarrow \gamma\gamma$ decay is noted, compatible with a statistical fluctuation. No evidence for new BSM Higgs bosons is observed.

Acknowledgments

We wish to congratulate our colleagues in the CERN accelerator departments for the excellent performance of the LHC machine. We thank the technical and administrative staff at CERN and other CMS institutes, and acknowledge support from: FMSR (Austria); FNRS and FWO (Belgium); CNPq, CAPES, FAPERJ, and FAPESP (Brazil); MES (Bulgaria); CERN; CAS, MoST, and NSFC (China); COLCIENCIAS (Colombia); MSES (Croatia); RPF (Cyprus); Academy of Sciences and NICPB (Estonia); Academy of Finland, ME, and HIP (Finland); CEA and CNRS/IN2P3 (France); BMBF, DFG, and HGF (Germany); GSRT (Greece); OTKA and NKTH (Hungary); DAE and DST (India); IPM (Iran); SFI (Ireland); INFN (Italy); NRF and WCU (Korea); LAS (Lithuania); CINVESTAV, CONACYT, SEP, and UASLP-FAI (Mexico); PAEC (Pakistan); SCSR (Poland); FCT (Portugal); JINR (Armenia, Belarus, Georgia, Ukraine, Uzbekistan); MST and MAE (Russia); MSTD (Serbia); MICINN and CPAN (Spain); Swiss Funding Agencies (Switzerland); NSC (Taipei); TUBITAK and TAEK (Turkey); STFC (United Kingdom); DOE and NSF (USA).

References

[1] CMS Collaboration, doi:10.1088/1748-0221/3/08/S08004 JINST **3** (2008) S08004
[2] CMS Collaboration, CMS-PAS-HIG-12-001
[3] CMS Collaboration, arXiv:1202.1487v1 [hep-ex], submitted to PLB
[4] CMS Collaboration, arXiv:1202.1997 [hep-ex]
[5] CMS Collaboration, arXiv:1202.3478 [hep-ex]
[6] CMS Collaboration, arXiv:1202.1489 [hep-ex]
[7] CMS Collaboration, arXiv:1202.4083 [hep-ex]
[8] CMS Collaboration, CMS-PAS-HIG-12-007
[9] CMS Collaboration, CMS-PAS-HIG-12-006
[10] CMS Collaboration, arXiv:1202.4195 [hep-ex]
[11] CMS Collaboration, CMS-PAS-HIG-12-008
[12] CMS Collaboration, arXiv:1202.4083 [hep-ex], submitted to PLB
[13] CMS Collaboration, CMS-PAS-HIG-11-019
[14] CMS Collaboration, CMS-PAS-HIG-12-004
[15] CMS Collaboration, CMS-PAS-HIG-12-005
[16] CMS Collaboration, CMS-PAS-HIG-12-002

Dijet imbalance in 2.76 TeV PbPb collisions in CMS

Sevil Salur for the CMS Collaboration

Department of Physics and Astronomy Rutgers, the State University of New Jersey 136 Frelinghuysen Road Piscataway, NJ 08854-8019

E-mail: salur@physics.rutgers.edu

Abstract. We present the measurement of dijet production in PbPb collisions at a nucleon-nucleon center-of-mass energy of 2.76 TeV that is studied with the CMS detector at the LHC. We use data corresponding to an integrated luminosity of 150 μb^{-1}. Jets are reconstructed using the anti-k_T algorithm on particle flow objects. The dijet momentum balance and angular correlations are studied as a function of collision centrality and leading jet momentum. For the most peripheral PbPb collisions, good agreement of the dijet momentum balance distributions with pp data and reference calculations at the same collision energy is found. However, more central collisions show a strong imbalance of leading and subleading jet transverse momenta, confirming the previous observations made using a smaller dataset. The extended jet transverse momentum range significantly increases the discriminatory power of the measurement for different models of parton energy loss in hot QCD matter.

1. Introduction

Well identified decay products of partonic interactions at large momentum transfers, also called hard probes, are used to study the structure and dynamics of the Quark Gluon Plasma [1, 2, 3]. These probes are well calibrated, as their expected yields are calculable using the perturbative QCD theoretical framework and as their propagation through the medium are affected via strong interactions. At RHIC, indirect measurements of energy loss in the medium ("jet quenching") have been made via observables of leading fragments of jets and their correlations [4, 5, 6]. Recent results from LHC confirmed these observations and expanded our knowledge of the jet-quenching effects by utilizing fully reconstructed jets, correlations between jets and single particles, and charged particle measurements [7, 8, 9, 10, 11].

By utilizing back to back jets, the LHC analyses revealed significant jet quenching, consistent with theoretical expectations that involve differential energy loss of back-to-back hard-scattered partons [12, 13, 14]. However, angular correlations between the jets are found to be almost unchanged, ruling out single-hard-gluon radiation as the leading energy loss mechanism. The jet fragmentation pattern also showed no change due to the existence of the hot QCD medium [16]. Similar quenching was also observed in isolated photon+jet pairs [15]. While these results constrain the mechanism of parton energy loss [17, 18], further understanding requires the measurement of the p_T dependence of the observed effects. The goal of this analysis is to characterize possible modifications of dijet event properties as a function of centrality and leading jet transverse momentum in PbPb collisions. Further details of this analysis can be found in [19].

2. Analysis

For the di-jet analyses, a total integrated luminosity of 150 μb^{-1} PbPb collisions and a total integrated luminosity of 231 $nb^{-1}$of pp collisions at a nucleon-nucleon center-of-mass energy of 2.76 TeV, collected with the Compact Muon Solenoid (CMS) detector are used. The CMS experiment at the LHC is a general multi-purpose detector designed to explore physics at the large TeV energy scales [20]. The two most important detector requirements for a successful reconstruction of jets in heavy ion collisions are a good energy measurement of large species of particles and an efficient jet trigger. With its high quality electromagnetic and hadronic calorimeters covering a wide pseodorapidity and a full azimuthal range, by design CMS is extremely well suited to measure hard scattering processes [21]. Sub-detectors such as the high precision silicon tracker, which has very good momentum resolution, complement the calorimeters for jet studies.

Jet reconstruction in heavy-ion collisions in CMS is performed with anti-k_T jet algorithm that is encoded in the FastJet framework [22]. A small value of 0.3 for resolution parameter R is selected to reduce the deterioration of the jet energy resolution in PbPb collisions due to fluctuations of the background from soft interactions. An algorithm that is a variant of an iterative "noise/pedestal subtraction" technique is used to estimate the heavy ion background event-by-event [23]. The input particles of the jet reconstruction are Particle Flow (PF) objects that are reconstructed by combining information from various sub-detectors, most importantly by combining tracks with clusters in electromagnetic and hadronic calorimeters [24]. The reconstructed jet energies are corrected by using a factorized multi-step approach used for all jet analyses in CMS [25]. Jet energy corrections are derived from PYTHIA [26] simulations without PbPb underlying events.

3. Di-Jet Asymmetry Results

The distribution of the angle $\Delta\phi$ between the leading and subleading jets as a function of leading jet p_T is shown in Figure 1. The results from the most central (0-20% central) PbPb events are shown as black points. The red histogram shows the results for PYTHIA dijets embedded into HYDJET PbPb simulated events [27]. The error bars represent only the statistical uncertainties. The distribution around the $\Delta\phi = \pi$ peak reflects the back-to-back dijet production and although this distribution changes across the various leading-jet p_T bins, there is no significant difference between PbPb data and the simulation. Compared with the simulated events, the two most central bins have an apparent excess of jets at low $\Delta\phi$ which is likely due to the matching of the leading jet with a random underlying event fluctuation instead of the true subleading jet partner. The difference in the rate of such events between the PbPb data and the simulated sample is compatible with the effect of quenching.

To characterize the jet energy balance or imbalance quantitatively, we define the asymmetry ratio $A_J = (p_{T,1} - p_{T,2})/(p_{T,1} + p_{T,2})$ where the subscript 1 always refers to the leading jet so that A_J is positive by construction. This asymmetry value was calculated for pairs of jets where the leading jet has p_T >120 GeV/c and the subleading jet has p_T >30 GeV/c. Jets are also required to be separated by $\Delta\phi > 2\pi/3$ to ensure that di-jets are approximately back-to-back). It is important to note that the use of A_J to a large extent removes uncertainties due to possible constant shifts of the jet energy scale.

The fraction of the events as a function of A_J for six centrality bins is shown in Figure 2. The data is compared to the reconstructed PYTHIA dijets embedded in HYDJET data events as shown as the red histograms. Only statistical errors are shown. For the most peripheral bins, the data point values are somewhat similar to those of the simulations. However, as the events become more central, a shift in data points becomes quite significant. Dijets found in the most central events show a very significant deficit of events in which the two jets are balanced and a significant excess of unbalanced pairs.

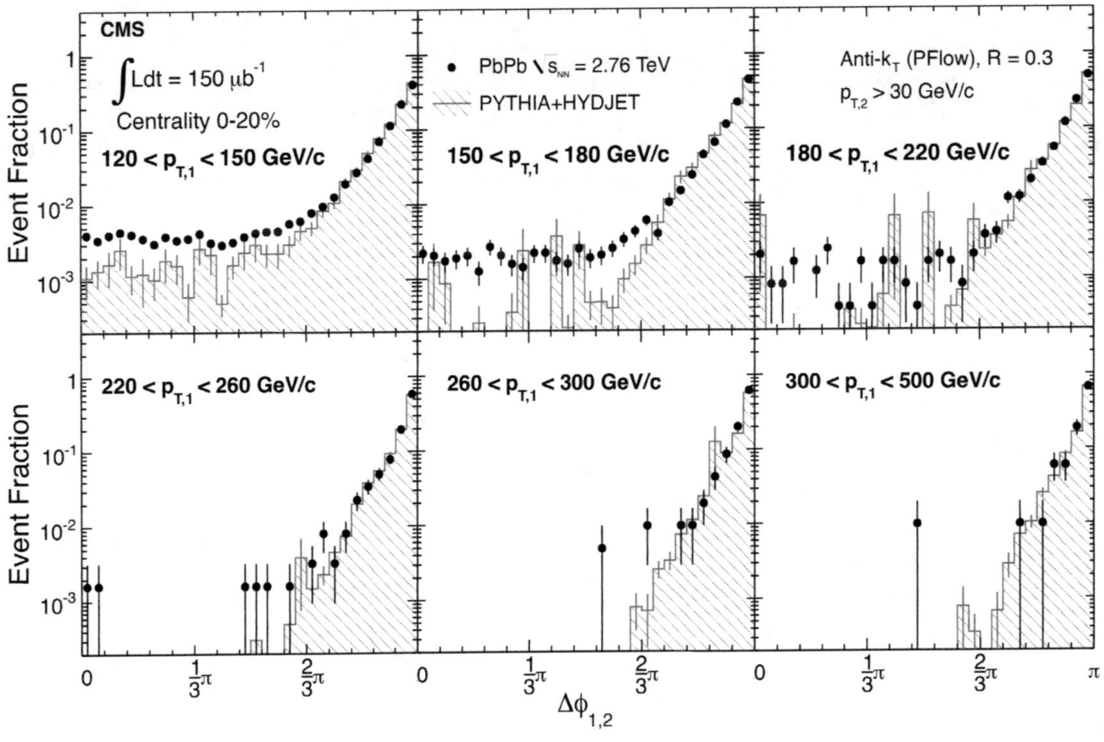

Figure 1. Distribution of the angle $\Delta\phi$ between the leading and subleading jets in bins of leading jet transverse momentum described in the figures for subleading jets of $p_T^{sub} > 30\text{GeV}/c$. Results for 0-20% central PbPb events are shown as points while the histogram shows the results for PYTHIA dijets embedded into HYDJET PbPb simulated events.

The A_J, distributions in various leading jet p_T selections are constructed for the most central 20% of the events and is presented in Figure 3. The shape of the distributions evolves in a different fashion in data and PYTHIA embedded simulations. This evolution is influenced by both the change in the relative jet resolution and splitting probability as present in pp collisions, and the medium induced energy loss which is present only in PbPb collisions. In order to extract the properties of medium induced energy loss from this phenomenon, one has to model the energy loss with these effects in mind.

The distributions of the p_T^{sub}/p_T^{lead}, is shown in Figure 4 to provide a more intuitive way of quantifying the energy loss. The arrows show the mean values of the distributions with equivalent shifts of data and simulations for all leading jet p_T. Both A_J and p_T^{sub}/p_T^{lead} distributions shows a strong evolution in the shapes across the various jet p_T bins, while a significant difference between PbPb data and simulations persists.

Average dijet momentum ratio p_T^{sub}/p_T^{lead} as a function of leading jet p_T for three bins of collision centrality is shown in Figure 5. Results for PbPb data are shown as points with vertical bars and brackets indicating the statistical and systematic uncertainties. Results for the PYTHIA dijets embedded into HYDJET PbPb simulated events are shown as squares. In the 50-100% centrality bin, results are also compared with pp data, which is shown as the open circles. The difference between the PbPb measurement and the simulation expectations is shown in the bottom panels. Both the data and the simulations reveal an increasing trend for the mean value of the jet transverse momentum ratio, as a function of the leading jet p_T. This is mostly due to the improvement of the energy resolution and the reduction of the jet splitting in higher

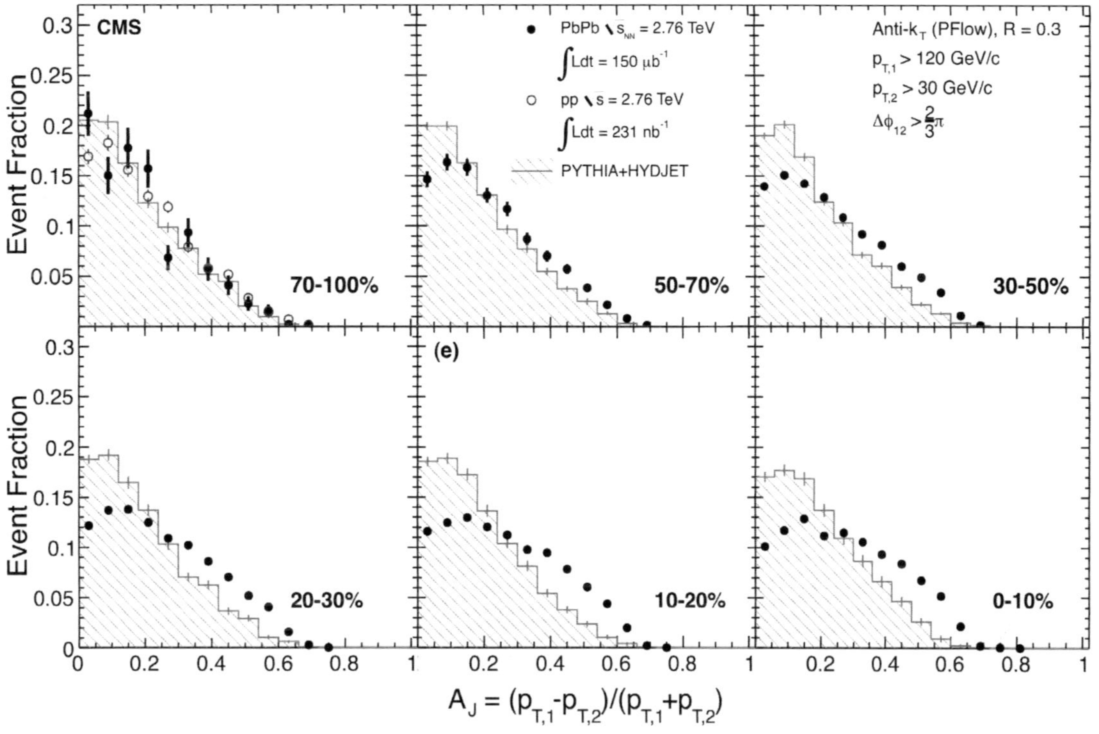

Figure 2. Dijet asymmetry ratio, A_J, for leading jets of p_T >120 GeV/c and subleading jets of p_T >30 GeV/c with a selection of $\Delta\phi > 2\pi/3$ between the two jets. Plots are shown for (a) 70-100% (b) 50-70%, (c) 30-50%, (d) 20-30%, (e) 10-20%, and (f) for the most central 0-10%. Results from PbPb data are shown as black and pp data as blue points. The red histogram shows the reconstruction of PYTHIA dijets embedded into data events.

jet momenta. The central data points lie consistently below the simulations for all jet p_T The difference between the pp data and the simulation is of the order of the systematic uncertainty of the measurement, whereas the difference between PbPb data and the reference is much larger. This observed difference is related to the parton energy loss for central PbPb collisions.

4. Results from Missing p_T

Studies of jet-hadron correlations involving vector summation of charged hadron momenta, find that the energy balance in events with large di-jet asymmetry is recovered on average by an excess of low-momentum particles in the hemisphere of the away-side jet, at large angles relative to the jet axes [7]. Complementary information about the overall momentum balance in the di-jet events can be obtained using the projection of missing p_T of reconstructed charged tracks onto the leading jet axis. For each event, this projection was calculated as

$$\not{p}_{\mathrm{T}}^{\parallel} = \sum_{\mathrm{i}} -p_{\mathrm{T}}^{\mathrm{i}} \cos\left(\phi_{\mathrm{i}} - \phi_{\mathrm{Leading\ Jet}}\right), \qquad (1)$$

where the sum is over all tracks with $p_T > 0.5$ GeV/c and $|\eta| < 2.4$. The results were then averaged over events to obtain $\langle \not{p}_{\mathrm{T}}^{\parallel} \rangle$. No background subtraction was applied, which allows this study to include the $|\eta_{jet}| < 0.8$ and $0.5 < p_{\mathrm{T}}^{\mathrm{Track}} < 1.0$ GeV/c regions.

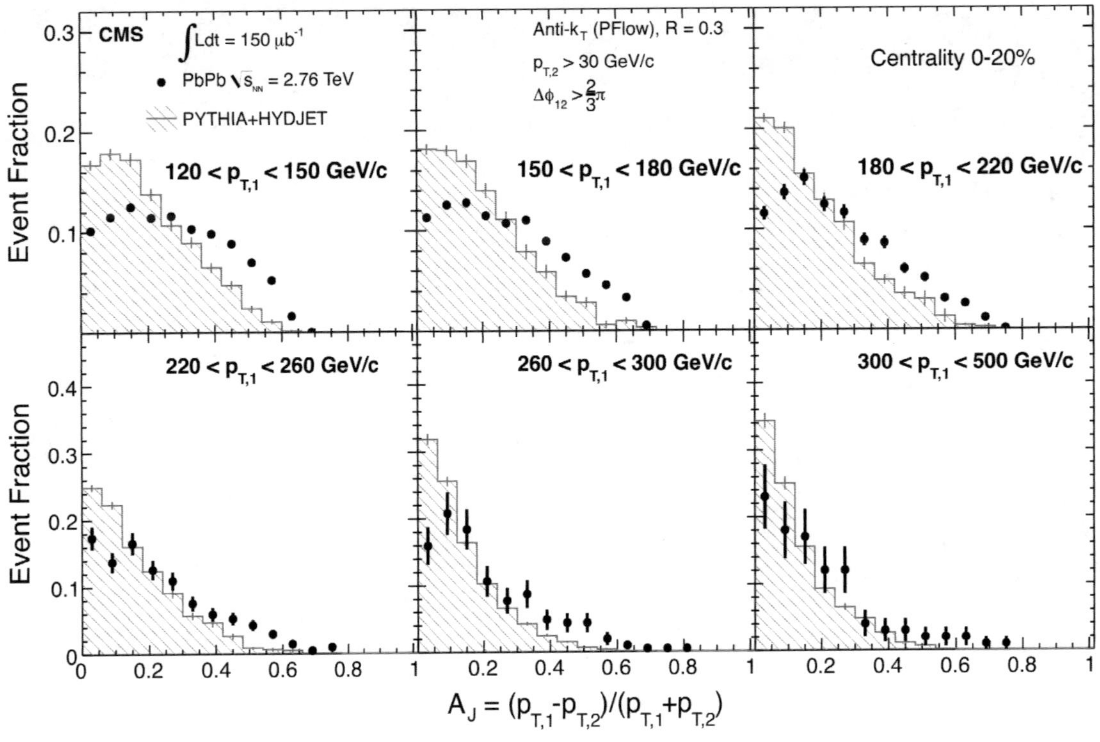

Figure 3. Dijet asymmetry ratio, A_J, in bins of leading jet p_T with subleading jets of p_T >30 GeV/c with a selection of $\Delta\phi > 2\pi/3$ between the two jets, for 0-20% central events. The effect of energy loss is visible at high leading jet p_T bins as well as lower ones, however the shape of the distribution evolves as one goes to higher p_T, both in MC and data. This is mostly because of a different amount of jet energy resolution being involved, and it should be taken into account when inferring about the change in the energy loss.

The $\langle \not{p}_T^{\parallel} \rangle$ values are shown as a function of dijet asymmetry A_J for 0-30% centrality in the right panel of Figure 6. Colored bands show the contribution to $\langle \not{p}_T^{\parallel} \rangle$ for various ranges of track p_T. For the solid circles, vertical bars and brackets represent the statistical and systematic uncertainties, respectively and for the individual p_T ranges, the statistical uncertainties are shown as vertical bars. Using tracks with $|\eta| < 2.4$ and $p_T > 0.5$ GeV/c, one sees the momentum balance of the events, shown as solid circles, is recovered within uncertainties for all events including with large observed di-jet asymmetry events. This shows that, the dijet momentum imbalance is not related to undetected activity in the event due to instrumental (e.g. gaps or inefficiencies in the calorimeter) or physics (e.g. neutrino production) effects. The radial dependence of the momentum balance can be studied with $\langle \not{p}_T^{\parallel} \rangle$ for tracks inside cones of size $\Delta R = 0.8$ around the leading and subleading jet axes, and for tracks outside of these cones. The results of this study are shown in the middle and the right panel of the Figure 6 for the in-cone balance and out-of-cone balance for central events respectively. One observes that for an in-cone imbalance of $\langle \not{p}_T^{\parallel} \rangle \approx -20$ GeV/c is found for the $A_J, > 0.33$ selection. This is balanced by a corresponding out-of-cone imbalance of $\langle \not{p}_T^{\parallel} \rangle \approx 20$ GeV/c. The out-of-cone contribution is carried almost entirely by tracks with $0.5 < p_T < 4$ GeV/c.

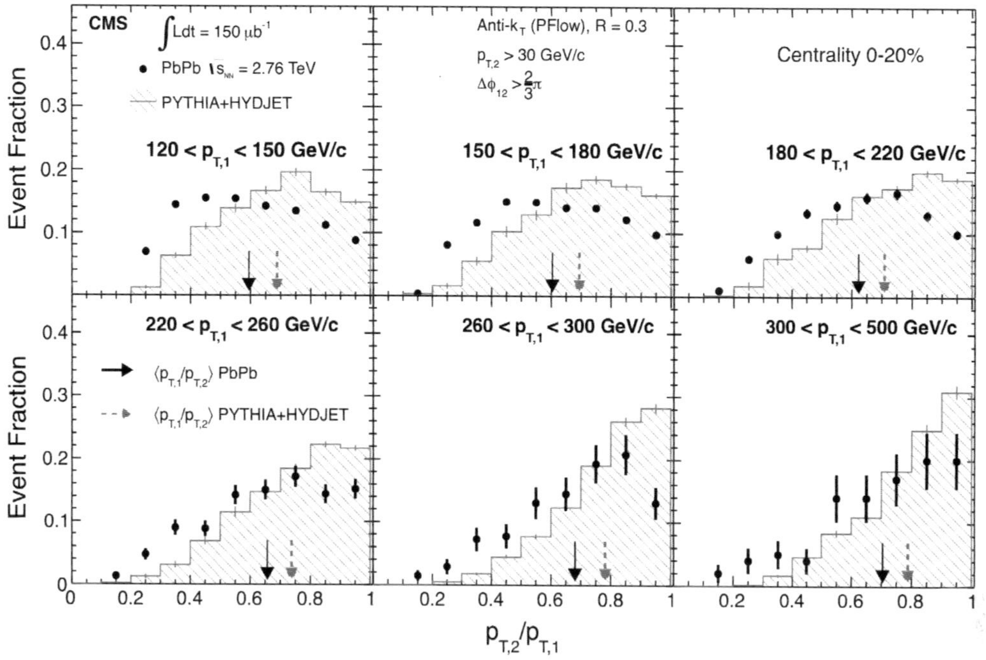

Figure 4. Subleading jet transverse momentum fraction (p_T^{sub}/p_T^{lead}), in bins of leading jet transverse momentum as described in the figures. A $\delta\phi > 2\pi/3$ cut is required between leading and subleading jets. Results for 0-20% central PbPb events are shown as points, while the red histogram shows the results for PYTHIA dijets embedded into HYDJET PbPb simulated events. The arrows show the mean values of the distributions and the error bars represent the statistical uncertainties.

5. Conclusions

Quenching of jets in PbPb collisions is studied by utilizing the di-jets measured with the CMS detector. The anti-k_T algorithm is used to reconstruct jets based on combined tracker and calorimeter information. The data is compared to the reconstructed PYTHIA dijets embedded in HYDJET data events tuned to reproduce the observed underlying event fluctuations. A good agreement between data and simulations is observed for the most peripheral collisions. This agreement starts to fail with increased centrality and for the most central collisions, the dijet momentum imbalance in the data appears to be significantly larger than as seen in the simulation. Across the entire range of jet momenta studied, no significant broadening of the dijet angular correlations is observed with respect to the reference distributions.

The dijet momentum imbalance is studied as a function of the leading jet p_T for different centrality ranges in comparison to simulations. For mid-central (30-50%) and more central PbPb event selections, a significantly lower average dijet momentum ratio (p_T^{sub}/p_T^{lead}) is observed than in the pp data and in the simulations. The mean values of the distributions of the p_T^{sub}/p_T^{lead} show a downward shift with respect to the simulation reference independent of the leading jet p_T. The missing energy was also recovered in the form of soft particles that are far away in phase space from the back to back jets.

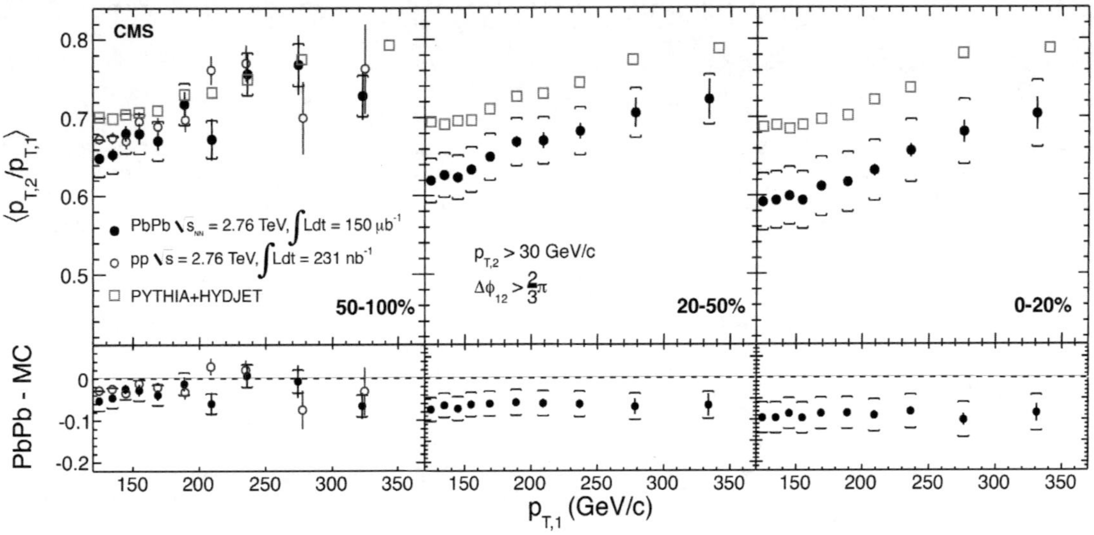

Figure 5. Data and MC simulation comparison of the centrality dependent $< p_{T,2}/p_{T,1} >$ as a function of leading jet p_T. The difference between PbPb and MC results are also shown in the bottom panels.

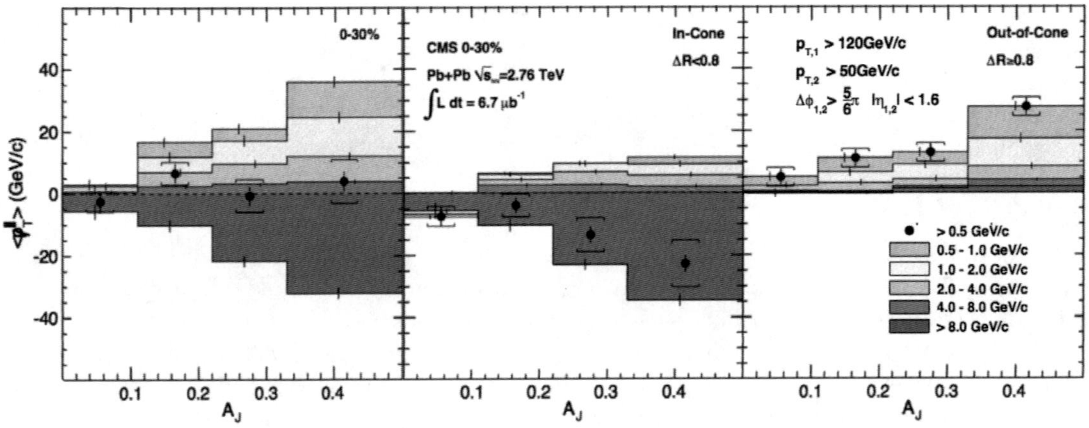

Figure 6. Left: Average missing transverse momentum, $\langle \not{p}_{\mathrm{T}}^{\parallel} \rangle$, for tracks with $p_T > 0.5\mathrm{GeV}/c$, projected onto the leading jet axis (solid circles) for the 0-30% central collisions. Colored bands show the contribution to $\langle \not{p}_{\mathrm{T}}^{\parallel} \rangle$ for various ranges of track p_T. The $\langle \not{p}_{\mathrm{T}}^{\parallel} \rangle$ values are shown inside ($\Delta R < 0.8$) one of the leading or subleading jet cones (middle) and outside ($\Delta R < 0.8$) the leading and subleading jet cones (right) [7].

6. Acknowledgments

This material is based upon work supported by the National Science Foundation under Grant No. 1067907.

References

[1] M. Gyulassy, I. Vitev, X. N. Wang, B. W. Zhang, "Jet quenching and radiative energy loss in dense nuclear matter", Quark Gluon Plasma P123-191 (2003), [nucl-th/0302077].
[2] M. Gyulassy and M. Plumer, "Jet quenching as a probe of dense matter", Nucl. Phys. A 527 (1991) 641.
[3] X. N. Wang and M. Gyulassy, "Gluon shadowing and jet quenching in A + A collisions at $\sqrt{s_{NN}}$ = 200-GeV", Phys. Rev. Lett. 68 (1992) 1480.
[4] J. Adams et al., [STAR Collaboration], "Evidence from d + Au measurements for final state suppression of high p(T) hadrons in Au+Au collisions at RHIC", Phys. Rev. Lett. 91 (2003) 072304, [arXiv:nucl-ex/0306024].
[5] J. Adams et al., [STAR Collaboration], "Direct observation of dijets in central Au+Au collisions at $\sqrt{s_{NN}}$ = 200-GeV", Phys. Rev. Lett. 97 (2006) 162301, [arXiv: nucl-ex/0604018].
[6] S. S. Adler et al., [PHENIX Collaboration], "Common suppression pattern of eta and pi0 mesons at high transverse momentum in Au+Au collisions at $\sqrt{s_{NN}}$ = 200-GeV", Phys. Rev. Lett. 96 (2006) 202301, [arXiv: nucl-ex/0601037].
[7] CMS Collaboration, "Observation and studies of jet quenching in PbPb collisions at nucleon-nucleon center-of-mass energy = 2.76 TeV", Phys. Rev. C 84 (2011) 024906, [arXiv:1102.1957].
[8] ATLAS Collaboration, "Observation of a Centrality-Dependent Dijet Asymmetry in Lead-Lead Collisions at sqrt(S(NN))= 2.76 TeV with the ATLAS Detector at the LHC", Phys. Rev. Lett. 105 (2010) 252303, [arXiv:1011.6182].
[9] ALICE Collaboration, "Particle-yield modification in jet-like azimuthal di-hadron correlations in Pb-Pb collisions at $\sqrt{s_{NN}}$ = 2.76 TeV", (2012), [arXiv:1012.1004].
[10] CMS Collaboration, "Study of high-pT charged particle suppression in PbPb compared to pp collisions at $\sqrt{s_{NN}}$ = 2.76 TeV", (2012), [arXiv:1202.2554].
[11] ALICE Collaboration, "Suppression of Charged Particle Production at Large Transverse Momentum in Central PbPb Collisions at $\sqrt{s_{NN}}$ = 2.76 TeV", Phys. Lett. B 696 (2011) 30, [arXiv:1012.1004].
[12] Y. He, I. Vitev, and B.-W. Zhang, "Next-to-leading order analysis of inclusive jet and di-jet production in heavy ion reactions at the Large Hadron Collider", (2011), [arXiv:1105.2566].
[13] C. Young, B. Schenke, S. Jeon, "Dijet asymmetry at the energies available at the CERN Large Hadron Collider", Phys. Rev. C 84 (2011) 024907, [arXiv:1103.5769].
[14] G.-Y. Qin and B. Muller, "Explanation of Di-jet asymmetry in Pb+Pb collisions at the Large Hadron Collider", Phys. Rev. Lett. 106 (2011) 162302, [arXiv:1012.5280].
[15] CMS Collaboration,"Studies of jet quenching using isolated-photon+jet correlations in PbPb and pp collisions at $\sqrt{s_{NN}}$ = 2.76 TeV", [arXiv:1205.0206]. (Also these proceedings.)
[16] CMS Collaboration, "Measurement of jet fragmentation into charged particles in pp and PbPb collisions at $\sqrt{s_{NN}}$ = 2.76 TeV", [arXiv:1205.5872]. (Also these proceedings.)
[17] J. Casalderrey-Solana, J. G. Milhano, and U. A. Wiedemann, "Jet Quenching via Jet Collimation", J. Phys. G 38 (2011) 035006, [arXiv:1012.0745].
[18] P. M. Chesler, Y.-Y. Ho, and K. Rajagopal, "Shining a Gluon Beam Through Quark-Gluon Plasma", (2011) arXiv:1111.1691
[19] CMS Collaboration,"Jet momentum dependence of jet quenching in PbPb collisions at $\sqrt{s_{NN}}$ = 2.76 TeV", Phys. Lett. B, vol. 712, (2012), p. 176.
[20] CMS Collaboration, "The CMS experiment at the CERN LHC", JINST S08004 (2008).
[21] CMS Collaboration, "CMS Physics Technical Design Report: Addendum on High Density QCD with Heavy Ions", J. Phys. G: Nucl. Part. Phys. 34 (2007) 2307.
[22] M. Cacciari, G. P. Salam and G. Soyez, "The anti-kt jet clustering algorithm", JHEP 0804 (2008) 063, [arXiv:0802.1189].
[23] O. Kodolova, I. Vardanian, A. Nikitenko et al., "The performance of the jet identification and reconstruction in heavy ions collisions with CMS detector", Eur. Phys. J. C 50 (2007) 117.
[24] CMS Collaboration, "Commissioning of the Particle-Flow Reconstruction in Minimum-Bias and Jet Events from pp Collisions at 7 TeV", CMS Physics Analysis Summary CMS-PAS-PFT-10-002, (2010).
[25] CMS Collaboration, "Determination of Jet Energy Calibration and Transverse Momentum Resolution in CMS", JINST 06 (2011) P11002, [arXiv:1107.4277].
[26] T.Sjostrand, S.Mrenna, and P.Skands, "PYTHIA6.4 Physics and Manual", JHEP 05 (2006) 026, [arXiv:hep-ph/0603175].
[27] I.P. Lokhtin, A.M. Snigirev, "A model of jet quenching in ultrarelativistic heavy ion collisions and high-pt hadron spectra at RHIC", Eur. Phys. J. C 45 (2006) 211.

Journal of Physics: Conference Series **389** (2012) 012004 doi:10.1088/1742-6596/389/1/012004

Connecting the underlying event with jet properties in pp collisions at $\sqrt{s} = 7$ TeV with the ALICE experiment

Hermes León Vargas (for the ALICE collaboration)
Institut für Kernphysik, Goethe-Universität Frankfurt am Main, Max-von-Laue-Str. 1, Frankfurt am Main, 60438, Germany

E-mail: hleon@ikf.uni-frankfurt.de

Abstract. A preliminary study of the fragmentation properties of charged particle jets as a function of the Underlying Multiplicity (UM) is presented. The UM is defined such that it can be related to the global soft event characteristics by excluding the contribution due to jet fragmentation. The measurement of jet properties as a function of the UM might be connected to the impact parameter dependence of the transverse nucleon structure as described via Generalized Parton Distributions. The results from the studies are compared to Monte Carlo (MC) models.

1. Introduction

The standard procedure to study the Underlying Event (UE) characteristics starts with defining regions of the azimuthal detector acceptance based on the direction of the leading charged jet [1] or the leading track [2] in an event. These regions, each of them covering an equal angle of $|\Delta\phi| = 2\pi/3$ are called Towards, Transverse and Away. The Toward and Away regions are expected to contain the tracks that are produced by the fragmentation of the partons that took part in the hard scattering while the Transverse region is expected to collect tracks from the UE. ALICE [3] has characterized the UE in pp collisions at $\sqrt{s} = 0.9$ and 7 TeV in [2] by measuring the charged particle density, summed $p_{\rm t}$ and azimuthal correlations (between the leading track and the remaining tracks from the event) in the above mentioned regions. This study presents an estimator for the multiplicity of the underlying event of proton-proton collisions and then uses it to characterize the properties of charged jets in different ranges of the underlying activity. Thus, this study allows to characterize the interplay between the low momentum transfers related to the underlying activity and the hard partonic interactions that produce jets.

2. Data analysis

The experimental results presented in this work are based on a sample of 172 million minimum bias pp collisions at $\sqrt{s} = 7$ TeV. Charged tracks were reconstructed using the Inner Tracking System (ITS) and the Time Projection Chamber (TPC). The combined information from the ITS and the TPC provide a uniform acceptance of tracks as well as good momentum resolution. Details about the track selection used in the charged jet analysis in ALICE can be found in [4]. The jet activity is identified using a jet finder algorithm. For this work we use the SISCone jet finder [5] which is a seedless infrared-safe cone algorithm. The jet finding is performed using charged particles with $p_t > 1$ GeV/c and a cone radius of $R_{\rm Jet}$=0.4 as input parameters for the algorithm. The minimum jet transverse momentum was set to 5 GeV/c.

3. Underlying Multiplicity

The UE definition is that it comprises all the activity in the event but that related to the hardest scattering in a pp collision. In this work we propose to use a jet finding algorithm to identify the orientation in the $\eta - \phi$ space of the products of the jet fragmentation and then characterize the underlying activity in the regions where the jet finder had not identified jet fragmentation activity. We developed a multiplicity estimator to characterize the UM following a suggestion by Frankfurt, Strikman and Weiss who discuss the transverse nucleon structure in [6]. In that paper it is suggested to associate the multiplicity to the impact parameter of the pp collision. Therefore we propose in this work to use a multiplicity estimator, that is not biased by the multiplicity created by the jet fragmentation, to access experimentally the centrality of the pp collision. This provides also a test of the implementation of pp centrality in modern MC generators, which contain simplifications such as the assumption of factorization of the parton x and the impact parameter as well as the lack of information of the flavor and x dependence of the proton's transverse composition as described in [7].

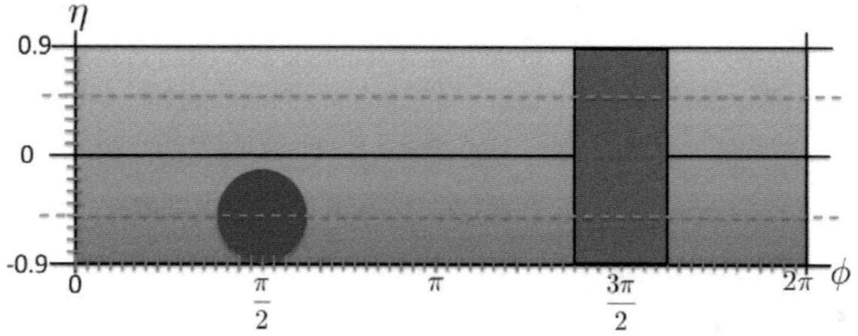

Figure 1. Definition of soft event multiplicity in jet events. The drawing represents the acceptance of the ALICE TPC in the η-ϕ space. The dark blue (colors in the online version) circle shows the jet area covered by the jet cone with radius $R_{\rm Jet}$. The grey rectangular area is the di-jet area. The width of the excluded di-jet area in the ϕ direction is $2R_{\rm Jet}$ and covers the whole acceptance in the η direction. The dotted lines represent the maximum limits for the η coordinate of the accepted jet axis, so the complete jet area falls inside the TPC acceptance.

The UM estimation is based on the separation of the event phase space into two regions: one that contains jet activity and the complementary region. The UM is calculated using the charged tracks measured within the TPC acceptance ($|\eta| < 0.9$ and $0 < \phi \leq 2\pi$). The first classification of the tracks from the events is done by their transverse momentum. The p_t cutoff of the jet finder (1 GeV/c) has been chosen to allow for a gap between this cutoff and the transverse momentum of the tracks that are used in the multiplicity estimation. Therefore we define as the total TPC multiplicity those tracks that have 150 MeV/$c < p_t^{\rm Soft-Track} <$ 900 MeV/c. The second requirement to select tracks for the UM comes from the geometry of the event, this selection can be easily explained using figure 1. The figure shows the η and ϕ acceptance of the ALICE TPC. To calculate the soft TPC multiplicity, that corresponds to the UM measured in the ALICE central barrel, the tracks that fall into two areas of the acceptance are not taken into account in the multiplicity estimation[1]. The first area is shown in figure 1 as a circle that

[1] The analysis is restricted to events where there is only one reconstructed jet in the whole TPC acceptance.

corresponds to the jet cone with axis coordinates $\eta^{\rm Jet}$=0.5 and $\phi^{\rm Jet}$=$\pi/2$ in this example. The second area that is excluded is the di-jet area[2], centred at $\phi^{\rm Exc.}$=$\phi^{\rm Jet}$+π, having a width in the η axis of $2R_{\rm Jet}$ and covering the whole acceptance in the ϕ direction. Figure 2 shows the distribution of the mean values of the soft TPC multiplicity, shown with black markers, as a function of the total TPC multiplicity defined above. The figure shows as well, using a dashed line, the values of the soft TPC multiplicity that would be expected in the case that the reduction in the multiplicity is caused only by a geometrical exclusion[3]. The experimental result shows that the soft TPC multiplicity is a fraction of the geometry expectation and increases linearly as a function of the total TPC multiplicity.

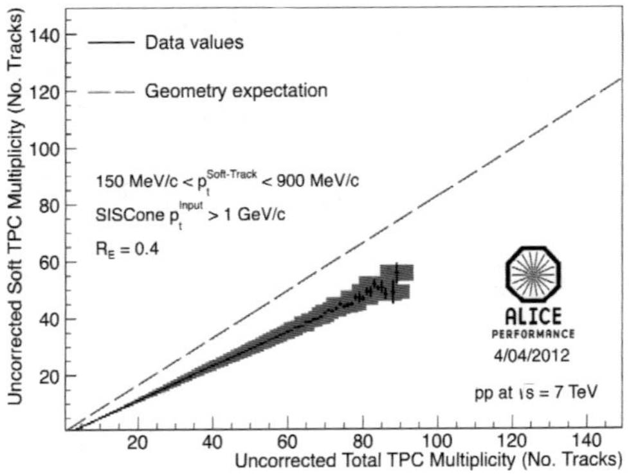

Figure 2. Soft TPC multiplicity as a function of the total TPC multiplicity. The figure shows the experimental measurements as well as the expectation values from a pure geometrical exclusion.

4. The *NT90* jet observable

The jet observable that is used in this work to characterize the fragmentation properties of the charged jets is called *NT90*. *NT90* was proposed in [8] as an observable that can be used to discriminate jets originating from quark or gluon fragmentation. *NT90* is constructed by selecting the tracks that are contained in the jet cone and that fulfil the same $p_{\rm t}$ cutoff used during the jet finding. This set of tracks is then sorted according to their transverse momentum going from the highest to the lowest in magnitude. Then one adds the $p_{\rm t}$ of the sorted tracks until 90% of the total charged jet transverse momentum ($p_{\rm t}^{\rm Ch.Jet}$) is recovered. *NT90* is the minimum number of tracks necessary to recover 90% of the total $p_{\rm t}^{\rm Ch.Jet}$. Figure 3 shows the mean values of *NT90* as a function of the $p_{\rm t}^{\rm Ch.Jet}$ obtained with Pythia6 minimum bias events that include the detector simulation of ALICE. The figure shows the different distributions obtained for quark and gluon initiated jets separately as well as for the inclusive jet sample. One can notice that the mean value of *NT90* at a fixed value of the $p_{\rm t}^{\rm Ch.Jet}$ is larger for gluon initiated jets than for quark initiated jets. This shows the known feature that the gluon jets have on average a softer fragmentation than quark jets at the same parton $p_{\rm t}$.

This is done in order to increase the statistical significance of the multiplicity estimator avoiding the subtraction of larger areas. The number of jet events after making this restriction is reduced by 10% from the total number of jet events in the sample.

[2] Even though the analysis is restricted to events where there is only one reconstructed jet the di-jet area is considered in order to make sure that even if the di-jet was not identified by the jet finder the multiplicity estimation is not biased.

[3] The geometrical exclusion is obtained by making the assumption that the tracks are isotropically distributed over the ALICE central barrel acceptance.

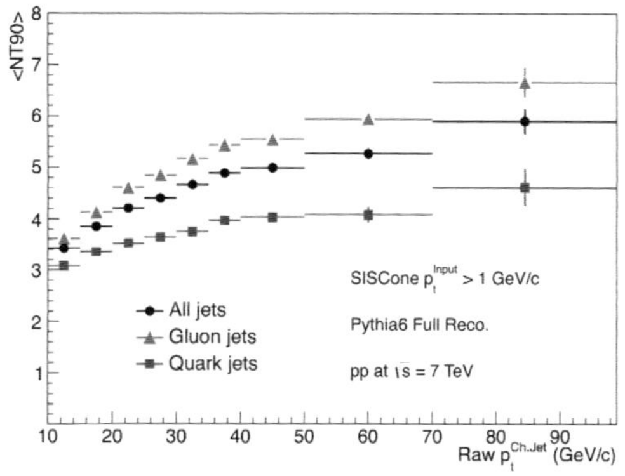

Figure 3. Mean values of *NT90* as a function of the $p_t^{\mathrm{Ch.Jet}}$ for a Pythia6 simulation of pp collisions at $\sqrt{s} = 7$ TeV. The triangles show the distribution obtained for gluon initiated jets while the squares show the distributions for quark initiated jets. The black circle distribution shows the results obtained for the inclusive jet sample.

5. *NT90* and the UM

Figure 4 shows the ratio of the mean values of *NT90* from jet samples corresponding to two bins of the UM as a function of the raw $p_t^{\mathrm{Ch.Jet}}$. The ratio is between a bin with low UM, $\langle NT90\rangle_{\mathrm{LM}}$, and a bin with high UM, $\langle NT90\rangle_{\mathrm{HM}}$. The bin with low UM has a soft TPC multiplicity smaller than 0.4 times the mean soft TPC multiplicity[4] and the one with high UM is associated to a soft TPC multiplicity larger than 1.2 times the mean soft TPC multiplicity. From the figure one can see that the jet properties from the two jet samples are different. The mean values of the *NT90* from the jets associated to high UM are larger than those associated to low values of the UM. This means that the jets in events with large UM have a softer fragmentation compared to the jets associated to low values of the UM.

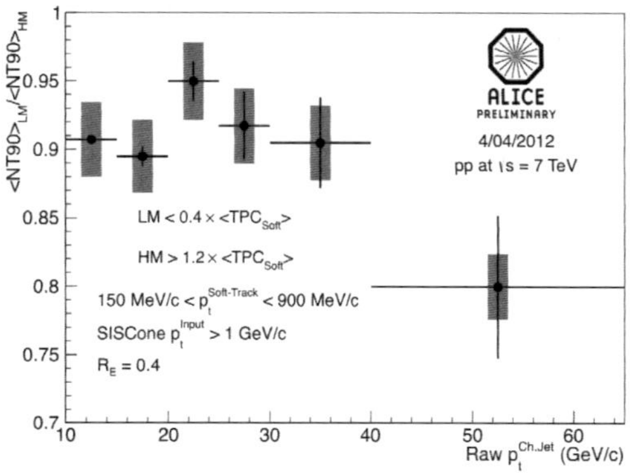

Figure 4. Ratio of the mean values of *NT90* obtained from two jet samples associate to different values of the UM as a function of the raw $p_t^{\mathrm{Ch.Jet}}$. One jet sample is associated to values of the UM smaller than the average soft TPC multiplicity (LM) while the other is from jets associated to values of the UM larger than the average soft TPC multiplicity (HM). The data is presented at the detector level.

[4] The mean soft TPC multiplicity is the average number of tracks with 150 MeV/$c < p_t^{\mathrm{Soft-Track}} <$ 900 MeV/c that are outside the areas related with jet fragmentation.

After looking at the differences in the properties of the charged jets from the two bins of UM, figure 5 shows the results corresponding to those from figure 4 but now obtained from MC generators including full detector simulation. The left panel of figure 5 shows the results obtained with Phojet [9], and the right panel shows the results obtained with a Pythia6 [10] simulation using the tune Perugia 0 [11]. As one can see from the figure the MC generators are able to reproduce a general trend observed in the data, i.e. a softer fragmentation of the jets associated to high values of the UM compared to those associated to low UM.

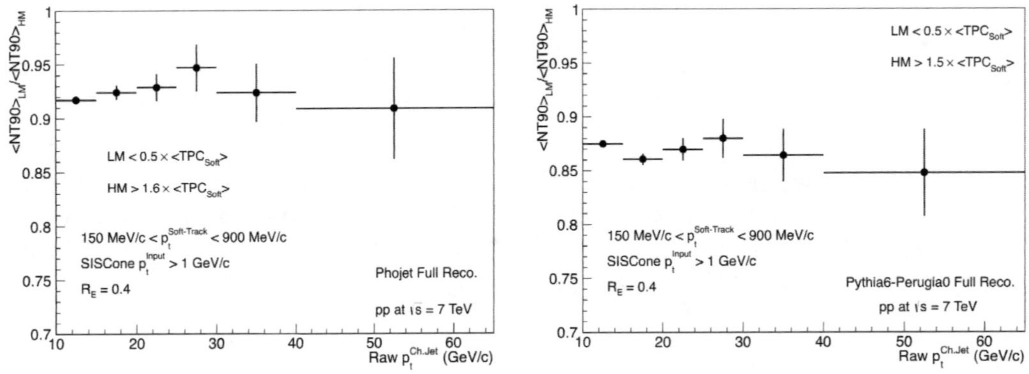

Figure 5. Ratio of the mean values of *NT90* from two bins of the UM as a function of the raw $p_{\mathrm{t}}^{\mathrm{Ch.Jet}}$. Left panel: Results obtained with a Phojet simulation including full detector simulation. Right panel: Results obtained with a Pythia6 simulation with the Perugia 0 tune including full detector simulation.

A comparison of the experimental results and the predictions from the MC generators is presented in figure 6, which shows the mean values of *NT90* as a function of the normalized UM measured in the ALICE central barrel.

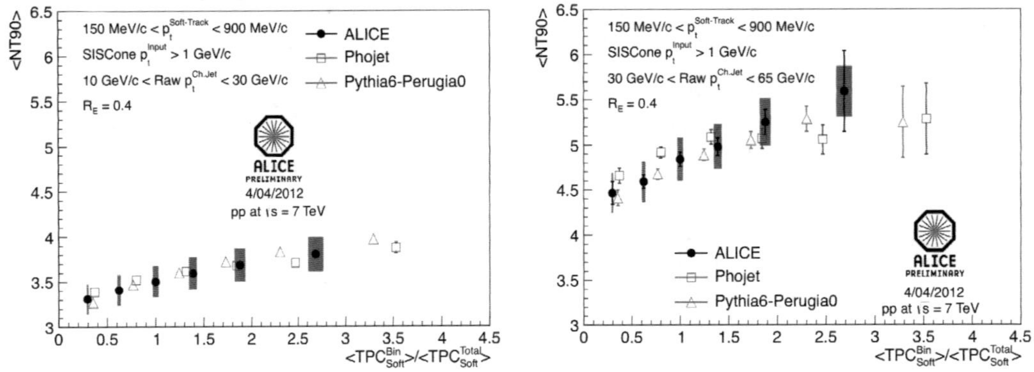

Figure 6. Mean values of *NT90* as a function of the normalised UM. Left panel: Results obtained for jets with 10 GeV/c < $p_{\mathrm{t}}^{\mathrm{Ch.Jet}}$ < 30 GeV/c. Right panel: Results for jets with 30 GeV/c < $p_{\mathrm{t}}^{\mathrm{Ch.Jet}}$ < 65 GeV/c.

The standard way to present results as a function of multiplicity is by plotting the observable as a function of fractions of the mean multiplicity. In this case the results are presented for different bins of UM, $\langle \mathrm{TPC}^{\mathrm{Bin}}_{\mathrm{Soft}} \rangle$, normalized by the mean value of UM, $\langle \mathrm{TPC}^{\mathrm{Total}}_{\mathrm{Soft}} \rangle$, measured in the central barrel for each data sample. The data was separated into two ranges of $p_{\mathrm{t}}^{\mathrm{Ch.Jet}}$. The left panel of figure 6 shows the results obtained for 10 GeV/c < $p_{\mathrm{t}}^{\mathrm{Ch.Jet}}$ < 30 GeV/c and in the right panel for jets with 30 GeV/c < $p_{\mathrm{t}}^{\mathrm{Ch.Jet}}$ < 65 GeV/c. The results obtained from the MC generators are presented after the full ALICE detector simulation so they are directly comparable with the experimental results. For jets with 10 GeV/c < $p_{\mathrm{t}}^{\mathrm{Ch.Jet}}$ < 30 GeV/c Pythia6 shows a similar dependence with the multiplicity as the one observed in the experimental data, while Phojet shows a flatter dependency. For jets with 30 GeV/c < $p_{\mathrm{t}}^{\mathrm{Ch.Jet}}$ < 65 GeV/c one can observe a monotonic increase of the mean values of *NT90* with increasing multiplicity for the experimental data. A similar trend is observed in the Pythia6 results while Phojet on the other hand shows an increase of the mean values of *NT90* that seems to saturate after reaching the mean value of the UM.

6. Conclusions

We have presented an estimator of the underlying soft multiplicity in pp collisions. The characterization is done with the UM, that uses the information from a jet finder to localize the regions of the detector acceptance that contain most of the jet fragmentation products. Then we separated the collision data in bins of UM and studied the charged jet fragmentation using the *NT90* observable. We observe that events with high activity of the UM are correlated to jets that undergo a softer fragmentation. The MC models Pythia6 and Phojet are able to reproduce the experimental observations for low momentum jets, while for intermediate jet transverse momentum Phojet seems to fail for UM above the mean values. The fact that the MC generators are able to describe the dependence of the jet fragmentation as a function of the UM needs further studies given that the models contain a simplified approach to describe the nucleon-nucleon collision centrality.

Acknowledgments

The author thanks C. Blume for the collaboration in the analysis presented in this work and C. Klein-Bösing and M. van Leeuwen for their comments and support. This work was supported by the Helmholtz Research School for Quark Matter Studies.

References

[1] CDF Collaboration 2002 *Phys. Rev.* D **65** 092002
[2] The ALICE Collaboration 2011 *Preprint* arXiv:1112.2082v2
[3] The ALICE Collaboration 2008 *JINST* **3** S08002
[4] The ALICE Collaboration 2012 *JHEP* **3** 053
[5] Salam G P and Soyez G 2007 *Preprint* arXiv:0704.0292v2
[6] Frankfurt L, Strikman M and Weiss C 2011 *Phys. Rev.* D **83** 054012
[7] Buckley A *et al.* 2011 *Phys. Rep.* **504** 054012 145-233
[8] Pumplin J 1991 *Phys. Rev.* D **44** 7 2025-32
[9] Roesler S, Engel R and Ranft J 2000 *Preprint* arXiv:hep-ph/0012252v1
[10] Sjöstrand, Mrenna S and Skands P 2006 *JHEP* **05** 026
[11] Skands P 2012 *Phys. Rev.* D **82** 074018

Charged particle jet measurements with the ALICE experiment in proton-proton collisions at the LHC

S. K. Prasad for the ALICE Collaboration[1]

Wayne State University, 666 West Hancock Street, Detroit, MI, USA, 48201

E-mail: sprasad@cern.ch

Abstract. We present preliminary results of measurements of charged particle jet properties in proton-proton collisions at $\sqrt{s}$ = 7 TeV using the ALICE detector. Jets are reconstructed using anti $-$ k_T, k_T and SISCone jet finding algorithms with resolution parameter $R = 0.4$ in the range of transverse momentum from 20 to 100 GeV/c in the midrapidity region ($|\eta| < 0.5$). The uncorrected charged jet spectra obtained using the three different jet finders show good agreement. The data are compared to predictions from PYTHIA-Perugia0, PYTHIA-Perugia2011, and PHOJET. The mean charged particle multiplicity in leading jets increases with increasing jet p_T and is consistent with model predictions. The radial distributions of transverse momentum about the jet direction and the distributions of the average radius containing 80% of the total jet p_T found in the jet cone ($R = 0.4$ in this analysis), indicate that high p_T jets are more collimated than low p_T jets.

1. Introduction

Jets are collimated sprays of particles originating from the fragmentation of hard scattered partons (quarks and gluons) produced in pp (or A–A) collisions [2]. As such, jets serve as a proxy for the high p_T partons produced in elementary hard scatterings. In pp collisions, they provide a unique tool to test perturbative quantum chromodynamics (pQCD) [3]. Measurements of jet shapes in particular, provide the details of the jet fragmentation process [2]. Jet shapes are also sensitive to the type of partons (quarks or gluons) that fragment into hadrons. In addition, measurements in pp collisions provide a baseline for similar measurements in more complex A–A collisions where various nuclear effects are expected to take place. Jet shape observables have previously been measured by the CDF [4, 5, 6] and D0 [7] collaborations in $p\bar{p}$ collisions and more recently by the ATLAS [8] and CMS [9] collaborations in pp collisions at $\sqrt{s}$ = 7 TeV. In this paper we report preliminary results of charged particle jet properties measured by A Large Ion Collider Experiment (ALICE) in pp collisions at $\sqrt{s}$ = 7 TeV.

2. Data analysis

The data used in this analysis were collected during the Large Hadron Collider (LHC) [10, 11] run in the year 2010 with the ALICE detector [12]. The main detector subsystems used for tracking are the Time Projection Chamber (TPC) [13] and the Inner Tracking System (ITS) [14]. The V-ZERO (V0) counters [15] and the ITS are used for online trigger to select the minimum bias events. Only events with primary vertex within ± 10 cm along the beam axis from the nominal

[1] for a full list of authors and the acknowledgements see [1].

Journal of Physics: Conference Series **389** (2012) 012005 doi:10.1088/1742-6596/389/1/012005

interaction point are analyzed to ensure uniform acceptance in pseudo-rapidity (η) and minimize the need for complex acceptance corrections. This analysis is based on 161 M minimum bias events. Tracks are reconstructed using combined information from the TPC and the ITS. These tracks have uniform high tracking efficiency for charged particles, good momentum resolution and uniform ϕ distribution. Tracks are included in the analysis if $p_{\mathrm{T,track}} > 0.150$ GeV/c and $\mid \eta_{\mathrm{track}} \mid < 0.9$.

3. Jet reconstruction method and observable definitions

Charged particle jets are reconstructed using the sequential recombination algorithms anti-$\mathrm{k_T}$ [16] and $\mathrm{k_T}$ [17, 18, 19] from the FastJet package [20] and a seedless infrared-safe cone based SISCone [21] algorithm. Jets are reconstructed with the resolution parameter $R = 0.4$. Only jets within $\mid \eta \mid < 0.5$ and in the transverse momentum range from 20 to 100 GeV/c are considered in this analysis. The jet properties are measured for leading jets only i.e. the jet with the highest p_{T} in each event. The jet shape observables studied in this analysis are defined in the following subsections.

3.1. Charged particle multiplicity in leading jet

The charged particle multiplicity of the leading jet is defined as the number of charged particles within the jet cone. The mean is computed in bins of the jet p_{T}.

3.2. Leading charged jet size (R_{80})

The size of the jet, R_{80}, is defined as the radius in $\eta - \phi$ space that contains 80% of the total transverse momentum found in the jet cone ($R = 0.4$ in this analysis). We report the mean value of R_{80} for leading jets as a function of jet p_{T}.

3.3. Radial distribution of transverse momentum within the leading charged jet

We also study the distribution of transverse momentum within a leading charged jet as a function of distance $r = \sqrt{(\Delta\eta)^2 + (\Delta\phi)^2}$ from the jet direction. We construct annular regions as illustrated in Fig. 1 at given r values. The scalar sum of the transverse momenta ($p_{\mathrm{T}}^{\mathrm{sum}}$) of all charged particles produced in each annulus is calculated jet by jet as a function of distance r. We report the mean value of this sum (Eq. 1) as a function of jet p_{T}.

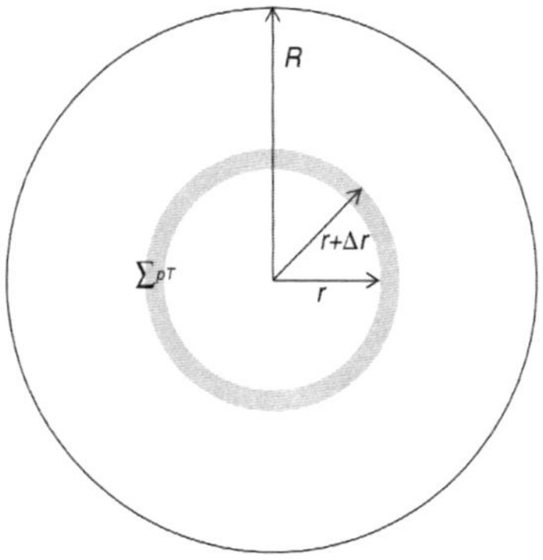

Figure 1. Illustration of radial distribution of scalar $p_{\mathrm{T}}^{\mathrm{sum}}$ about the jet axis in the leading charged jet. Scalar $p_{\mathrm{T}}^{\mathrm{sum}}$ is obtained for each annular region as a function of distance $r = \sqrt{(\Delta\eta)^2 + (\Delta\phi)^2}$ from the jet axis.

Journal of Physics: Conference Series **389** (2012) 012005 doi:10.1088/1742-6596/389/1/012005

$$< \frac{dp_{\mathrm{T}}^{\mathrm{sum}}}{dr} > (r) = \frac{1}{\Delta r} \frac{1}{\mathrm{N_{jets}}} \sum p_{\mathrm{T}}(r - \Delta r/2, r + \Delta r/2) \tag{1}$$

where, $p_{\mathrm{T}}(r - \Delta r/2, r + \Delta r/2)$ denotes the summed p_{T} of all tracks inside the annular ring between $r - \Delta r/2$ and $r + \Delta r/2$. Δr is the radial width of the annulus. The analysis was carried out with $\Delta r = 0.02$. $\mathrm{N_{jets}}$ denotes the total number of jets.

4. Instrumental effects and systematic uncertainties

We used a bin-by-bin correction procedure similar to those used by the CDF [5] and ATLAS [8] collaborations to correct the measured jet shape observables for detector effects. PYTHIA 6.4 [22] Monte Carlo (MC) with tune Perugia0 [23] describes the measured jet shapes reasonably well for $R = 0.4$ and therefore is used for the correction. Correction factors are computed for each bin in jet p_{T}. They are defined as the ratio between the jet shape observable at particle level to that at detector level in their respective jet p_{T} bin as described by Eq. 2,

$$\mathrm{CF(p_t)} = \frac{\mathrm{Obs_{mc}^{part}}}{\mathrm{Obs_{mc}^{det}}} \tag{2}$$

where, $\mathrm{Obs_{mc}^{part}}$ is the observable at particle level and $\mathrm{Obs_{mc}^{det}}$ is that at the detector level. Jet shape observables at particle level are obtained using MC jets at particle level. A full GEANT [24] simulation implementing the detector effects has been performed to obtain the observables at detector level. We have not corrected for underlying events (UE).

The main contributors to the systematic uncertainties are the uncertainties in efficiency and momentum resolution in the track reconstruction. These effects were estimated using MC (PYTHIA) events. Uncertainties on the measured observables are estimated by varying the efficiency and momentum resolution by 5% and 20% respectively.

5. Results

5.1. Comparative performance analysis of jet finders

Figure 2 (Top panel) shows the uncorrected transverse momentum distributions of charged particle jets obtained using the anti $-$ $\mathrm{k_T}$ (filled black circles), $\mathrm{k_T}$ (filled red triangles), and SISCone (open blue circles) jet finding algorithms with $R = 0.4$ within $\mid \eta \mid < 0.5$, in the range of transverse momentum from 20 to 100 GeV/c for pp collisions at $\sqrt{s} = 7$ TeV. The ratios between the uncorrected yields obtained with the $\mathrm{k_T}$, and SISCone to that obtained with the anti-$\mathrm{k_T}$ are shown in the bottom panel. We observe that the three different jet reconstruction algorithms agree well with each other at detector level.

5.2. Charged particle multiplicity in leading jet

The mean charged particle multiplicity ($< \mathrm{N_{ch}} >$) distributions for leading jets and their dependence on jet p_{T} are shown in Fig.3. One observes that the mean charged multiplicity increases with increasing jet p_{T} as expected based on prior measurements by the CDF [5] collaboration. This behavior is observed also in the MC predictions. The ratios of measured data to MC predictions suggest that models reproduce the data rather well.

5.3. Leading charged jet size

The distributions of the average radius ($< R_{80} >$) containing 80% of the total jet p_{T} found in the jet cone ($R = 0.4$) are shown in Fig. 4 as a function of jet p_{T}. At low jet p_{T} (20 GeV/c), $< \mathrm{R_{80}} >$ is contained in a very small cone of radius ~0.16, which decreases further down to ~0.08 for high p_{T} jets (100 GeV/c).

Journal of Physics: Conference Series **389** (2012) 012005 doi:10.1088/1742-6596/389/1/012005

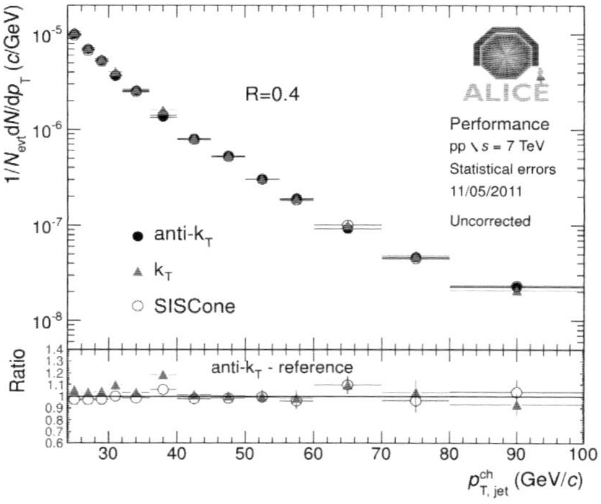

Figure 2. (Top panel) Uncorrected p_{T}-distributions of jets for pp collisions at $\sqrt{s} = 7$ TeV obtained with the anti $-$ $\mathrm{k_T}$ (filled black circles), $\mathrm{k_T}$ (filled red triangles), and SISCone (open blue circles) jet finding algorithms. (Bottom panel) The ratios between the yields obtained with the $\mathrm{k_T}$, and SISCone to that with the anti $-$ $\mathrm{k_T}$. The solid line at unity is to guide the eye. Error bars indicate the statistical uncertainties.

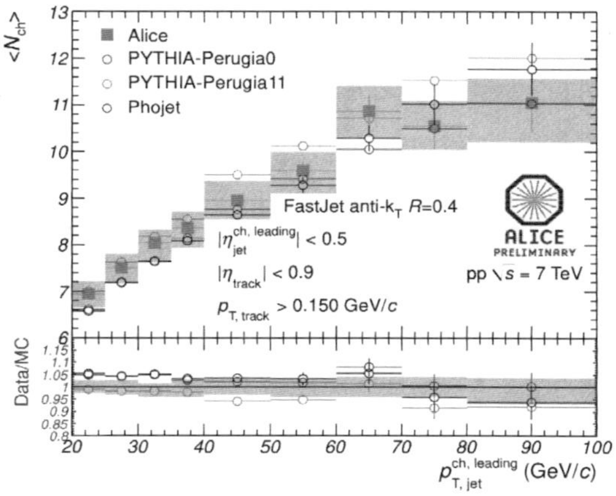

Figure 3. (Top panel) Mean charged particle multiplicity in the leading jet as a function of jet p_{T} for pp collisions at $\sqrt{s} = 7$ TeV (filled magenta squares) compared to predictions from PYTHIA-Perugia0 [23] (open blue circles), PYTHIA-Perugia2011 [23] (open red circles) and PHOJET [25] (open black circles). (Bottom panel) The ratios between data to MC predictions. Error bars indicate the statistical uncertainties while the gray bands indicate the systematic uncertainties.

5.4. Radial distribution of transverse momentum within the leading charged jet

Figure 5 shows the radial distributions of transverse momentum in the leading jet about the jet axis for jet p_{T} between 20 to 30 GeV/c (left panel) and 60 to 80 GeV/c (right panel). It can be clearly seen in the figure that the transverse momentum density is largest near the jet axis and decreases with r. The higher slop for this decrease in case of high p_{T} jets (right panel) indicates that high p_{T} jets are more collimated than low p_{T} jets.

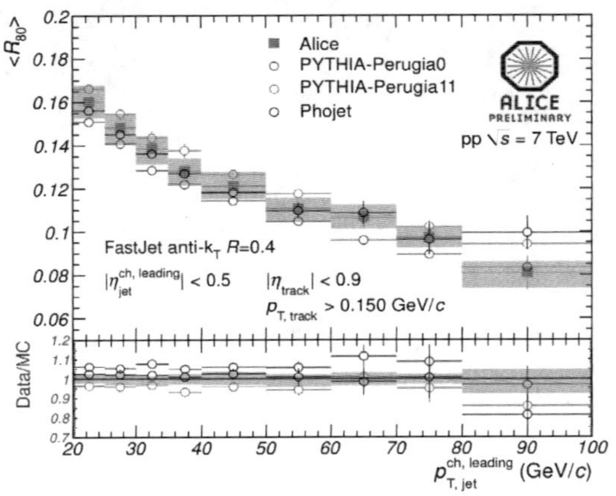

Figure 4. (Top panel) Distributions of the average radius R_{80} containing 80% of jet p_{T} found in the jet cone ($R = 0.4$) as a function of jet p_{T} for pp collisions at $\sqrt{s} = 7$ TeV (magenta squares) compared to predictions from PYTHIA-Perugia0 [23] (open blue circles), PYTHIA-Perugia2011 [23] (open red circles) and PHOJET [25] (open black circles). (Bottom panel) The ratios between the data to the predictions from simulations. Error bars indicate the statistical uncertainties while the gray bands indicate the systematic uncertainties.

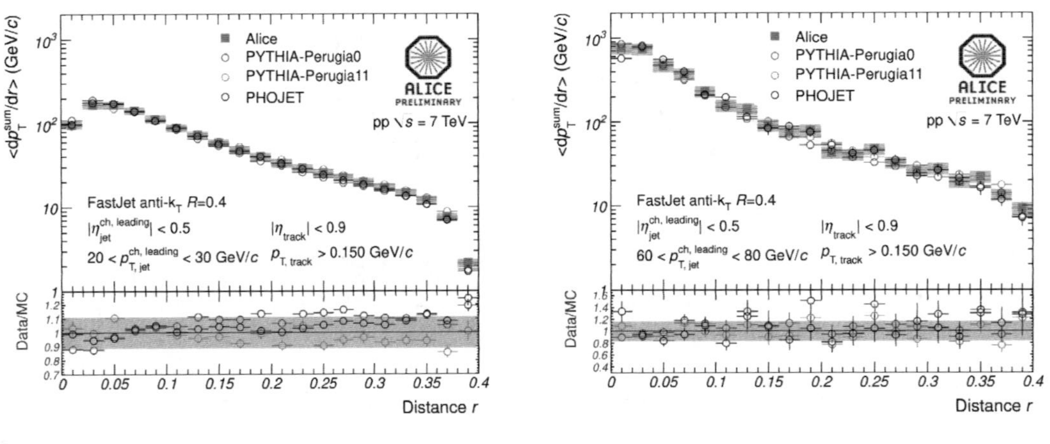

Figure 5. (Left panel) (Top) Radial distributions of transverse momentum about the jet axis in the leading charged jet (magenta squares) for pp collisions at $\sqrt{s} = 7$ TeV for $20 <$ jet $p_{\mathrm{T}} <$ 30 GeV/c compared to predictions from PYTHIA-Perugia0 [23] (open blue circles), PYTHIA-Perugia2011 [23] (open red circles) and PHOJET [25] (open black circles). (Bottom) The ratios between the data to the predictions from simulations. (Right panel) Results for $60 <$ jet $p_{\mathrm{T}} <$ 80 GeV/c. Error bars indicate the statistical uncertainties while the gray bands indicate the systematic uncertainties.

6. Conclusions

We reported preliminary measurements of charged particle jet properties in pp collisions at $\sqrt{s} = 7$ TeV using the ALICE detector. Jets were reconstructed using the anti $-$ $\mathrm{k_T}$, $\mathrm{k_T}$, and SISCone jet finding algorithms with resolution parameter $R = 0.4$ in the p_{T} range from 20 to

Journal of Physics: Conference Series **389** (2012) 012005 doi:10.1088/1742-6596/389/1/012005

100 GeV/c. A comparison between the uncorrected charged jet spectra obtained using three different jet finding algorithms show a good agreement. The mean charged particle multiplicity in the leading charged jet increases with increasing jet p_{T}. Similar to the previous observations reported by the CDF [5] and ATLAS [8] collaborations, results on R_{80} and $p_{\mathrm{T}}^{\mathrm{sum}}$ show that high p_{T} jets are more collimated than low p_{T} jets. Predictions from simulations (PYTHIA-Perugia0, PYTHIA-Perugia2011, and PHOJET) describe the jet shape observables reasonably well.

References

[1] ALICE Collaboration 2012 *arXiv [nucl-ex/1207.0900]*
[2] Ellis S D, Kunszt Z and Soper D E 1992 *Phys. Rev. Lett.* **69** 3615
[3] Gross D J and Wilczek F 1973 *Phys. Rev.* D **8** 3633
[4] CDF Collaboration 1993 *Phys. Rev. Lett.* **70** 713
[5] CDF Collaboration 2001 *Phys. Rev.* D **65** 092002
[6] CDF Collaboration 2005 *Phys. Rev.* D **71** 112002
[7] D0 Collaboration 1995 *Phys. Lett.* B **357** 500
[8] ATLAS Collaboration 2011 *Phys. Rev.* D **83** 052003
[9] CMS Collaboration 2011 *Jet transverse structure and momentum distribution in pp collisions at 7 TeV* (Technical Report: CMS-PAS-QCD-10-014 CERN Geneva)
[10] Pettersson T S (ed.) and Lefèvre (ed.) P 1995 *The Large Hadron Collider : conceptual design* (CERN-AC-95-05 LHC)
[11] *LHC Design Report Volume I+II+III* 2004 (CERN-2004-003-V-1, CERN-2004-003-V-2, CERN-2004-003-V-3)
http://ab-div.web.cern.ch/ab-div/Publications/LHC-DesignReport.html
[12] ALICE Collaboration 2008 *JINST* **3** S08002
[13] Alme J *et al* 2010 *Nucl. Instrum. Meth.* A **622** 316-367
[14] Aamodt K *et al* 2010 *JINST* **5** P03003
[15] ALICE Collaboration 2004 *ALICE forward detectors: FMD, TO and VO: Technical Design Report* (CERN-LHCC-2004-025)
http://cdsweb.cern.ch/record/781854
[16] Cacciari M, Salam G P and Soyez G 2008 *JHEP* **04** 63
[17] Catani S, Dokshitzer Y L, Seymour M H and Webber B R 1993 *Nucl. Phys.* B **406** 187
[18] Catani S, Dokshitzer Y L, Olsson M, Turnock G and Webber B R 1991 *Phys. Lett.* B **269** 432
[19] Ellis S D and Soper D E 1993 *Phys. Rev.* D **48** 31603166
[20] Cacciari M and Salam G P 2006 *Phys. Lett.* B **641** 57-61
[21] Salam G P and Soyez G 2007 *JHEP* **05** 086
[22] Sjostrand T, Mrenna S and Skands P Z 2006 *JHEP* **0605** 026
[23] Skands P Z 2010 *Phys. Rev.* D **82** 074018
[24] Brun R, Bruyant F, Maire M, McPherson A C and Zanarini P 1985 *GEANT3 user guide, CERN data handling division* (DD/EE/84-1)
[25] Roesler S, Engel R and Ranft J 2000 *The Monte Carlo event generator DPMJET-III* (Lisbon 2000-Advanced Monte Carlo) 1033-1038 *[hep-ph/0012252]*

28th Winter Workshop on Nuclear Dynamics 2012 IOP Publishing
Journal of Physics: Conference Series **389** (2012) 012006 doi:10.1088/1742-6596/389/1/012006

Reconstructed Jet Results in p+p, d+Au and Cu+Cu collisions at 200 GeV from PHENIX

D.V. Perepelitsa on behalf of the PHENIX Collaboration
Columbia University in the City of New York, New York, NY 10027, USA
Nevis Laboratories, Irvington, NY 10533, USA
E-mail: dvp@bnl.gov

Abstract. Jet reconstruction in heavy ion collisions at RHIC and the LHC is becoming a popular tool for exploring medium effects including the energy loss and modified fragmentation of hard-scattered partons. In p+A and d+A collisions, reconstructed jets are important for evaluating cold nuclear matter effects such as the impact parameter dependence of nuclear parton distribution functions and initial state energy loss. We present current PHENIX results from p+p, d+Au, and Cu+Cu collisions at 200 GeV using the Gaussian filter and anti-k_T algorithms. The systematic study of direct jet reconstruction across a variety of collisions systems at PHENIX will help to tell a coherent story of jet physics at RHIC.

1. Introduction

Jet reconstruction in heavy ion collisions at RHIC and the LHC is evolving into a sophisticated method to probe suppression, quenching and other hot and cold nuclear medium effects on fragmenting hard-scattered partons. Because reconstructed jets can better determine the full kinematics of a fragmenting parton at leading order (LO), they may be more sensitive to medium effects than inclusive measurements of leading hadrons from parton fragmentation. Together, jet measurements at RHIC and the LHC can provide a complementary set of high-statistics heavy ion results to investigate energy loss mechanics at two different temperatures.

The flexibility of colliding species in the RHIC facility in particular allows experimental control over the system size, energy density and geometry of heavy ion collisions, all of which can be exploited with a developing jet physics program.

An important signature of cold and hot nuclear medium effects on reconstructed jets is the deviation of jet yields from a naive geometry-dependent scaling of jets in $p+p$ collisions. The nuclear modification factor R_{AA} is defined as the ratio of jet yields in Cu+Cu collisions of a given centrality to the scaled $p+p$ reference,

$$R_{\mathrm{AA}}^{cent} = \frac{(1/N_{\mathrm{evt}}^{cent})(dN^{\mathrm{cent}}/dp_{\mathrm{T}})}{< T_{\mathrm{AB}}^{cent} > (d\sigma^{\mathrm{p+p}}/dp_{\mathrm{T}})}, \tag{1}$$

where $< T_{AB}^{cent} >$ is the nuclear overlap function derived from a Glauber simulation of Cu+Cu collisions. The R_{AA} is sensitive to effects from the hot medium formed in heavy ion collisions, such as partonic energy loss or jet quenching.

Published under licence by IOP Publishing Ltd

Journal of Physics: Conference Series **389** (2012) 012006 doi:10.1088/1742-6596/389/1/012006

In lieu of a p+p reference, peripheral collisions, where the medium effects are expected to be small, can be used instead. The nuclear modification factor $R_{\rm CP}$ is defined as the ratio of jet yields relative to that in peripheral collisions,

$$R_{\rm CP}^{cent} = \frac{(1/N_{\rm coll}^{cent})(1/N_{\rm evt}^{cent})(dN^{\rm cent}/dp_{\rm T})}{(1/N_{\rm coll}^{60-88\%})(1/N_{\rm evt}^{60-88\%})(dN^{60-88\%}/dp_{\rm T})}, \tag{2}$$

where $N_{\rm coll}^{cent}$ is the mean number of binary collisions. In d+Au collisions, the $R_{\rm CP}$ measures the scale of the differences in jet production between the centralities that arise from cold nuclear matter effects (CNM) such as initial and final state energy loss or modification of nuclear parton distribution functions (nPDF's). It does not, however, measure the absolute scale of the nuclear effect relative to p+p collisions, which is necessary for a full understanding of CNM effects and their implications for suppression/quenching in A+A (ion-ion) collisions.

In these proceedings, we present the latest jet reconstruction results in p+p and Cu+Cu[1] and in d+Au collisions[2] at PHENIX.

2. Experimental setup

2.1. PHENIX Detector

This analysis uses the two PHENIX central arm spectrometers, which cover a pseudorapidity range of $-0.35 < \eta < +0.35$ and have a azimuthal acceptance of $\Delta\phi = \pi/2$ in each arm.

Charged tracks from charged hadrons and electrons are reconstructed using the drift chamber (DC), pad chamber (PC) and ring-imaging Čerenkov detector (RICH). Electromagnetic energy from photons, π^0's and other neutral hadrons is clustered by the electromagnetic calorimeter (EMCal). Clusters are required to have a time of flight within a $\pm 5\sigma_{tof}$ window of a light-speed particle, where σ_{tof} is the timing resolution in calorimeter towers.

Reconstructed tracks and clusters with $p_{\rm T} > 400$ MeV/c are used as the inputs to the jet reconstruction algorithms. For the case when charged tracks are associated with a cluster in the EMCal, only the track (which better captures the momentum of charged hadrons) is ultimately used.

2.2. Dataset

The measurements presented here are based on data taken by the PHENIX detector in p+p and Cu+Cu collisions in 2005 and d+Au collisions in 2008, all at a center of mass energy per nucleon of $\sqrt{s_{\rm NN}} = 200$ GeV.

The high rate of the PHENIX data acquisition system (> 6 kHz in p+p collisions) allows the complementary recording of a triggered dataset for high-statistics high-$p_{\rm T}$ results as well as a large minimum bias dataset to understand the trigger efficiency in a data-driven way. Data from minimum bias collisions is used to evaluate and extrapolate the performance of the high-$p_{\rm T}$ Electromagnetic/RICH Trigger (ERT), which requires > 2.1(1.6) GeV within a 4×4 calorimeter tower window in the PbSc (PbGl) sectors of the EMCal. Efficiency-corrected, trigger-selected data is used to construct the jet yields used in the final result.

3. Methodology

3.1. Jet Reconstruction

In p+p and Cu+Cu collisions, jets are reconstructed with the Gaussian filter algorithm (parameter $\sigma = 0.3$), a seedless, cone-like, infrared- and collinear-safe algorithm with a continuous angular weighing of energy deposits[3]. The Gaussian filter jet axis (η, ϕ) is defined by the position at which the convolution of the event energy density $p_{\rm T}(\eta', \phi')$ with a gaussian kernel is a local maximum:

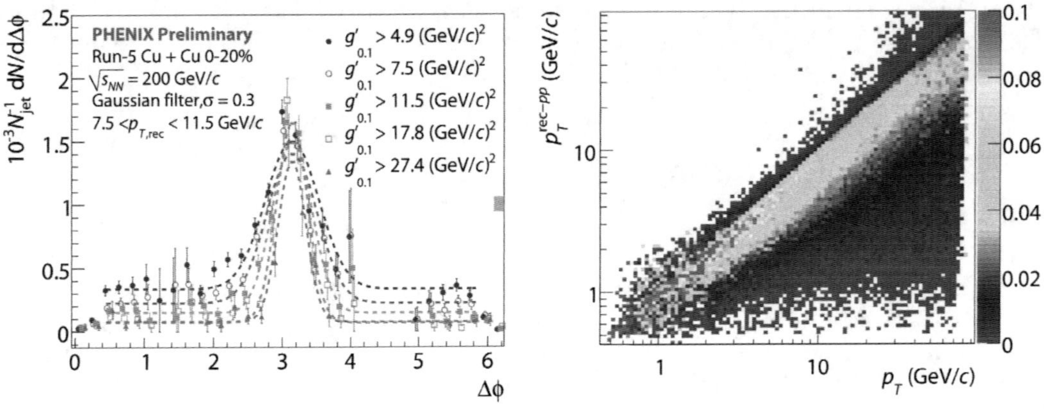

Figure 1. Left: $\Delta\phi$ distributions for jets in Cu+Cu 0-20% collisions that pass successively harder fake jet rejection cuts. Right: transfer matrix $P(p_T^{\rm rec}|p_T^{\rm true})$ for reconstructed jets in p+p collisions at PHENIX.

$$p_T^{\rm jet} \equiv \max\left\{\int\int d\eta' d\phi' p_T\left(\eta', \phi'\right) e^{-(\Delta\eta^2+\Delta\phi^2)/2\sigma^2}\right\}, \tag{3}$$

where $\Delta\eta = \eta - \eta'$, $\Delta\phi = \phi - \phi'$ and $p_T^{\rm jet}$ is the local maximum value which is taken to be the transverse momentum of the jet. The continuous angular weighing stabilizes the jet axis in the presence of background and optimizes the signal to background ratio (S/B) by focusing on the energetic core of the jet.

In d+Au collisions, the anti-k_T jet algorithm[4] with a cone size of $R = 0.3$ is used. This is the first time anti-k_T, a sequential recombination algorithm which is popular in other heavy ion jet analyses, was tested for use in the PHENIX experiment.

Reconstructed jets are required to have three of more constituents within a 30° solid angle of the jet axis. Additionally, jets must pass a fiducial cut requiring that they lie $\Delta\eta, \Delta\phi > 0.05$ units away from any edge of the PHENIX central arm acceptance. Finally, a cut on the charged fraction of the jet momentum is applied to eliminate residual background from mis-identified charged tracks.

3.1.1. Fake jet rejection Two different methods are used to handle the influence of fake jets (jets reconstructed from a combinatorial collection of nearby underlying heavy ion event particles). In d+Au collisions, we only consider the jet yields > 15 GeV/c, where the fake rate is determined from data to be negligible.

In Cu+Cu collisions, PHENIX has developed a technique to separate real low-p_T jets from UE fluctuations on a jet by jet basis, which is implemented by forming the g-discriminant for each jet:

$$g \equiv \sum_{i\in\rm fragment} (p_T)_i^2 \exp\left(-(\Delta\eta^2 + \Delta\phi^2)/2\sigma_{\rm dis}^2\right), \tag{4}$$

where $\Delta\eta$ and $\Delta\phi$ are the distances in η/ϕ space between the jet axis and fragment, and $\sigma_{\rm dis} = 0.1$ is chosen to provide a tighter angular weighing than the Gaussian filter jet reconstruction. Requiring a minimum value of g is similar to an angularly weighted p_T cut which is efficient for collimated jets with a harder fragmentation pattern, and inefficient for jets consistent with underlying event fluctuations.

A data-driven method examining the angular correlation of reconstructed jets is used to determine the point at which point successively harder cuts on the g-discriminant saturate. In Figure 1, the combinatorial component in the $\Delta\phi$ distribution for reconstructed di-jets (indicative of fake jets) decreases for harder cuts on g, until the effect of the cut saturates and only a back-to-back peak (indicative of angularly correlated QCD dijets) remains. A fake jet rejection cut of $g > 17.8$ GeV^2 is used for the Cu+Cu results here.

3.2. Performance and Spectrum Corrections

The detector response to partonic jets in p+p collisions is evaluated with a Monte Carlo (MC) simulation. 14×10^6 total `PYTHIA` events were generated with values of the minimum transverse momentum transfer in the QCD hard scattering ranging from $\hat{p}_T = 0.5$ to 64 GeV/c, and Gaussian jet reconstruction is run on final state particles with $p_{\mathrm{T}}^{\mathrm{true}} > 400$ MeV/c. Other `PYTHIA` tunes, as well as the `HERWIG` generator, are used as a cross-check. Then, the jet reconstruction performance is determined with a `GEANT3` simulation of the PHENIX detector geometry. The transfer matrix $P(p_{\mathrm{T}}^{\mathrm{rec}}|p_{\mathrm{T}}^{\mathrm{true}})$, which relates jets reconstructed at the detector scale to the truth jets is shown in Figure 1. In PHENIX, the energy scale is driven mainly by the loss of neutral energy (primarily n and K_L^0), tracking inefficiency and edge of acceptance effects rather than fluctuations of the underlying event or the intrinsic DC and EMCal resolutions.

Using the transfer matrix and jet reconstruction efficiency, the p+p reconstructed jet spectrum is unfolded to the truth spectrum using a singular value decomposition (SVD) unfolding method[5]. Future work in this part of the analysis now includes an additional hadronization correction that associates the hadronic and partonic energy scales, allowing for a better comparison to NLO theory calculations.

In d+Au collisions, the effects of the mild underlying event on the jet p_{T} are evaluated with an embedding analysis in which jet reconstruction is run on MC jet events embedded into minimum bias data. From this, the reconstructed jet yields are corrected to an equivalent yield in p+p collisions though a bin by bin unfolding correction.

In Cu+Cu collisions, we subtract an event-averaged underlying event background which is parameterized in centrality and z-vertex position. The effects of the underlying event fluctuations on the jet p_{T} are determined through an embedding procedure similar to d+Au, and the reconstructed jet yields are corrected to the equivalent p+p yield through a bin by bin unfolding.

4. Results

The invariant cross section for inclusive jet production in p+p collisions is shown in Figure 2. Comparisons to Gaussian filter reconstruction run on `PYTHIA` final-state hadrons, an NLO calculation, and a jet measurement by the STAR Collaboration[6] are shown as well. The fragmentation function for charged particles $(1/N_{\mathrm{jet}})(dN/dz)$ is shown for different selections of jet p_{T}, where $z = p_{||}^{\mathrm{charged}}/p_{\mathrm{T}}^{\mathrm{jet}}$ is the longitudinal fraction of the jet energy in the fragment.

The nuclear modification factor R_{CP} for anti-k_{T} jets in d+Au collisions is shown in Figure 3. Different centrality selections (relative to the most peripheral collisions) are plotted in different colors. We observe an R_{CP} flat in p_{T} that is suppressed in the most central collisions ($R_{\mathrm{CP}} \sim 0.75$-$0.80$) and is close to unity ($R_{\mathrm{CP}} \sim 0.95$-$1.00$) in the mid-peripheral collisions. The 0-20%/60-88% R_{CP} is also compared to that of single π^0's from PHENIX 2003 data[7]. This comparison emphasizes the increased kinematic reach of reconstructed jets compared to using leading single hadrons. The R_{CP} for π^0's suggests a suppression at high-p_{T} but has growing statistical errors after > 10 GeV/c. The R_{CP} for reconstructed jets confirms this behavior out to 35 GeV/c.

The $p_{\mathrm{out}} \equiv (p_{\mathrm{T}})_{\mathrm{low}} \cdot \sin\Delta\phi$ for reconstructed dijets can be used to search for possible k_{T}-broadening from cold nuclear matter effects. Figure 4 shows the p_{out} in different centralities,

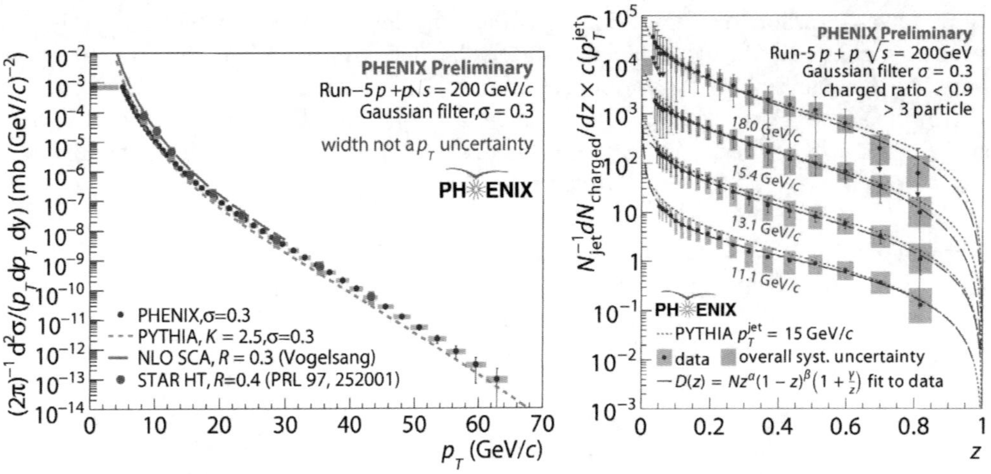

Figure 2. Left: Jet production cross section in p+p collisions, corrected to the $p_{\mathrm{T}}^{\mathrm{true}}$ hadron energy scale. Shown compared with theory. Right: Fragmentation function for charged particles in p+p jets. Curves for different trigger jets are shown.

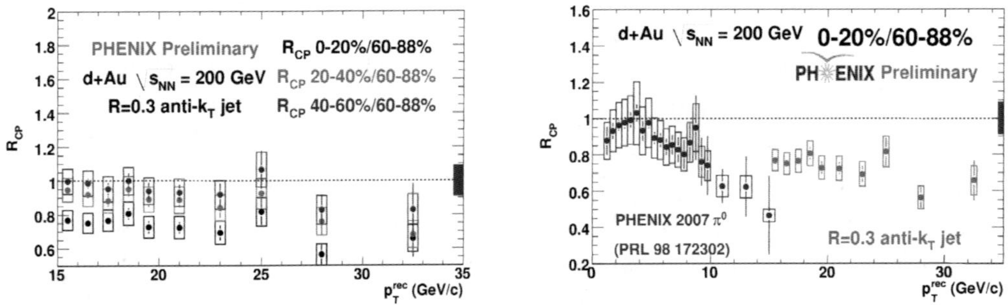

Figure 3. Left: Nuclear modification factor R_{CP} for jets in d+Au collisions as a function of the p+p-reconstructed energy scale p_{T}^{rec} for three different centrality bins (0-20%/60-88%, 20-40%/60-88%, and 40-60%/60-88%). Right: The 0-20%/60-88% R_{CP} for reconstructed jets compared to that of π^0's[7].

with kinematic and away-side peak cuts to select real QCD dijets. The distribution are very nearly centrality-independent.

The nuclear modification factor R_{AA} for Gaussian filter jets in Cu+Cu collisions, as well as the $\Delta\phi$ distributions for reconstructed dijets, are shown in Figure 5. At high-p_{T}, the nuclear modification for peripheral collisions shows no effect ($R_{\mathrm{AA}} \sim 1.0$) while for central collisions it indicates a strong suppression ($R_{\mathrm{AA}} \sim 0.45 - 0.50$). As the $\Delta\phi$ distributions shows, the suppressed jets do not become angularly de-correlated in central events.

5. Jets at RHIC Study

Recently, the PHENIX Collaboration has proposed a re-imagining of the PHENIX experiment as "sPHENIX"[11], a dedicated jet detector with the acceptance, hermiticity and hadronic

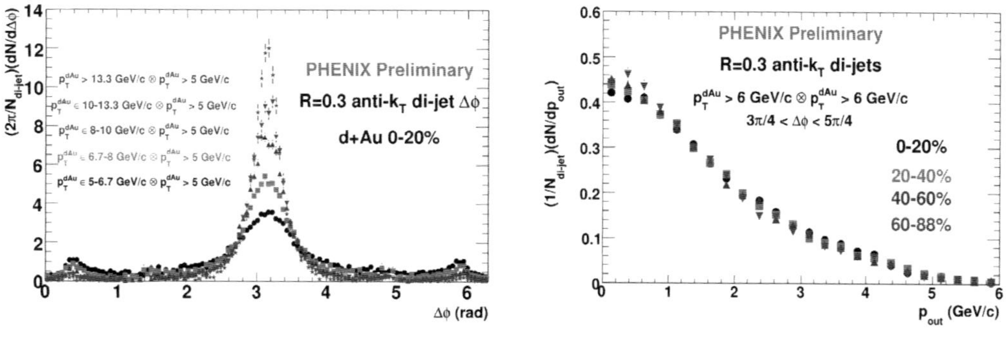

Figure 4. Left: Acceptance-corrected $\Delta\phi$ distribution for d+Au 0-20% jets, different p_T selections. Right: Self-normalized $p_{\rm out}$ distributions for different d+Au centrality selections.

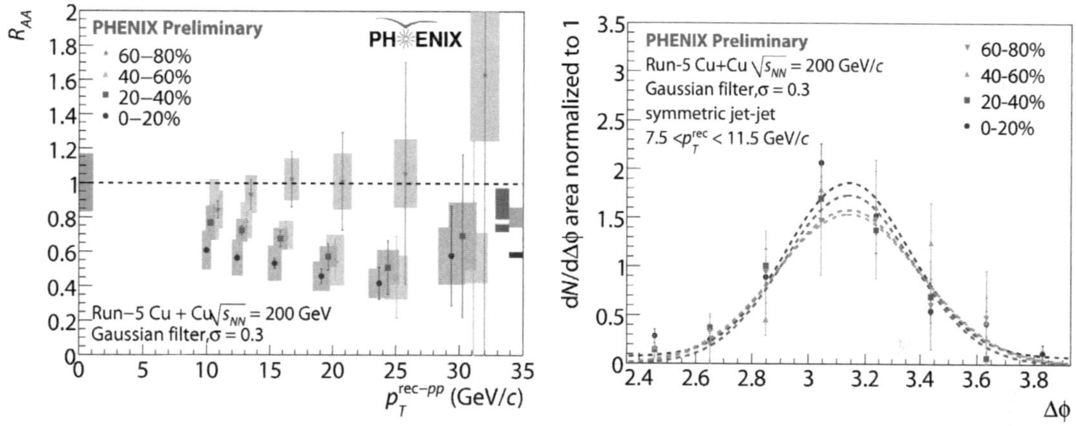

Figure 5. Left: Nuclear modification factor $R_{\rm AA}$ for jets in Cu+Cu collisions as a function of the p+p-reconstructed energy scale $p_{\rm T}^{rec}$ for four different centrality bins (0-20%, 20-40%, 40-60% and 60-80%). Right: Reconstructed dijet away-side $\Delta\phi$ distributions for the different Cu+Cu centrality bins (each distribution has been given unit normalization).

calorimetry needed to support a robust jet physics program at RHIC. To this end, a proof of principle study[10] was undertaken to demonstrate the feasibility of separating real high-$p_{\rm T}$ reconstructed jets from fakes in Au+Au collisions at $\sqrt{s_{\rm NN}} = 200$ GeV, without any fake jet rejection or subtraction scheme. The anti-$k_{\rm T}$ algorithm is used exclusively throughout, with cone sizes $R = 0.2, 0.3, 0.4$.

The total Monte Carlo dataset was 0.75×10^9 minimum bias HIJING events. The study assumes an "ideal" calorimetric detector spanning $|\eta| < 1$ by $\phi \in [0, 2\pi)$ with a segmentation of $\Delta\eta \times \Delta\phi = 0.1 \times 0.1$ towers. The HIJING code was modified to run jet reconstruction at every call of the HIJING fragmentation subroutine, and the fragmentation jets with $E_{\rm T}^{true} > 5$ are defined as the truth jets used to evaluate the reconstruction performance.

Figure 6 outlines the underlying event subtraction procedure, which builds upon the one described by the ATLAS Collaboration[12]. In the first pass of the procedure, "seed jets" ($R = 0.2$ jets containing an energetic tower with > 3 times the mean tower energy) are excluded

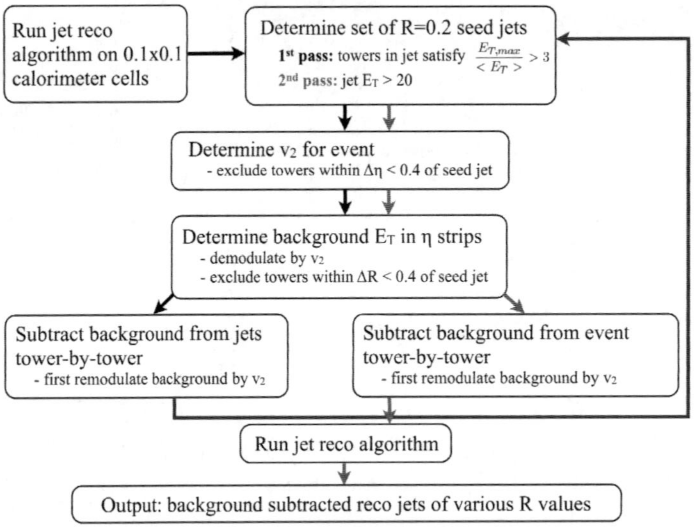

Figure 6. Left: Flowchart of underlying event subtraction procedure.

from the determination of the background, which is corrected for v_2 flow modulation in the event and done separately in $\Delta\eta = 0.1$ strips. This zeroth-order estimate of the background is subtracted from the towers in seed jets. The background determination procedure is repeated, this time only excluding seed jets with $E_{\rm T} > 20$ from the background determination. In the second iteration, the v_2-modulated, η-dependent background is subtracted from all towers in the event.

Jet reconstruction with the different R values is run on the background-subtracted towers. The reconstruction efficiency to have a reconstructed jet within $\Delta R < 0.25$ of a truth jets rises to 100% above $E_{\rm T}^{true} > 15$, > 20 and > 25 GeV for the $R = 0.2$, 0.3 and 0.4 cone sizes, respectively. The jet energy scale $\langle E_{\rm T}^{reco} - E_{\rm T}^{true}\rangle / E_{\rm T}^{true}$ is within 5% of zero for jets above 20 GeV. The jet energy resolution $\sigma(E_{\rm T}^{reco} - E_{\rm T}^{true})/E_{\rm T}^{true}$ is $< 25\%$ and $< 15\%$ GeV for $R = 0.2$ and $R = 0.4$ jets above 20 GeV (and decreases with increasing truth jet energy).

Figure 7 shows the distribution of $E_{\rm T}^{true}$ for truth jets matched to a given R value and $E_{\rm T}^{reco}$ bin. We observe a dominant peak with $E_{\rm T}^{reco} \sim E_{\rm T}^{true}$ with a ~ 5 GeV width well separated from the low-$E_{\rm T}^{true}$ exponential background. This is precisely the behavior necessary for a well-controlled unfolding from the reconstructed to true energy scales. Figure 7 also shows the distribution of $E_{\rm T}$ for $R = 0.3$ truth jets, and reconstructed jets matched and unmatched to truth jets. Reconstructed jets resulting from a hard scattering dominate above $E_{\rm T}^{reco} > 30$ GeV (> 20 and > 40 GeV for $R = 0.2$ and $R = 0.4$ jets, respectively).

Taken together, the results from this study demonstrate a clear separation of jets resulting from a hard scattering from underlying event fluctuations at RHIC energies. As a reminder, this jet-background separation is achieved before any fake jet rejection or subtraction scheme is implemented, suggesting that it may be improved even further with additional work.

6. Conclusions

The direct jet cross section in p+p collisions at $\sqrt{s} = 200$ GeV, and its consistency with theory calculations over ten orders of magnitude, demonstrates the PHENIX capability for performing jet reconstruction measurements to $p_{\rm T}^{\rm true} > 60$ GeV/c and validates the Gaussian filter algorithm for general use.

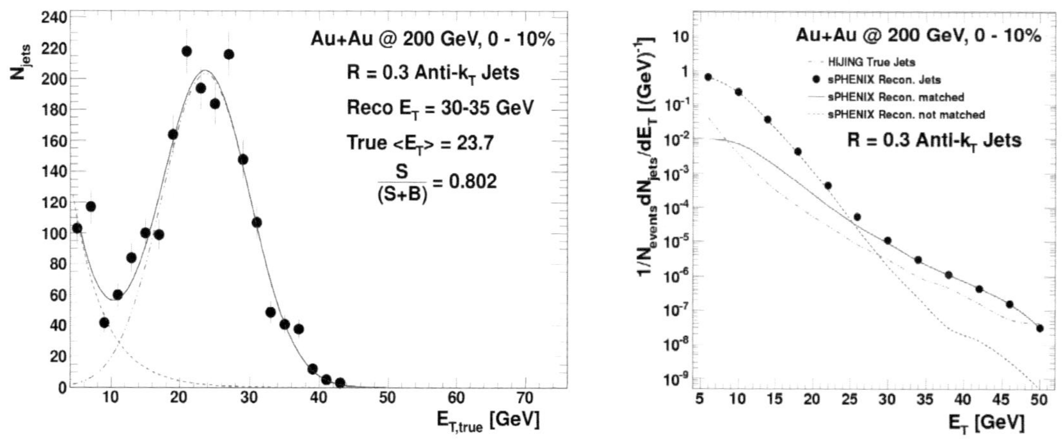

Figure 7. Left: Distribution of E_{T}^{true} for truth jets matched to $R = 0.3$ jets with $E_{\mathrm{T}}^{reco} \in 30-35$ GeV. Right: Distribution of E_{T} for $R = 0.3$ truth jets and reconstructed jets matched and unmatched to truth jets.

Reconstructed jets have also been measured in d+Au and Cu+Cu collisions at $\sqrt{s_{\mathrm{NN}}} = 200$ GeV for different centrality selections and over a wide p_{T} range, and are starting to produce physics results. In d+Au collisions, we observe a suppression of high-p_{T} jets in central collisions relative to peripheral ones. Future work, such as the extracting the $R_{d\mathrm{A}}$ (the nuclear modification factor relative to p+p collisions) and controlling the fake rate to extend the kinematic reach to lower p_{T}, is needed to better understand the source of this effect. These results are important for the recent developments in modeling the impact parameter dependence of nPDFs[8]. Finally, we can derive a strong constraint on any centrality-dependent angular broadening in d+Au collisions.

In Cu+Cu collisions, we observe a strong suppression of jet production relative to the p+p baseline, consistent with the level of suppression for high-p_{T} π^0's in Cu+Cu collisions also measured by PHENIX[9]. These preliminary results for reconstructed jets in different collision systems are currently being finalized.

The sPHENIX proposal, which re-envisions PHENIX as a dedicated jet detector, in tandem with the proposed RHIC luminosity upgrade would open the door to high-statistics inclusive jet, di-jet and γ-jet measurements. The study summarized here demonstrates that separation of reconstructed jets from background is feasible at RHIC energies with such an experimental setup.

References

[1] Y.S. Lai (PHENIX Collaboration), Nucl. Phys. A, 855 (2010), p. 295
[2] N.C. Grau (PHENIX Collaboration), J. Phys. G: Nucl. Part. Phys. 38 (2011) 124090
[3] Y.S. Lai, B.A. Cole, arXiv:0806.1499
[4] M. Cacciari, G. P. Salam and G. Soyez, JHEP 0804 (2008) 063
[5] A. Höcker and V. Kartvelishvili, Nucl. Instrum. o Methods A 372, 469 (1996)
[6] B. I. Abelev et al. (STAR Collaboration), Phys. Rev. Lett. 97, 252001 (2006)
[7] S. S. Adler et al. (PHENIX Collaboration), Phys. Rev. Lett. 98, 172302 (2007)
[8] I. Helenius, K.J. Eskola, H. Honkanen, C. A. Salgado, arXiv:1205.5359
[9] A. Adare et al. (PHENIX Collaboration), Phys. Rev. Lett. 101, 162301 (2008)
[10] J. A. Hanks et al. Phys. Rev. C 86, 024908
[11] A. Adare et al. (PHENIX Collaboration) nucl-ex/1207.6378
[12] ATLAS (ATLAS Collaboration) (2011), `cdsweb.cern.ch/record/1353220/files/ATLAS-CONF-2011-075.pdf`

A Very High Momentum Particle Identification Detector for the ALICE Experiment at the LHC

Edmundo Garcia[1] for the ALICE VHMPID Collaboration[2]

Department of Chemistry and Physics, Chicago State University, Chicago IL 60304

E-mail: edmundo.garcia@csu.edu

Abstract. The purpose of the ALICE experiment is the study of the quark-gluon plasma (QGP) in heavy ion collisions at LHC. To this end ALICE has unique PID capability to identify a wide variety of particles in the low to intermediate momentum range. We propose to enhance these capabilities with the Very High Momentum Particle Identification Detector (VHMPID), improving the identification of charged pions, kaons, protons and anti-protons in the momentum range $5 < p < 25$ GeV/c on a track-by-track basis. This upgrade provides ALICE with new opportunities to study parton-medium interactions. The underlying physics and the relevant degrees of freedom in the QGP are not known. Only specific experimental evidence of the modification of the harmonization process in medium will enable us to gain a deeper understanding of the fundamental process of hadron formation during the evolution of the deconfined matter. Since the low momentum region of the hadron spectrum is populated by many competing processes, only the high momentum part of the spectrum can be interpreted unambiguously. The VHMPID is a Focusing Ring Imaging Cerenkov with C_4F_{10} gaseous radiator coupled to a CsI-based photon detector. The VHMPID presents sufficient acceptance for triggered- and tagged-jet studies, allowing for the first time, identified charged hadron measurements in jets.

1. Introduction

ALICE (A Large Ion Collider Experiment) at the LHC (Large Hadron Collider) [1] consists of a variety of detector systems for measuring hadrons, leptons, and photons. ALICE is designed to perform measurements of high energy nucleus-nucleus collisions in order to study quark matter (QCD matter) [2] under extreme conditions. The detector saw the first heavy-ion (Pb-Pb) collisions produced at the LHC in November of 2010, figure 1 is a display of one of the first events. ALICE is a collaboration of more than 1000 physicists, engineers and technicians from 105 physics institutions in 30 countries across the world. A discussion of ALICE's physics program can be found in [3].

[1] Corresponding author

[2] See Appendix I for the full list of collaborators

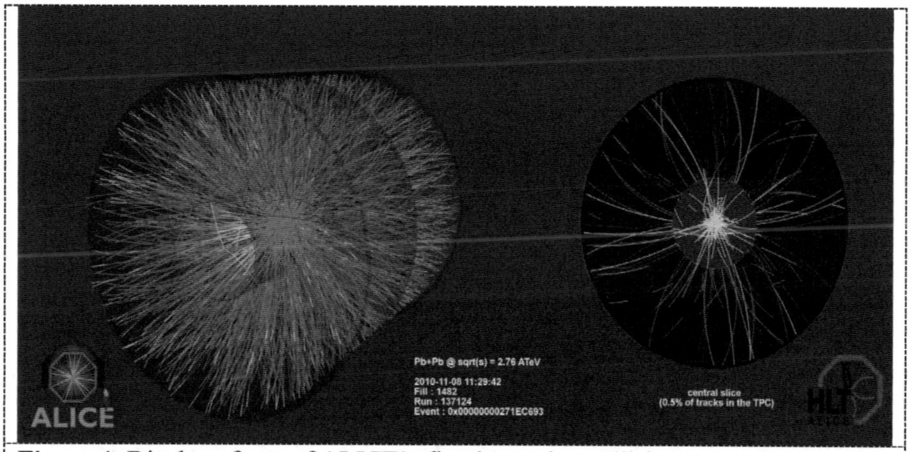

Figure 1. Display of one of ALICE's first heavy ion collisions.

As confirmed by early ALICE results, one of the most interesting features of the relativistic heavy ion collisions at the LHC is the study of particle yields in the available transverse momentum (p_T) range. Today, in ALICE, the reach of proton identification on a track-by-track basis is up to 5 GeV/c, and statistical identification is possible up to 30 GeV/c using the relativistic rise of the TPC [4]. In this paper we describe the development of a gas Cherenkov detector for track-by-track charged particle identification in the p_T range of 5 – 25 GeV/c the VHMPID.

At intermediate p_T the direct identification of jets is very difficult, given the large background from soft particles. Thus there is the need to use the correlation methods to study jets' energy loss and production mechanisms at p_T between 10 and 30 GeV/c. Given the underlying background present at these energies, the correlation studies will require tracking and exact track-by-track identification capabilities. The VHMPID will provide these capabilities, filling the existing gap between low and high p_T in the ALICE's acceptance [4]. The use of the relativistic rise in the ALICE's TPC allows for statistical separation of the baryons and mesons. However, the track-by-track identification will give a better interpretation in the details of measurements such as: the baryon-antibaryon content of individual jets, the identification of baryon in jets as a leading particle, and the details of the near and far side jet correlations.

The challenge is to use a relatively small detector covering about 30% of the acceptance of ALICE's central barrel with a suitable trigger. This detector is based on the combination of a Cherenkov radiator with a large area photon-to-electron converter (CsI), a charged particle detector with a pad segmented cathode (multi-wire proportional chamber) and a spherical mirror. With these elements the further characteristics of the VHMPID are based on the experimental constraints: the space available in ALICE, the 0.5 T magnetic fields surrounding it, and the granularity that allows the discrimination of background in a central heavy-ion collision at the LHC (2% occupancy). A schematic view of the geometry for the VHMPID is shown in figure 2 (a) and the proposed location for the detector with respect to ALICE's barrel is in figure 2 (b). At its present design this modular detector will be located in front of ALICE's calorimeter modules (EMCal, DCal, PHOS detectors [3]).

2. Physics Motivation

The primary purpose of the VHMPID is to investigate the hadro-chemistry of the fragmentation process in vacuum and in medium. This detector will allow a unique view of the hadronization process itself. Since the low momentum region of the hadron spectrum is populated by many competing processes, such as thermal production and recombination in addition to fragmentation, only the high momentum part of the spectrum can be interpreted unambiguously. Only specific experimental evidence of the modification of the hadronization process in medium will enable us to gain deeper understanding of the fundamental process of hadron formation during the evolution of deconfined matter.

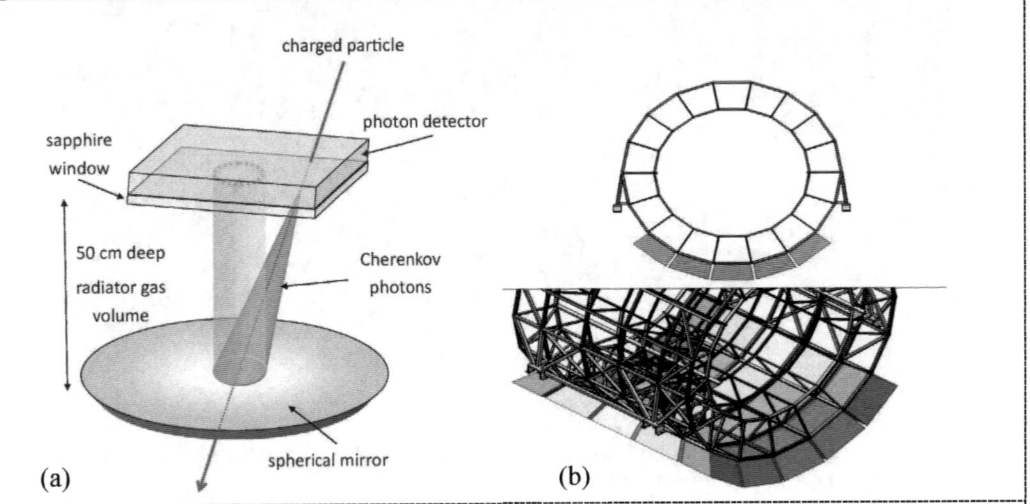

Figure 2. (a) VHMPID concept. The red line represents the incident track. (b) The proposed location within the ALICE detector for the VHMPID volumes.

Recent theory predictions [5] have hinted at significant flavor and baryon effects in the medium-modified fragmentation of jets in the deconfined matter. This is in part confirmed by recent results from RHIC [6]. Furthermore even the basic measurements of baryon fragmentation in vacuum have not been explored and none of the competing LHC experiments has the necessary detector capabilities to measure protons at high momentum. The precise measurement of parton fragmentation to protons and anti-protons [7], which has never been done before out to the relevant momentum transfer scale, will significantly constrain fragmentation functions and enable us to distinguish between quark and gluon fragmentation as well as shed light on the issue of di-quark formation during the fragmentation process [8]. Recent theory papers (e.g. [9]) have suggested that the quarkonium production mechanism (color singlet vs. color octet) can be determined by mapping the sub-leading jet structure (jet shape, hadron composition, etc.) in triggered heavy flavor jets in proton-proton collisions. The VHMPID, in conjunction with the central barrel tracking and the calorimeter jet trigger, enables the following new physics measurements in proton-proton collisions at the LHC.

The measurement of strong hadron suppression up to 20 GeV/c at RHIC and up to 100 GeV/c at the LHC has highlighted the importance of the high-p_T regime. This regime may have unexpected particle species dependent features. Certain theories, based on recombination, have postulated strong particle effects, in particular in the light flavor baryon sector (enhanced formation) out to 20 GeV/c [10], whereas theories based on enhanced gluon splitting or early formation time predict effects also in the meson sector [11]. Thus PID out to high p_T will unambiguously test certain recombination and fragmentation predictions. The VHMPID, together with the central barrel tracking and a calorimeter jet trigger, will enable the following physics measurements in heavy ion collisions at the LHC:

- Study of the baryon enhancement at intermediate to high p_T through measurement of hadro-chemistry in tagged jets.
- Detailed mapping of gluon splitting process (energy loss in medium) through measurement of hadro-chemistry in tagged jets.
- Study of medium modification of quarkonium production and gluon/quark fragmentation through measurement of sub-leading identified hadron distribution in tagged jets
- Determination of baryon/anti-baryon imbalance through momentum dependent proton/anti-proton measurement in tagged jets in medium.
- Determination of hadronic resonance modification in medium at high p_T.

Journal of Physics: Conference Series **389** (2012) 012007 doi:10.1088/1742-6596/389/1/012007

The main advantage of the proposed PID/Calorimeter combination is that all these flavor and baryon number dependent measurements will be based on triggered and identified jets. The analysis of unidentified back-to back jets in heavy ion collisions via di-hadron correlations have highlighted the role of the non-Abelian jet energy loss (gluon splitting) [12] , however fragmentation properties cannot be extracted from unidentified hadron correlation measurements alone.

3. Detector Performance Studies

In this section the resulting PID capabilities will be discussed by means of Monte-Carlo simulations [13] and preliminary results from recent beam tests performed at the CERN/PS.

The efficiency of identifying particles with Cherenkov detectors depends on the number of produced Cherenkov photons, their absorption in the traversed media, the detection efficiency and the corresponding angular resolution. The first two properties depend on the refraction index, this is known for C_4F_{10} at a per mil level so this radiator gas was used in the simulations. Another parameter used was the Cherenkov angle resolution that was estimated via theoretical calculations with a resulting value of 2.6 mrad. For the detector efficiencies, figure 3(a) shows the optical properties of all media used in the Monte-Carlo simulation. These characteristics and the full description of the photon detector were implemented in GEANT [14].

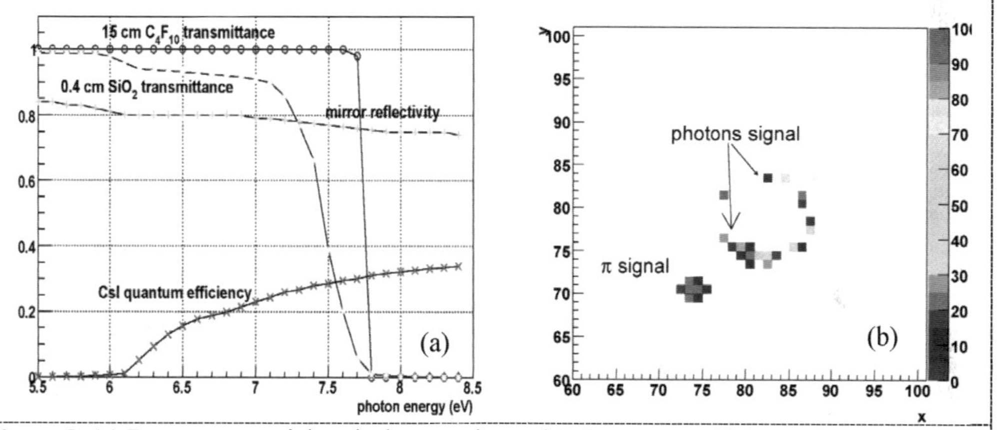

Figure 3. (a) Detector material optical properties and CsI quantum efficiency.(b) Cherenkov event of single pion at saturation with incident angle of 15°. Axes *x* and *y* are given in pad units, the color coding represents the integrated charge in ADC units.

The results of the simulations are shown in the following figures. Figure 3(b) is an example of the simulated detection single pion events for 15° incidence, in absence of background. Figure 4(a) is the distribution of reconstructed single photon Cherenkov angle, from pions at saturation in 50 cm C_4F_{10} at 3.5 atm in presence of Pb-Pb background. The distribution was obtained by embedding Cherenkov rings from pions, kaons and protons, at different momenta, with background Pb-Pb HIJING events at LHC energies. In figure 4(b) are presented the ring-averaged Cherenkov angle distributions obtained for pions, kaons and protons at 25 GeV/c, after the Hough Transform processing [15].

4. Test beam results

Preliminary results from very recent beam tests at the CERN/PS are discussed. The prototype detector for the test consisted of a 100 cm C_4F_8O radiator and the baseline photo-detector (MWPC with CsI photo-cathode segmented in 8 *x* 8.4 mm^2 pads). The event display for 6 GeV/c pions at atmospheric pressure and room temperature is shown in figure 4(a). The radiator gas was used without any cleaning and the transparency was measured using a UV-monochromator based system under commissioning, indicating a lower transmission. Thus, the observed number of photon cluster is about

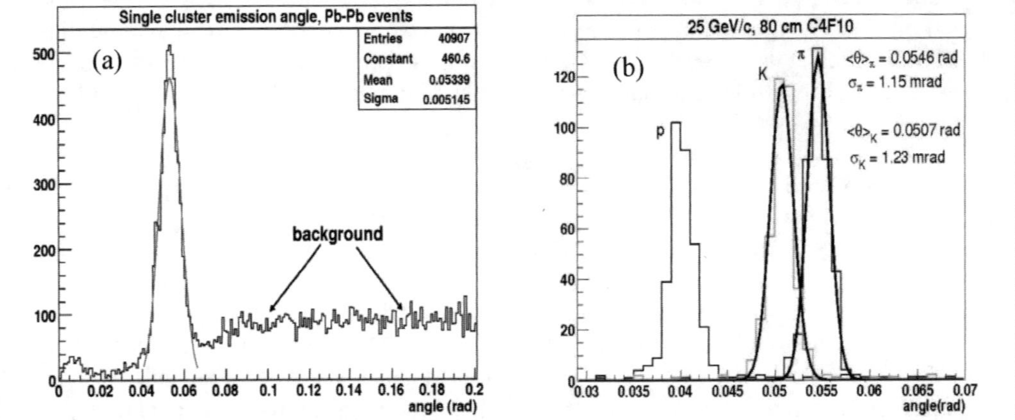

Figure 4. (a) Distribution of reconstructed single photon Cherenkov angle, from pions at saturation in 50 cm C_4F_{10} at 3.5 atm in presence of Pb-P b background (b) Distribution of ring-averaged Cherenkov angles for pions, kaons, and protons after the Hough Transform method at 25 GeV/c at 3.5 atm gas radiator pressure.

half of those expected from simulations using nominal gas transparency. The most relevant result is the observed single photon angular resolution of 2.9 mrad as shown in figure 4(b).(missing a verb)

A MWPC prototype with variable anode-cathode gap (in the range 0.5-2 mm) equipped with a CsI photo-cathode with smaller pads (0.4 *x* 0.8 cm) is under test with the aim of reducing the photon overlap probability (and thus maximizing the number of detected photons). In a recent test beam run the detector performance was checked with a new prototype with a pressurized radiator (3.5 atm), successful preliminary results from this run indicate the feasibility of the pressurized design.

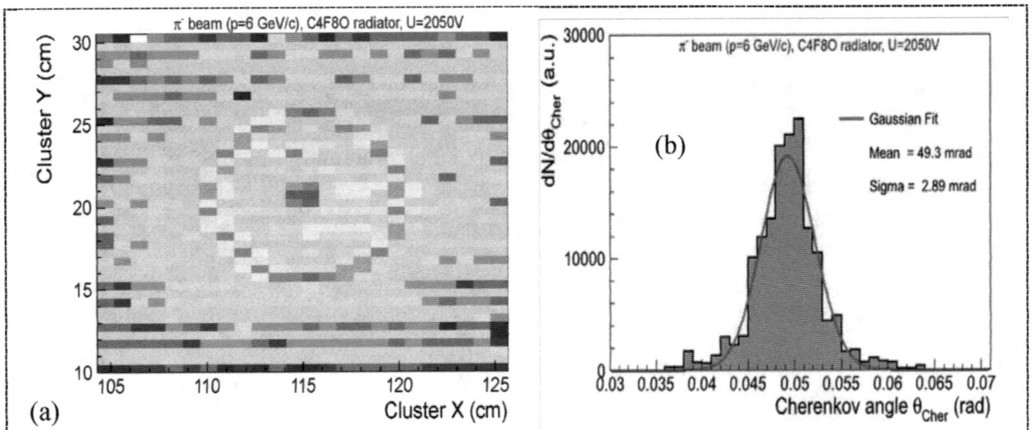

Figure 5. (a) Event display for pion beam. (b) Distribution of detected photons in the ring fiducial (yellow orange ring zone in (a)).

5. Physics Performance Studies

The combined VHMPID/Calorimeter coverage will allow full jet reconstruction with particle identification back to back with the existing EMCal [16]. The main measurement which will benefit from such an integrated detector system is the determination of particle identified fragmentation functions in heavy ion and proton-proton collisions, since the jet energy can be determined unambiguously and thus lead to a strict analysis of the fractional momentum. Furthermore the proposed detector combination enables us to do actual jet-jet correlation measurements with particle identification. Here we present Monte-Carlo simulation examples of the physics capabilities of the VHMPID.

For this study we have assumed the full coverage of the VHMPID implementation which leads to around 25% coverage of the TPC volume. About half of all EMCal triggered di-jets have their partner reconstructed in the combined acceptance. For jet-jet correlations we set the number of EMCal triggered di-jets with 100 GeV jet energy each. Figure 6 shows the accuracy to which a medium modification measurement of the proton fragmentation function can be performed. The anticipated medium effect is based on a comparison between PYTHIA and qPYTHIA predictions for 50 GeV jets [5]. The VHMPID makes a significant contribution to this plot by enabling the proton identification in the range from $\xi = 0.7$ to 2.3 ($\xi = \ln(E_{jet}/p_{had})$. In this range the measurement is clearly not statistics-limited. The systematic error bars on the proton yield (y-axis) and the jet energy (x-axis) are included.

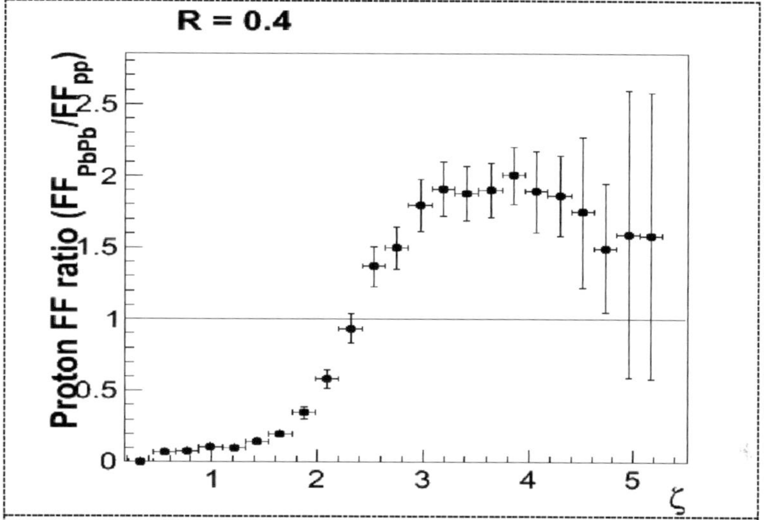

Figure 6. Ratio of proton fragmentation functions simulated for p-p and Pb-Pb collisions for 50 GeV jets using the PID uncertainties for TPC, TOF and VHMPID and the jet energy uncertainties for the EMCal. Jets are reconstructed with a cone algorithm using R=0.4 for the cone size. The statistics are based on one PbPb run at present luminosities.

Measuring identified particle ratios based on the spectra is a unique way to test medium modifications of the fragmentation process, as was suggested in [18]. Differences between the production of mesons/baryons, particles/anti-particles, and/or strange and heavy- flavor content hadrons are predicted for higher energies and low-x values. Predictions differ strongly, and the need for precise track-by-track measurements is obvious. In figure 7 we show the prediction by [5] for a specific identified particle ratio in reconstructed 100 GeV energy jets. The jet is fully reconstructed in the calorimeter first, which leads to a reliable determination of the jet energy. Then the VHMPID determines the particle species of all high momentum hadrons in the same jet cone track by track. The anticipated statistical and systematic error bars for the jet energy and the particle ratios are shown in the data points in the figure.

In addition to the identification of ground state hadrons, the VHMPID also enables the reconstruction of high momentum hadronic resonances. Since many of these excited states decay in the medium they have been suggested as a signature for chiral restoration as well as a measure of medium properties such as the medium lifetime and the modification of the fragmentation process in the medium. As an example we have studied the decay of the ϕ-meson into K^+K^- pairs. The right panel of figure 8 shows a simulation of the ϕ-meson peak reconstructed in the VHMPID from identified kaon pairs with both kaons above the $p_T > 5$ Gev/c threshold. This result is based on 5 million unquenched central Pb-Pb HIJING events. The signal to background ratio is 10 to 1, a 10-fold improvement upon the present ALICE measurement in the TPC. A good portion of this improvement stems from the tight p_T cut on the decay daughters, but the unambiguous identification of the decay daughters not only improves the signal to background ratio it also allows the correlation of these resonances with jets or other hadrons.

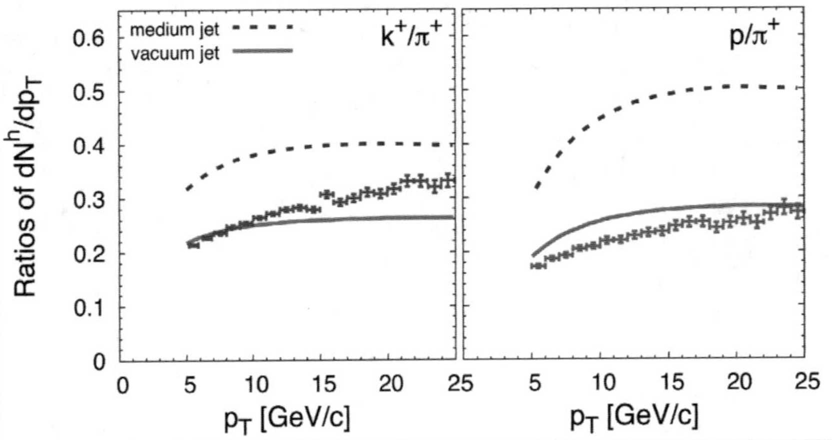

Figure 7. Calculated K^+/π^+ (left panel) and p/π^+ (right panel) ratios based on PYTHIA calculations using the VHMPID eefficiencies and systematic uncertainties (based on the annual yield of triggered 50 GeV jets) compared to theoretical pQCD predictions for vacuum and medium jets [5].

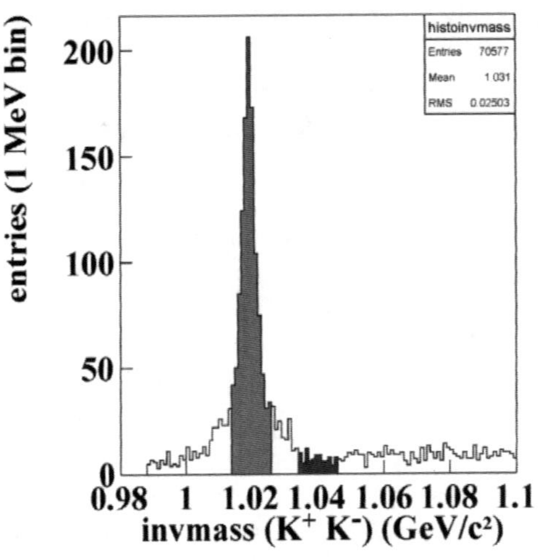

Figure 8. Reconstructed ϕ-meson peak in VHMPID acceptance based on 5 Million central HIJING events (unquenched). Red signal, blue background.

Acknowledgments
This work has been supported by the grant NSF-PHY-0968903. The work was also supported by Hungarian OTKA grants NK62044, NK778816, CK77719, CK77815 and National Research Foundation (NRF) of Korea. A. Di Mauro, P. Martinengo, F. Piuz and J. B. Van Beelen are warmly acknowledged for their important contributions to the VHMPID project.

Appendix A.

A. Harton, E. Garcia, *Chicago State University, USA* - D.A. Finogeev, T.L. Karavicheva, A.B. Kurepin, A.N. Kurepin, A.I. Maevskaya, Yu.V. Musienko, V.I. Razin, A.I. Reshetin, N.S. Topilskaya, E.A. Usenko, *Institute for Nuclear Research of the Russian Academy of Sciences, Moscow* - L. Olah, D. Varga, *Eotvos University, Budapest, Hungary* - D.W. Kim, J.S. Kim, *Gangneung-Wonju National University, Dept. of Physics, Gangneung, South Korea* - R.T. Jimenez, D. Mayani, G. Paic, M.E. Patino, V. Peskov, *Instituto de Ciencias Nucleares Universidad Nacional Autonoma de Mexico, Mexico* - J. Song, J. Yi, I.-K. Yoo, *Pusan National University, Dept. of Physics, South Korea* - S. Chattopadhyay, Saha Institute, Kolkata, India - F. Barile, G. De Cataldo, D. Di Bari, E. Nappi, C. Pastore, D. Perrino, I. Sgura, G. Volpe, *Universita degli Studi di Bari, Dipartimento Interateneo di Fisica "M. Merlin" & INFN Sezione di Bari, Italy* - T. V. Acconcia, A. K. Dash, J. Takahashi, *UNICAMP, University of Campinas, Brazil* - R. Bellwied, D. D. Chinellato, L. Pinsky, A. Timmins, M. Weber, *University of Houston, USA* - S. D'Ambrosio, F. Cindolo, S. De Pasquale, G. Iannone, G. Patimo, *University of Salerno, Italy* - A. Knospe, C. Markert, L. Xaplanteris, *University of Texas at Austin, USA* - A.G. Agocs, G.G. Barnafoldi, Gy. Bencedi Gy. Bencze, D. Berenyi, L., Boldizsar, E. Futo, G. Hamar, L. Kovacs, P. Levai, L. Molnar, S. Pochybova, *Wigner RCP, RMKI, HAS, Budapest, Hungary.*

References

[1] T L S Group, "The Large Hadron Collider Conceptual Design", CERN-AC-95-05 (1995), hep-ph/0601012
[2] B Müller "Quark Matter 2005 - Theoretical Summary", arxiv.org:nucl-th/0508062
[3] F Carminati et al. (ALICE Collaboration), J. Phys. G30, 1517 (2004).
[4] TPC Technical Design Report , CERN/LHCC-2000-001
[5] S. Sapeta and U. A. Wiedemann, Eur. Phys. J. C55, 293 (2008).
[6] Y. Xu (for the STAR collaboration), J. Phys. G37 094059 (2010)
[7] L. Grigoryan, Phys.Rev. C83 014904 (2011)
[8] Z. Nagy and D.E. Soper, arXiv:1202.4496
[9] A.C. Kraan, AIP Conference Procedings 1038 45 (2008)
[10] J. Zuo [STAR Collaboration], J.Phys. G35 044027 (2008)
[11] R. Bellwied and C. Markert, Phys. Lett. B691, 208 (2010)
[12] M. Gyulassy, P. Levai, and I. Vitev, Phys. Rev. Lett. 85 5535 (2000)
[13] G. Volpe Ph. D. Thesis: Study of Cherenkov detectors for high momentum charged particles identifi cation in the ALICE experiment at LHC. Universita degli Studi di Bari, Facolta di Scienze Matematiche, Fisiche e Naturali. Dottorato di ricerca in Fisica XIX ciclo. Anno accademico 2005-2006.
[14] S. Agostinelli et al., Nucl. Inst. and Meth. A 506 (2003) 250–303
[15] Hough, P.V.C., "Method and Means for Recognizing Complex Patterns," U.S. Patent No. 3069654, 1962.
[16] EMcal Technical Proposal CERN-LHCC-2006-014 and CERN-1996-032-add.3
[17] P. Aurenche and B.G. Zakharov, arXiv:1109.6819

Journal of Physics: Conference Series **389** (2012) 012008 doi:10.1088/1742-6596/389/1/012008

Recent Flow Results from PHENIX

Paul Stankus for the PHENIX Collaboration
Oak Ridge National Lab
E-mail: paul.w.stankus@gmail.com

Abstract. The phenomenon of collective flow, as revealed through azimuthal anisotropies of the "bulk" of produced hadrons, has been one of the most information-rich sectors of observables measured in RHIC and LHC heavy-ion collisions. A new wealth of data on bulk azimuthal anisotropies is now becoming available, including measurements involving (i) higher-order harmonics, (ii) identified hadrons, (iii) forward and backward rapidities, and (iv) system and beam energy dependencies. The latest compendium of flow results from PHENIX is reviewed and their range of physics implications will be discussed.

1. Introduction

The hydrodynamic paradigm has been quite successful in describing and explaining the global patterns of the large majority of light hadrons produced in high-energy A+A collisions at RHIC (and also at the LHC), and is now a prominent feature of the emerging "standard picture" of these collisions. Within this paradigm we can identify three basic stages of the collision's evolution:

(i) The energy of the incoming nuclei is transmuted into a localized energy density present at many different points over some three-dimensional volume.

(ii) The localized energy density exhibits some sufficiently well-defined pressure at each point, and thereafter evolves according to the laws of hydrodynamics along pressure gradients.

(iii) Whatever its composition, the fluid eventually expands and, with positive pressure, cools to the point where further expansion requires it to decay into color singlets, i.e. hadrons, which can spread out to a larger volume and eventually free-stream into their final states.

The hydro paradigm was able to explain the observation that the final states of RHIC collisions exhibited a large quadrupolar anisotropy in momentum, aka "elliptic flow", as a consequence of a quadrupolar azimuthal spatial asymmetry in the initial energy deposit pattern, and so in the pattern of initial pressure gradients. Quantitative consideration of the magnitude of elliptic flow then implied that (a) the time for stage (i) leading to the hydro stage (ii) must be very short, $<\sim 1$ fm/c, and so occur at a high density, and (b) the evolution during the hydro stage (ii) must exhibit low dissipation, expressed as shear viscosity, in order to preserve the quadrupolar anisotropy so efficiently. These implications matched with the theoretical description, that thermalized nuclear matter at these densities should be described as a Quark Gluon Plasma (QGP) and that the QGP should have very low dissipation.

Thus, signatures of hydrodynamic ("flow") behavior in the patterns of final-state hadron momenta now form the basis for arguing that the QGP as described by QCD is formed in

RHIC (and LHC) collisions, and for diagnosing both the hydrodynamic parameters of the QGP fluid and the shapes of the initial energy deposit pattern. This paper will exhibit a selection of recently published flow measurements made by the PHENIX experiment, and consider what they reveal about the hydro paradigm. To do this, we will address results which are relevant to the different stages described above, but in reverse time ordering: final-state description, diagnosing the hydro evolution, and investigating the initial-state energy deposit mechanism.

2. The Flowing Final State

Focusing on stage (iii) of the evolution as described above, the simplest picture is that, when a certain threshold density and temperature are reached, each element of the fluid is simply converted into a locally equilibrated gas of hadrons in accordance with the local temperature and chemical potentials. In this picture, the fact that the different species of hadrons all emerge from fluid elements with particular velocities will imprint particular "mass splitting" pattern on the elliptic flow strength for different species: stronger for the light mesons at lower transverse momenta then switching to be stronger for baryons at higher transverse momenta. When this pattern was observed at RHIC [1] it was taken as strong evidence for the hydro paradigm. The PHENIX experiment has now [2] extended the measurements of elliptic flow for identified charged hadrons out to much higher transverse momenta (p_T), as seen in Figure 1. The basic mass splitting is clearly seen, as well as the interesting turn down in the (anti-)proton elliptic flow at the higher p_T in mid-central collisions.

An alternative is the *coalescence picture*, in which the hadrons emerge from the final state of the fluid through the re-combination of valence/constituent quarks, which are themselves following the fluid's flow. This picture predicts a pattern in the v_2's of hadrons not just according to mass but according to constituent quark number; roughly speaking, the elliptic flow modulation of a two- or three-quark hadron will be two or three times that of the flowing quarks. The lower panels of Fig. 1 illustrate the observables which would be predicted to show a common behavior across hadrons in two coalescence mechanisms; but with the new higher-p_T data we can now see a unmistakeable departure from the trend for the (anti-)baryons.

3. The Evolution of Flow

If, as in the standard picture, the fluid is comprised of locally-equilibrated QGP during all or most of the hydro stage, then being able to diagnose or constrain what goes on in that stage offers the prospect for learning directly about the QGP itself, including its equation of state and transport properties.

A traditional probe for diagnosing the thermal properties of the early stages of nuclear collisions is *direct photons*, since photons radiated during all stages can survive nearly undisturbed to the final state. The PHENIX experiment has recently published – see Figure 2 – unique results showing a non-zero elliptic anisotropy of direct photons down to low $p_T \sim$ 1 GeV/c, where thermal emission sources are expected [4] to dominate their production. In principle, then, these data should encode the history of when the elliptic flow was being built up during the hydrodynamic stage, or even at its start. However, the theoretical interpretation of the direct photon v_2 in terms of thermal fluid radiation is unclear at present, since the currently available predictions based on hydrodynamical pictures are seen to greatly underestimate the magnitude of the direct photon anisotropy. It remains to be determined whether this discrepancy can be resolved within the hydro paradigm, or whether a new ingredient will be called for, such as substantial momentum anisotropy at the start of the hydro stage (aka "pre-equilibrium flow"); or whether the identification of direct photons in this p_T range with QGP radiation should be questioned.

Originally it was assumed that the quadrupolar anisotropy of elliptic flow would be aligned, collision by collision, with the direction of the reaction plane Ψ_{RP} connecting the two nuclei:

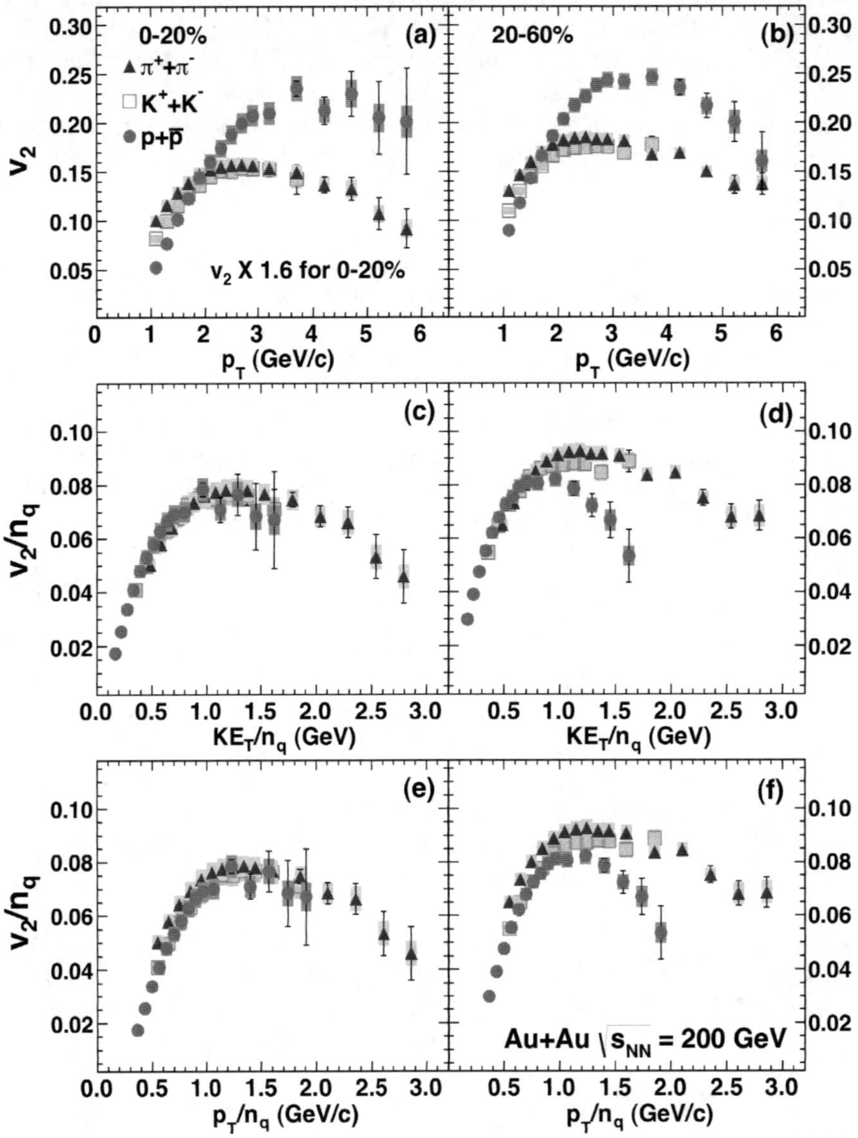

Figure 1. Upper: Elliptic flow strengths for identified charged hadrons, with charge signs combined, in Au+Au collisions as measured by PHENIX [2]. Middle: observation of "quark number scaling", where the v_2 amplitude is divided by the hadron's constituent quark number, plotted agains the hadron's transverse kinetic energy per constituent quark. Bottom: same, but plotted against transverse momentum per constituent quark. A common behavior for all the hadrons is often taken as validating a quark coalescence picture of hadron formation, but a clear departure can now be seen for the (anti-)baryons with the higher-p_T data.

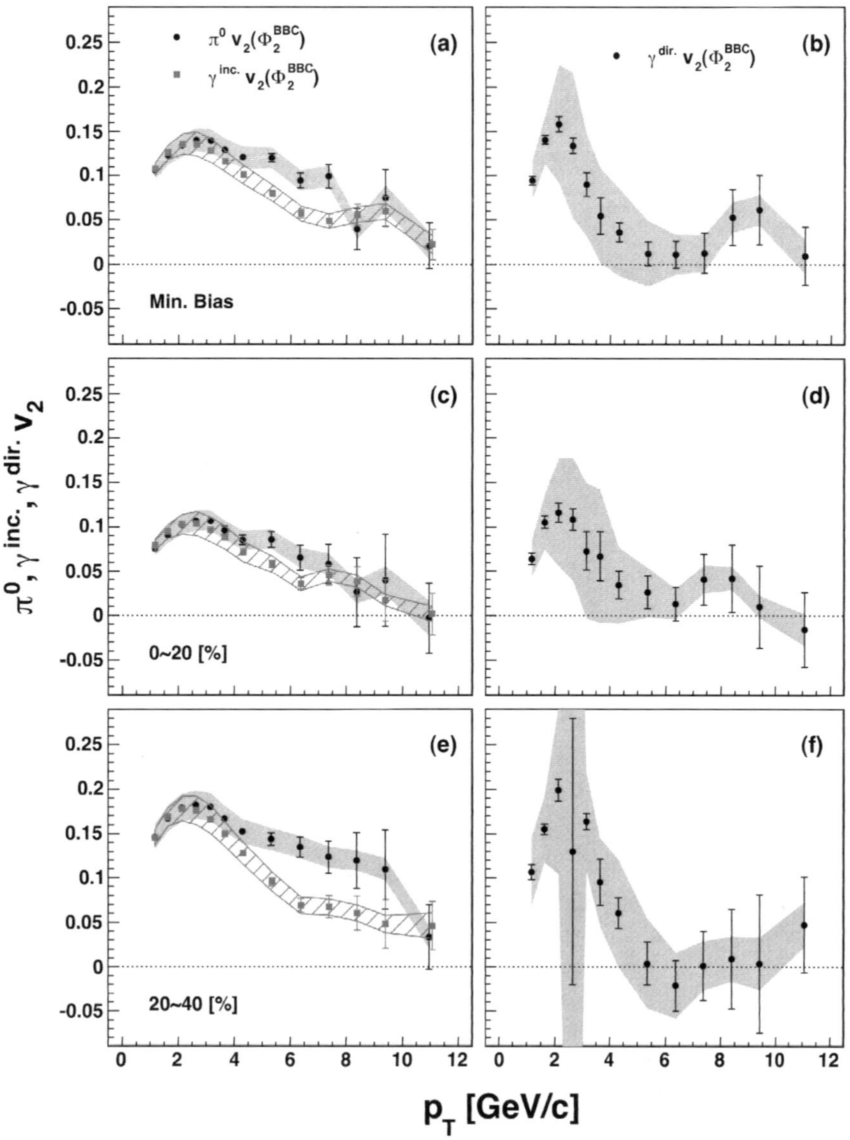

Figure 2. Elliptic flow strengths for π^0's, inclusive photons (left) and direct photons (right) in Au+Au collisions, as measured by PHENIX [3]. The v_2 parameters are measured for pions and inclusive photons at mid-rapidity with reference to a second-order event plane Φ_2^{BBC} in the PHENIX Beam-Beam Counters (BBC) at high rapidities; the direct photon v_2 is then derived from a decomposition of the inclusive photons into decay and direct components.

$$\frac{dN}{d\phi}(\phi; \Psi_{\rm RP}) \propto 1 + 2v_2 \cos(2(\phi - \Psi_{\rm RP})) \quad (1)$$

But if we allow for more general geometries, such as might be produced by initial-state spatial fluctuations, then we can in principle expect azimuthal modulations $\{v_n\}$ at all harmonic orders, referenced to a set $\{\psi_n\}$ of event plane directions at each order:

$$\frac{dN}{d\phi}(\phi; \{\psi_n\}) \propto 1 + 2v_1 \cos(\phi - \psi_1) + 2v_2 \cos(2(\phi - \psi_2)) + 2v_3 \cos(3(\phi - \psi_3)) + ... \quad (2)$$

As mentioned in Section 1, the fact (within the hydro paradigm) that an initial-state quadrupolar spatial anisotropy can faithfully drive a corresponding final-state quadrupolar momentum anisotropy was interpreted as evidence for low dissipation in the hydro phase, and could quantitatively constrain transport/dissipation properties of the QGP such as the shear viscosity. If higher-order azimuthal momentum anisotropies are also being driven by higher-order initial spatial anisotropies, then these should show the effects off dissipation/shear viscosity even more prominently due to their shorter spatial wavelengths. As such, beyond the $n = 2$ "elliptic flow" there is now considerable interest in measuring the $n = 3$ "triangular" and $n = 4$ "cloverleaf" flow patterns of final-state hadrons.

Figure 3 shows recent results from the PHENIX experiment [5] for elliptic (v_2) and triangular (v_3) flow of inclusive charged hadrons, across centrality for two p_T bins. Also plotted are the predictions of several different theoretical hydrodynamical calculations, with different models of initial-state fluctuations and choices of shear viscosity; systematic comparison reveals that, as expected, the new v_3 data provide significant additional constraint on the models, beyond that of the v_2 data alone. Systematic comparisons of this kind may offer the best opportunity for detailed and quantitative characterization of the QGP state of nuclear matter.

4. The Initial Deposit of Energy

In a symmetric A+A collision the distribution of initial energy deposit in the transverse plane across mid-rapidity must, on average, be "almond-shaped": mirror- and 180°- rotationally symmetric. Such an average distribution cannot, by symmetry, drive a final-state anisotropy of odd harmonic order at mid-rapidity, and will only drive relatively small ($v_4 \sim (v_2)^2$) anisotropies at higher even orders. Within the hydro paradigm, then, the observation of measurable triangular flow at mid-rapidity in RHIC collisions provides strong evidence that the event-by-event initial states do *not* all conform to the average distribution, but must exhibit substantial fluctuations.

Specific models for initial-state fluctuations, coupled with simulations of the hydro and hadronization phases, can be compared to data across centralities and transverse momentum as in Fig. 3. But even without a specific model we can deduce some general trends directly from the data. The fuller set of PHENIX data on v_2, v_3 and v_4 for inclusive hadrons across centrality and p_T are shown in Figure 4. The elliptic flow strength v_2 grows dramatically from central to peripheral collisions, consistent with the average geometry of the nuclear overlap region becoming less round and more elliptical. However, even the "roundest" central collisions exhibit a substantial v_2, consistent with a contribution to event-by-event non-roundness from fluctuations becoming visible. The v_3 strengths are nearly independent of centrality, which is qualitatively consistent with their origin lying entirely in fluctuations and having little or no connection to the average geometry. The v_4 trend, growing slowly from central to peripheral, fits the same picture, where it is driven largely by fluctuations but with some component tracking the average geometry as would be expected for an even-order moment.

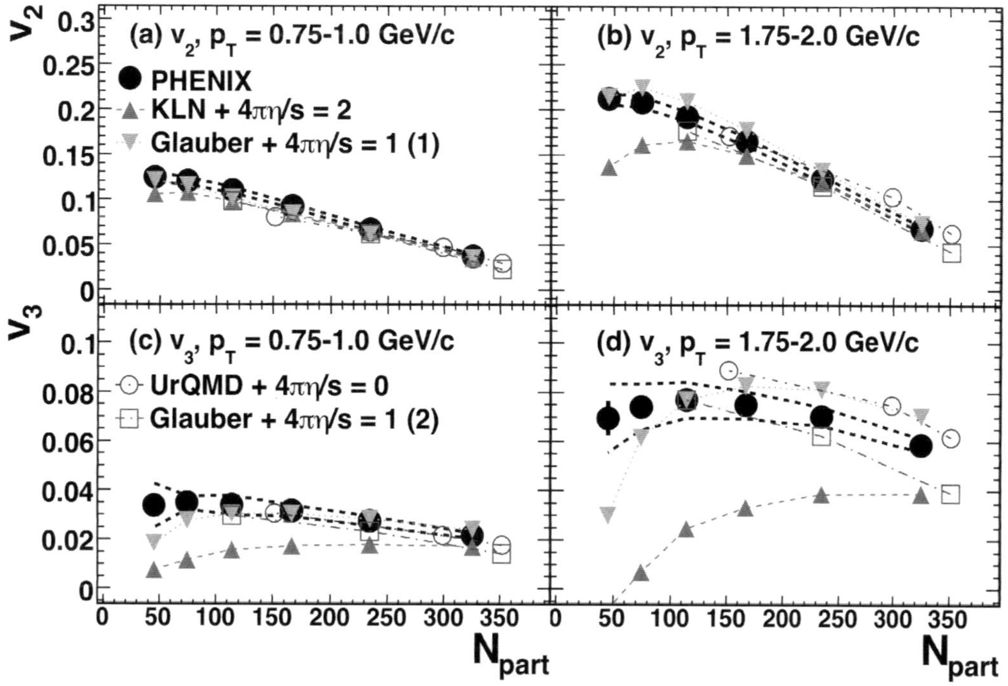

Figure 3. Elliptic (upper) and triangular (lower) flow strengths of inclusive charged hadrons, compared to specific hydrodynamical model calculations with the indicated choices of sQGP viscosity [5]. Within this set of models, the measurement of triangular flow strength provides substantial additional discrimination power in constraining the viscosity parameter.

The success of a picture with substantial initial-state fluctuations, which could in principle be measured and diagnosed, opens the way to a systematic investigation of how the initial deposit of energy density over space is achieved. The mechanism for this phase remains unknown at present, but understanding it would shed light on a whole new frontier of non-perturbative, non-thermal QCD physics.

Figure 5 shows another new PHENIX data set, which can bear directly on the question of what the initial hydro state looks like across three dimensions. On each event the PHENIX experiment measures, in several different detectors, the observable event plane orientation angles Φ_n, at each order n; these are experimental proxies for the true event plane angles ψ_n described in Eq. 2 above. What is shown in Fig. 5 is the azimuthal correlation between these measured event planes angles for combinations of detectors that span various ranges in rapidity. Each of these correlation strengths, measured as a function of centrality, provides information on the patterns of fluctuations in the initial state and how those are correlated across longitudinal space (initially) and across rapidity (finally).

To choose just one example, the blue points in panel (b) of Fig. 5 indicate that the triangular components of the final, and so presumably initial, state are positively correlated over a very wide separation in pseudorapitidy, $\Delta\eta \sim 7$, which is to say that their triangular orientations are parallel on average. This is actually quite a surprising result! in that the average distribution over three dimensions would produce triangles whose axes should be anti-correlated between forward and backward rapidities; and so this may indicate the dominance of initial fluctuations in the transverse plane which are correlated over long initial distances longitudinally. The new

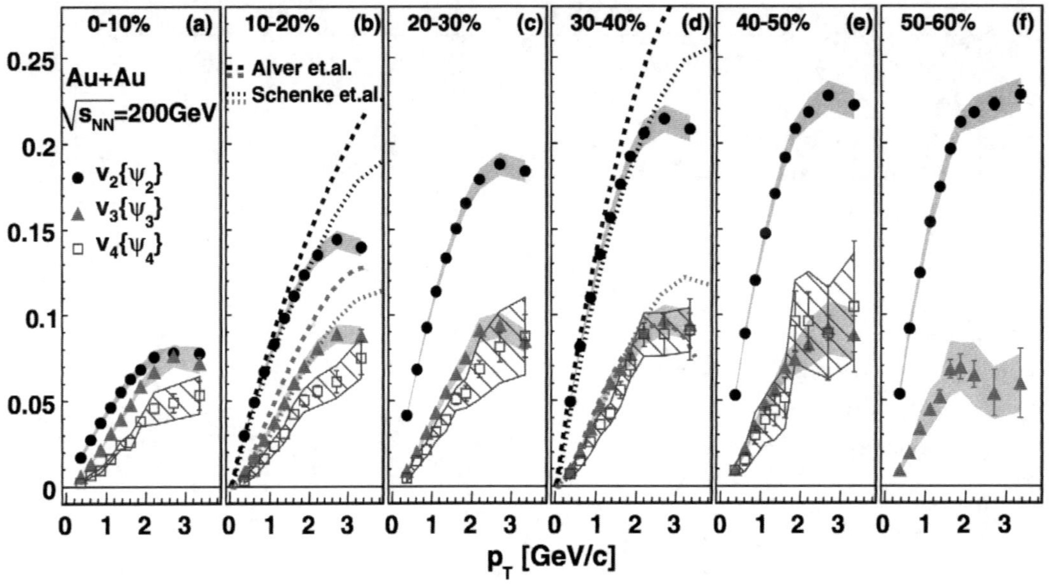

Figure 4. Elliptic (v_2), triangular (v_3) and "cloverleaf" (v_4) flow strengths for inclusive charged particles in Au+Au collisions, for different collision centralities [5]. The elliptic flow strength increases steadily in magnitude from central to peripheral, consistent with an origin primarily in geometry; while the higher-order moment strengths are nearly constant with centrality, consistent with an origin primarily in initial-state fluctuations.

wealth of this kind of data holds great potential for understanding the initial state in three dimensions, and awaits detailed theoretical attention at present.

5. Conclusion

The PHENIX experiment has recently published a range of unique flow results with intriguing physics implications. Identified hadron elliptic flow out to very high transverse momentum ($\sim$ 6 GeV/c) reveals a breakdown in the quark coalescence scaling behavior seen at lower momenta. An elliptic flow pattern has been observed in direct photons in an energy range thought to have been dominated by thermal radiation from a QGP phase, but hydrodynamical calculations of QGP radiation greatly under-predict the measured strengths. Measurement of higher-order flow moments v_3 and v_4 confirm a picture in which the initial energy density pattern is dominated by fluctuations, and also allow tighter constraints to be placed on transport properties of QGP-phase matter.

6. References

[1] S. S. Adler *et al.* [PHENIX Collaboration], "Elliptic flow of identified hadrons in Au+Au collisions at $\sqrt{s_{NN}} = 200$ GeV," Phys. Rev. Lett. **91**, 182301 (2003)

[2] A. Adare *et al.* [PHENIX Collaboration], "Deviation from quark-number scaling of the anisotropy parameter v_2 of pions, kaons, and protons in Au+Au collisions at $\sqrt{s_{NN}} = 200$ GeV," Phys. Rev. **C85**, 064914 (2012)

[3] A. Adare *et al.* [PHENIX Collaboration], "Observation of direct-photon collective flow in $\sqrt{s_{NN}} = 200$ GeV Au+Au collisions," arXiv:1105.4126 [nucl-ex], in press at Phys. Rev. Lett.

[4] A. Adare *et al.* [PHENIX Collaboration], "Enhanced production of direct photons in Au+Au collisions at $\sqrt{s_{NN}} = 200$ GeV and implications for the initial temperature," Phys. Rev. Lett. **104**, 132301 (2010)

[5] A. Adare *et al.* [PHENIX Collaboration], "Measurements of Higher-Order Flow Harmonics in Au+Au Collisions at $\sqrt{s_{NN}} = 200$ GeV," Phys. Rev. Lett. **107**, 252301 (2011)

Journal of Physics: Conference Series **389** (2012) 012008 doi:10.1088/1742-6596/389/1/012008

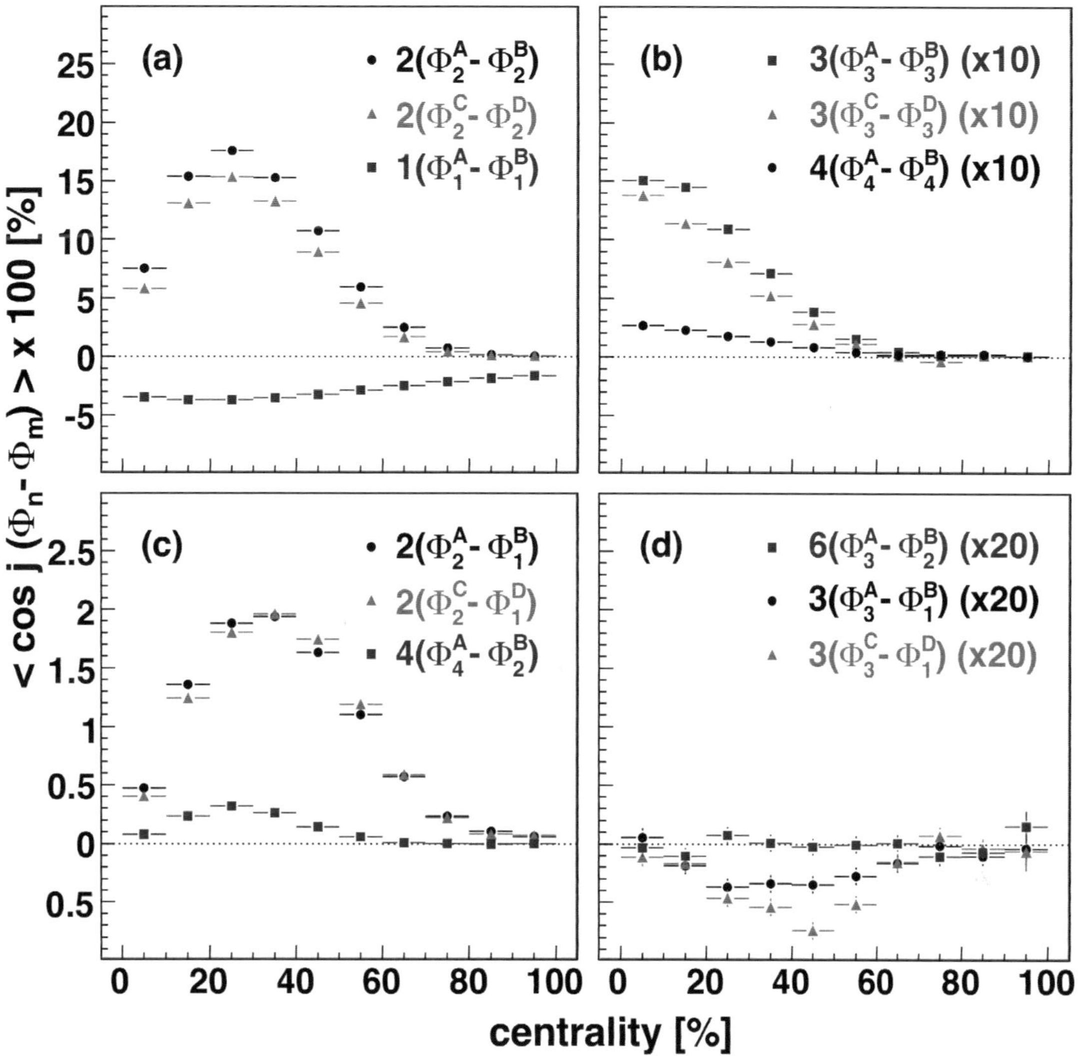

Figure 5. Correlation strengths between various combinations of measured event planes, as reconstructed in different PHENIX detectors at different harmonic orders (see [5] for full details). In particular, (i) the red triangles in panel (b) indicate that the third-order "triangular" event planes separated over a wide range in rapidity ($\Delta\eta \sim 7$) are positively correlated, i.e. they tend to have parallel orientations; while (ii) the blue squares in panel (d) indicate that the second-and third-order event planes are essentially uncorrelated over a slightly smaller separation.

Journal of Physics: Conference Series **389** (2012) 012009 doi:10.1088/1742-6596/389/1/012009

Probe the partonic/hadronic matter with elliptic flow in STAR Beam Energy Scan

Xiaoping Zhang for the STAR Collaboration

Department of Engineering Physics, Tsinghua University, Beijing 100084, China
Key Laboratory of Particle & Radiation Imaging (Tsinghua University), Ministry of Education, Beijing 100084, China

E-mail: xpzhangnju@gmail.com

Abstract. One of the goal of RHIC Beam Energy Scan program is to search for the phase transition boundary between partonic and hadronic matters. Azimuthal anisotropy is expected to be sensitive to the degree of freedom of the produced matter in the early stage of high energy nuclear collisions, which makes it an important probe to QCD phase transition. In this paper, we present the results of mid-rapidity elliptic flow measurements in Au + Au collisions at $\sqrt{s_{\rm NN}}$ = 7.7, 11.5 and 39 GeV from the STAR experiment at RHIC. The energy evolution of Number-of-Constituent-Quark (NCQ) scaling will be tested with the v_2 of identified hadrons ($\pi^{\pm}$, $K^{\pm}$, K^0_S, p, $\overline{p}$, ϕ, Λ, $\overline{\Lambda}$, Ξ^- and $\overline{\Xi}^+$). A significant difference in v_2 is observed between particles and corresponding anti-particles at the lowest energies of the beam energy scan. The implications on partonic-hadronic phase transition are discussed.

1. Introduction

One of the goal of RHIC Beam Energy Scan program is to search for the phase transition boundary between partonic and hadronic matters [1]. By systematic study of Au+Au collisions at $\sqrt{s_{\rm NN}} = 5 - 200$ GeV, one could access a wide region of temperature T and baryon chemical potential μ_B in the QCD phase diagram [1]. The elliptic flow (v_2) is a good probe to characterize the partonic or hadronic properties in the early stage of high energy nuclear collisions [2–4]. v_2 describes the azimuthal momentum space anisotropy of particle emission from non-central heavy-ion collisions in the plane transverse to the beam direction. It can be characterized by the second harmonic coefficient (v_2) of a Fourier decomposition of azimuthal momentum distribution of final state hadrons with respect to the reaction plane [5, 6], that is, $v_2 = \langle \cos 2(\varphi - \psi_R) \rangle$, where φ is the azimuthal angle of a produced particle in the laboratory frame, ψ_R is the reaction plane angle (for the details of ψ_R estimation from experimental data, see Ref. [6]), and the brackets denote the arithmetic average over all interested particles in corresponding events.

v_2 originates in the spatial anisotropy of the system when it is created in a non-central collision and in interactions in the evolving system which convert the spatial anisotropy to momentum anisotropy. This anisotropy is driven by the azimuthal dependent pressure gradient developed from the rescatterings. The frequency and intensity of such rescatterings depends on the partonic or hadronic properties of the system, which makes v_2 especially sensitive to the degrees of freedom of the system in the early time. From transport model calculations, v_2 in the partonic dominated system is expected to be significantly larger than that in hadronic case [2–4]. In Au+Au $\sqrt{s_{\rm NN}} = 200$ GeV collisions, RHIC has discovered the Number-of-Constituent-Quark

Journal of Physics: Conference Series **389** (2012) 012009 doi:10.1088/1742-6596/389/1/012009

(NCQ) scaling of v_2 for identified light and strange hadrons, which is an important evidence for the discovery of a new kind of deconfined matter with partonic collectivity [7]. Because of the possible transition from partonic dominated phase to hadronic dominated phase, it is expected that the established paradigm for partonic degrees of freedom at top RHIC energy may break at given low collision energy. Especially, the absence or reduction of collective flow and the breaking of NCQ scaling could indicate the system in hadronic phase [2–4]. In this paper, we study the energy dependence of v_2 in Au + Au collisions at $\sqrt{s_{\rm NN}}$ = 7.7, 11.5 and 39 GeV, based on data taken from STAR experiment at RHIC in the year 2010.

2. Experimental data analysis

The Au+Au collision events collected by the minimum bias trigger are used in this analysis. STAR's Time Projection Chamber (TPC) [8] and Time-of-Flight (TOF) detector [9,10] are used for tracking, particle identification and event plane determination. The events are required to have a primary Z vertex (along beam direction) within 70, 50, and 40 cm of the center of the TPC for Au+Au collisions at $\sqrt{s_{\rm NN}}$ = 7.7, 11.5, and 39 GeV, respectively, to ensure nearly uniform detector acceptance. After the event selection, we obtain about 4, 12, and 130 million Au+Au minimum bias triggered events at $\sqrt{s_{\rm NN}}$ = 7.7, 11.5, and 39 GeV, respectively. The collision centrality is determined by the measured raw charged hadron multiplicity from the TPC within a pseudorapidity window $|\eta| < 0.5$ [11]. The v_2 measurement is based on the η-sub event plane method to reduce the non-flow effect from short range η-correlations. The η-sub event plane method is similar to the event plane method, except one defines the flow vector for each particle based on the particles measured in the opposite hemisphere in pseudorapidity [12]. An η gap of $|\Delta\eta| > 0.05$ is guaranteed between the measured particle and the particles used for event plane reconstruction.

The TPC can identify particles by the specific energy loss ($\mathrm{d}E/\mathrm{d}x$) of charged particles traversing the TPC gas at low momentum. In the momentum range about 1–3 GeV/c, the pions, kaons and protons could not be identified by TPC $\mathrm{d}E/\mathrm{d}x$ only. With the fully installed TOF detector [9, 10] in the year 2010, one can identify event-by-event the $\pi^{\pm}$, $K^{\pm}$, proton, and anti-proton with the momentum above 0.2 GeV/c. The multi-strange hadron signals are obtained from the invariant mass distribution reconstructed by their hadronic decay channels: $\phi \rightarrow K^{+}+K^{-}$, $\Xi^{-}(\overline{\Xi}^{+}) \rightarrow \Lambda(\overline{\Lambda})+\pi^{-}(\pi^{+})$. The decay daughters $\Lambda(\overline{\Lambda})$ are reconstructed through $\Lambda(\overline{\Lambda}) \rightarrow p(\overline{p})+\pi^{-}(\pi^{+})$. K^0_S is reconstructed through its decay channel $K^0_S \rightarrow \pi^{+}+\pi^{-}$. Detailed descriptions on analysis cuts about geometry, kinematics as well as particle identification can be found in Refs. [12–15].

3. Results and discussions

In figure 1 we show the v_2 of Λ and $\overline{\Lambda}$ as a function of transverse momentum (p_T) in minimum bias (0-80%) Au+Au collisions at $\sqrt{s_{\rm NN}}$ = 7.7, 11.5 and 39 GeV. The v_2 gradually increases with the collision energy. One can see from figure 1 that Λ v_2 is systematically larger than $\overline{\Lambda}$ v_2. This difference increases with decreasing beam energy as shown for the data at 7.7 and 11.5 GeV.

Figure 2 shows the integrated v_2 difference between particles (P) and their corresponding anti-particles ($\overline{P}$) relative to the particle v_2. The difference is less than 5% between $K^{+}(\pi^{+})$ and $K^{-}(\pi^{-})$ at 39 GeV. For baryons such as protons(Λ) and anti-protons($\overline{\Lambda}$), the differences are a little bit larger than those of pions and kaons, about 10% at 39 GeV. At 7.7 and 11.5 GeV, a significant v_2 difference (up to 60%) between protons(Λ) and anti-protons($\overline{\Lambda}$) is observed. For mesons, the v_2 difference between $K^{+}(\pi^{+})$ and $K^{-}(\pi^{-})$ is up to 15%, much smaller compared to the baryons. Color bands in figure 2 are calculations from a multi-phase transport model (AMPT) [16]. The mean-field potentials in the hadronic phase are included in the AMPT model, which lead to the splittings of the elliptic flows of particles and their anti-particles. In this way, it

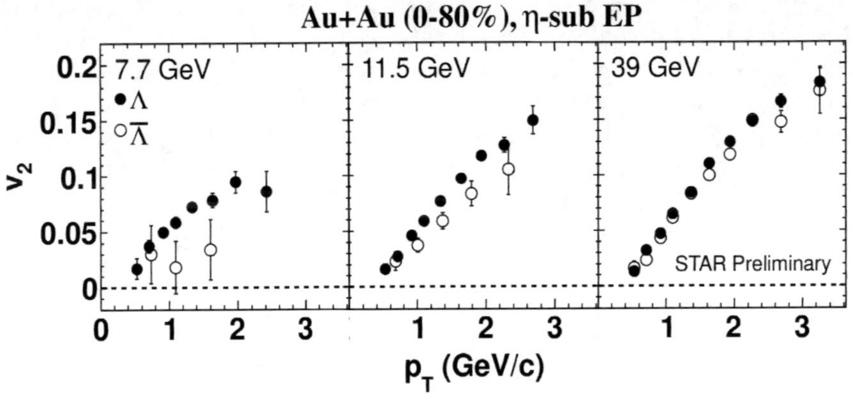

Figure 1. v_2 of Λ and $\overline{\Lambda}$ as a function of p_T in minimum bias (0–80%) Au+Au collisions at $\sqrt{s_{\rm NN}}$ = 7.7, 11.5 and 39 GeV.

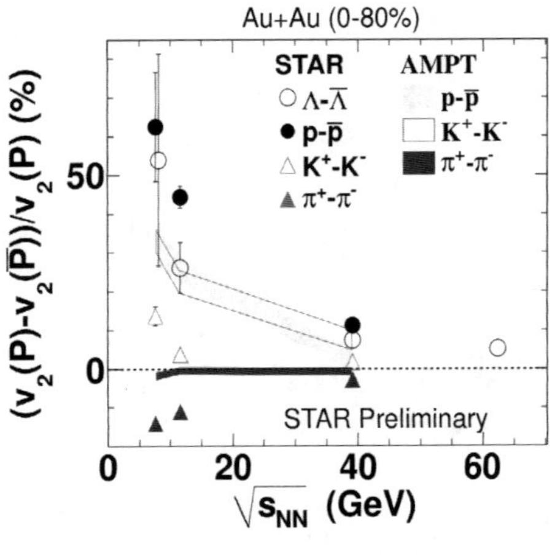

Figure 2. The relative difference of integrated v_2 between particles (P) and their corresponding anti-particles ($\overline{P}$) (see the legend) as a function of $\sqrt{s_{\rm NN}}$ for 0–80% Au+Au collisions. Color bands are calculations from a multi-phase transport model (AMPT) [16].

could describe qualitatively the v_2 differences between particles and their anti-particles. Without including the hadronic mean-field potentials, the model fails to reproduce the v_2 differences. Even with the hadronic mean-field potentials included, the AMPT model still underestimates the v_2 difference between protons(π^+) and anti-protons(π^-), while overestimates that between K^+ and K^-. Another model considers the different v_2 between transported quarks (u, d) and produced quarks ($\overline{u}$, $\overline{d}$) due to baryon transportation to mid-rapidity [17]. It further assumes a smaller strange quark v_2 compared to produced light quarks. With above assumptions and a parton coalescence scenario, the model could qualitatively describe the v_2 difference between different hadrons. To distinguish different models, further quantitative comparisons with model calculations on p_T, centrality and rapidity dependent v_2 should be performed.

In figures 3(a)-(c), we compare the v_2 of ϕ-meson to those of proton and Λ in minimum bias Au+Au collisions at $\sqrt{s_{\rm NN}}$ = 11.5, 39, and 200 GeV. We use the TPC event plane for v_2 measurement at 200 GeV [18, 19], and the TPC η-sub event plane at 11.5 and 39 GeV. One can see from figure 3(a) that ϕ-meson v_2 is close to those of proton and Λ for p_T below

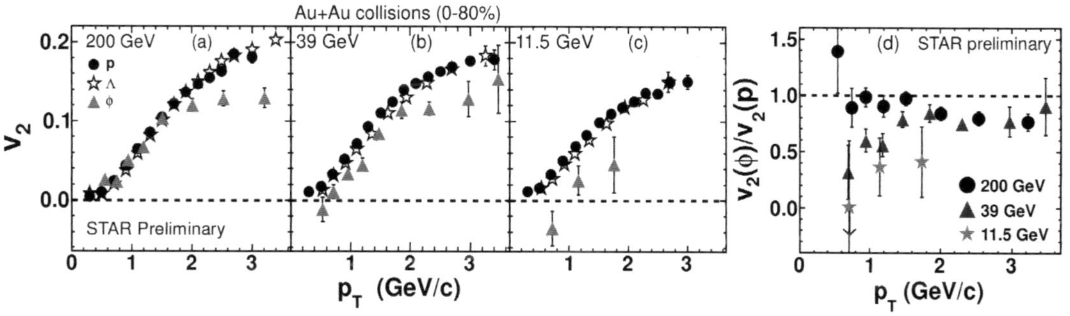

Figure 3. (a)-(c): ϕ-meson v_2 versus p_T compared to proton and Λ in minimum bias Au+Au collisions at 11.5, 39 and 200 GeV, respectively. The proton and Λ v_2 at 200 GeV are from Refs. [18, 19]. (d) The ratio between ϕ-meson v_2 and proton v_2 versus p_T in above energies. The central value of $v_2(\phi)/v_2(p)$ is far below 0 for the first low p_T point in Au+Au 11.5 GeV collisions. To suppress the y-axis scale, we place the central value of this point at $v_2(\phi)/v_2(p) = 0$ and use a down arrow to show that the real central value of this point is below 0. Errors are statistical only.

1.5 GeV/c in Au+Au collisions at $\sqrt{s_{\rm NN}} = 200$ GeV. This is expected from the hydrodynamic model since the mass of ϕ-meson is close to those of the proton and Λ [20]. For $p_T > 2$ GeV/c, there is baryon-meson separation which could be understood by quark coalescence in deconfined matter [21].

In Au+Au collisions at $\sqrt{s_{\rm NN}} = 39$ GeV, we observe the similar baryon-meson separation for $p_T > 2$ GeV/c, indicating that there is a partonic degree of freedom in the collision system. However, the ϕ-meson v_2 at low p_T is smaller than the proton and Λ v_2 at 11.5 and 39 GeV. One can see from figure 3(d) that the ratios between ϕ-meson v_2 and proton v_2 show a decreasing trend with decreasing collision energy at low p_T, indicating that relative to light quarks, the strange quark collectivity becomes weaker. Possible reasons could be that the strange quark are not fully thermalized in 11.5 and 39 GeV Au+Au collisions [17] and/or the hadronic interactions start to play an important role in driving ϕ-meson elliptic flow [2–4]. Note that ϕ-meson has small hadronic cross sections, a small v_2 is expected in hadronic dominated system [2–4], which is seen in the present experimental data.

In figure 4, we test the transverse kinetic energy $(m_T - m_0)$ scaling of v_2 with various particles π^+, K^0_S, ϕ, proton, Λ, and Ξ^-. Most of the particles follow one common curve except for the ϕ mesons at 11.5 GeV. The ϕ meson v_2 shows a different trend away from the scaling of other hadrons. The mean deviation of the ϕ-meson v_2 from the π^+ v_2 is 0.02 ± 0.008 at 11.5 GeV. This observation shows that ϕ meson has a smaller v_2 than the expectation from $m_T - m_0$ scaling of other hadrons at 11.5 GeV.

4. Summary

In summary, we present the results of mid-rapidity elliptic flow measurements in Au + Au collisions at $\sqrt{s_{\rm NN}} = 7.7$, 11.5 and 39 GeV from the STAR experiment at RHIC. A significant difference in v_2 is observed between particles and corresponding anti-particles in Au+Au 7.7 and 11.5 GeV, suggesting that NCQ-scaling between particles and anti-particles is broken at low beam energies. In Au+Au collisions at $\sqrt{s_{\rm NN}} = 39$ GeV, the baryon-meson separation between ϕ-meson and proton (or Λ) v_2 at $p_T > 2$ GeV/c supports the quark coalescence picture. At low p_T, the ϕ-meson v_2 is smaller than those of the proton and Λ, thus violates the mass ordering observed in Au+Au collisions at $\sqrt{s_{\rm NN}} = 200$ GeV. This might be related to incomplete thermalization of strange quarks and/or the increasing contributions from hadronic interactions

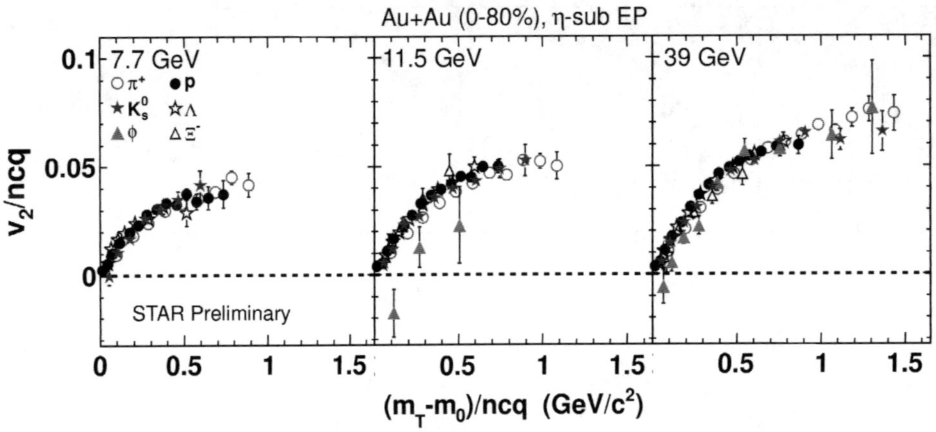

Figure 4. v_2 per number-of-constituent quarks (ncq) in 0−80% minimum bias Au+Au collisions for π^+, K_S^0, ϕ, proton, Λ and Ξ^- as a function of $(m_T - m_0)/ncq$ for the three collision energies 7.7, 11.5 and 39 GeV. Errors are statistical only.

to the elliptic flow with decreasing collision energies. Coming results from Au+Au collisions at 19.6 and 27 GeV measured in year 2011 will provide further information.

Acknowledgments

X. Zhang thanks the support by the National Natural Science Foundation of China (Grant Nos. 10865004, 10905029, 11035009, 11065005, and 11105079), by the China Postdoctoral Science Foundation (Grant No. 20100480017), and by the Foundation for the Authors of National Excellent Doctoral Dissertation of P. R. China (FANEDD) (No. 201021).

References

[1] M. M. Aggarwal *et al.* (STAR Collaboration.), arXiv:1007.2613.
[2] B. Mohanty and N. Xu, J. Phys. G **36**, 064022 (2009).
[3] J. Tian *et al.*, Phys. Rev. C **79**, 067901 (2009).
[4] K. J. Wu, F. Liu, and N. Xu, J. Phys. G **37**, 094029 (2010).
[5] S. Voloshin and Y. Zhang, Z. Phys. C **70**, 665 (1996).
[6] A. M. Poskanzer and S. A. Voloshin, Phys. Rev. C **58**, 1671 (1998).
[7] B. I. Abelev *et al.* (STAR Collaboration.), Phys. Rev. Lett. **99**, 112301 (2007).
[8] K. H. Ackermann *et al.* (STAR Collaboration), Nucl. Instrum. Meth. A **499**, 624 (2003).
[9] M. Shao *et al.*, Nucl. Instrum. Meth. A **558**, 419 (2006).
[10] W. J. Llope, Nucl. Instrum. Meth. B **241**, 306 (2005).
[11] B. I. Abelev *et al.* (STAR Collaboration), Phys. Rev. C **79**, 034909 (2009).
[12] B. I. Abelev *et al.* (STAR Collaboration), Phys. Rev. C **77**, 054901 (2008).
[13] B. I. Abelev *et al.* (STAR Collaboration), Phys. Rev. C **79**, 064903 (2009).
[14] A. Schmah (STAR Collaboration), J. Phys. G **38**, 124049 (2011).
[15] X. Zhang (STAR Collaboration), Acta Phys. Pol. B Proc. Suppl. **5**, 509 (2012).
[16] J. Xu, L.-W. Chen, C. M. Ko, and Z.-W. Lin, Phys. Rev. C **85**, 041901(R) (2012).
[17] J. C. Dunlop, M. A. Lisa, and P. Sorensen, Phys. Rev. C **84**, 044914 (2011).
[18] S. Shi (STAR Collaboration), Nucl. Phys. A **830**, 187c (2009).
[19] X. Zhang (STAR collaboration), Plasma Sci. and Technol. **14**, 449 (2012).
[20] P. Huovinen *et al.*, Phys. Lett. B **503**, 58 (2001).
[21] D. Molnar and S. A. Voloshin, Phys. Rev. Lett. **91**, 092301 (2003); R. C. Hwa, C. B. Yang, Phys. Rev. C **67**, 064902 (2003); R. J. Fries *et al.*, Phys. Rev. Lett. **90**, 202303 (2003).

28th Winter Workshop on Nuclear Dynamics 2012 IOP Publishing
Journal of Physics: Conference Series **389** (2012) 012010 doi:10.1088/1742-6596/389/1/012010

Energy scan of π^0 Suppression and Flow in Au+Au Collisions at PHENIX

Norbert Novizky for the PHENIX collaboration
University of Jyväskylä, Jyväskylä, Finland
Helsinki Institute of Physics, Helsinki, Finland
E-mail: Norbert.Novitzky@cern.ch

Abstract. The neutral pion spectra were measured by the PHENIX detector in $|y| < 0.35$ in Au+Au system at $\sqrt{s_{NN}} = 39$, 62.4 and 200 GeV. The nuclear modification factor (R_{AA}) shows significant suppression and strong centrality dependence in Au+Au collisions. The azimuthal anisotropy is measured with respect to the reaction plane, determined at large rapidity. Here we present both R_{AA} and azimuthal anisotropy of the π^0 in Au+Au for three different energies.

1. Introduction

A large suppression of the inclusive π^0 production, up to a factor of 5, observed for the first time in Au+Au at $\sqrt{s_{NN}} = 130$ GeV collisions [1] and $\sqrt{s_{NN}} = 200$ GeV collisions [2], is one of the most significant manifestations of the opaque QCD medium produced in heavy ion collisions. Absence of this suppression in d+Au at $\sqrt{s_{NN}} = 200$ GeV collisions [3] indicates that this phenomenon cannot be attributed to initial state effects or suppression by the cold nuclear matter [4].

Event anisotropy is expected to be sensitive to the early stages of heavy ion collisions. A possible formation of the quark-gluon plasma (QGP) could affect the way how the initial anisotropy in coordinate space is transferred into momentum space in the final state. At high-p_{T}, where the hard scattering processes are dominant over the soft processes, the event anisotropy provides us deeper understanding of the energy loss mechanism.

We show the new results from the latest RHIC low energy program from 2010, namely Au+Au collisions at $\sqrt{s_{NN}} = 62.4$ and 39 GeV. The aim of the low energy scan is to search for the critical point of the phase transition and to see how the suppression of high-p_{T} particles evolves. We present the π^0 elliptical flow and the nuclear modification factor.

2. Nuclear Modification Factor R_{AA}

The suppression of high-p_{T} particles in medium is studied in terms of the nuclear modification factor:

$$R_{AA}(p_{\mathrm{T}}) = \frac{(1/N_{\mathrm{evt}}^{AA})d^2N^{AA}/dydp_{\mathrm{T}}}{\langle N_{\mathrm{coll}}\rangle\,(1/N_{\mathrm{evt}}^{pp})d^2N^{pp}/dydp_{\mathrm{T}}}, \quad (1)$$

where $d^2N/dydp_{\mathrm{T}}$ is the yield in the corresponding system (p+p or Au+Au) and $\langle N_{\mathrm{coll}}\rangle$ is the average number of binary collisions obtained from the Glauber model Monte-Carlo [5]. The observed suppression in Au+Au at $\sqrt{s_{NN}}$ = 200 GeV [2] ($R_{AA} \approx 0.2$) was interpreted to be caused by the energy loss of partons traversing the medium created in Au+Au collisions.

There are four main classes of theoretical models describing the energy-loss mechanism: opacity expansion (GLV) [6], multiple soft scattering (BDMPS-Z-ASW) [7], higher-twist (HT) [8], thermal field theory (AMY) [9]. Although all models provided quite accurate description of the measured ϕ-integrated nuclear modification factor of π^0 at $\sqrt{s_{NN}}$ = 200 GeV [2], we are still lacking a detailed understanding of the jet quenching phenomena [10].

3. Azimuthal Anisotropy v_2

The elliptic flow v_2 is the second Fourier coefficient of the azimuthal distribution of selected particles. Neglecting the higher harmonics, the azimuthal yield can be written as

$$\frac{dN}{d\phi} \propto 1 + 2 \cdot v_2 \cos 2(\phi - \psi_{\mathrm{RP}}), \tag{2}$$

where ϕ is the azimuthal angle of an outgoing particle and ψ_{RP} is the reaction plane angle. In the optical Glauber model ψ_{RP} coincides with the direction of the nuclear impact parameter. However, in the real collision the ψ_{RP} does not necessarily coincide with the impact parameter due to the event-by-event nuclear density fluctuations. The measurements of the parameter v_2 in PHENIX at RHIC were performed for charged particles [11, 12] as well as for identified particles [13, 14].

The nuclear modification factor with respect to the reaction plane ($\Delta\phi = \phi - \psi_{\mathrm{RP}}$)[15] allows to study an effect of the initial geometry on the observed suppression:

$$R_{AA}\left(\Delta\phi_i, p_{\mathrm{T}}\right) = R_{AA}^{\mathrm{incl}}\left(p_{\mathrm{T}}\right) \times \frac{N\left(\Delta\phi_i, p_{\mathrm{T}}\right)}{(1/N_{bin})\sum_{i=1}^{N_{bin}} N\left(\Delta\phi_i, p_{\mathrm{T}}\right)}, \tag{3}$$

where $R_{AA}^{\mathrm{incl}}(p_{\mathrm{T}})$ is the nuclear modification factor as defined in Eq. (1), $N\left(\Delta\phi_i, p_{\mathrm{T}}\right)$ is the number of particles observed in the given $(\Delta\phi_i, p_{\mathrm{T}})$ bin and N_{bin} is the number of the reaction plane bins. However, up to the contribution from higher harmonics, the high-p_{T} v_2 and the corresponding

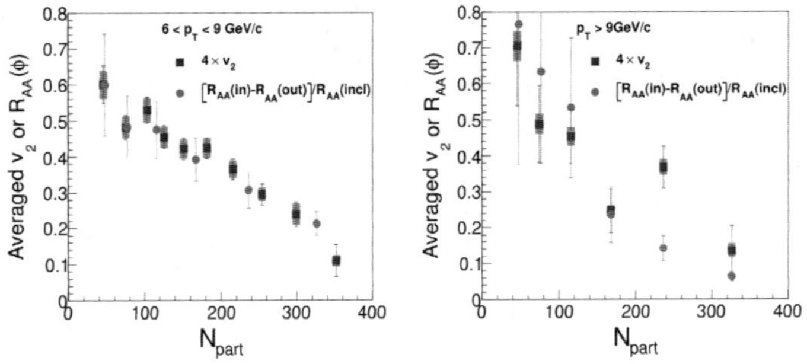

Figure 1. The integrated v_2 and the in- and out-of-plane difference of the nuclear modification factor defined on the left hand side of the Eq. (5). The data are from the π^0 measurement in Au+Au at $\sqrt{s_{NN}}$=200 GeV [14, 15].

inclusive R_{AA} contain the same information as $R_{AA}(\Delta\phi, p_{\mathrm{T}})$. This can be seen from:

$$\begin{aligned} R_{AA}\left(\Delta\phi, p_{\mathrm{T}}\right) &= \frac{d^3N^{\mathrm{AuAu}}/d\Delta\phi dy dp_{\mathrm{T}}}{\langle N_{\mathrm{coll}}\rangle\, d^3N^{pp}/d\Delta\phi dy dp_{\mathrm{T}}} \approx \frac{d^2N^{\mathrm{AuAu}}/dy dp_{\mathrm{T}}\left(1 + 2\cdot v_2\cos\left(2\Delta\phi\right)\right)}{\langle N_{\mathrm{coll}}\rangle\, d^2N^{pp}/dy dp_{\mathrm{T}}} \\ &= R_{AA}^{\mathrm{incl}}(p_{\mathrm{T}})\left(1 + 2\cdot v_2\cos\left(2\Delta\phi\right)\right). \end{aligned} \tag{4}$$

Note that contributions from the higher harmonics (v_3, v_4, ...) are negligible at high-p_T [16]. From Eq. (4) it follows that the often shown difference between the in- and the out-of-plane R_{AA} is simply

$$\frac{R_{AA}(0,p_T)-R_{AA}(\pi/2,p_T)}{R_{AA}^{\text{incl}}(p_T)} \approx 4\cdot v_2(p_T). \tag{5}$$

PHENIX data [15] for $R_{AA}(\Delta\phi, p_T)$ and high-p_T v_2 are shown in Fig. 1 to check how well Eq. (5) holds for the measured data. The different p_T regions of the particle spectra are dominated by different physics processes, while the low-p_T is considered for flow and hydrodynamical behavior, the production of high-p_T particles are dominated by hard processes.

4. Energy Scan

In 2005 RHIC provided Cu+Cu collisions at three different center-of-mass energies, namely $\sqrt{s_{NN}}$ = 22.4, 62.4 and 200 GeV [17]. The large suppression of π^0 at high-p_T was observed at $\sqrt{s_{NN}}$ = 62.4 and 200 GeV and moderate enhancement at $\sqrt{s_{NN}}$ = 22.4 GeV, see Fig. 2 left. The p_T averaged R_{AA} showed a clear centrality dependence at $\sqrt{s_{NN}}$ = 62.4 and 200 GeV and no centrality dependence at $\sqrt{s_{NN}}$ = 22.4 GeV, see Fig. 2 right.

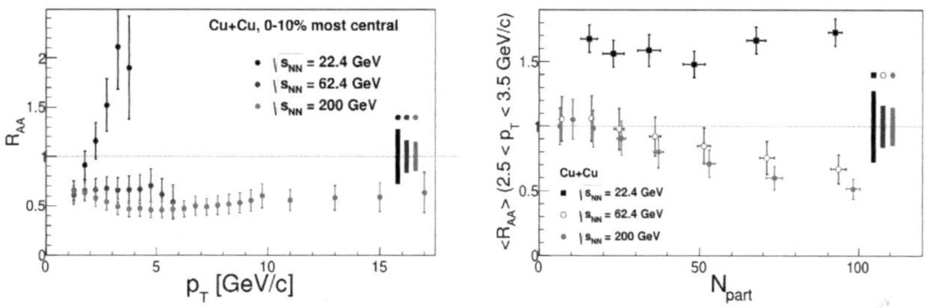

Figure 2. Measured π^0 nuclear modification factor in Cu+Cu collisions [17]. The error bars show the statistical errors and the error boxes around the dashed line show the systematical uncertainties. The left panel shows the R_{AA} in a function of p_T, while the right panel shows the p_T averaged R_{AA} in function of the number of participants.

In 2010, PHENIX collected 700M events at $\sqrt{s_{NN}}$ = 62.4 GeV and 250M events at $\sqrt{s_{NN}}$ = 39 GeV Au+Au collisions. The π^0 was reconstructed using the $\pi^0 \to \gamma\gamma$ channel. The π^0 invariant yield was reconstructed in the EMCal calorimeter; results are shown in Fig. 3 [18]. From fitting the minimum bias spectra with a power law function ($\propto p_T^n$) for $p_T > 4$ GeV/c, we obtained the powers $n_{39} = -12.07 \pm 0.18$ and $n_{62.4} = -10.60 \pm 0.09$ which are steeper than at $\sqrt{s_{NN}}$ = 200 GeV, where $n_{200} = -8.06 \pm 0.08$ for minimum bias collisions. The slopes of corresponding p+p data are $n_{39}^{pp} = -12.02 \pm 0.31$, $n_{62.4}^{pp} = -9.82 \pm 0.18$ and $n_{200}^{pp} = -8.22 \pm 0.09$ [2].

To determine the v_2 coefficient of π^0 , the relative azimuthal emission angle w.r.t. reaction plane was divided into six bins and the corresponding yield was fitted with Eq. (2). The reaction plane was determined combining the inner and outer ring of reaction plane detector (RXNin, $1.0 < |\eta| < 1.5$, and RXNout, $1.5 < |\eta| < 2.8$), the beam-beam counter (BBC, $3.1 < |\eta| < 3.9$) and muon piston calorimeter (MPC, $3.1 < |\eta| < 3.9$). Fig. 4 shows the extracted π^0 v_2 as a function of transverse momentum for three center of mass energies, where the event plane was defined by combination of the RXNin and RXNout for $\sqrt{s_{NN}}$ = 39 and 62.4 GeV and the combination of RXNin and MPC for $\sqrt{s_{NN}}$ = 200 GeV. Within the studied range of center of

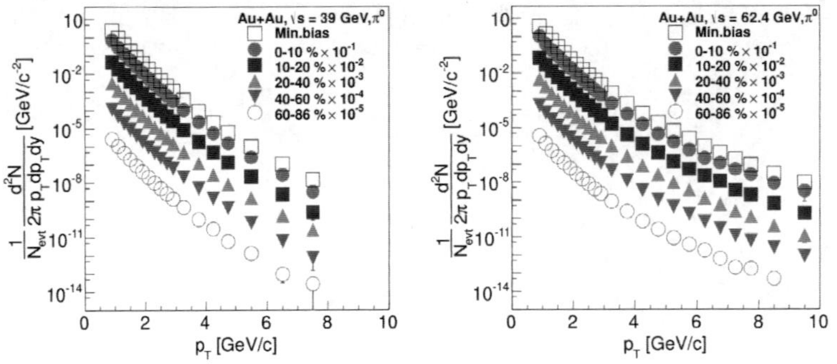

Figure 3. π^0 invariant yield spectra in Au+Au collisions at $\sqrt{s_{NN}}$ = 39 GeV (left) and 62.4 GeV (right) in different centrality bins. The minimum bias data are shown as the open squares. Only the statistical uncertainties are shown.

mass energies, $\sqrt{s_{NN}}$ = 39-200 GeV, the data suggest either no or a rather week dependence of the measured v_2 on $\sqrt{s_{NN}}$.

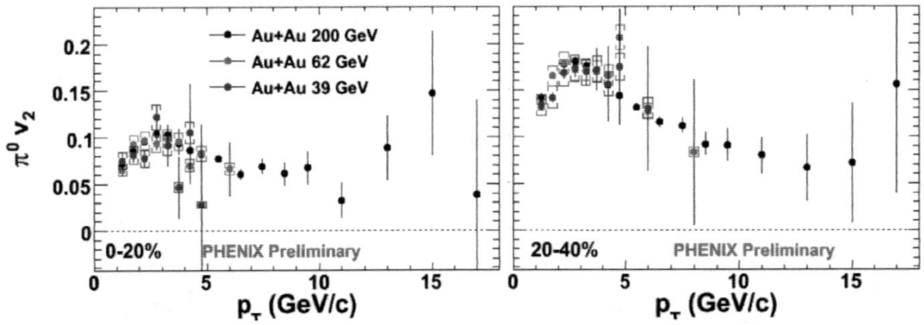

Figure 4. The v_2 of π^0 as a function of p_T in $\sqrt{s_{NN}}$ = 39, 62.4 and 200 GeV. Left panel shows the most central collisions (0-20%), right panel corresponds to more peripheral collisions (20-40%).

Evaluation of R_{AA} requires knowledge of the corresponding p+p reference. A p+p reference with the same experimental set-up as the heavy ion data reduces a significant part of the systematic uncertainties.

In the case of $\sqrt{s}$ = 62.4 GeV p+p reference, PHENIX measured π^0 cross-section up to $p_T <$ 7 GeV/c (see Fig 1. in [17]). However, the corresponding Au+Au data reach up to $p_T <$ 10 GeV/c (see Fig. 3 right). In the range $7 < p_T < 10$ GeV/c, the reference data were estimated by means of a power-law extrapolation of the existing data points. The systematic uncertainties resulting from the extrapolation are assessed based on varying the fit range used for the power-law fit. The extrapolated reference is compatible with the data points measured by the CCOR [20] experiment at ISR in the range $7 < p_T < 10$ GeV/c within the systematic uncertainties (see Fig. 5).

The p+p reference at $\sqrt{s}$ = 39 GeV has not been measured by PHENIX yet and therefore we used the π^0 yield measured by the Tevatron E706 experiment [21]. The acceptance correction coming from the different rapidity coverage was estimated based on the PYTHIA8 simulation [22] (see Fig. 6 right).

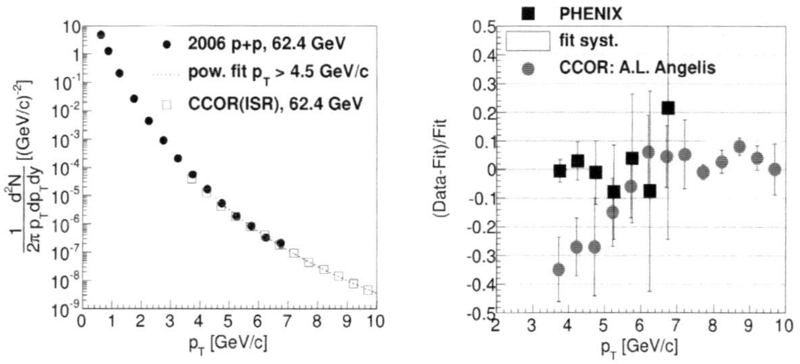

Figure 5. Left: The p+p reference at $\sqrt{s} = 62.4$ GeV from PHENIX [19] fitted with a power-law function. The data are compared with the data from the CCOR experiment at ISR [20]. Right: Relative difference between the measured data and the powerlaw fit which was used as the reference in the range $7 < p_{\rm T} < 10$ GeV/c. The yellow band gives the estimated systematic uncertainties following from the extrapolation procedure.

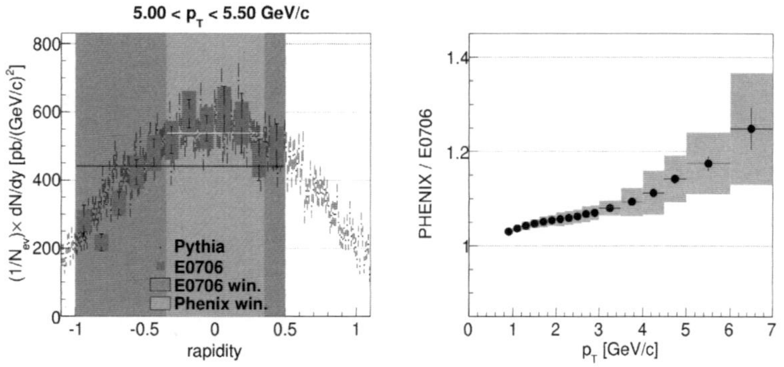

Figure 6. Left: An example of the π^0 rapidity distribution obtained from E706 [21] and PYTHIA8 [22] in p+p collisions at $\sqrt{s_{NN}} = 39$ GeV. The $p_{\rm T}$ range of selected π^0 s was $5.0 < p_{\rm T} < 5.5$ GeV/c. The green and the blue box represent the rapidity coverage of the PHENIX and the E706 experiment respectively. Right: The ratio of PYTHIA8 simulated π^0 invariant yields in PHENIX and E706 acceptance. The estimated systematic uncertainty of this correction is given by the gray boxes around the data points.

The resulting π^0 R_{AA} [18] was in Au+Au at $\sqrt{s_{NN}} = 39$ and 62.4 GeV is shown in Fig. 7. The left panel corresponds to the most central collisions where all three energies shows large suppression indicating that an opaque medium is produced even at the $\sqrt{s_{NN}} = 39$ GeV Au+Au collisions. The right panel shows a significant suppression of the $\sqrt{s_{NN}} = 200$ and 62.4 GeV Au+Au collisions for $p_{\rm T} > 4$ GeV/c and for the same $p_{\rm T}$ region for $\sqrt{s_{NN}} = 39$ GeV it is close to unity. In both centralities, the R_{AA} for $\sqrt{s_{NN}} = 62.4$ and 200 GeV are comparable for $p_{\rm T} > 6$ GeV/c.

The $p_{\rm T}$ averaged R_{AA} was calculated in two ranges, $4 < p_{\rm T} < 6$ GeV/c and $p_{\rm T} > 6$GeV/c (see Fig. 8). In the lower $p_{\rm T}$ region we observe a suppression of π^0 at $\sqrt{s_{NN}} = 62.4$ and 200 GeV and for the $\sqrt{s_{NN}} = 39$ GeV data the suppression starts to be pronounced only when $N_{\rm part} > 100$. In the high $p_{\rm T}$ range, $\sqrt{s_{NN}} = 62.4$ and 200 GeV data seem to have a similar magnitude of R_{AA} at all centralities.

Journal of Physics: Conference Series **389** (2012) 012010 doi:10.1088/1742-6596/389/1/012010

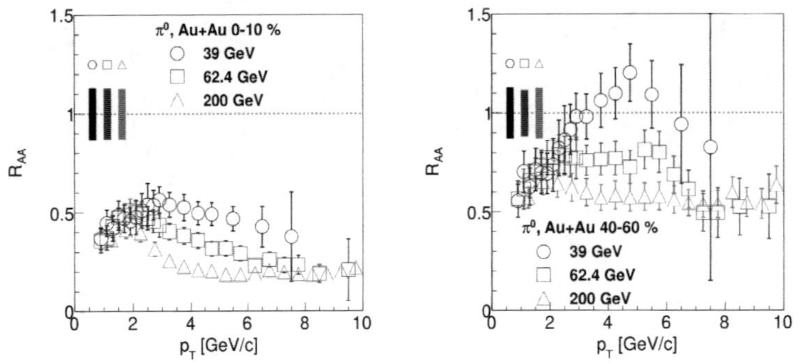

Figure 7. The nuclear modification factor measured in Au+Au collisions at $\sqrt{s_{NN}} = 39$, 62.4 and 200 GeV. The error bars shown represent a quadratic sum of the statistical uncertainties and the point-to-point uncorrelated and correlated systematic uncertainties. The boxes around unity indicate uncertainties related to $\langle N_{\rm coll}\rangle$ and absolute normalization. Left: The most central collisions 0-10%. Right: The mid-peripheral collisions 40-60%.

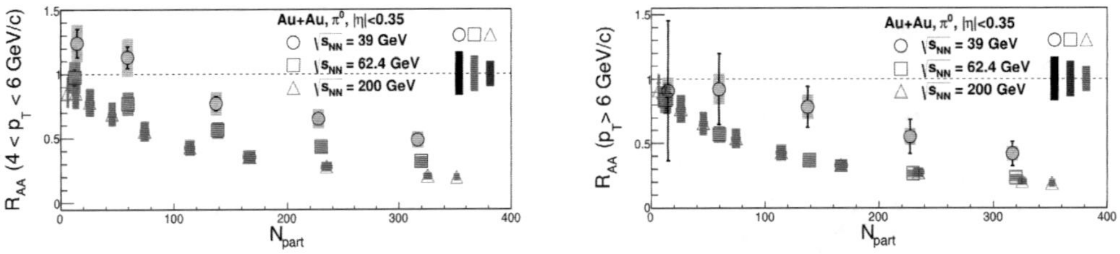

Figure 8. The $p_{\rm T}$ averaged R_{AA} of π^0 in Au+Au at $\sqrt{s_{NN}} = 39$, 62.4 and 200 GeV as a function of the number of participants ($N_{\rm part}$). Data corresponding to two π^0 $p_{\rm T}$ ranges are shown: $4 < p_{\rm T} < 6$ GeV/c (left) and $p_{\rm T} > 6$ GeV/c (right).

5. x_T Scaling and Fractional Energy Loss

Inclusive single-particle spectra at sufficiently high-$p_{\rm T}$ and collision energy were predicted to exhibit scaling in $\sqrt{s}$ with the variable $x_T = 2p_{\rm T}/\sqrt{s}$ such that the production cross section can be written in a form [23, 24]

$$E\frac{d^3\sigma}{dp^3} = \frac{1}{\sqrt{s}^{n(x_T,\sqrt{s})}}G(x_T), \tag{6}$$

where $G(x_T)$ is a general function and the effective exponent $n(x_T, \sqrt{s})$ characterizes the process [24]. The scaling variable $n_{\rm eff}(x_T, \sqrt{s})$ is evaluated using the linear variation of the logarithm of the ratio of the invariant cross-section in p+p collisions at different center of mass energies. In case of Au+Au data the total cross-section can be defined via the Glauber model where we observe the difference in collision energy range $\sqrt{s_{NN}} = 39$, 62.4 and 200 GeV is very small. Therefore, we obtain the effective exponent $n_{\rm eff}(x_T, \sqrt{s})$ in case of Au+Au , from invariant yields instead of cross-section:

$$n_{\rm eff}^{AA}(x_T) = \frac{\log\left(yield(x_T,\sqrt{s_1})/yield(x_T,\sqrt{s_2})\right)}{\log\left(\sqrt{s_2}/\sqrt{s_1}\right)}. \tag{7}$$

Fig. 9 left panel shows the x_T scaling power $n_{\text{eff}}(x_T)$ for pairing the π^0 invariant cross-sections in p+p and invariant yields in Au+Au collisions. In case of 62.4 and 200 GeV both for the p+p and Au+Au data show similar shape and they are comparable within the uncertainties. The p+p data for 39 and 200 GeV is comparable with the higher energy results. However, the $n_{\text{eff}}(x_T)$ for 39 and 200 GeV in Au+Au collisions behave differently when compared other data pairs. The rise of the effective exponent in lower x_T values is attributed with the dominance of soft processes [25], while at higher x_T values they deviate strongly from leading-twist scaling predictions [24, 26]. The data on Fig. 9 suggest that the transition from soft processes to hard processes occurs at higher p_{T} in Au+Au collisions than in p+p collisions at 39 GeV.

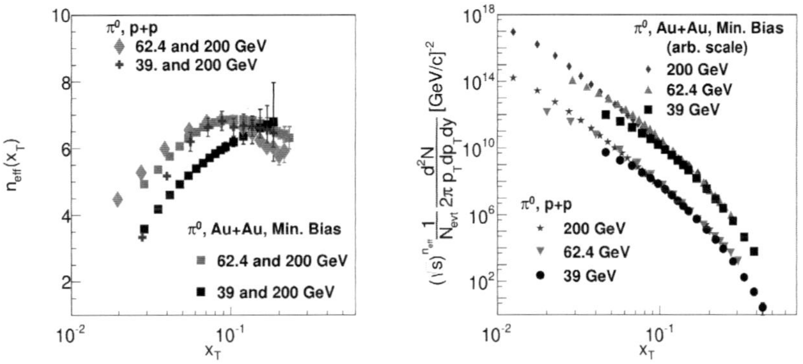

Figure 9. Left: The x_T scaling power $n_{\text{eff}}(x_T)$ for p+p and Au+Au (minimum bias) calculated with different collision energy pairs. Right: Using the n_{eff} we scaled the Au+Au and the p+p invariant yields. The Au+Au yield have an additional arbitrary scaling.

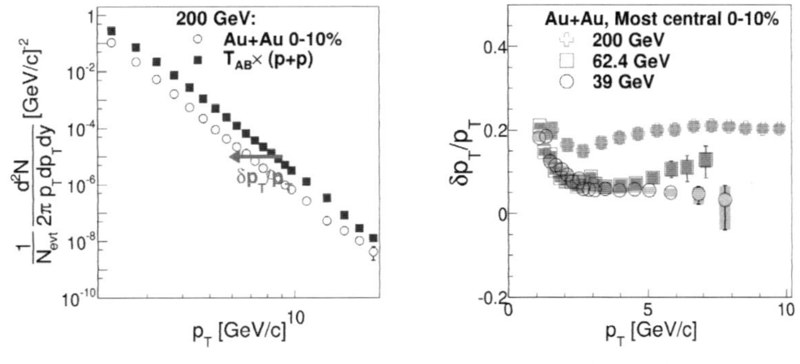

Figure 10. Left panel shows the spectra of π^0 in the Au+Au and T_{AB} scaled p+p spectra at 200 GeV collisions. The fractional momentum loss $\delta p_{\text{T}}/p_{\text{T}}$ is illustrated as the horizontal shift between the two spectra. Right panel shows the fractional momentum loss result for the most central collisions (0-10%) at 39, 62.4 and 200 GeV collisions.

The parton created from hard scattering is propagating through the medium and looses its energy. Final state fragments of this parton have also decreased magnitude of momentum, $p'^{\pi^0}_T = p^{\pi^0}_T - \delta p^{\pi^0}_T$, where $p^{\pi^0}_T$ and $p'^{\pi^0}_T$ are the original and final momentum of the π^0 , respectively. The average momentum loss of the neutral pions can be studied as the momentum shift between the Au+Au and $T_{\rm AB}$ scaled p+p spectra. On Fig. 10 left panel we show the spectra of π^0 in Au+Au and scaled p+p collisions. The fractional momentum loss $\delta p_{\rm T}/p_{\rm T}$ is defined as a horizontal shift of the spectrum. On Fig. 10 right panel we show the point-by-point fractional momentum loss results for π^0 at three energies, namely 39, 62.4 and 200 GeV. The data shows the energy loss is the largest at 200 GeV collisions and approximately flat in this range

6. Summary

The energy scan program at RHIC provides information about the evolution of the high-$p_{\rm T}$ suppression. The R_{AA} of π^0 was measured at three different energies in Au+Au at $\sqrt{s_{NN}} = 39$, 62.4 and 200 GeV. The most central collisions (0-10%) shows factor of 2 suppression at $\sqrt{s_{NN}} =$ 39 GeV, while in mid-peripheral collision (40-60%) the high-$p_{\rm T}$ particles are not suppressed. A strong suppression is shown at $\sqrt{s_{NN}} = 62.4$ GeV and 200 GeV both in the most central and a still significant suppression in mid-peripheral collisions. The results suggest a smooth transition of the energy-loss in $\sqrt{s_{NN}}$ and system size (centrality). The measurement of the flow (v_2) of the π^0 at same colliding energies shows no significant dependence on the $\sqrt{s_{NN}}$, indicating the soft processes vary slowly with the collision energy. The x_T scaling exponent suggest the transition from soft processes to hard processes occurs at higher $p_{\rm T}$ in Au+Au collisions at $\sqrt{s_{NN}} = 39$ GeV than in p+p collisions at same collision energies.

References

[1] K. Adcox *et al.*, Phys.Rev.Lett. **88**, 022301 (2002).
[2] A. Adare *et al.*, Phys.Rev.Lett. **101**, 232301 (2008).
[3] S. S. Adler *et al.*, Phys. Rev. Lett. **98**, 172302 (2007).
[4] A. Accardi and M. Gyulassy, Phys.Lett. **B586**, 244 (2004).
[5] M. L. Miller, K. Reygers, S. J. Sanders, and P. Steinberg, Ann.Rev.Nucl.Part.Sci. **57**, 205 (2007).
[6] M. Gyulassy, P. Levai, and I. Vitev, Phys.Lett. **B538**, 282 (2002).
[7] U. A. Wiedemann, Nucl.Phys. **B588**, 303 (2000).
[8] X.-f. Guo and X.-N. Wang, Phys.Rev.Lett. **85**, 3591 (2000).
[9] P. B. Arnold, G. D. Moore, and L. G. Yaffe, JHEP **0011**, 001 (2000).
[10] W. Horowitz and B. Cole, Phys.Rev. **C81**, 024909 (2010).
[11] K. Adcox *et al.*, Phys. Rev. Lett. **89**, 212301 (2002).
[12] A. Adare *et al.*, Phys.Rev.Lett. **105**, 062301 (2010).
[13] S. S. Adler *et al.*, Phys. Rev. Lett. **91**, 182301 (2003).
[14] A. Adare *et al.*, Phys.Rev.Lett. **105**, 142301 (2010).
[15] S. Afanasiev *et al.*, Phys. Rev. **C80**, 054907 (2009).
[16] A. Adare *et al.*, Phys.Rev.Lett. **107**, 252301 (2011).
[17] A. Adare *et al.*, Phys.Rev.Lett. **101**, 162301 (2008).
[18] A. Adare *et al.*, nucl-ex/1204.1526 (2012).
[19] A. Adare *et al.*, Phys. Rev. **D79**, 012003 (2009).
[20] A. Angelis *et al.*, Phys.Lett. **B79**, 505 (1978).
[21] L. Apanasevich *et al.*, Phys.Rev. **D68**, 052001 (2003).
[22] T. Sjostrand, S. Mrenna, and P. Z. Skands, Comput.Phys.Commun. **178**, 852 (2008).
[23] R. Blankenbecler, S. J. Brodsky, and J. Gunion, Phys.Lett. **B42**, 461 (1972).
[24] S. Brodsky, H. Pirner, and J. Raufeisen, Phys.Lett. **B637**, 58 (2006).
[25] S. Berman, J. Bjorken, and J. B. Kogut, Phys.Rev. **D4**, 3388 (1971).
[26] F. Arleo, S. J. Brodsky, D. S. Hwang, and A. M. Sickles, Phys.Rev.Lett. **105**, 062002 (2010).

Journal of Physics: Conference Series **389** (2012) 012011 doi:10.1088/1742-6596/389/1/012011

Measurement of elliptic and higher order flow harmonics in 2.76 TeV Pb-Pb collisions with the ATLAS detector

Soumya Mohapatra on behalf of the ATLAS Collaboration
Departments of Physics & Astronomy and Chemistry, Stony Brook University, NY 11794, USA
E-mail: soumya@cern.ch

Abstract. Measurements of flow harmonics $v_1 - v_6$ are presented in a broad range of transverse momentum and centrality for $\sqrt{s_{NN}} = 2.76$ TeV Pb-Pb collisions with the ATLAS experiment. The fourier coefficients of two-particle correlations, $v_{n,n} = \langle\cos(n\Delta\phi)\rangle$ are shown to factorize as products of the single particle v_n for $n = 2-6$. However, the factorization breaks for $n = 1$ due to effects of global momentum conservation A scaling relation in the transverse momentum dependence ofthe v_n : $v_n(p_T)^{1/n} \propto v_2(p_T)^{1/2}$ is shown to approximately hold. For central and mid-central collisions and at low and intermediate p_T, the long range structures in two-particle correlations are found to be entirely accounted for by the single particle v_n.

1. Introduction

A central goal of the heavy ion program at the Large Hadron Collider (LHC) is to understand the properties of the hot and dense matter produced in relativistic heavy ion collisions, commonly termed as the quark gluon plasma. One of the ways to study its properties is to perform high accuracy measurements of the azimuthal yields of the produced particles. Due to anisotropic pressure gradients, the "fireball" produced in the collisions expands differently in different directions with more particles emitted along the larger gradients. This anisotropy in the particle yields can be expressed as a Fourier series:

$$\frac{dN}{d\phi} \propto 1 + 2\sum_{n=1}^{\infty} v_n(p_T, \eta)\cos(n(\phi - \Phi_n)) \tag{1}$$

where ϕ is the azimuthal angle of the produced particles, v_n is the magnitude of the nth order harmonic flow and Φ_n is the orientation of the corresponding harmonic plane [1].

The elliptic flow v_2 has been studied extensively as it usually dominates other coefficients, being strongly influenced by the elliptic geometry of the overlap region between the colliding nuclei. However, the initial geometry of the produced fireball may have angular moments of several orders: elliptic, triangular and higher orders due to fluctuations in the spatial positions of the nucleons in the colliding nuclei. These higher order moments in the initial geometry can give rise to measurable higher order flow ($n \geq 3$).

A better understanding of the higher order v_n can explain the origin of the ridge, an elongated structure along $\Delta\eta$ at $\Delta\phi \sim 0$ [2] and the so-called "mach-cone", a double-hump structure on

Journal of Physics: Conference Series **389** (2012) 012011 doi:10.1088/1742-6596/389/1/012011

the away-side [3] seen in two-particle correlations. These were initially interpreted as response of the medium to the energy deposited by the quenched jets. However, recent studies [4] have shown that higher order flow harmonics make a significant contribution to these structures. The v_n can also be used to obtain information about the initial geometry and transport properties of the medium, for instance the viscosity to entropy-density ratio [5, 6, 7].

The v_n can be measured by the event plane (EP) method by correlating the single particle azimuthal distributions with the harmonic planes Φ_n. They can also be measured by the two-particle correlation (2PC) method where the particle pair distribution in relative azimuthal angle $\Delta\phi = \phi_a - \phi_b$ are measured (the subscripts a and b label the two particles, commonly called trigger particle and partner particle). The 2PC correlation function can be expanded in a Fourier series in $\Delta\phi$ as:

$$C(\Delta\phi) \propto 1 + 2\sum_{n=1}^{\infty} v_{n,n}(p_{\mathrm{T}}^{\mathrm{a}}, p_{\mathrm{T}}^{\mathrm{b}}) \cos(n\Delta\phi) \qquad (2)$$

If the correlations are dominated by single particle anisotropies (Eq. 1), then the Fourier coefficients $v_{n,n}$ are equal to the product of the individual single particle v_n [8] :

$$v_{n,n}(p_{\mathrm{T}}^{\mathrm{a}}, p_{\mathrm{T}}^{\mathrm{b}}) = v_n(p_{\mathrm{T}}^{\mathrm{a}}) v_n(p_{\mathrm{T}}^{\mathrm{b}}) \qquad (3)$$

Using Eq. 3, one can obtain the v_n from the 2PC. The above relation is violated if non-flow effects, such as jets and resonance decays are large. Thus the non-flow effects must be suppressed before extracting the flow harmonics from the correlations.

The v_n results presented are for charged hadrons reconstructed in the ATLAS inner detector [9] covering $|\eta| < 2.5$. The Φ_n planes for the v_2-v_6 measurements via the event plane method were determined using the forward calorimeter covering $|\eta| \in (3.3, 4.9)$. All measurements were done using $8\mu b^{-1}$ of Minimum Bias Pb-Pb data at $\sqrt{s_{NN}}$ of 2.76 TeV. Details of the results presented here are published in [10].

2. Event plane analysis

Figure 1 shows the v_n from EP method as a function of centrality in 5% centrality bins plus a (0-1)% super-central bin, for two different p_{T} ranges. The v_2 has a much stronger centrality dependence as compared to the other harmonics. For non-central events, v_2 becomes much larger than the other v_n due to the elliptical geometry of the fireball. The other v_n, mainly driven by event-by-event fluctuations in the initial geometry have a weaker centrality dependence. In most central events (top 5%) and at sufficiently high p_{T}, v_3, v_4 and even v_5 can become larger than or comparable to v_2.

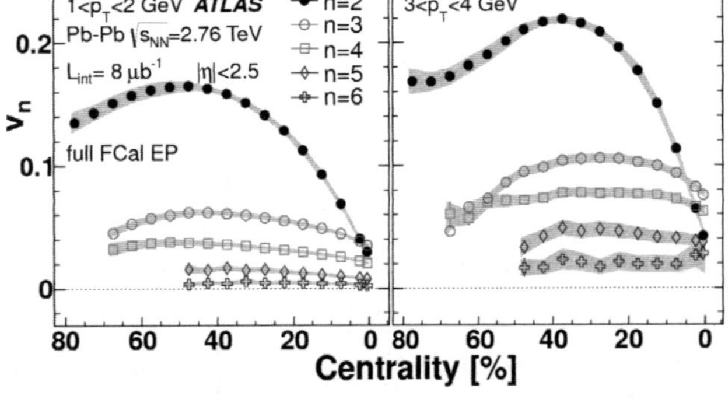

Figure 1. Flow harmonics v_n as a function of centrality in 5% centrality bins plus a (0-1)% most central bin (the right most point) for two p_{T} bins. The shaded bands indicate systematic uncertainties.

Journal of Physics: Conference Series **389** (2012) 012011 doi:10.1088/1742-6596/389/1/012011

Figure 2 shows v_n as a function of p_T for two different centralities. For all centralities the v_n have a similar trend, they increase to reach a maximum between 3 GeV $< p_T <$ 4 GeV and then decrease. This is interpreted as the driving mechanism behind the v_n changing from collective flow at low p_T to path-length dependent suppression at high p_T. In the (0-5)% central events, v_3 and v_4 become larger than v_2 at $p_T \sim 2$ GeV and 3 GeV respectively.

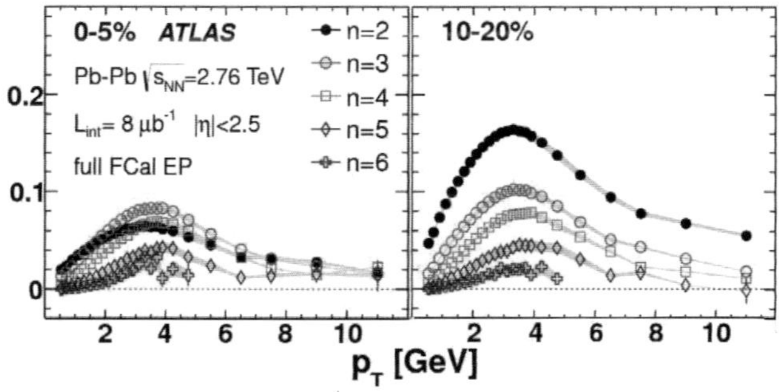

Figure 2. Flow harmonics v_n as a function of p_T for two centrality bins. The shaded bands indicate the systematic uncertainties.

A scaling relation is seen in the p_T dependence of v_n: $v_n^{1/n} = k_n v_2^{1/2}$ where k_n are only weakly dependent on p_T. This is shown in in the left panels of Fig. 3 for several different centralities. This scaling is also observed at RHIC for n=2-4 [11]. The centrality dependence of the k_n is shown in the right panel of Fig. 3 for $p_T \in (2, 3)$ GeV. They vary weakly with centrality in mid-central and peripheral events, but increase strongly in central events.

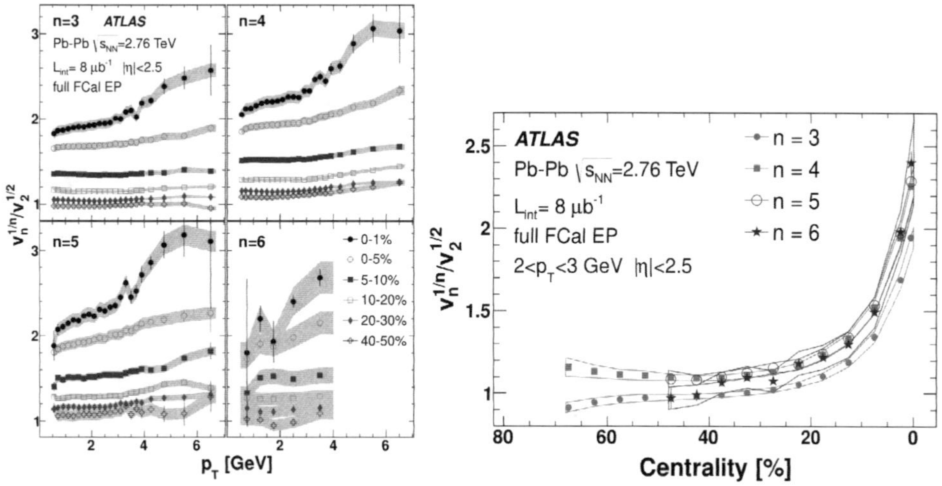

Figure 3. Left panels: $v_n^{1/n}/v_2^{1/2}$ as function of p_T in different bins of centrality. Right panel: $v_n^{1/n}/v_2^{1/2}$ vs centrality for $p_T \in (2, 3)$GeV.

In Fig. 4 the η dependence of the v_n is shown for $p_T \in (2, 3)$ GeV. In this p_T interval, v_2 decreases by at most 5% from $\eta = 0$ to $\eta = 2.5$, for $n > 2$ the decrease is slightly larger. In general, there is a weak dependence of the v_n on η in central and mid-central collisions. Eq. 3 stated that ignoring non-flow effects, the 2PC $v_{n,n}$ is the product of the single particle v_n. This relation is only true if there is no η dependence of the v_n (or approximately true if the η dependence is weak). The weak η dependence justifies the use of Eq. 3.

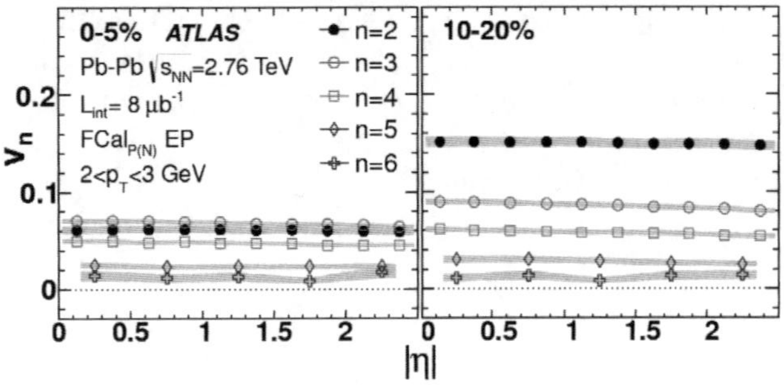

Figure 4. Pseudorapidity dependence of v_n for $p_{\mathrm{T}} \in (2, 3)$GeV. Each panel is a different centrality bin.

3. Two-particle correlations

Figure 5 shows the 2PC in $\Delta\eta$-$\Delta\phi$ space for $p_{\mathrm{T}}^{\mathrm{a}}, p_{\mathrm{T}}^{\mathrm{b}} \in (2, 3)$ GeV for three centrality bins. Such correlations, where both trigger and partner particle have same p_{T} range, are termed as fixed-p_{T} correlations. The near-side jet peak reported in [2] is clearly visible at $(\Delta\eta, \Delta\phi) \sim (0, 0)$ across all centralities and is more prominent in peripheral events. The near-side ridge has a weak $\Delta\eta$ dependence in central and mid-central collisions. The strength of the ridge increases from central to mid-central collisions and then decreases for peripheral collisions, completely disappearing in the (80-90)% centrality class. On the away-side, a double-hump structure is seen in the (0-5)% central collisions which becomes a ridge-like structure in the (20-30)% mid-central collisions. In the (80-90)% peripheral collisions, it is replaced by the away-side jet. The presence of the near-side jet biases the values of the flow harmonics obtained from the two-particle correlations. This bias can be minimized by requiring a $|\Delta\eta| > 2.0$ cut between the trigger and the associated particle in the correlations.

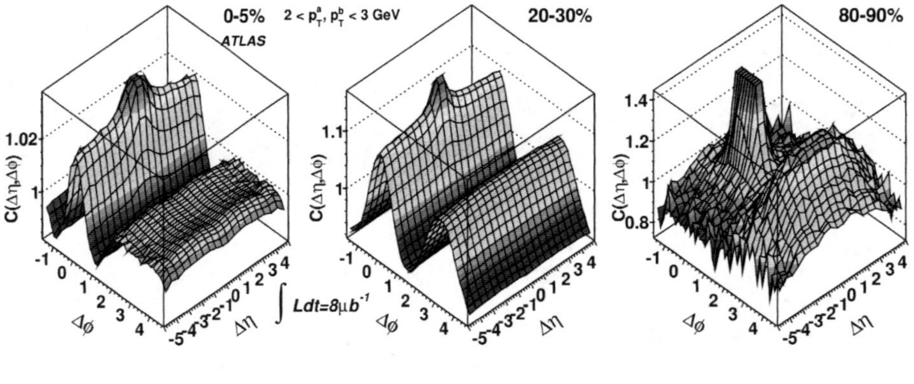

Figure 5. Two-particle $\Delta\eta - \Delta\phi$ correlations for 2 GeV< $p_{\mathrm{T}}^{\mathrm{a}}, p_{\mathrm{T}}^{\mathrm{b}}$ <3 GeV. Each panel is a different centrality bin. The (80-90)% peripheral events have the near-side jet peak truncated to better reveal the long-range structures in $\Delta\eta$.

The left panel in Fig. 6 shows shows the one-dimensional $\Delta\phi$ correlation corresponding to the first panel of Fig. 5 for $2 < |\Delta\eta| < 5$. As the trigger and partner particle p_{T} ranges are chosen to be the same, Eq. 3 becomes:

$$v_{n,n}(p_{\mathrm{T}}^{\mathrm{a}}, p_{\mathrm{T}}^{\mathrm{a}}) = v_n^2(p_{\mathrm{T}}^{\mathrm{a}}) \tag{4}$$

Using this relation, the values of v_n and its uncertainties are calculated from $v_{n,n}$ and are plotted as a function of n in the panel c) of Fig. 6. The v_n values are plotted up to $n = 15$. However, the analysis is limited to $n \leq 6$ because for higher n the systematic and statistical uncertainties are large compared to the extracted value of v_n.

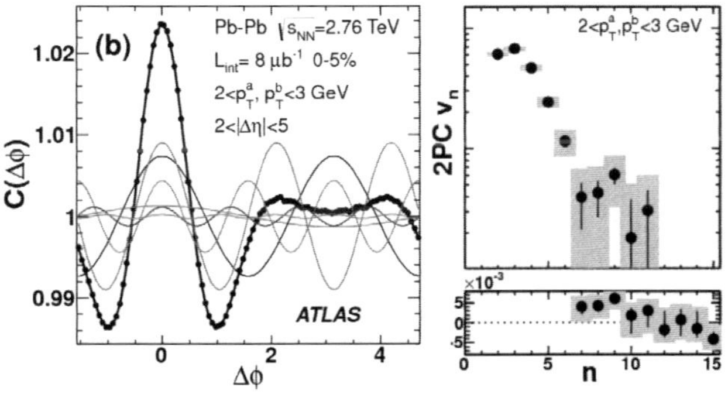

Figure 6. Left Panel: The $\Delta\phi$ correlation function (black points) for 2< $|\Delta\eta|$ <5, overlaid with contributions from individual Fourier components and their sum. Right Panel: v_n vs. n (The bars and bands indicate statistical and systematic uncertainties respectively). The lower sub panel shows v_n for $n \geq 7$ in a linear scale

The 2PC results can be used to check where the factorization (Eq. 3) breaks. If collective flow dominates the 2PC, then the near-side peak must be larger than the away-side peak. From Fig. 7 it is seen that this is roughly true up to 50% centrality (for p_T^a,$p_T^b \in (3, 4)$ GeV) beyond which the away-side peak becomes larger indicating a break-down of the $v_{n,n}$ factorization.

In Fig. 8, the p_T evolution of the correlations is shown for $(0-10)\%$ central collisions. At low p_T the correlation is driven by flow and the near-side peak is larger than the away-side. However, for p_T^a,$p_T^b > 6$ GeV, the 2PCs are dominated by the away-side jet peak. Thus it is clear that at high p_T (> 6 GeV) and in peripheral collisions (> 50% centrality), non-flow effects become important and Eq. 3 is not expected to hold.

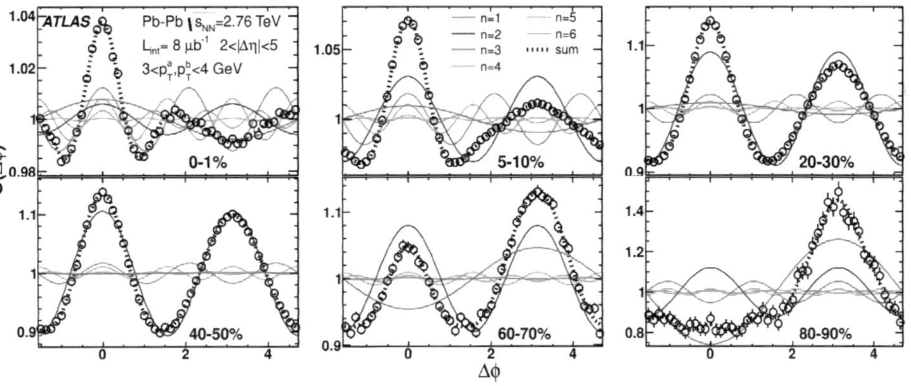

Figure 7. Centrality dependence of $\Delta\phi$ correlations for $p_T^a, p_T^b \in (3, 4)$GeV and $\Delta\eta \in (2, 5)$. The superimposed solid lines (thick-dashed lines) indicate contributions from individual $v_{n,n}$ components (sum of the first six components).

4. Comparison of v_n obtained from 2PC and EP methods

The left panels of Fig. 9, compare the v_2 from fixed-p_T 2PC method with those from the EP method for (0-10)% central events. The two methods agree within 5% to 15% for v_2 for $p_T < 4$ GeV. Deviations are observed for $p_T >$ 4-5 GeV, presumably due to contributions from the away-side jet.

The v_n values obtained from fixed-p_T correlations can be cross-checked using correlations where the trigger and partner have different p_T^a and p_T^b values. Such correlations are termed as mixed-p_T correlations. The $v_n(p_T^b)$ of the partner can be obtained as:

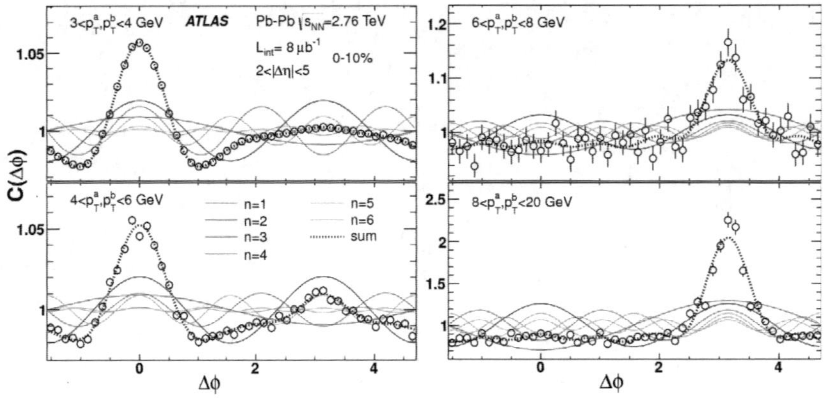

Figure 8. Fixed-p_{T} correlation functions in the (0-10)% centrality interval for several p_{T} ranges with $2 < |\Delta\eta| < 5$.

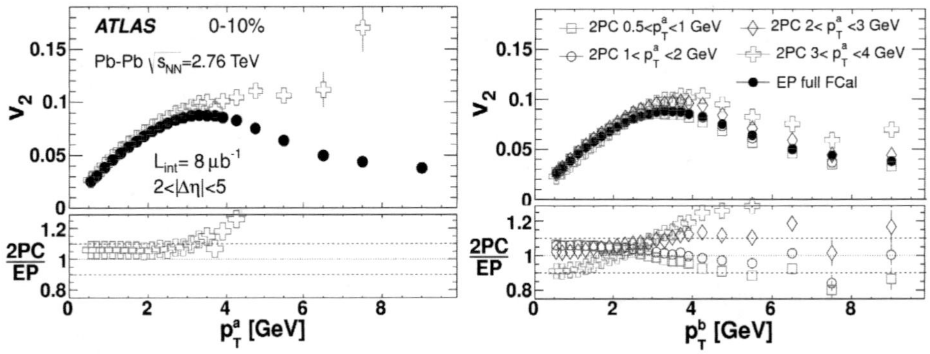

Figure 9. Left panel: comparision between the fixed-p_{T} $v_2(p_{\mathrm{T}})$ and EP $v_2(p_{\mathrm{T}})$. Right panel: Comparison of $v_2(p_{\mathrm{T}}^{\mathrm{b}})$ obtained for four different reference $p_{\mathrm{T}}^{\mathrm{a}}$ ranges (0.5-1, 1-2, 2-3, 3-4 GeV) with the EP values. The error bars indicate the statistical uncertainties only. The dashed lines in the ratio plots indicate a $\pm 10\%$ band to guide the eye. All results are for the (0-10)% centrality interval.

$$v_n(p_{\mathrm{T}}^{\mathrm{b}}) = v_{n,n}(p_{\mathrm{T}}^{\mathrm{a}}, p_{\mathrm{T}}^{\mathrm{b}})/v_n(p_{\mathrm{T}}^{\mathrm{a}}) \tag{5}$$

Equation 5 can be checked by measuring the same $v_n(p_{\mathrm{T}}^{\mathrm{b}})$ for different $v_n(p_{\mathrm{T}}^{\mathrm{a}})$. This is illustrated for v_2 in the right panel of Fig. 9. It is seen that factorization of $v_{n,n}$ works well for $n = 2$ and the values of $v_2(p_{\mathrm{T}}^{\mathrm{b}})$ are reasonably independent of $p_{\mathrm{T}}^{\mathrm{a}}$. Further it is seen that when $p_{\mathrm{T}}^{\mathrm{a}}$ is below 3 GeV, the agreement of the 2PC values with the EP method extend out to much higher $p_{\mathrm{T}}^{\mathrm{b}}$. This shows that as long as the reference $p_{\mathrm{T}}^{\mathrm{a}}$ is low, the factorization relation is valid and hence the v_n can be measured to high p_{T} via the 2PC method. This factorization between hard and soft particles is expected since the high p_{T} particles, due to path-length dependent energy loss, are correlated with the same geometry that drives the collective expansion at low p_{T} [13].

In Fig. 10 the validity of the factorization is checked for all harmonics as a function of $|\Delta\eta|$ for four different $p_{\mathrm{T}}^{\mathrm{a}}, p_{\mathrm{T}}^{\mathrm{b}}$ combinations. For harmonics v_2 to v_6 and $|\Delta\eta| > 1.0$, all trigger-partner combinations give nearly identical values of $v_n(p_{\mathrm{T}}^{\mathrm{b}})$ even though the trigger v_n values vary over a large range. This validates the factorization for $n = 2 - 6$.

The first panel of Fig. 10 shows that for $v_{1,1}$ the factorization does not hold for any value of

$\Delta\eta$. This is because the $v_{1,1}$ is strongly influenced by global momentum conservation (GMC), which modifies the factorization for $v_{1,1}$. A detailed analysis of $v_{1,1}$ to extract the v_1 has been done in ATLAS and is presented in a separate talk in this conference by J. Jia [12], details of which can be found in [10].

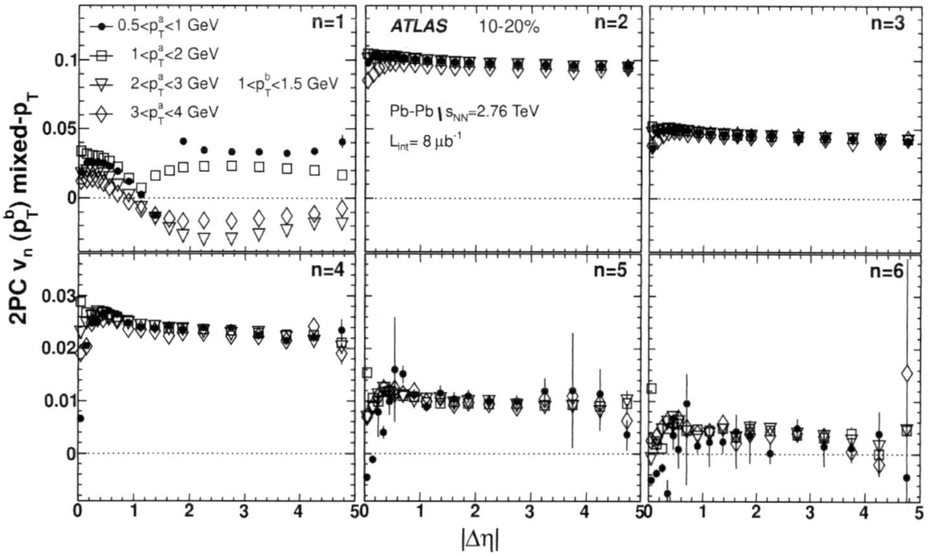

Figure 10. Flow harmonic $v_n(p_T^b)$ as a function of $|\Delta\eta|$ for $p_T^b \in (1, 1.5)$ GeV, calculated for four different reference p_T^a ranges (0.5-1, 1-2, 2-3, and 3-4 GeV). The error bars indicate the statistical uncertainties only.

The left panel of Fig. 11 shows the centrality dependence of the v_n obtained from the two methods. For clarity only the statistical errors are shown. The ratio plots in the bottom panels show that the values for v_2, v_3 and v_4 agree within 5% for central and mid-central collisions. For v_5 and v_6 they agree within 10% and 15% respectively and are consistent within the systematic uncertainties quoted in [10].

The agreement between the 2PC and EP methods implies that the structures of the two-particle correlation at low p_T and large $\Delta\eta$ mainly reflect collective flow. This is verified explicitly by reconstructing the correlation function as:

$$C^{\rm reco}(\Delta\phi) = N_0^{\rm 2PC}\left(1 + 2v_{1,1}^{\rm 2PC}cos(\Delta\phi) + 2\sum_{n=2}^{6} v_n^{\rm EP}v_n^{\rm EP}cos(n\Delta\phi)\right) \quad (6)$$

where $N_0^{\rm 2PC}$ and $v_{1,1}^{\rm 2PC}$ are the average of the correlation function and first harmonic coefficient from the 2PC analysis, and the remaining coefficients are calculated from the v_n measured from the event plane method. The right panel of Fig. 11 compares a measured 2PC to the corresponding reconstructed correlation (Eq. 6) showing excellent agreement between the two. This demonstrates that the ridge and cone are a manifestation of single particle v_1-v_6.

5. Summary

Measurements of the flow harmonics v_1-v_6 over a large p_T, η and centrality range are presented. The 2PC $v_{n,n}$ are shown to factorize into products of single particle v_n for $n = 2-6$ in central and mid-central collisions as long as one particle has low p_T ($<$ 3GeV). The factorization breaks

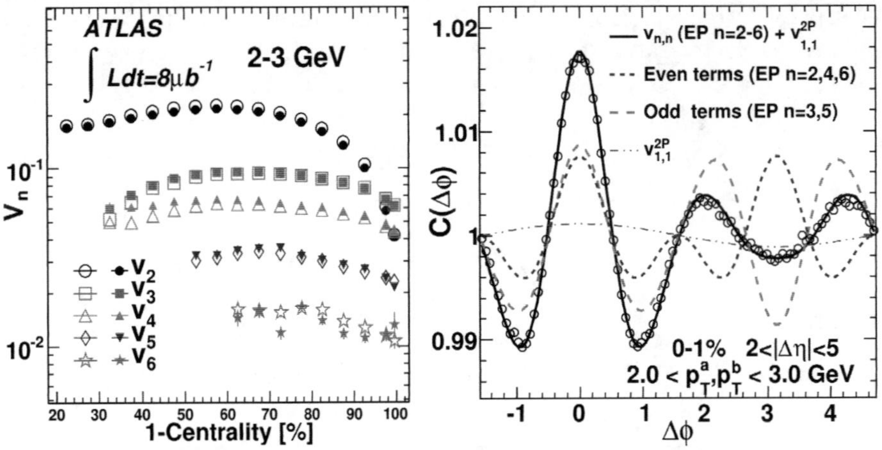

Figure 11. Left panel: Comparison of v_n obtained from the 2PC method (solid symbols) to the EP results (open symbols) as a function of centrality. Right panel: Measured 2PC (circles) compared with the one reconstructed from EP method for (0-1)% centrality.

for $v_{1,1}$ due to contribution from momentum conservation effects. It is shown that the features in two-particle correlations for $|\Delta\eta| > 2$ at low and intermediate p_T (< 3 GeV) such as the ridge and double-hump are accounted for by the v_1-v_6.

This work is in part supported by NSF under award number PHY-1019387.

References

[1] A. M. Poskanzer and S. A. Voloshin, Phys. Rev. C 58, 1671 (1998).
[2] PHENIX Collaboration, A. Adare *et al.*, Phys. Rev. C 78, 014901 (2008).
[3] STAR Collaboration, B. I. Abelev *et al.* Phys. Rev. C 80, 064912 (2009).
[4] B. Alver and G. Roland, Phys. Rev. C 81, 054905 (2010) [Erratum Phys. Rev. C 82, 039903 (2010)]
[5] Z. Qiu and U. W. Heinz, Phys. Rev. C 84, 024911 (2011).
[6] B. Schenke, S. Jeon, C. Gale, Phys. Lett. B702, 59 (2011).
[7] G. Y. Qin, H. Petersen, S. A. Bass and B. Muller, Phys. Rev. C 82, 064903 (2010).
[8] R. A. Lacey, Nucl. Phys. A698, 559 (2002); R. J. M. Snellings, Nucl. Phys. A698, 193 (2002).
[9] ATLAS Collaboration, JINST 3, S08003 (2008).
[10] ATLAS Collaboration, Phys. Rev. C 86, 014907 (2012).
[11] R. Lacey *et al*, Scaling of the higher-order flow harmonics: implications for initial-eccentricity models and the "viscous horizon", arXiv:1105.3782 [nucl-ex].
[12] J.Jia for the ATLAS collaboration, Measurement of the rapidity-even dipolar flow in Pb-Pb collisions with the ATLAS detector, arXiv:1208.1874v1 [nucl-ex].
[13] J. Jia, Azimuthal anisotropy in a jet absorption model with fluctuating initial geometry in heavy ion collisions, arXiv:1203.3265v2 [nucl-th].

Journal of Physics: Conference Series **389** (2012) 012012 doi:10.1088/1742-6596/389/1/012012

Azimuthal Anisotropy in Charged Hadron Production in 2.76 TeV PbPb Collisions Measured by CMS

Eric Appelt on behalf of the CMS Collaboration

Vanderbilt University, Nashville, United States

E-mail: eric.appelt@gmail.com

Abstract. We report on the CMS measurements of charged hadron anisotropic azimuthal distributions from PbPb collisions at a center-of-mass energy of 2.76 TeV per nucleon pair and their decomposition into a Fourier series up to the 6th coefficient. The results are presented as a function of transverse momentum, centrality and pseudorapidity and cover a broad kinematic range. These results can provide constraints on the theoretical description of the early dynamics in the hot and dense medium and its transport properties.

1. Introduction

The azimuthal anisotropy of charged hadrons produced in PbPb collisions are typically characterized by the coefficients (v_n) of Fourier harmonics in an azimuthal distribution (ϕ) of the charged particle yield, $dN/d\phi \propto 1 + \sum_{n=1}^{\infty} 2v_n \cos(n(\phi - \Psi_R))$, where Ψ_R is a reference angle determined event-by-event that corresponds to a "participant plane". The coefficient v_2, commonly referred to as the elliptic flow coefficient, is dominant in non-central events due to the typically lenticular shape of the overlap region of the colliding nuclei. Particularly for the measurement of v_2, the participant plane is defined by the minor axis of the overlap region and the beam direction. The results presented here include detailed measurements of the elliptic flow of charged particles produced in PbPb collisions at $\sqrt{s_{NN}} = 2.76$ TeV in a broad kinematic range of $|\eta| < 2.4$, and $0.3 < p_T < 60$, in the centrality range of 0-80% of the total inelastic PbPb cross-section. Additionally, preliminary results for higher flow harmonic coefficients ranging from v_3 to v_6 are presented.

In the low-p_T region, these coefficients are generally associated with the hydrodynamic flow of the medium produced in the collision [1]. Event-by-event fluctuations in the value of v_2 may be driven by fluctuations in the shape of the initial overlap area [2], and different experimental methods for measuring v_2 can respond differently to these fluctuations [3]. In addition, measurements of v_2 at low-p_T may be affected by "non-flow" particle correlations from jets and resonance decays. Multiple methods are therefore employed to measure the flow harmonics in order to better understand the initial state of the collision and the hydrodynamic properties of the medium.

At high-p_T, a non-zero value of v_2 can be induced if there is a stronger suppression of the hadron yield along the long axis than the short axis of the overlap region. The measurement of v_2 at high-p_T can quantitatively constrain models of partonic energy loss in the medium that

Published under licence by IOP Publishing Ltd

take account of the influence of the path length and shape of the overlap region of the colliding nuclei.

2. Experimental method

These measurements were taken with the Compact Muon Solenoid (CMS) detector using $\sqrt{s_{NN}} = 2.76$ TeV data obtained during the 2010 and 2011 heavy-ion runs at the Large Hadron Collider (LHC). The CMS detector primarily consists of a silicon tracker, lead-tungstenate electromagnetic calorimeter, brass/scintillator hadronic calorimeter, and gas-ionization muon chambers. The tracker and calorimeters are housed within a 3 m radius superconducting solenoid that provides a 3.8T magnetic field. In addition, the detector includes extensive forward calorimetry including two steel/quartz Čerenkov Hadron Forward (HF) calorimeters, and Beam Scintillation Counters (BSC) placed along the beamline. The silicon tracker is comprised of a inner silicon pixel detector and outer silicon strip detector. A more detailed description of the CMS detector can be found elsewhere [4].

2.1. Event selection and tracking

Minimum bias PbPb events were triggered via coincident signals from both ends of the detector in either the BSC or HF calorimeters. Additional offline event selection was performed in order to remove events due to noise, beam backgrounds, electromagnetic interactions leading to the breakup of nuclei, and other non-collision events [5]. The minimum bias trigger has an acceptance of $(97 \pm 3)\%$ for inelastic hadronic PbPb collisions. For the 2011 heavy-ion run, only a small fraction of all minimum-bias triggered events could be recorded due to hardware limits on the data acquisition rate. In order to maximize the sample of events containing high-p_T particles emitted in PbPb collisions, a dedicated high-p_T single track trigger was implemented, utilizing a tracking algorithm identical to the offline track reconstruction which identified high-p_T charged particles starting with a candidate trajectory from the three-layer silicon pixel detector with $p_T > 11$ GeV/c. [6] The efficiency of this high-p_T track trigger rises from about 75% at $p_T \approx 12$ GeV/c to nearly 100% above $p_T \approx 20$ GeV/c.

The algorithm used to reconstruct charged particle tracks for the 2010 analyses presented here employed a two-iteration tracking algorithm, with one iteration using only the pixel detector to reconstruct charged particles with a p_T as low as 0.3 GeV/c, and is described in detail in Ref. [5]. The charged particle reconstruction algorithm used for the 2011 data used multiple iterations to provide good efficiency at very high-p_T, and is described in detail in Ref. [7].

2.2. Centrality and Glauber modeling

Centrality determination was performed using the total energy in both of the HF detectors. Using this total energy distribution, the event samples were divided into 40 centrality bins, each representing 2.5% of the total inelastic nucleus-nucleus cross section. These small bins are then regrouped into centrality classes ranging from 0% to 80% for the various analyses.

The Monte Carlo Glauber model is a multiple collision model that treats each nucleus-nucleus collision as an independent sequence of nucleon-nucleon collisions. From this model, quantities of interest characterizing the events of each centrality class that are not directly observable are calculated, such as the number of participating nucleons, or N_{part}. The eccentricity of the overlap region of the colliding nuclei as determined by this model, called the participant eccentricity, is denoted here as ϵ_{part} or just ϵ. Additionally, the cumulant moments, defined as $\epsilon\{2\}^2 \equiv \langle\epsilon^2_{\text{part}}\rangle$ and $\epsilon\{4\}^4 \equiv 2\langle\epsilon^2_{\text{part}}\rangle^2 - \langle\epsilon^4_{\text{part}}\rangle$, are calculated from this model.

2.3. Methods for measuring the anisotropy parameters

The value of v_n measured by the event-plane method, denoted here as $v_n\{\text{EP}\}$, is defined as $v_n\{\text{EP}\} = \langle\cos n(\phi - \psi_{\text{EP}})\rangle/R$. Here ψ_{EP} denotes a reconstructed event-plane, in this analysis determined from the azimuthal energy distribution in the HF calorimeters. R represents a correction factor for the resolution of the reconstructed event-plane, and the average is taken over all particles in all events [8].

The cumulant methods measure flow by a cumulant expansion of the multiparticle azimuthal correlations, without determining the orientation of the event-plane. The value of v_n measured by the two-particle cumulant method, denoted here as $v_n\{2\}$, is defined as $v_n\{2\} = \sqrt{\langle \cos n(\phi_1 - \phi_2)\rangle}$ where ϕ_1 and ϕ_2 represent the azimuthal angles of two charged particles and the average is taken over all pairs of particles in all events. The two-particle cumulant method is particularly sensitive to non-flow. The value of v_n measured by the four-particle cumulant method, denoted here as $v_n\{4\}$, is similarly defined as $v_n\{4\} = \left(2\langle \cos n(\phi_1 - \phi_2)\rangle^2 - \langle \cos n(\phi_1 + \phi_2 - \phi_3 - \phi_4)\rangle\right)^{1/4}$. The four-particle cumulant method removes lower order non-flow effects from correlations between pairs of particles. For both the two- and four-particle cumulant methods, $v_n\{2\}$ and $v_n\{4\}$ are estimated using a generating function of the multiparticle correlations in the complex plane [9].

The Lee–Yang zeros method, from which the measured value of v_2 is denoted here as $v_2\{\mathrm{LYZ}\}$, is based on multiparticle correlations involving all particles in the event. Using the asymptotic behavior of the cumulant expansion, the location of the zeros of a complex function are related to the magnitude of the integrated flow in the system [10]. The Lee–Yang zeros method approximates a cumulant of very high order and removes the influence of non-flow from the measured v_2 signal.

A much more detailed description of the methods used in these analyses is given in Ref. [5] and the references contained therein.

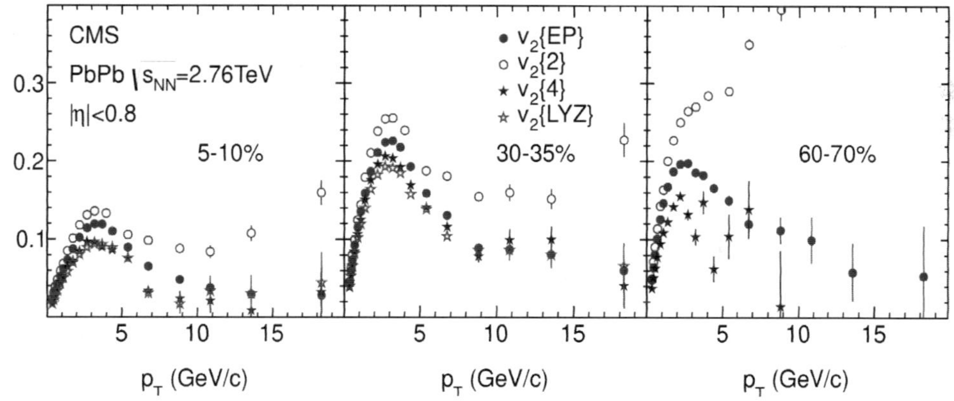

Figure 1. Comparison of the four different methods for measuring v_2 as a function of p_T at mid-rapidity ($|\eta| < 0.8$) for three representative centrality classes given in the figures [5]. The error bars show the statistical uncertainties only.

3. Elliptic flow at midrapidity

In Fig. 1, the v_2 measurement as a function of p_T is given at midrapidity ($|\eta| < 0.8$) for the event-plane, two-particle cumulant, four-particle cumulant, and Lee–Yang zeros methods, denoted as $v_2\{\mathrm{EP}\}$, $v_2\{2\}$, $v_2\{4\}$, and $v_2\{\mathrm{LYZ}\}$, respectively. Excepting the $v_2\{2\}$ results in peripheral events which are dominated by non-flow effects at high-p_T, the transverse momentum dependence shows a rise of v_2 up to $p_T \approx 3$ GeV/c and then a decrease.

In the left panel of Fig. 2, the v_2 obtained from the event-plane and cumulant methods at midrapidity is integrated over the range $0.3 < p_T < 3.0$, scaled by the participant eccentricity as determined from the Glauber-model simulation, and plotted as a function of centrality. These data show a near-linear decrease in the eccentricity scaled v_2 from central to peripheral collisions. The values of v_2 from the cumulant methods were scaled with their respective cumulant moments of the participant eccentricity, which following the definition of $v_2\{2\}$ and $v_2\{4\}$ may be written [3]:

$$\frac{v_2\{2\}}{\epsilon\{2\}} = \left(\frac{\langle v_2^2\rangle}{\langle \epsilon^2\rangle}\right)^{1/2}, \quad \text{and} \quad \frac{v_2\{4\}}{\epsilon\{4\}} = \left(\frac{2\langle v_2^2\rangle^2 - \langle v_2^4\rangle}{2\langle \epsilon^2\rangle^2 - \langle \epsilon^4\rangle}\right)^{1/4}, \tag{1}$$

where $\langle v_2\rangle$ represents the event-by-event average of the v_2 value from each event. If the hydrodynamic response to the participant eccentricity is linear, i.e. $v_2 \sim \epsilon$, and nonflow effects are negligible, then the identity $v_2\{2\}/\epsilon\{2\} = v_2\{4\}/\epsilon\{4\}$ should hold [2]. In the right panel of Fig. 2, $v_2\{2\}/\epsilon\{2\}$ and $v_2\{4\}/\epsilon\{4\}$ are plotted and agree in the centrality range of 15-40%, suggesting that the difference between these methods can be attributed to their respective sensitivities to event-by-event flow fluctuations.

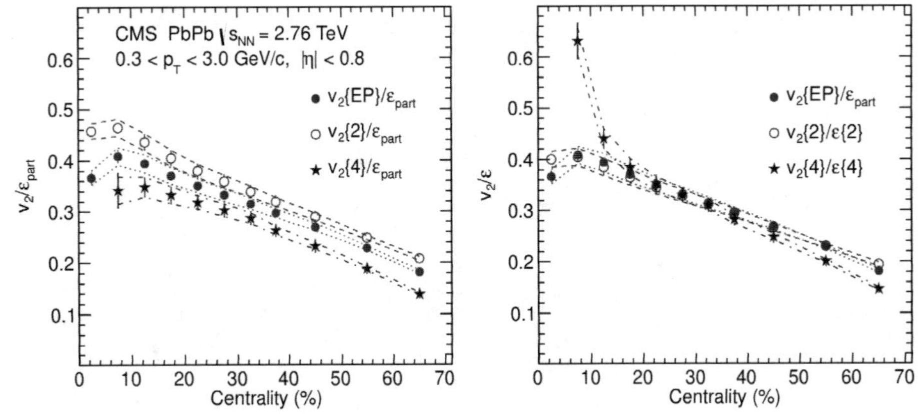

Figure 2. Left panel: Centrality dependence of the p_T-integrated v_2 divided by the participant eccentricity, $\varepsilon_{\text{part}}$, measured at mid-rapidity ($|\eta| < 0.8$) for the event-plane, two-particle cumulant, and four-particle cumulant methods [5]. Right panel: The same measurements as in the left panel, but here the two-particle and four-particle cumulant results are divided by their corresponding moments of the participant eccentricity, $\varepsilon\{2\}$ and $\varepsilon\{4\}$. In both panels the error bars show the sum in quadrature of the statistical and systematic uncertainties in the measurement of v_2, and the lines show the systematic uncertainties in the eccentricity determination.

4. Elliptic flow at forward rapidity

The measurement of $v_2(\eta)$ integrated over the range $0.3 < p_T < 3.0$ for the four methods is shown in three representative centrality classes in Fig. 3. The values for $v_2\{4\}$ and $v_2\{\text{LYZ}\}$ are in agreement and smaller than $v_2\{\text{EP}\}$ and $v_2\{2\}$. The value of $v_2(\eta)$ is greatest at midrapidity and slowly decreases at higher values of $|\eta|$. This decrease is more pronounced in the more peripheral collisions.

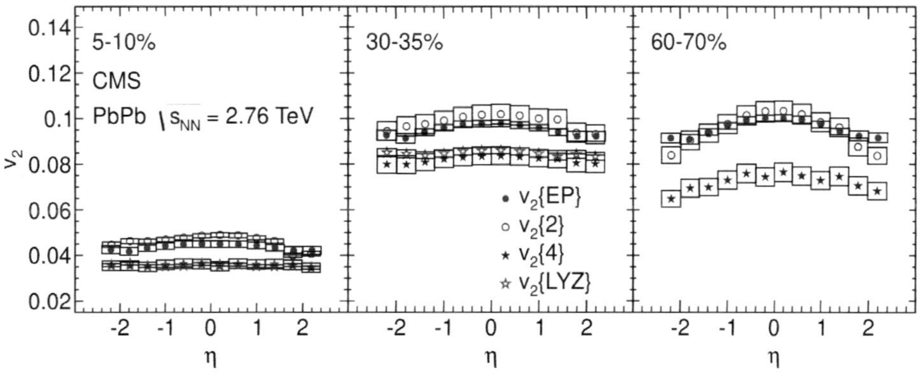

Figure 3. Pseudorapidity dependence of v_2 integrated over the range $0.3 < p_T < 3$ GeV/c for all four methods in three representative centrality classes [5]. The error bars show the statistical uncertainties, and the boxes show the systematic ones.

To ascertain whether the decrease in $v_2(\eta)$ at high $|\eta|$ is due to a change in the underlying spectra, or if it is due to a pseudorapidity dependence of $v_2(p_T)$, a measurement of $v_2(p_T)$ with the event-plane method in three pseudorapidity intervals is displayed in Fig. 4. For the events from 0-40% centrality, the $v_2(p_T)$ distribution is not seen to change with pseudorapidity within statistical uncertainties, indicating that any change in $v_2(\eta)$ is due to the underlying charged particle spectra. For more peripheral events, a slight decrease in $v_2(p_T)$ is observed at high $|\eta|$, indicating that a pseudorapidity dependence of the $v_2(p_T)$ distribution contributes to the decrease in $v_2(\eta)$ with increasing $|\eta|$ in peripheral events.

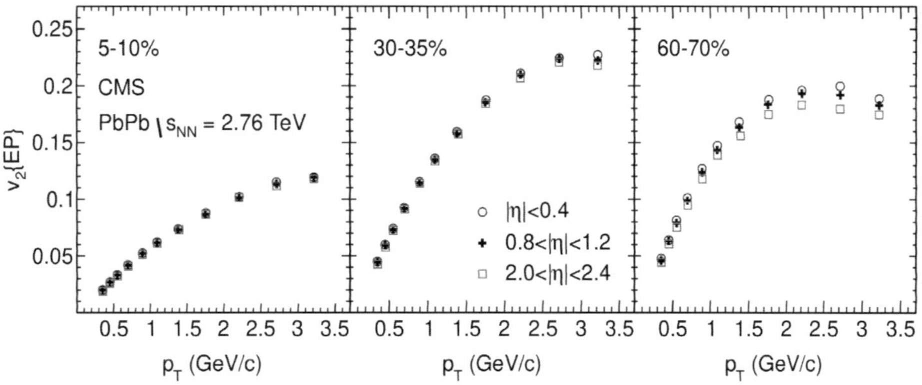

Figure 4. $v_2(p_T)$ as measured by the event-plane method in three pseudorapidity regions and in three representative centrality classes [5]. The error bars show the statistical uncertainties.

The PHOBOS experiment at the Relativistic Heavy Ion Collider (RHIC) has observed that the v_2 over a large range of collision energies ranging from $\sqrt{s_{NN}} = 19.6$ GeV to 200 GeV exhibited extended longitudinal scaling when viewed from the rest frame of one of the colliding nuclei [11]. As the measured particles are unidentified in both the CMS and PHOBOS measurements, the pseudorapidity η^+ (η^-) of the particles in the rest frame of the nucleus moving in the positive (negative) direction is given by $\eta^{\pm} = \eta \pm y_{\text{beam}}$ where η is the pseudorapidity of the particles in the laboratory or center-of-mass frame, and $y_{\text{beam}} \approx \ln \sqrt{s_{NN}}$[GeV]. In Fig. 5, the left (right) half of each plot shows v_2 in the rest frame of the beam moving in the positive (negative) direction. The CMS results plotted here are obtained from the event-plane method integrated over the range $0 < p_T < 3.0$, where values of $v_2(p_T)$ below $p_T = 0.3$ GeV/c are estimated by extrapolating the charged particle spectrum and $v_2(p_T)$ distribution to $p_T = 0$ by assuming that $v_2(p_T = 0) = 0$. Although the CMS data cover 4.8 units of pseudorapidity, they do not overlap the PHOBOS data when plotted in the rest frames of the colliding nuclei due to the extreme difference in beam energies. The CMS data show a weaker pseudorapidity dependence than the PHOBOS data, and suggest a nearly boost-invariant region which is wider in η for more central events.

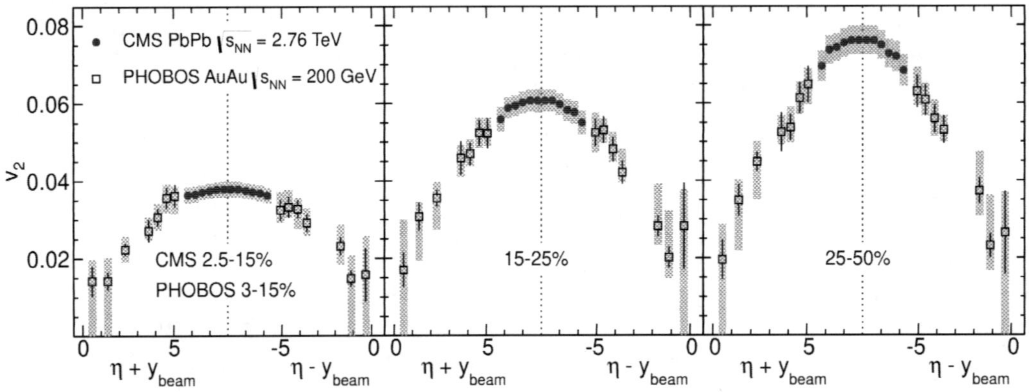

Figure 5. Measured $v_2(\eta)$ from CMS [5] (closed symbols) and PHOBOS [12] (open symbols) in three centrality intervals. The left (right) half of each plot shows v_2 in the rest frame of the beam moving in the positive (negative) direction. The error bars show the statistical uncertainties, and the boxes show the systematic ones.

5. Higher order flow

Preliminary measurements of higher order flow harmonics from v_3 to v_6 have been taken from the 2010 heavy-ion data using the two-particle cumulant, three-particle cumulant, five-particle cumulant, and Lee–Yang zeros methods [13]. In Fig. 6, measurements the $v_2\{4\}$, $v_3\{2\}$, $v_4\{3\}$, $v_4\{5\}$, $v_4\{\text{LYZ}\}$, and $v_6\{\text{LYZ}\}$ methods have been integrated over the range $0.3 < p_T < 3.0$ and are given a function of the number of participant nucleons corresponding to each centrality bin as estimated by MC Glauber simulations. [13] The value of v_3 increases slowly from central to peripheral collisions, reaches a maximum at around $N_{\text{part}} = 100$, then decreases slightly. This is in contrast to v_2, which increases rapidly with centrality and saturates at approximately the same value of N_{part} as the measured v_3.

6. Azimuthal anisotropy at extremely high transverse momentum

The results for $v_2(p_T)$ for p_T ranging from 1 GeV/c to 60 GeV/c taken from the 2011 heavy-ion data using the event plane method in six centrality ranges from 0-10% to 50-60% are shown in Fig. 7. These new CMS results significantly extend the p_T reach of previous v_2 measurements achieved by the ALICE [14], ATLAS [15], and CMS [5] Collaborations. Beyond $p_T = 10$ GeV/c

Journal of Physics: Conference Series **389** (2012) 012012 doi:10.1088/1742-6596/389/1/012012

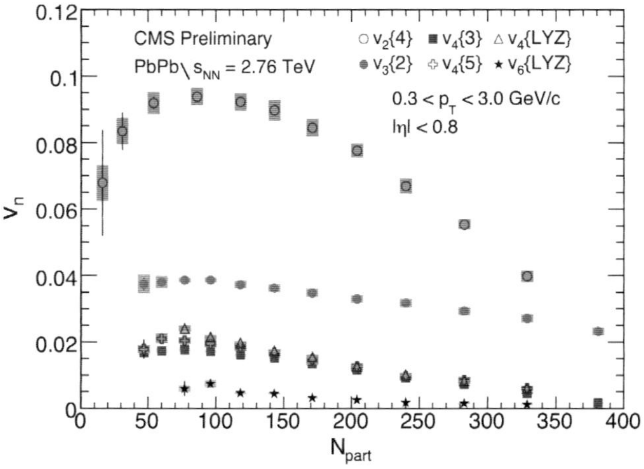

Figure 6. Integrated $v_2\{4\}$, $v_3\{2\}$, $v_4\{3\}$, $v_4\{5\}$, $v_4\{\mathrm{LYZ}\}$, and $v_6\{\mathrm{LYZ}\}$ as a function of the number of participant nucleons corresponding to each centrality bin as estimated by MC Glauber simulations. [13]. The error bars show the statistical uncertainties and the colored bands show the systematic ones

the value of v_2 shows a much weaker dependence on p_T than at lower transverse momenta, and the values of v_2 continue to gradually decrease but remain greater than zero up to $p_T = 40$ GeV/c.

This result indicates a clear dependence of v_2 on p_T in this extremely high-p_T region where the charged particle production is dominated by parton fragmentation and energy-loss models are most applicable. This suggests that the path-length dependence of partonic energy loss in the medium is itself dependent on the initial energy of the parton.

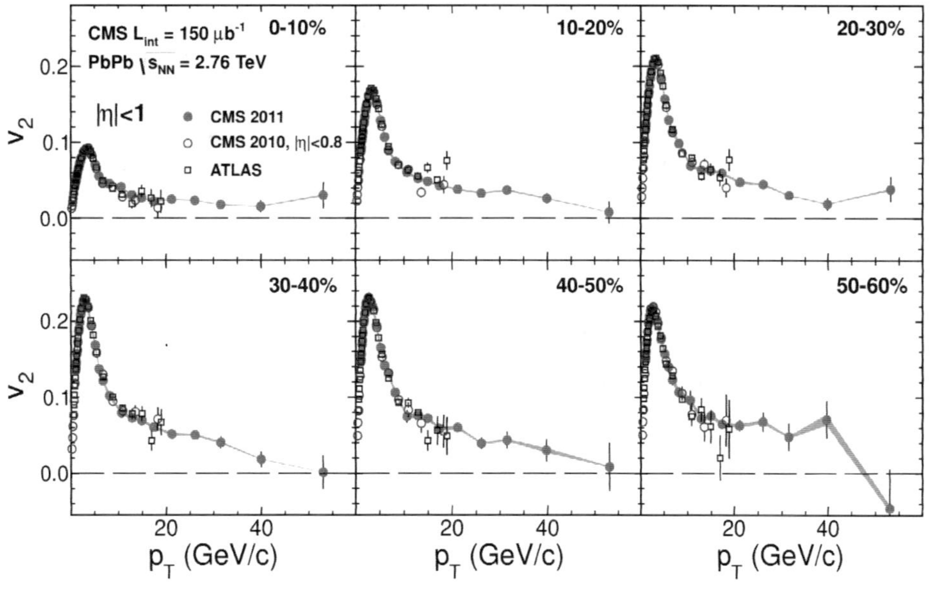

Figure 7. $v_2(p_T)$ from 1.0 to 60.0 GeV/c with $|\eta| < 1$ for six centrality ranges in PbPb collisions at $\sqrt{s_{NN}} = 2.76$ TeV, Solid markers indicate the results from CMS using data collected in 2011 [6]. For comparison, results from the ATLAS [15] experiment are shown in open squares, and earlier CMS results from data collected in 2010 are shown in open circles [5]. Error bars show the statistical uncertainties, while the grey bands correspond to the systematic uncertainties.

7. Summary

Detailed measurements of the azimuthal anisotropies for charged particles in $\sqrt{s_{NN}} = 2.76$ TeV PbPb collisions have been presented. Four methods have been used to determine the value of v_2 as a function of transverse momentum, pseudorapidity, and centrality in the broad kinematic range of $0.3 < p_T < 20$ and $|\eta| < 2.4$. The value of $v_2(p_T)$ has been found to increase up to $p_T \approx 3$ GeV/c and then slowly decrease. The measurement of $v_2(p_T)$ with the event-plane method was greatly expanded to $p_T = 60$ GeV/c using the enhanced statistics of the 2011 heavy-ion data produced by the single-track trigger. A non-zero value of v_2 was observed at a value of at least $p_T = 40$ GeV/c. The value of $v_2(\eta)$ was found to have a weak dependence on pseudorapidity in central collisions, and as $|\eta|$ increases the value of $v_2(\eta)$ gradually drops in peripheral collisions. Comparisons of $v_2(\eta)$ to PHOBOS data suggested that extended longitudinal scaling may hold to LHC energies. Preliminary measurements of higher order flow harmonics ranging from v_3 to v_6 have also been presented. The precision data over a wide kinematic range presented here will provide important constraints on the initial state of the collision, the hydrodynamic properties of the medium formed in the collision, and models of partonic energy loss in the medium.

Acknowledgments

We congratulate our colleagues in the CERN accelerator departments for the excellent performance of the LHC machine. We thank the technical and administrative staff at CERN and other CMS institutes, and acknowledge support from: BMWF and FWF (Austria); FNRS and FWO (Belgium); CNPq, CAPES, FAPERJ, and FAPESP (Brazil); MES (Bulgaria); CERN; CAS, MoST, and NSFC (China); COLCIENCIAS (Colombia); MSES (Croatia); RPF (Cyprus); MoER, SF0690030s09 and ERDF (Estonia); Academy of Finland, MEC, and HIP (Finland); CEA and CNRS/IN2P3 (France); BMBF, DFG, and HGF (Germany); GSRT (Greece); OTKA and NKTH (Hungary); DAE and DST (India); IPM (Iran); SFI (Ireland); INFN (Italy); NRF and WCU (Korea); LAS (Lithuania); CINVESTAV, CONACYT, SEP, and UASLP-FAI (Mexico); MSI (New Zealand); PAEC (Pakistan); MSHE and NSC (Poland); FCT (Portugal); JINR (Armenia, Belarus, Georgia, Ukraine, Uzbekistan); MON, RosAtom, RAS and RFBR (Russia); MSTD (Serbia); SEIDI and CPAN (Spain); Swiss Funding Agencies (Switzerland); NSC (Taipei); TUBITAK and TAEK (Turkey); STFC (United Kingdom); DOE and NSF (USA).

This research is supported in part by the Department of Energy Office of Science Graduate Fellowship Program (DOE SCGF), made possible in part by the American Recovery and Reinvestment Act of 2009, administered by ORISE-ORAU under contract no. DE-AC05-06OR23100.

References

[1] Voloshin S, Poskanzer A and Snellings R 2008 (*Preprint* `nucl-ex/0809.2949`)
[2] Qiu Z and Heinz U 2011 *Phys. Rev. C* **84** 024911
[3] Ollitrault J Y, Poskanzer A and Voloshin S 2009 *Phys. Rev. C* **80** 014904
[4] CMS Collaboration 2008 *JINST* **3** S08004
[5] CMS Collaboration 2012 Submitted for publication in *Phys. Rev. C* (*Preprint* `nucl-ex/1204.1409`)
[6] CMS Collaboration 2012 *Phys. Rev. Lett.* **109** 022301
[7] CMS Collaboration 2012 *Eur. Phys. J. C.* **72** 1945
[8] Poskanzer A and Voloshin S 1998 *Phys. Rev. C* **58** 1671
[9] Borghini N, Dinh P and Ollitrault J Y 2001 *Phys. Rev. C* **64** 054901
[10] Bhalerao R, Borghini N and Ollitrault J Y 2003 *Nucl. Phys. A* **727** 373
[11] PHOBOS Collaboration 2005 *Phys. Rev. Lett.* **94** 122303
[12] PHOBOS Collaboration 2005 *Phys. Rev. C* **72** 051901
[13] CMS Collaboration 2011 *CMS PAS HIN-11-005* http://cdsweb.cern.ch/record/1361385
[14] ALICE Collaboration 2010 *Phys. Rev. Lett.* **105** 252302
[15] ATLAS Collaboration 2012 *Phys. Lett. B* **707** 330

Measurement of the rapidity-even dipolar flow in Pb-Pb collisions with the ATLAS detector

Jiangyong Jia on behalf of the ATLAS Collaboration

Chemistry Department, Stony Brook University, Stony Brook, NY 11794, USA
Physics Department, Brookhaven National Laboratory, Upton, NY 11796, USA

E-mail: `jjia@bnl.gov`

Abstract. The rapidity-even dipolar flow v_1 associated with dipole asymmetry in the initial geometry is measured over a broad range in transverse momentum 0.5 GeV< $p_{\rm T}$ <9 GeV, and centrality (0-50)% in Pb-Pb collisions at $\sqrt{s_{\rm NN}}$ = 2.76 TeV, recorded by the ATLAS experiment at the LHC. The v_1 coefficient is determined via a two-component fit of the first order Fourier coefficient $v_{1,1} = \langle\cos\Delta\phi\rangle$ of two-particle correlations in azimuthal angle $\Delta\phi = \phi_{\rm a} - \phi_{\rm b}$ as a function of $p_{\rm T}^{\rm a}$ and $p_{\rm T}^{\rm b}$. This fit is motivated by the finding that the $p_{\rm T}$ dependence of $v_{1,1}(p_{\rm T}^{\rm a}, p_{\rm T}^{\rm b})$ data are consistent with the combined contributions from a rapidity-even v_1 and global momentum conservation. The magnitude of the extracted momentum conservation component suggests that the system conserving momentum involves only a subset of the event (spanning about 3 units in η in central collisions). The extracted v_1 is observed to cross zero at $p_{\rm T} \approx 1.0$ GeV, reaches a maximum at 4–5 GeV with a value comparable to that for v_3, and decreases at higher $p_{\rm T}$. Interestingly, the magnitude of v_1 at high $p_{\rm T}$ exceeds the value of the v_3 in all centrality interval and exceeds the value of v_2 in central collisions. This behavior suggests that the path-length dependence of energy loss and initial dipole asymmetry from fluctuations corroborate to produce a large dipolar anisotropy for high $p_{\rm T}$ hadrons, making the v_1 a valuable probe for studying the jet quenching phenomena.

Recently, the measurements of harmonic flow coefficients v_n have provided valuable insights into the properties of the hot and dense matter created in heavy ion collisions at the Relativistic Heavy Ion Collider (RHIC) and the Large Hadron Collider (LHC). These coefficients are obtained from a Fourier expansion of the particle production in azimuthal angle ϕ:

$$\frac{dN}{d\phi} \propto 1 + 2\sum_{n=1}^{\infty} v_n \cos n(\phi - \Phi_n) \tag{1}$$

where Φ_n represents the phase of v_n. These v_n coefficients are closely related to the Fourier coefficients $v_{n,n}$ of the two-particle correlation (2PC) in relative azimuthal angle $\Delta\phi = \phi_{\rm a} - \phi_{\rm b}$:

$$\frac{dN^{pairs}}{d\Delta\phi} \propto 1 + 2\sum_{n=1}^{\infty} v_{n,n} \cos n(\Delta\phi), \tag{2}$$

which in general contains contributions from both harmonic flow v_n and other non-flow correlations:

$$v_{n,n}(p_{\rm T}^{\rm a}, p_{\rm T}^{\rm b}) = v_n(p_{\rm T}^{\rm a}) v_n(p_{\rm T}^{\rm b}) + \text{non-flow} \tag{3}$$

The v_2 coefficient, known to be associated with the elliptic shape of the overlap region, dominates in the non-central collisions. However, large v_n values have also been measured for $n > 2$ [1–4], which have been interpreted as the hydrodynamic response to the fluctuations of the matter in the initial states. The exotic ridge and Mach-cone like structures first observed at RHIC in the 2PC in $\Delta\phi$ and $\Delta\eta$ [5–7] are naturally explained by the sum of these harmonics. The main sources of non-flow correlations include contributions from resonance decays, Bose-Einstein correlation, jets, and the global momentum conservation (GMC), a source that is very important for the $n = 1$ term. This proceedings focuses on the v_1 results extracted from $v_{1,1}$ by taking into account the contribution of GMC in Eq. 3, using a method similar to Ref. [8]. The main results have been published in Ref. [4].

During the Quark Matter 2011 conference, all three LHC experiments demonstrated that the non-flow effects are small for $n = 2 - 6$ and the factorization:

$$v_{n,n}(p_T^a, p_T^b) = v_n(p_T^a)v_n(p_T^b) \tag{4}$$

is valid within $\sim 5 - 10\%$ (see S. Mohapatra's talk in this workshop [9]). They further show that this factorization is valid as long as one of the particle has $p_T \lesssim 4$ GeV, beyond that the influence of dijet correlation becomes large. In contrast, the factorization breaks down for $n = 1$ over the entire p_T and η range. This is illustrated in Figure 1, which shows the v_n values calculated from $v_{n,n}$ via Eq. 4, assuming no non-flow effect; They are obtained for $1 < p_T^b < 1.5$ GeV using four different reference p_T^a ranges. The extracted $v_2 - v_6$ are independent of reference p_T^a for $|\Delta\eta| > 1$, while the $v_1(p_T^b)$ values calculated this way clearly change with the choice of p_T^a, indicating a breakdown of the factorization over the measured p_T and $\Delta\eta$ range.

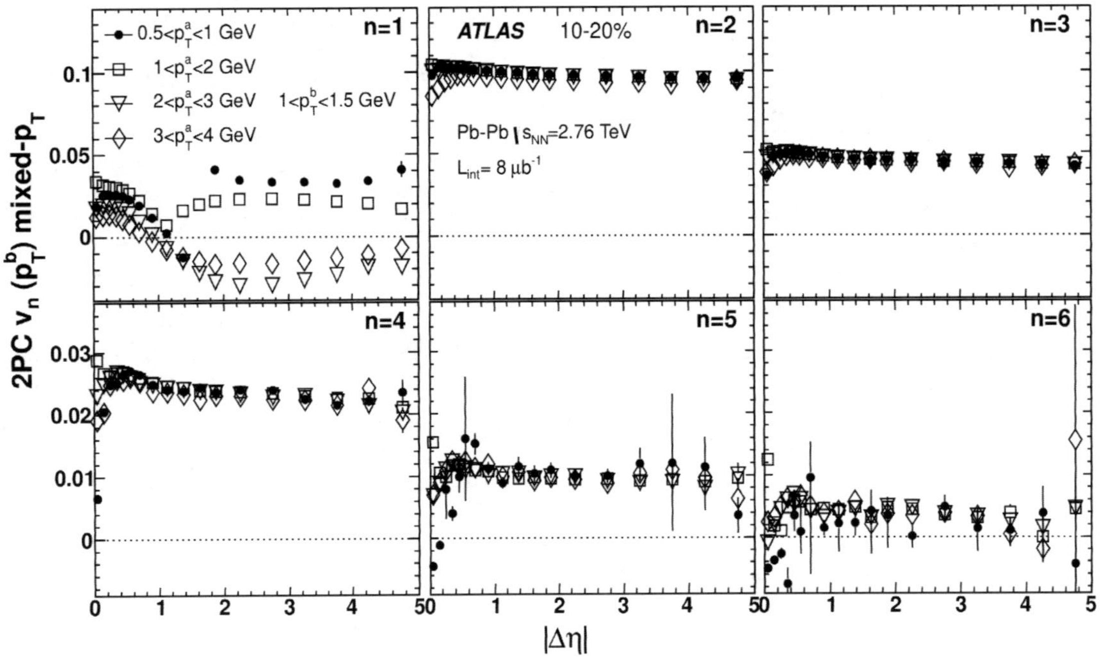

Figure 1. The values of $v_n(p_T^b) = \frac{v_{n,n}(p_T^a, p_T^b)}{v_n(p_T^a)}$ vs. $|\Delta\eta = \eta_a - \eta_b|$ for $1 < p_T^b < 1.5$ GeV, calculated from a reference $v_n(p_T^a)$ in four p_T^a ranges (0.5–1, 1–2, 2–3, and 3–4 GeV) [4]. The error bars indicate the statistical uncertainties.

The main reason for the lack of factorization for $v_{1,1}$ is that it is strongly influenced by GMC, while all higher-order coefficients conserve momentum due to their multi-fold symmetries. One example of GMC effects comes from dijet fragmentation in peripheral events, which tend to give a negative $v_{1,1}$ at large $|\Delta\eta|$. The influence of GMC on the two-particle correlation was studied extensively [10, 11], and was shown to explicitly break the factorization relation Eq. 4 [10]:

$$v_{1,1}(p_{\mathrm{T}}^{\mathrm{a}}, p_{\mathrm{T}}^{\mathrm{b}}, \eta_{\mathrm{a}}, \eta_{\mathrm{b}}) \approx v_1(p_{\mathrm{T}}^{\mathrm{a}}, \eta_{\mathrm{a}})v_1(p_{\mathrm{T}}^{\mathrm{b}}, \eta_{\mathrm{b}}) - \frac{p_{\mathrm{T}}^{\mathrm{a}}p_{\mathrm{T}}^{\mathrm{b}}}{M\langle p_{\mathrm{T}}^2\rangle}, \tag{5}$$

where M and $\langle p_{\mathrm{T}}^2\rangle$ are the multiplicity and average squared transverse momentum for the whole event, respectively. The second term is a leading-order approximation for the GMC. This approximation is expected to be valid when the correction term is much smaller than one. The GMC effect is important in peripheral events and for high p_{T} particles, but is diluted in central events due to the large multiplicity.

The dependence of the v_1 on η can be generally separated into a rapidity-odd component and a rapidity-even component. The rapidity-odd v_1 is thought to arise from the deflection of the colliding ions [12], and changes sign from negative η to positive η. However, the rapidity-odd v_1 is generally small < 0.005 for $|\eta| < 2$ at the RHIC and LHC. The corresponding contribution to $v_{1,1}$ ($< 2.5 \times 10^{-5}$) is negligible compared to the typical $v_{1,1}$ values (see below). The rapidity-even v_1 signal is argued to be associated to the dipole anisotropy of the pressure gradient arising from the dipole asymmetry of the nuclear overlap; this dipole asymmetry is mainly due to fluctuations in the initial geometry [13]. If the rapidity-even v_1 depends weakly on η, similar to the higher-order v_n, Eq. 5 can be simplified to:

$$v_{1,1}(p_{\mathrm{T}}^{\mathrm{a}}, p_{\mathrm{T}}^{\mathrm{b}}) \approx v_1(p_{\mathrm{T}}^{\mathrm{a}})v_1(p_{\mathrm{T}}^{\mathrm{b}}) - \frac{p_{\mathrm{T}}^{\mathrm{a}}p_{\mathrm{T}}^{\mathrm{b}}}{M\langle p_{\mathrm{T}}^2\rangle}. \tag{6}$$

The top-left panel of Figure 1 indeed suggests that the $v_{1,1}$ values are independent of $\Delta\eta$ for $|\Delta\eta| > 2$. Hence the $v_{1,1}$ values are integrated over $2 < |\Delta\eta| < 5$ to obtain one value for each $p_{\mathrm{T}}^{\mathrm{a}}$ and $p_{\mathrm{T}}^{\mathrm{b}}$ combination, and the resulting $v_{1,1}(p_{\mathrm{T}}^{\mathrm{a}}, p_{\mathrm{T}}^{\mathrm{b}})$ functions are shown in Figure 2. In peripheral events ($>$ 40% centrality), the $v_{1,1}$ values are always negative and their magnitudes increase nearly linearly with $p_{\mathrm{T}}^{\mathrm{a}}$ and $p_{\mathrm{T}}^{\mathrm{b}}$, due to the contribution from the away-side jet. In more central events, the contribution of this negative $v_{1,1}$ component is smaller, reflecting a dilution of the GMC term by the large event multiplicity. Furthermore, Figure 2 clearly suggests that a positive $v_{1,1}$ component sets in for $2 \lesssim p_{\mathrm{T}} \lesssim 6$ GeV. Its magnitude increases with p_{T} and eventually drives $v_{1,1}$ into the positive region.

To extract the rapidity-even v_1, a least-squares fit of the $v_{1,1}$ data is performed for each centrality interval with the following form:

$$\chi^2 = \sum_{\mathrm{a,b}} \frac{\left(v_{1,1}(p_{\mathrm{T}}^{\mathrm{a}}, p_{\mathrm{T}}^{\mathrm{b}}) - [v_1^{\mathrm{Fit}}(p_{\mathrm{T}}^{\mathrm{a}})v_1^{\mathrm{Fit}}(p_{\mathrm{T}}^{\mathrm{b}}) - cp_{\mathrm{T}}^{\mathrm{a}}p_{\mathrm{T}}^{\mathrm{b}}]\right)^2}{\left(\sigma_{\mathrm{a,b}}^{\mathrm{stat}}\right)^2 + \left(\sigma_{\mathrm{a,b}}^{\mathrm{sys,p2p}}\right)^2}, \tag{7}$$

where $\sigma_{\mathrm{a,b}}^{\mathrm{stat}}$ and $\sigma_{\mathrm{a,b}}^{\mathrm{sys,p2p}}$ denote the statistical and point-to-point systematic uncertainties for $v_{1,1}(p_{\mathrm{T}}^{\mathrm{a}}, p_{\mathrm{T}}^{\mathrm{b}})$, respectively. The $v_1^{\mathrm{Fit}}(p_{\mathrm{T}})$ function is defined via a smooth interpolation of its values at 15 discrete p_{T} points $v_1^{\mathrm{Fit}}(p_{\mathrm{T},i})|_{i=1}^{15}$ (see Table 1), and these together with the parameter c constitute a total of 16 fit parameters. This interpolation procedure is motivated by the fact that the function $v_1^{\mathrm{Fit}}(p_{\mathrm{T}})$ is unknown but is expected to vary smoothly with p_{T}. In the peripheral collisions where the v_1 contribution is small, the fit is also repeated including only the GMC contribution (i.e. removing the $v_1^{\mathrm{Fit}}(p_{\mathrm{T}}^{\mathrm{a}})v_1^{\mathrm{Fit}}(p_{\mathrm{T}}^{\mathrm{b}})$ in Eq. 7).

Journal of Physics: Conference Series **389** (2012) 012013 doi:10.1088/1742-6596/389/1/012013

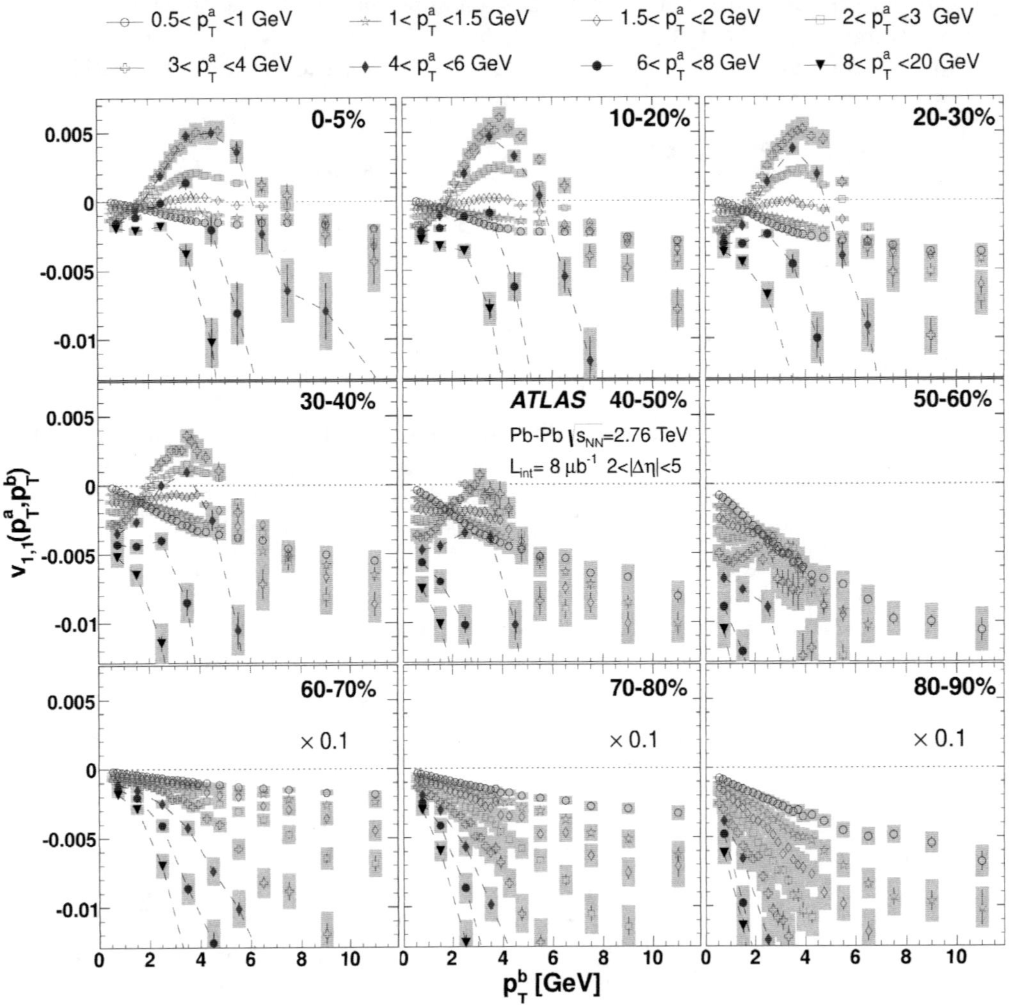

Figure 2. The values of $v_{1,1}(p_{\mathrm{T}}^{\mathrm{a}}, p_{\mathrm{T}}^{\mathrm{b}})$ for $2 < |\Delta\eta| < 5$ vs. $p_{\mathrm{T}}^{\mathrm{b}}$ for different $p_{\mathrm{T}}^{\mathrm{a}}$ ranges. Each panel presents results in one centrality interval. The error bars and shaded bands represent statistical and systematic uncertainties, respectively. The data points for the three highest $p_{\mathrm{T}}^{\mathrm{a}}$ intervals have coarser binning in $p_{\mathrm{T}}^{\mathrm{b}}$, hence are connected by dashed lines to guide the eye. The data in the bottom three panels are scaled down by a factor of ten to fit within the same vertical range.

The top panel of Figure 3 shows the fit to the $v_{1,1}$ data from Figure 2 for the (0–5)% centrality interval. The $v_{1,1}$ data in Figure 2 are plotted as a function of $p_{\mathrm{T}}^{\mathrm{b}}$ for six intervals of $p_{\mathrm{T}}^{\mathrm{a}}$. The seemingly-complex patterns of the $v_{1,1}$ data in (0-5)% centrality interval are well described by the two-component fit across broad ranges of $p_{\mathrm{T}}^{\mathrm{a}}$ and $p_{\mathrm{T}}^{\mathrm{b}}$, with the dot-dashed lines indicating the estimated contributions from the GMC. The typical χ^2/DOF of the fit is between one and two depending on the centrality, as shown in Table 1. The deviations of the data from the fit, as shown in the bottom section of each panel, are less than 10^{-4} for $p_{\mathrm{T}} < 5$ GeV. Above that p_{T} and in more peripheral events the deviations increase, possibly reflecting the limitation of the leading-order approximation for the momentum conservation term in Eq. 6, or the two-component assumption in general.

The bottom panel of Figure 3 shows the fit to the $v_{1,1}$ data from Figure 2 for the (80–90)%

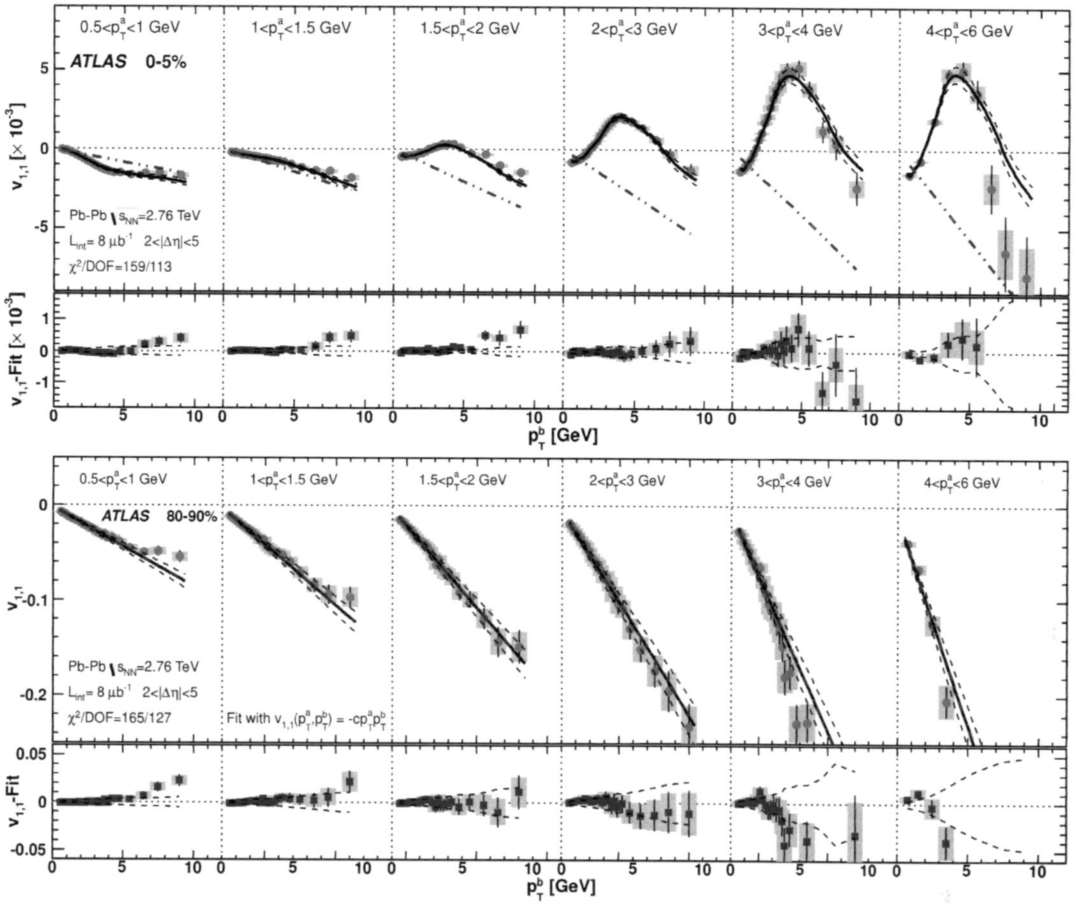

Figure 3. Top panel: Global fit to the $v_{1,1}$ data for the (0–5)% centrality interval via Eq. 7. The fit is performed simultaneously over all $v_{1,1}$ data points in a given centrality interval, organized as a function of $p_{\mathrm{T}}^{\mathrm{b}}$ for various $p_{\mathrm{T}}^{\mathrm{a}}$ ranges, with shaded bars indicating the correlated systematic uncertainties. The fit function and the associated systematic uncertainties are indicated by the thick-solid lines and surrounding dashed lines, respectively. The dot-dashed lines intercepting each dataset (with negative slope) indicate estimated contributions from the momentum conservation component. The lower part of the panel shows the difference between the data and fit (solid points), as well as the systematic uncertainties of the fit (dashed lines). Bottom panel: Global fit for the (80–90)% centrality interval considering only the GMC contribution.

centrality interval, including only the CMC term. The data are consistent with the GMC up to 10 GeV in both $p_{\mathrm{T}}^{\mathrm{a}}$ and $p_{\mathrm{T}}^{\mathrm{b}}$, well into the region where the contribution from away-side jet is expected to dominate the 2PC. This behavior suggests that in the very peripheral collisions, the system that conserves the momentum involves mainly the particles from dijets.

Figure 4 shows the extracted $v_1^{\mathrm{Fit}}(p_{\mathrm{T}})$ for various centrality intervals together with the v_2 and v_3 data measured using the event plane method [4]. The $v_1^{\mathrm{Fit}}(p_{\mathrm{T}})$ is negative for $p_{\mathrm{T}} \lesssim 1.0$ GeV (decrease to about 0.9 GeV in (40-50)% centrality interval), confirming a generic feature expected for collective v_1 as suggested by hydrodynamic model calculations [11,13]. The crossing point extracted from ALICE data [2,8] is significantly higher ($\sim 1.3 - 1.4$ GeV), which could be related to the fact the $|\Delta\eta|$ gap of 0.8 in ALICE data is not sufficient to fully suppress the near-

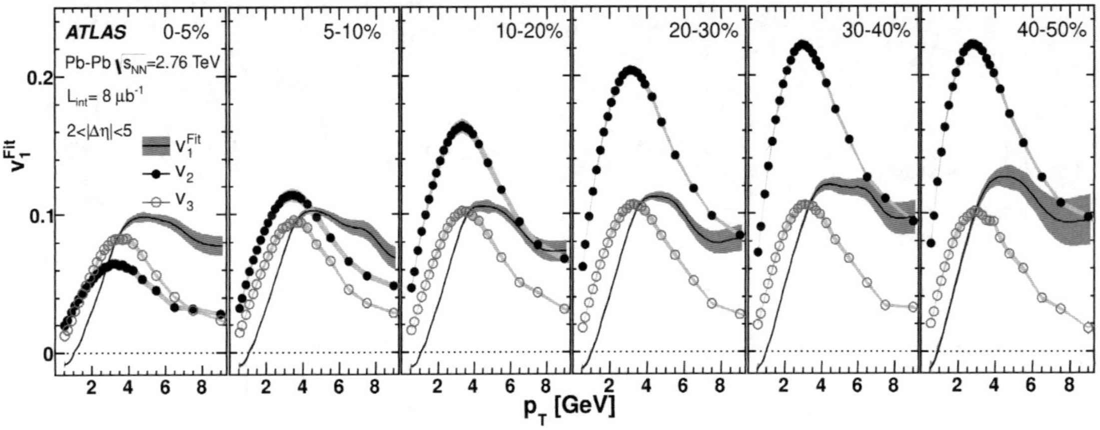

Figure 4. $v_1^{\rm Fit}$ vs. $p_{\rm T}$ for various centrality intervals. The shaded bands indicate the total uncertainty. The uncertainty bands are reproduced on their own at the bottom of the figure for clarity.

side peak (see top left panel of Figure 1 or Figure 2c of [4]). Nevertheless, that calculation clearly demonstrates that both the zero crossing point and the peak value are sensitive to the shear viscosity to entropy ratio, making our precision v_1 measurement a valuable tool for constraining the transport properties of the medium.

At higher $p_{\rm T}$, the measured $v_1^{\rm Fit}$ values reach a maximum between 4 GeV and 5 GeV and then fall. The magnitude of the $v_1^{\rm Fit}$ is large: its peak value is comparable or slightly larger than that for the v_3. The peak value increases by about 20% over the measured centrality range, also significantly larger than the increase for v_3 (about $\sim$ 10% over the same centrality range). The falloff of $v_1^{\rm Fit}(p_{\rm T})$ at higher $p_{\rm T}$ may indicate the onset of path-length dependent jet energy loss, which correlates with the dipole asymmetry in the initial geometry to produce the large v_1 coefficient at high $p_{\rm T}$. This falloff is slower than that for the v_2 and v_3, with an magnitude that is surprisely larger than v_2 in central collisions, consistent with a finding based on jet absorption model calculation [14]. These high $p_{\rm T}$ behaviors suggest that our $v_1^{\rm Fit}(p_{\rm T})$ results also provide valuable input for understanding jet quenching phenomena.

Finally, results in Figure 4 also imply that the rapidity-even collective v_1 is an important component of the two-particle correlation at intermediate $p_{\rm T}$. For example, the large positive $v_{1,1}$ harmonic in Figure 5, responsible for the mach-cone like structure in the the bottom-right panel, is mainly due to the rapidity-even v_1: its contribution to $v_{1,1}$ can be three times larger than the negative momentum conservation term estimated by the global fit (top panel of Figure 3).

If the two-component ansatz of Eq. 6 is valid, the fit parameter c should be inversely proportional to multiplicity M and $\langle p_{\rm T}^2\rangle$ of the whole event. This ansatz is checked by calculating the product of c and the charged hadron multiplicity at mid-rapidity $\frac{dN}{d\eta}_{|\eta=0}$ from [15], assuming $\frac{dN}{d\eta}_{|\eta=0} \propto M$. The results are summarized in Table 1 for each centrality interval. Since $\langle p_{\rm T}^2\rangle$ for the whole event is expected to vary weakly with centrality, the product is also expected to vary weakly with centrality. Table 1 shows that this is indeed the case, supporting the assumptions underlying Eq. 6.

It has been argued that the system that conserves momentum may only involves a subset of the event [17]. If the value of $\langle p_{\rm T}^2\rangle$ is known, one can estimate the effective size of the system that is correlated due to the GMC. For $\langle p_{\rm T}^2\rangle \sim 1$ GeV^2 (estimated from charged hadron spectrum from ALICE [16]), the system size is estimated to be $M \sim 5000$ in (0-5)% most central collisions,

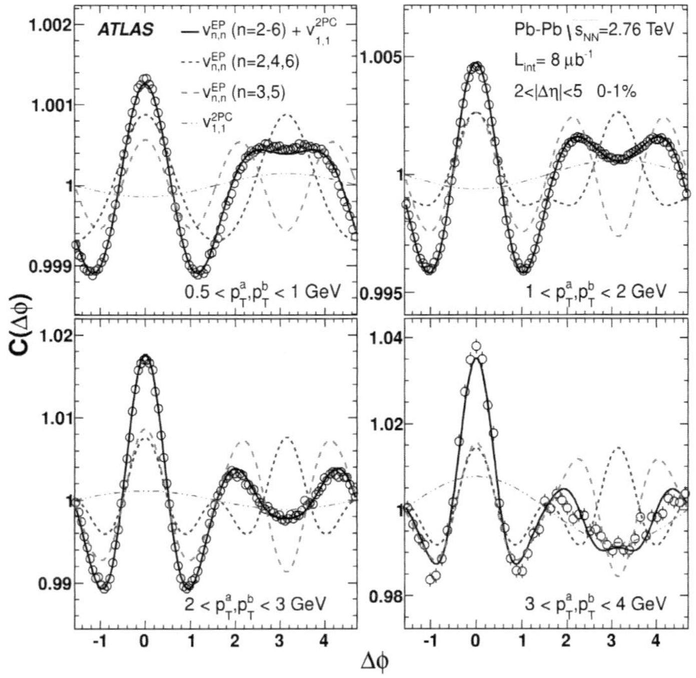

Figure 5. Measured correlation functions compared with those reconstructed from v_2–v_6 measured from the $\mathrm{FCal}_{\mathrm{P(N)}}$ method and $v_{1,1}$ from the 2PC method in (0–1)% centrality interval for four p_{T} ranges. The contributions from $n = 1$, $n = 3, 5$ and $n = 2, 4, 6$ are shown separately. The error bars indicate the statistical uncertainties.

covering about 3 units in η; the system size increases to about 4 units for (40-50)% centrality interval.

Centrality	χ^2/DOF	$c[\times 0.001\mathrm{GeV}^{-2}]$	$c\frac{dN}{d\eta}_{\vert\eta=0}[\times\mathrm{GeV}^{-2}]$
0-5%	159/113	0.24 ± 0.02	0.39 ± 0.04
5-10%	133/113	0.28 ± 0.02	0.37 ± 0.04
10-20%	165/113	0.35 ± 0.03	0.36 ± 0.04
20-30%	134/113	0.50 ± 0.04	0.34 ± 0.03
30-40%	188/113	0.75 ± 0.05	0.33 ± 0.03
40-50%	181/113	1.16 ± 0.09	0.32 ± 0.03
15 interpolation points used in the default fit: 0.5, 0.7, 0.9, 1.1, 1.3, 1.5, 2.0, 2.5, 3.0, 3.5, 4.5, 5.5, 6.5, 7.5, 9.0 GeV			

Table 1. Quality of the fit χ^2/DOF, fit parameter c, and corresponding multiplicity scaled values $c\frac{dN}{d\eta}_{\vert\eta=0}$ for various centrality intervals. The uncertainty of $c\frac{dN}{d\eta}_{\vert\eta=0}$ is calculated as the quadrature sum of uncertainties from c and $\frac{dN}{d\eta}_{\vert\eta=0}$ of Ref. [15]. The bottom row lists the 15 p_{T} interpolation points used in the default fit.

Equation 6 was derived based on fairly general grounds, hence it is expected to be valid for any Monte Carlo model that contains the flow physics and respects momentum conservation. One example is the AMPT (A Multi Phase Transport) model [18]. This model combines the initial condition from HIJING and final state interaction via a parton and hadron transport model. The initial condition from HIJING is seeded by a Glauber model for nucleon-nucleon scatterings, hence it contains the fluctuations which generate event-by-event spatial asymmetries; the parton

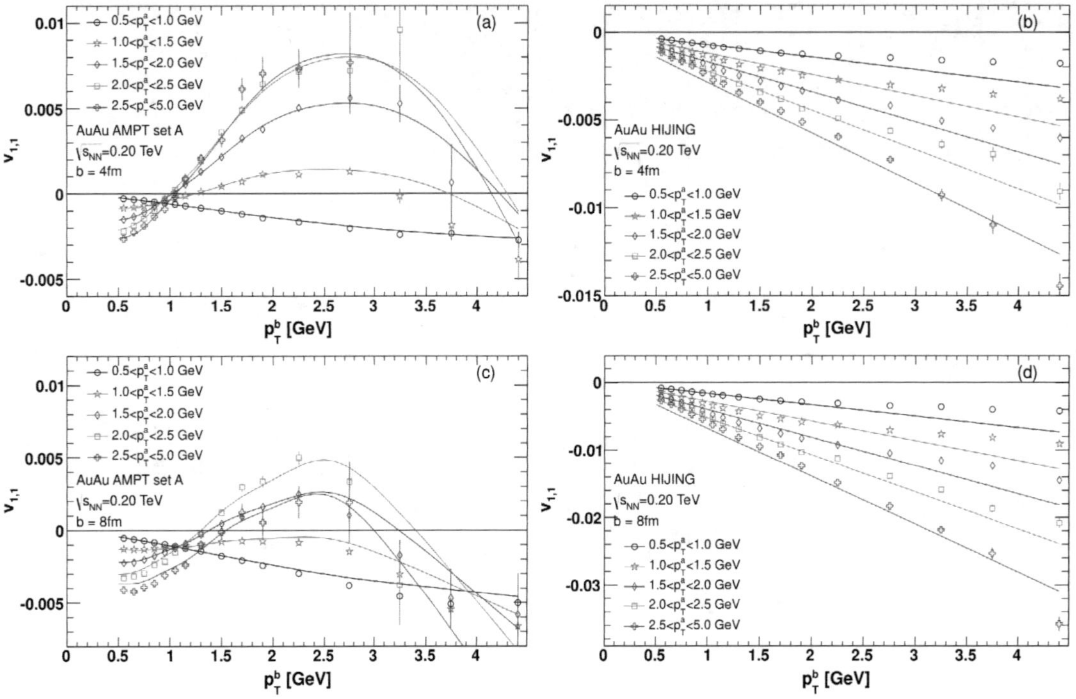

Figure 6. $v_{1,1}(p_{\rm T}^{\rm a}, p_{\rm T}^{\rm b})$ (symbols) from AMPT (left panels) and HIJING (right panels) for $b = 4$ fm (top panels) and $b = 8$ fm in AuAu collisions at $\sqrt{s_{NN}} = 0.2$ TeV. Lines in the left (right) panels are result of a global fit to Eq. 6 (a fit to pure momentum conservation component: $cp_{\rm T}^{\rm a}p_{\rm T}^{\rm b}$). Taken from [19].

and hadron transport is responsible for transforming the initial asymmetries into the momentum anisotropy. Figure 6 shows the factorization behavior from both HIJING and AMPT at the RHIC and LHC energies obtained from [19]. The $v_{1,1}(p_{\rm T}^{\rm a}, p_{\rm T}^{\rm b})$ values from HIJING model, which has no final state interactions, exhibit linear dependence on $p_{\rm T}^{\rm a}$, $p_{\rm T}^{\rm b}$, consistent with correlations coming from only GMC. However once the final state interactions are switched on as is in the AMPT model, the $v_{1,1}(p_{\rm T}^{\rm a}, p_{\rm T}^{\rm b})$ data develops crossing pattern at $p_{\rm T} \sim 1$ GeV and hump-back structures in $2 < p_{\rm T} < 6$ GeV ranges, very similar to what is observed in the data. A simple two-component fit via Eq. 6 is able to describe these structures simultaneously. This clearly demonstrates the first component of Eq. 6 is indeed associated with the dipolar flow arising from the initial geometry fluctuations and strong interactions in the final state.

In summary, the $v_{1,1} = \langle\cos(\Delta\phi)\rangle$ component of the two-particle correlation data are extracted for Pb-Pb collisions at $\sqrt{s_{\rm NN}} = 2.76$ TeV. The $p_{\rm T}$ and $\Delta\eta$ dependence of $v_{1,1}$ data are found to be consistent with contributions from rapidity-even dipolar flow v_1 and global momentum conservation. A two-component fit is used to extract the individual contributions from these two components. The extracted v_1 function crosses zero at $p_{\rm T} \sim 1$ GeV, reaches a value of 0.1-0.12 (comparable or slightly larger than v_3), and then decreases at higher $p_{\rm T}$. The $p_{\rm T}$ at which v_1 reaches maximum is about 1 GeV higher than other v_n. Extracted v_1 shows a mild but stronger increase with centrality ($\sim 20\%$) than the v_3. The magnitude of the extracted momentum conservation component suggests that the system conserving momentum involves only a subset of the event (spanning about 3 units in η in central collisions). Simulation based on AMPT model qualitatively reproduces the complex $p_{\rm T}$ dependence of the $v_{1,1}$. Together with HIJING model simulation, they supports the two-component interpretation of the $v_{1,1}$ data and

Journal of Physics: Conference Series **389** (2012) 012013 doi:10.1088/1742-6596/389/1/012013

strongly suggest that the presence of dipolar flow requires both initial geometry fluctuations and strong final state interactions.

This work is in part supported by NSF under award number PHY-1019387.

References

[1] PHENIX Collaboration, Phys. Rev. Lett. **107**, 252301 (2011)
[2] ALICE Collaboration, Phys. Lett. B **708**, 249 (2012).
[3] CMS Collaboration, Eur. Phys. J. C **72**, 2012 (2012).
[4] ATLAS Collaboration, Phys. Rev. C **86**, 014907 (2012), auxiliary plots can be found at http://atlas.web.cern.ch/Atlas/GROUPS/PHYSICS/PAPERS/HION-2011-01/
[5] PHENIX Collaboration, Phys. Rev. C **78**, 014901 (2008).
[6] STAR Collaboration, Phys. Rev. C **80**, 064912 (2009).
[7] PHOBOS Collaboration, Phys. Rev. Lett. **104**, 062301 (2010).
[8] E. Retinskaya, M. Luzum and J. -Y. Ollitrault, Phys. Rev. Lett. **108**, 252302 (2012)
[9] S. Mohapatra for the ATLAS Collaboration, "Measurement of elliptic and higher order flow harmonics in 2.76 TeV Pb-Pb collisions with the ATLAS detector", this proceedings.
[10] N. Borghini, P. M. Dinh, J. -Y. Ollitrault, Phys. Rev. C **62**, 034902 (2000).
[11] F. G. Gardim, F. Grassi, Y. Hama, M. Luzum, J. -Y. Ollitrault, Phys. Rev. C **83**, 064901 (2011).
[12] S. A. Voloshin, A. M. Poskanzer, R. Snellings, "Collective phenomena in non-central nuclear collisions," arXiv:0809.2949 [nucl-ex].
[13] D. Teaney and L. Yan, Phys. Rev. C **83**, 064904 (2011).
[14] J. Jia, "Azimuthal anisotropy in a jet absorption model with fluctuating initial geometry in heavy ion collisions," arXiv:1203.3265 [nucl-th].
[15] Phys. Lett. B **710**, 363 (2012)
[16] ALICE Collaboration, "Transverse momentum spectra of identified charged hadrons with the ALICE detector in Pb-Pb collisions at $\sqrt{s_{NN}} = 2.76$ TeV," arXiv:1111.0763 [nucl-ex].
[17] Z. Chajecki and M. Lisa, Phys. Rev. C **79**, 034908 (2009)
[18] Z. -W. Lin, C. M. Ko, B. -A. Li, B. Zhang and S. Pal, Phys. Rev. C **72**, 064901 (2005)
[19] J. Jia, S. K. Radhakrishnan and S. Mohapatra, "A study of the anisotropy associated with dipole asymmetry in heavy ion collisions," arXiv:1203.3410 [nucl-th].

28th Winter Workshop on Nuclear Dynamics 2012 IOP Publishing
Journal of Physics: Conference Series **389** (2012) 012014 doi:10.1088/1742-6596/389/1/012014

Anisotropic Flow and Viscous Hydrodynamics

Li Yan*

Department of Physics and Astronomy, Stony Brook University, Stony Brook, NY, 11794, USA

E-mail: li.yan@stonybrook.edu

Abstract. We report part of our recent work on viscous hydrodynamics with consistent phase space distribution $f(x, \mathbf{p})$ for freeze out. We develop the gradient expansion formalism based on kinetic theory, and with the constraints from the comparison between hydrodynamics and kinetic theory, viscous corrections to $f(x, \mathbf{p})$ can be consistently determined order by order. Then with the obtained $f(x, \mathbf{p})$, second order viscous hydrodynamical calculations are carried out for elliptic flow v_2.

1. Introduction

Relativistic hydrodynamics is an important theoretical tool in heavy ion collisions. Especially it successfully reproduces the observed elliptic flow v_2. In the past several years, with respect to the updated understanding of collective phenomena in heavy-ion collisions, two major progresses have been made in hydrodynamical models. One is the realization of fluctuations in initial state[1], which leads to the practically used event-by-event hydrodynamics[2, 3]. And correspondingly many new types of anisotropic flow, such as dipole flow v_1[4] and triangular flow v_3[1, 5], are studied. The other one is the inclusion of viscous corrections[6]. In practice, although ideal hydrodynamics has been found effective in characterizing the collective medium expansion, viscous corrections are not negligible[7]. One particular example of viscous corrections to hydrodynamics is the viscous damping of elliptic flow[8, 9, 10].

Many attempts have been made in the investigation of the viscous corrections to hydrodynamics[8, 9, 10]. Since first order viscous hydrodynamics suffers from the causality problem(see for example[11]), most of the established viscous hydrodynamical models have contained second order viscous terms in the equations of motion, such as Israel-Stewart theory. Recently, in an important paper, Baier and his collaborators(BRSSS) developed second order viscous hydrodynamical theory with conformal symmetry assumption[12]. Hydrodynamical simulation to the medium expansion in heavy ion collisions is achieved by solving hydrodynamical equations of motion, with respect to the specified initial conditions. Observables are obtained then by freeze out process. Due to the correspondence between hydrodynamics and kinetic theory, freeze out process also needs viscous corrections, and the corrections have to be constrained so that they are consistent with hydrodynamics itself. In particular, in Cooper-Fryer formula, this is reflected in the determination of the phase-space distribution function $f(x, \mathbf{p})$. However, to the knowledge of authors, this constraints have not been systematically discussed in the literature through second order, and this will be the main subject of this paper. For more details, refer to [13].

*In collaboration with Derek Teaney.

The paper is organized as follows. The theoretical formalism is constructed by reviewing viscous effects in kinetic theory and hydrodynamics in section 2. In section 3 the determination of phase-space distribution function is discussed. Hydrodynamical simulations with consistent form of $f(x, \mathbf{p})$ is carried out, and elliptic flow v_2 is calculated at RHIC energy, in section 4. Our conclusions are summarized in section 5.

2. Kinetic Theory and Hydrodynamics

A liquid system out of equilibrium can be approached through either transport theory or hydrodynamics, provided that the non-equilibrium part can be seen as perturbations. As a result, in order to sustain the continuity around freeze out, there must be a determined correspondence between kinetic theory and hydrodynamics. This correspondence originates the constraints on distribution function $f(x, \mathbf{p})$. In terms of the dependence on transverse momentum p_T, one can generally fix δf by the moment method[14], in which δf is decomposed into moment expansion and cut at a certain order. The widely accepted form, for example,

$$\delta f_\pi = \frac{n_p(1 \pm n_p)}{2(e+\mathcal{P})T^2} p^\mu p^\nu \pi_{\mu\nu}, \tag{1}$$

is obtained by a fourteen-momentum method. However there are two flaws in this form for realistic simulations. First, in actual simulations $\pi_{\mu\nu}$ is taken from second order hydrodynamical calculations, while δf_π is computed using a first order approximation. Second, and more importantly, the p_T dependence in δf has been investigated in [15], and found to vary based on microscopic dynamics. Although the quadratic ansatz (i.e. Eq. (1)) is appropriate for collisional energy loss, it fails in the case where radiative energy loss dominates.

In kinetic theory, the viscous corrections $\delta f(x, \mathbf{p})$ corresponds to the viscous corrections to hydrodynamics, i.e., stress tensor $\pi^{\mu\nu}$, so it also depends on gradient expansion. In this way, following Chapman-Enskog method[14] where expansion in gradients is used to find asymptotic form of the solution to Boltzmann equation, $\delta f(x, \mathbf{p})$ can be determined order by order. When relaxation time approximation can be taken into account, form of δf is analytically solvable for the specified viscous hydrodynamics. In this paper, we take relaxation time approximation and determine $\delta f(x, \mathbf{p})$ through second order in gradients, with respect to BRSSS hydrodynamics.

The matrix convention we use throughout the paper is $(-, +, +, +)$. p^μ is used for four-momentum. And for the sake of convenience, we introduce a tilde on a quantity from time to time to indicate it as dimensionless. In the derivation of hydrodynamics, tensor index is always split into temporal and spatial parts. Since flow four-velocity $u^\mu = \gamma(1, \vec{v})$ is purely temporal in the local rest frame(LRF), the decomposition is then realized based on u^μ and the projection operator $\Delta^{\mu\nu} = u^\nu u^\nu + g^{\mu\nu}$. For derivative operator we have $\partial^\mu = u^\mu D + \nabla^\mu$, with $D = u^\mu \partial_\mu$ and $\nabla^\mu = \Delta^{\mu\nu}\partial_\nu$ representing pure time and space derivatives in LRF respectively. A further decomposition according to rotation group is also considered comparing to the existed tensor form in hydrodynamics, so the indices of a tensor in brackets $\langle \ldots \rangle$ stand for being symmetric, (projection operator) traceless and transverse,

$$A^{\mu\ldots\langle\ldots\alpha\ldots\beta\ldots\rangle\ldots\nu} = A^{\mu\ldots\langle\ldots\beta\ldots\alpha\ldots\rangle\ldots\nu} \tag{2}$$

$$A^{\mu\ldots\langle\ldots\alpha\ldots\beta\ldots\rangle\ldots\nu}\Delta_{\alpha\beta} = 0 \tag{3}$$

$$A^{\mu\ldots\langle\ldots\alpha\ldots\rangle\ldots\nu}\Delta_{\alpha\beta} = 0 \tag{4}$$

More details of the decomposition can be found in [13].

2.1. Hydrodynamics

In Chapman-Enskog expansion, the so-called solubility condition[14] relates spatial gradients and temporal derivatives. This is actually equivalent to hydrodynamical equations of motion

$\partial_\mu T^{\mu\nu} = 0$. In the current section we will derive the viscous corrections through second order to BRSSS hydrodynamical equations of motion.

The energy-momentum tensor of viscous hydrodynamics reads

$$T^{\mu\nu} = eu^\mu u^\nu + \mathcal{P}\Delta^{\mu\nu} + \pi^{\mu\nu}, \tag{5}$$

where $\pi^{\mu\nu}$ is the viscous correction to stress tensor. In this paper, the bulk viscosity $\zeta = 0$ and baryon chemical potential $\mu_B = 0$. Up to second order viscous corrections, BRSSS[16] determined that the possible forms of the gradient expansion in the stress tensor in a conformal liquid are

$$\pi^{\mu\nu} = -\eta\sigma^{\mu\nu} + \eta\tau_\pi \left[\langle D\sigma^{\mu\nu} \rangle + \frac{1}{d-1}\sigma^{\mu\nu}\nabla\cdot u \right] + \lambda_1 \left\langle \sigma^\mu{}_\lambda \sigma^{\nu\lambda} \right\rangle + \lambda_2 \left\langle \sigma^\mu{}_\lambda \Omega^{\nu\lambda} \right\rangle + \lambda_3 \left\langle \Omega^\mu{}_\lambda \Omega^{\nu\lambda} \right\rangle, \tag{6}$$

where $\sigma^{\mu\nu} = 2\langle \nabla^\mu u^\nu \rangle$ and the vorticity tensor is defined as

$$\Omega^{\mu\nu} = \frac{1}{2}\Delta^{\mu\alpha}\Delta^{\nu\beta}\left(\nabla_\alpha u_\beta - \nabla_\beta u_\alpha\right). \tag{7}$$

$(\tau_\pi, \lambda_1, \lambda_2, \lambda_3)$ are corresponding second order transport coefficients. $d = 4$ is the number of space-time dimensions.† Then equations of motion according to Eq. (6) can be formulated as

$$De = -(e + \mathcal{P})\nabla\cdot u + \frac{\eta}{4}\sigma_{\mu\nu}\sigma^{\mu\nu} + O(\nabla^3), \tag{8}$$

$$Du_\mu = -\frac{\nabla_\alpha \mathcal{P}}{e + \mathcal{P}} + \frac{u_\mu D\sigma^{\mu\nu}\Delta_{\alpha\nu}}{e + \mathcal{P}} - \frac{\Delta_{\alpha\nu}\nabla_\mu \sigma^{\mu\nu}}{e + \mathcal{P}} + O(\nabla^3). \tag{9}$$

Note that the temporal derivatives of the flow velocity u^μ and the energy density e are written in terms of spatial gradients through second order, while higher order terms are neglected.

2.2. Kinetics

To determine the viscous corrections to the distribution function, the strategy is to solve the kinetic equations in a relaxation time approximation order by order in the gradient expansion.

$$f(x, \mathbf{p}) \equiv n_{\mathbf{p}}(x, \mathbf{p}) + \delta f(x, \mathbf{p}) = n_{\mathbf{p}}(x) + \delta f_{(1)} + \delta f_{(2)} + O(\nabla^3), \tag{10}$$

where $n_{\mathbf{p}}(x)$ is equilibrium distribution, which depends on the space-time coordinates through the dimensionless combination $p\cdot u/T = \tilde{p}\cdot u$. In classical limit, $n_{\mathbf{p}}(x) = \exp[p\cdot u(x)/T(x)]$. In a relaxation time approximation the Boltzmann equation reads

$$p^\mu \partial_\mu f_{\mathbf{p}}(x) = -\frac{T^2}{\tilde{C}_{\mathbf{p}}}\left[f - n(p\cdot u_*(x)/T_*(x))\right], \tag{11}$$

where the dimensionless coefficient $\tilde{C}_{\mathbf{p}}$ is related to the canonical momentum dependent relaxation time τ_R,

$$\tilde{C}_{\mathbf{p}} = -T^2\,\frac{\tau_R(-p\cdot u)}{p\cdot u}. \tag{12}$$

†In this paper, we will ignore the possible effects of higher space-time dimensions, so we write explicitly wherever $d = 4$. Also we ignore all curved space-time effects discussed in [16].

Following the convention used in [15], the canonical relaxation time is proportional to $(-\tilde{p}\cdot u)^{1-\alpha}$, where α lies generally between the quadratic ansatz limit($\alpha = 0$) and linear ansatz limit ($\alpha = 1$). We have

$$\tilde{C}_{\mathbf{p}} = c_0(-\tilde{p}\cdot u)^{-\alpha}, \tag{13}$$

with c_0 to be fixed by shear viscosity (see below). The parameters $u^{*\mu}(x)$ and $T^*(x)$ which appear in the Eq. (11) are equal at leading order to the temperature and flow velocities (T and u^μ). They satisfy the Landau matching conditions, i.e. ,

$$\left(\nu\int\frac{d^3p}{(2\pi)^3E_{\mathbf{p}}}p^\mu p^\nu\delta f(u^*,T^*)\right)u^*_\mu = 0, \qquad \left(\nu\int\frac{d^3p}{(2\pi)^3E_{\mathbf{p}}}p^\mu p^\nu n_{\mathbf{p}}(u^*,T^*)\right)u^*_\mu = eu^{*\nu}. \tag{14}$$

So there exist the expansion

$$T^*(X) \equiv T(X) + \delta T^*\,, \tag{15}$$
$$u^{*\mu}(X) \equiv u^\mu(X) + \delta u^{*\mu}\,, \tag{16}$$

with $\delta T^* = \sum_{n=1}\delta T_{(n)}$ and $\delta u^{*\mu} = \sum_{n=1}\delta u^\mu_{(n)}$ indicating corrections from all higher order gradients. Expanding n^*_p (with an obvious notation) we find correspondingly,

$$n^*_{\mathbf{p}} \equiv n_{\mathbf{p}} + \delta n^*_p \quad\text{and}\quad \delta n^*_{\mathbf{p}} = n'_{\mathbf{p}}\left[\frac{p\cdot\delta u_*}{T} - E_{\mathbf{p}}\frac{\delta T_*}{T^2}\right], \tag{17}$$

where here and in the following a prime stands for the derivative on $\tilde{p}\cdot u$. $\delta n^*_{\mathbf{p}}$ with higher order gradients is important for the determination of viscous correction δf, as will be shown in the next section. Eq. (14), together with the correspondence between kinetic theory and hydrodynamics,

$$\nu\int\frac{d^3p}{(2\pi)^3E_{\mathbf{p}}}p^\mu p^\nu\delta f(u^*,T^*) = \pi^{\mu\nu}\,, \tag{18}$$

constrains the solution of Boltzmann equation $f(x,\mathbf{p})$.

3. Determination of δf

We substitute the expansion Eq. (10) into the kinetic equation Eq. (11) and equate orders. In doing so we use the hydrodynamic equations of motion to write time derivatives of $T(x)$ and $u(x)$ in terms of spatial gradients of these fields, see Eq. (8). Indeed, with the help of these equations of motion, up to first order in gradient,

$$p^\mu\partial_\mu n_{\mathbf{p}} = +n'_{\mathbf{p}}\frac{p^\mu p^\nu}{2T}\sigma_{\mu\nu} + O(\nabla^2)\,. \tag{19}$$

The terms linearly proportional to the $\sigma_{\mu\nu}$ is ultimately responsible for shear viscosity.

Collecting all the terms of first order in gradient for δf we obtain the preliminary first order solution

$$\delta f_{(1)} = -\,\tilde{C}_{\mathbf{p}}\;n'_{\mathbf{p}}\;\frac{p^\mu p^\nu\sigma_{\mu\nu}}{2T^3} + \delta n^*_{\mathbf{p}(1)}. \tag{20}$$

But Eq. (20) is not completed until the constant coefficient c_0, the form of $\delta u^\mu_{(1)}$ and $\delta T_{(1)}$ are determined through constraint conditions Eq. (14) and Eq. (18). Obviously there exist the trivial solution $\delta u^\mu_{(1)} = 0$ and $\delta T_{(1)} = 0$ to Eq. (14), and then from Eq. (18) we find,

$$\eta = \frac{T^3}{15}B_1. \tag{21}$$

Journal of Physics: Conference Series **389** (2012) 012014 doi:10.1088/1742-6596/389/1/012014

B_1 is a dimensionless constant from the integral,

$$B_1(\alpha) = \nu \int \frac{d^3\tilde{p}}{(2\pi)^3 \tilde{E}_{\mathbf{p}}} (\tilde{p} \cdot u)^4 [n'_{\mathbf{p}} \tilde{C}_{\mathbf{p}}]. \tag{22}$$

It only depends on the underlying microscopic dynamics in terms of α. We will encounter three more similar constants,

$$B_2(\alpha) = \nu \int \frac{d^3\tilde{p}}{(2\pi)^3 \tilde{E}_{\mathbf{p}}} (\tilde{p} \cdot u)^6 [n'_{\mathbf{p}} \tilde{C}_{\mathbf{p}}]' \tilde{C}_p \,, \tag{23}$$

$$B_3(\alpha) = -\nu \int \frac{d^3\tilde{p}}{(2\pi)^3 \tilde{E}_{\mathbf{p}}} (\tilde{p} \cdot u)^5 [n'_{\mathbf{p}} \tilde{C}_{\mathbf{p}}] \tilde{C}_p \,, \tag{24}$$

$$B_4 = -\nu \int \frac{d^3\tilde{p}}{(2\pi)^3 \tilde{E}_{\mathbf{p}}} (\tilde{p} \cdot u)^3 [n'_{\mathbf{p}}] \,. \tag{25}$$

c_0 is then the solution to Eq. (21), and it is interesting to know that for a massless gas and in the quadratic ansatz limit, $c_0 = \eta/s$.

We extend the derivation to second order $\delta f_{(2)}$, then

$$\begin{aligned} \delta f_{(2)} = &- \tilde{C}_{\mathbf{p}}\, n'_{\mathbf{p}}\, \frac{\eta}{s} \frac{1}{T^4} \Big[p \cdot u\, p_\mu \left(2 \left\langle \sigma^{\mu\nu} \nabla_\nu ln T \right\rangle + \left\langle \Delta^\mu_\alpha \nabla_\beta \sigma^{\alpha\beta} \right\rangle \right) + \frac{(p \cdot u)^2}{4(d-1)} \sigma^2 \Big] \\ &- \frac{\tilde{C}_{\mathbf{p}}}{T^2}\, p^\mu \partial_\mu \delta f_{(1)} + \delta n^*_{\mathbf{p}(2)} \,. \end{aligned} \tag{26}$$

Clearly there are three sources of contributions, those generated from $\delta f_{(1)}$ in Boltzmann equation, from second order corrections to flow velocity and temperature and from second order hydrodynamics equations of motion. The dominant term is from $p^\mu \partial_\mu \delta f_{(1)}$, since in Boltzmann equation the derivative gives rise to one higher order dependence on p_T simultaneously. To write $\delta f_{(2)}$ explicitly we decompose the resulting tensors into irreducible tensors of the rotation group in the LRF ; [13] provides a few more details. Straightforward then from (26), with somewhat tedious algebra, we obtain,

$$\begin{aligned} \delta f_{(2)} = & \frac{[n'\tilde{C}_{\mathbf{p}}]'\tilde{C}_{\mathbf{p}}}{4T^2} \tilde{p}^\mu \tilde{p}^\nu \tilde{p}^\alpha \tilde{p}^\beta \left\langle \sigma_{\mu\nu} \sigma_{\alpha\beta} \right\rangle - \frac{[n'\tilde{C}_{\mathbf{p}}]\tilde{C}_{\mathbf{p}}}{2T^2} \tilde{p}^\mu \tilde{p}^\nu \tilde{p}^\alpha \left[3 \left\langle \sigma_{\mu\nu} \nabla_\alpha ln T \right\rangle + \left\langle \nabla_\alpha \sigma_{\mu\nu} \right\rangle \right] \\ & + \tilde{p}^\mu \tilde{p}^\nu \left\langle \sigma^\lambda_\mu \sigma_{\nu\lambda} \right\rangle \left[\frac{[n'\tilde{C}_{\mathbf{p}}]'\tilde{C}_{\mathbf{p}}}{7T^2} (\tilde{p} \cdot u)^2 + \frac{[n'\tilde{C}_{\mathbf{p}}]\tilde{C}_{\mathbf{p}}}{2T^2} \tilde{p} \cdot u \right] \\ & + \tilde{p}^\nu \tilde{p}^\nu \left\langle \sigma_{\mu\lambda} \Omega^\lambda_\nu \right\rangle \frac{[n'\tilde{C}_{\mathbf{p}}]\tilde{C}_{\mathbf{p}}}{T^2} \tilde{p} \cdot u - \tilde{p}^\mu \tilde{p}^\nu \left[\left\langle D\sigma_{\mu\nu} \right\rangle + \frac{\sigma_{\mu\nu}}{3} \nabla \cdot u \right] \frac{[n'\tilde{C}_{\mathbf{p}}]\tilde{C}_{\mathbf{p}}}{2T^2} \tilde{p} \cdot u \\ & + \tilde{p}_\mu \tilde{p} \cdot u \left[\frac{\eta}{s} \frac{n'\tilde{C}_{\mathbf{p}}}{T^2} + \frac{[n'\tilde{C}_{\mathbf{p}}]\tilde{C}_{\mathbf{p}}}{5T^2} \tilde{p} \cdot u \right] \left[2 \left\langle \sigma^{\mu\nu} \nabla_\nu ln T \right\rangle + \left\langle \Delta^\mu_\alpha \nabla_\beta \sigma^{\alpha\beta} \right\rangle \right] \\ & + \sigma^2 \left[-\frac{\eta}{s} \frac{n'\tilde{C}_{\mathbf{p}}}{T^2} \frac{(\tilde{p} \cdot u)^2}{12} + \frac{[n'\tilde{C}_{\mathbf{p}}]'\tilde{C}_{\mathbf{p}}}{4T^2} \frac{2(\tilde{p} \cdot u)^4}{15} + \frac{[n'\tilde{C}_{\mathbf{p}}]\tilde{C}_{\mathbf{p}}}{2T^2} \frac{(\tilde{p} \cdot u)^3}{3} \right] \\ & + \delta n^*_{\mathbf{p}(2)} \,. \end{aligned} \tag{27}$$

The first line in (27) does not contribute to $\pi^{\mu\nu}$, since all tensor structure from momentum integral are orthogonal to these terms in brackets. Besides the already fixed parameter c_0, we still need to determine second order corrections to flow velocity and temperature, by solving

Landau-Lifshitz matching condition. And we find that these following solutions satisfy all the constraint conditions, up to second order in gradient,

$$\delta u^{\mu}_{(2)} = -\frac{3}{B_4 T^2}\left[-\frac{\eta}{3s}B_1 + \frac{B_3}{12}\right]\left[2\left\langle\sigma^{\mu\nu}\nabla_{\nu} lnT\right\rangle + \left\langle\Delta^{\mu}_{\alpha}\nabla_{\beta}\sigma^{\alpha\beta}\right\rangle\right] \tag{28}$$

$$\delta T_{(2)} = -\frac{\sigma^2}{B_4 T}\left[-\frac{\eta}{s}\frac{B_1}{12} + \frac{B_2}{30} - \frac{B_3}{6}\right]. \tag{29}$$

As a by-product of the constraint condition Eq. (18), we also recognize that

$$\eta\tau_{\pi} = \frac{T^2 B_3}{15}, \quad \lambda_1 = \eta\tau_{\pi}\left[\frac{2B_2}{7B_3} + 1\right], \quad \lambda_2 = -2\eta\tau_{\pi}, \quad \lambda_3 = 0. \tag{30}$$

These relations are in the consistent range with those discussed in [17]. In particular, all these non-trivial second order transport coefficients are non-free parameters in our formalism, as long as η/s is fixed. This reduces the number of free parameters in hydrodynamical calculations, and thus quite significant for the recent study on the extraction of η/s[18].

4. Viscous Hydrodynamic Simulation and Elliptic Flow

The simulation of hydrodynamics in code needs solving hydrodynamical equations of motion. For viscous hydrodynamics, for the convenience of computation, the algorithm is introduced by approximation. In BRSSS formalism, up to second order in gradients, Eq. (6) can be rewritten as

$$\pi^{\mu\nu} = -\eta\sigma^{\mu\nu} - \tau_{\pi}\left[\langle D\pi^{\mu\nu}\rangle + \frac{4}{3}\pi^{\mu\nu}\nabla\cdot u\right] - \frac{\lambda_1}{\eta}\langle\pi_{\lambda}^{\ \mu}\sigma^{\lambda\nu}\rangle - \frac{\lambda_2}{\eta}\langle\pi_{\lambda}^{\ \mu}\Omega^{\lambda\nu}\rangle + O(\nabla^3), \tag{31}$$

such that a linearized equation for $\pi^{\mu\nu}$, which is approximately correct if all higher order gradients are neglected, is obtained. Then together with general hydrodynamical equations of motion, $\partial_{\mu}T^{\mu\nu}$=0, and appropriately selected equation of state, we have a complete set of equations for the unknown variables, $(\pi^{\mu\nu}, e, u^{\mu})$. There are two options for the equation of state in our calculations. For theoretical interest we can consider conformal equation of state(CEOS). Although CEOS does not have the accurate description for real medium expansion in heavy ion collisions, it has the conformal symmetry as assumed in BRSSS hydrodynamics. The previously used lattice equation of state(LEOS) by Romatschke and Luzum[9] is also tested. The details of the algorithm can be found in [13]. For the initial state, smooth initial condition is considered. As indicated by several groups[3, 19], event fluctuations in the calculations of v_2 can be safely ignored. A constant temperature freeze out scheme is taken, with T_{fo} = 150 MeV.

The effects of including consistent δf in freeze out procedure can be best seen in the results of elliptic flow. In Fig. 1 the results of a series of calculations are shown. Compare to ideal hydrodynamics, viscous hydrodynamics indeed damps the generated v_2, and a major fraction of the damping is from δf. This is reflected in the difference between the calculated v_2 with equilibrium distribution function n_p and that with $n_p + \delta f$'s. The complete and consistent calculation with respect to BRSSS hydrodynamics needs also $\delta f_{(2)}$ correction. However, in the quadratic ansatz limit and conformal equation of state the corresponding v_2(left panel in Fig. 1) quickly becomes unreliable when p_T goes up to 1.5 GeV. This is due to the dramatic increase of viscous corrections at large p_T region, and has already been discussed in the first order case with δf_{π}[10]. But now that $\delta f_{(2)}$ has higher $p_T(p_T^4)$ dependence, this effect is much stronger. In another way, we can estimate the magnitude of viscous corrections by calculating the spectrum. Then gradient expansion formalism fails when δf contributes to spectrum no

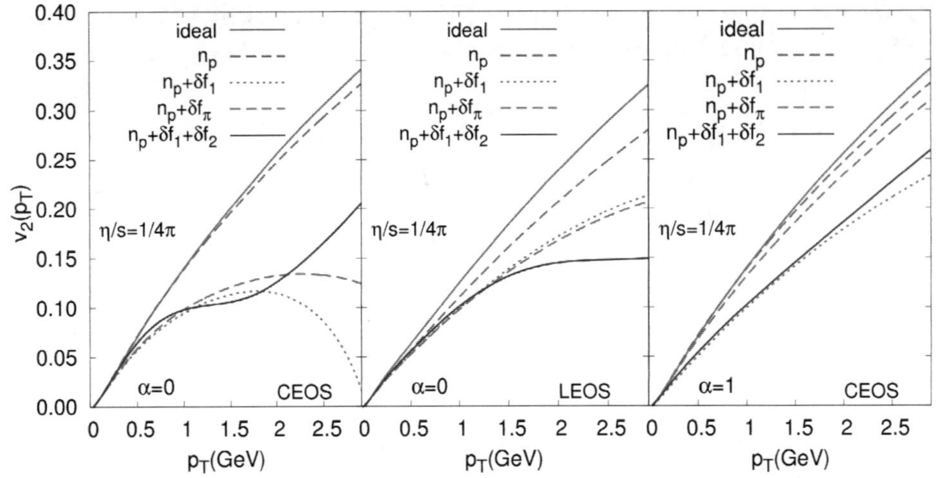

Figure 1. Differential v_2 results from various viscous hydrodynamical calculations, $b = 6.8$ fm.

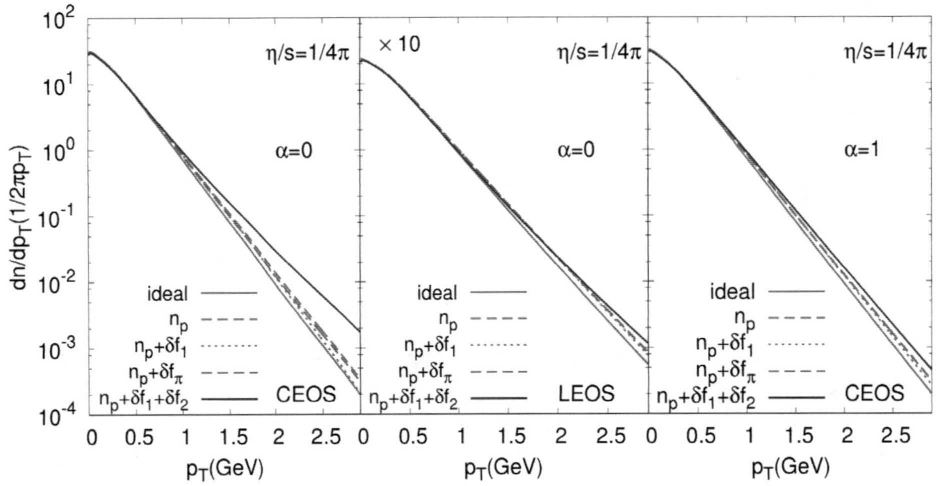

Figure 2. Spectrum from various viscous hydrodynamical calculations, $b = 6.8$ fm. Note that a factor of 10 has been multiplied in the case of LEOS to make it on the same scale.

longer perturbatively. As seen in the left panel in Fig. 2, we find similarly for CEOS and quadratic ansatz limit, $p_T > 1.5$GeV region is beyond the feasibility of our formalism.

Changing equation of state or the α dependence, however, can help solving the problem to some extent. As seen in the two other panels in Fig. 1 and Fig. 2, viscous corrections are smaller. On one hand these changes extend our reliable calculation to higher p_T region. On the other hand, realistic medium in heavy ion collisions has equation of state more close to LEOS, and also quadratic ansatz limit does not necessarily reflect the real microscopic dynamics. So as expected, second order corrections in δf lead to extra corrections to calculated observables.

5. Discussions and Future Work

We have presented a formalism for the derivation of viscous corrections to phase space distribution function at freeze out. With respect to BRSSS hydrodynamics, from this formalism we have obtained the consistent form of $\delta f_{(2)}$, and checked its impact on hydrodynamical simulations. $\delta f_{(2)}$ affects the results as corrections, and the corrections increase with p_T. But for LEOS or $\alpha = 1$, the corresponding corrections are as small as perturbations. This is expected from the gradient expansion of the formalism we are following. At last, this dependence on α verifies the claim in [15] that quadratic ansatz may be questionable in real calculations.

One important property of higher order viscous corrections in $f(x, \mathbf{p})$, is its higher p_T dependence. This plays a crucial role in the study of anisotropic flow for two aspects. 1. This makes the flow results more sensitive to η/s. 2. This makes higher order anisotropic flow v_n more sensitive. Both of these aspects have positive effects on the extraction of η/s from heavy ion collisions. And this will be the main subject of our future work.

Acknowledgements

This work is supported in part by the Sloan Foundation and by the Department of Energy through the Outstand Junior Investigator programm DE-FG-02-08ER4154.

References

[1] Alver B and Roland G 2010 *Phys.Rev.* **C81** 054905 (*Preprint* 1003.0194)
[2] Schenke B, Jeon S and Gale C 2011 *Phys.Rev.Lett.* **106** 042301 (*Preprint* 1009.3244)
[3] Qiu Z and Heinz U W 2011 *Phys.Rev.* **C84** 024911 (*Preprint* 1104.0650)
[4] Teaney D and Yan L 2011 *Phys.Rev.* **C83** 064904 (*Preprint* 1010.1876)
[5] Alver B H, Gombeaud C, Luzum M and Ollitrault J Y 2010 *Phys.Rev.* **C82** 034913 (*Preprint* 1007.5469)
[6] Teaney D 2003 *Phys.Rev.* **C68** 034913 (*Preprint* nucl-th/0301099)
[7] Danielewicz P and Gyulassy M 1985 *Phys.Rev.* **D31** 53–62
[8] Dusling K and Teaney D 2008 *Phys.Rev.* **C77** 034905 (*Preprint* 0710.5932)
[9] Luzum M and Romatschke P 2008 *Phys.Rev.* **C78** 034915 (*Preprint* 0804.4015)
[10] Song H and Heinz U W 2008 *Phys.Rev.* **C77** 064901 (*Preprint* 0712.3715)
[11] Romatschke P 2010 *Int.J.Mod.Phys.* **E19** 1–53 (*Preprint* 0902.3663)
[12] Baier R, Romatschke P, Son D T, Starinets A O and Stephanov M A 2008 *JHEP* **0804** 100 (*Preprint* 0712.2451)
[13] Teaney D and Yan L , In progress
[14] De Groot S, Van Leeuwen WA e and Van Weert C 1980
[15] Dusling K, Moore G D and Teaney D 2010 *Phys.Rev.* **C81** 034907 (*Preprint* 0909.0754)
[16] Policastro G, Son D and Starinets A 2001 *Phys.Rev.Lett.* **87** 081601 (*Preprint* hep-th/0104066)
[17] York M A and Moore G D 2009 *Phys.Rev.* **D79** 054011 (*Preprint* 0811.0729)
[18] Shen C, Bass S A, Hirano T, Huovinen P, Qiu Z *et al.* 2011 *J.Phys.G* **G38** 124045 (*Preprint* 1106.6350)
[19] Gardim F G, Grassi F, Luzum M and Ollitrault J Y 2012 *Phys.Rev.* **C85** 024908 (*Preprint* 1111.6538)

Journal of Physics: Conference Series **389** (2012) 012015 doi:10.1088/1742-6596/389/1/012015

Collective properties of nucleus-nucleus collisions from AGS to LHC energies

V P Konchakovski[1], V D Toneev[2], W Cassing[1], E L Bratkovskaya[3], S A Voloshin[4] and V Voronyuk[5]

[1] Institute for Theoretical Physics, University of Giessen, Giessen, Germany
[2] Joint Institute for Nuclear Research, Dubna, Russia
[3] Institute for Theoretical Physics, University of Frankfurt, Frankfurt, Germany
[4] Wayne State University, Detroit, Michigan, USA
[5] Bogolyubov Institute for Theoretical Physics, Kiev, Ukraine

E-mail: Wolfgang.Cassing@theo.physik.uni-giessen.de

Abstract. The azimuthal anisotropies of the collective transverse flow of charged hadrons are investigated in a wide range of heavy-ion collision energies within the microscopic Parton-Hadron-String Dynamics (PHSD) transport approach which incorporates explicit partonic degrees-of-freedom in terms of strongly interacting quasiparticles (quarks and gluons) in line with an equation-of-state from lattice QCD as well as the dynamical hadronization and hadronic collision dynamics in the final reaction phase. The experimentally observed increase of the elliptic flow v_2 of charged hadrons with collision energy is successfully described in terms of the PHSD approach. The analysis of higher-order harmonics v_3 and v_4 in the azimuthal angular distribution shows a similar tendency of growing deviations between partonic and purely hadronic models with increasing collision energy. This demonstrates that the excitation functions of azimuthal anisotropies reflect the increasing role of quark-gluon degrees of freedom in the early phase of relativistic heavy-ion collisions. Furthermore, the specific variation of the ratio $v_4/(v_2)^2$ with respect to bombarding energy, centrality and transverse momentum is found to provide valuable information on the underlying partonic dynamics.

1. Introduction

The discovery of large azimuthal anisotropic flow at the Relativistic-Heavy-Ion-Collider (RHIC) provides a conclusive evidence for the creation of dense partonic matter in ultra-relativistic nucleus-nucleus collisions. With sufficiently strong parton interactions, the medium in the collision zone can be expected to achieve local equilibrium and exhibit approximately hydrodynamic flow [1, 2, 3]. The momentum anisotropy is generated due to pressure gradients of the initial "almond-shaped" collision zone produced in noncentral collisions [1, 2]. The azimuthal pressure gradient extinguishes itself soon after the start of the hydrodynamic evolution, so the final flow is only weakly sensitive to later stages of the fireball evolution. The pressure gradients have to be large enough to translate an early asymmetry in density of the initial state to a final-state momentum-space anisotropy. In these collisions a new state of strongly interacting matter is created, being characterized by a very low shear viscosity η to entropy density s ratio, η/s, close to a nearly perfect fluid [4, 5, 6]. Lattice QCD (lQCD) calculations [7, 8, 9] indicate that a crossover region between hadron and quark-gluon matter should have been reached in these experiments.

An experimental manifestation of this collective flow is the anisotropic emission of charged particles in the plane transverse to the beam direction. This anisotropy is described by the different flow parameters defined as the proper Fourier coefficients v_n of the particle distributions in azimuthal angle ψ with respect to the reaction plane angle Ψ_{RP}. At the highest RHIC collision energy of $\sqrt{s_{NN}} = 200$ GeV, differential elliptic flow measurements $v_2(p_T)$ have been reported for a broad range of centralities or number of participants N_{part}. For N_{part} estimates, the geometric fluctuations associated with the positions of the nucleons in the collision zone serve as the underlying origin of the initial eccentricity fluctuations. These data are found to be in accord with model calculations that an essentially locally equilibrated quark gluon plasma (QGP) has little or no viscosity [10, 11, 12]. Collective flow continues to play a central role in characterizing the transport properties of the strongly interacting matter produced in heavy-ion collisions at RHIC. Particle anisotropy measurements are considered as key observables for a reliable extraction of transport coefficients.

It was shown before that higher-order anisotropy harmonics, in particular v_4, may provide a more sensitive constraint on the magnitude of η/s and the freeze-out dynamics, and the ratio $v_4/(v_2)^2$ might indicate whether a full local equilibrium is achieved in the QGP [13]. The role of fluctuations and so-called 'nonflow' correlations are important for such measurements. It is well established that initial eccentricity fluctuations significantly influence the magnitudes of $v_{2,4}$ [14, 15]. However, the precise role of nonflow correlations, which lead to a systematic error in the determination of $v_{2,4}$, is less clear. Recently, significant attention has been given to the study of the influence of initial geometry fluctuations on higher order eccentricities $\epsilon_n (n \geq 3)$ for a better understanding of how such fluctuations manifest themselves in the harmonic flow correlations characterized by v_n. Even more, it was proposed that the analysis of v_n^2 for all values of n can be considered as an analogous measurement to the Power Spectrum extracted from the Cosmic Microwave Background Radiation providing a possibility to observe superhorizon fluctuations [16].

A large number of anisotropic flow measurements have been performed by many experimental groups at SIS, AGS, SPS and RHIC energies over the last twenty years. Very recently, the azimuthal asymmetry has been measured also at the LHC [17]. However, the fact that these data have not been obtained under the same experimental conditions as at RHIC experiments, does not directly allow for a detailed and meaningful comparison in most cases. The experimental differences include: different centrality selection, different transverse momentum acceptance, different particle species, different rapidity coverage and different methods for flow analysis as pointed out in Ref. [18].

The Beam-Energy-Scan (BES) program proposed at RHIC [19] covers the energy interval from $\sqrt{s_{NN}} = 200$ GeV, where partonic degrees of freedom play a decisive role, down to the AGS energy of $\sqrt{s_{NN}} \approx 5$ GeV, where most experimental data may be described successfully in terms of hadronic degrees-of-freedom, only. Lowering the RHIC collision energy and studying the energy dependence of anisotropic flow allows to search for the possible onset of the transition to a phase with partonic degrees-of-freedom at an early stage of the collision as well as possibly to identify the location of the critical end-point that terminates the cross-over transition at small quark-chemical potential to a first order phase transition at higher quark-chemical potential [20, 21].

This contribution aims to summarize excitation functions for different harmonics of the charged particle anisotropy in the azimuthal angle at midrapidity in a wide transient energy range, *i.e.* from the AGS to the top RHIC energy. The first attempts to explain the preliminary STAR data with respect to the observed increase of the elliptic flow v_2 with the collision energy have failed since the traditional available models did not allow to clarify the role of the partonic phase [22]. In this contribution we investigate the energy behavior of different flow coefficients, their scaling properties and differential distributions (cf. Ref. [23, 24]). Our analysis of the

Journal of Physics: Conference Series **389** (2012) 012015 doi:10.1088/1742-6596/389/1/012015

STAR/PHENIX RHIC data – based on recent results of the BES program – will be performed within the Parton-Hadron-String Dynamics (PHSD) transport model [25] that includes explicit partonic degrees-of-freedom as well as a dynamical hadronization scheme for the transition from partonic to hadronic degrees-of-freedom and vice versa. For more detailed descriptions of PHSD and its ingredients we refer the reader to Refs. [26, 27, 28, 29].

2. Results for collective flows

We directly continue with the results from PHSD in comparison with other approaches and the available experimental data.

2.1. Elliptic flow

The largest component, known as elliptic flow v_2, is one of the early observations at RHIC [30]. The elliptic flow coefficient is a widely used quantity characterizing the azimuthal anisotropy of emitted particles,

$$v_2 =< cos(2\psi - 2\Psi) >=< \frac{p_x^2 - p_y^2}{p_x^2 + p_y^2} > \ , \tag{1}$$

where Ψ_{RP} is the azimuth of the reaction plane, p_x and p_y are the x and y component of the particle momenta and the brackets denote averaging over particles and events. This coefficient can be considered as a function of centrality, pseudorapidity η and/or transverse momentum p_T. We note that the reaction plane in PHSD is given by the $(x - z)$ plane with the z-axis in beam direction. The reaction plane is defined as a plane containing the beam axes and the impact parameter vector.

We recall that at high bombarding energies the longitudinal size of the Lorentz contracted nuclei becomes negligible compared to its transverse size. The forward shadowing effect then becomes negligible and the elliptic flow fully develops in-plane, leading to a positive value of the average flow v_2 since no shadowing from spectators takes place. In Fig. 1 (l.h.s.) the experimental v_2 data compilation for the transient energy range is compared to the results from HSD calculations and further available model results as included in Ref. [22]. The centrality selection is the same for the data and the various models.

In order to interpret the results in Fig. 1 we have to recall the various ingredients of the models employed for comparison. The UrQMD (Ultra relativistic Quantum Molecular Dynamics) model is a microscopic transport theory based on the relativistic Boltzmann equation [31, 32]. It allows for the on-shell propagation of all hadrons along classical trajectories in combination with stochastic binary scattering, color string formation and resonance decay. The model incorporates baryon-baryon, meson-baryon and meson-meson interactions based on experimental data (when possible). This Boltzmann-like hadronic transport model has been employed for proton-nucleus and nucleus-nucleus collisions from AGS to RHIC energies [32]. The comparison of the data on v_2 to those from the UrQMD model will thus essentially provide information on the contribution from the hadronic phase. As seen in Fig. 1, being in agreement with data at the lowest energy $\sqrt{s_{NN}}$ = 9.2 GeV, the UrQMD model results then either remain approximately constant or decrease slightly with increasing $\sqrt{s_{NN}}$; UrQMD thus does not reproduce the rise of v_2 with the collision energy as seen experimentally.

The HSD model [33, 34, 35] is also a hadron-string model including formally the same processes as UrQMD. However, being based on the off-shell generalized transport equation [29] followed from Kadanoff-Baym approach, the quasiparticles in the HSD model take into account in-medium modifications of their properties in the nuclear environment which is rather essential for many observables and in particular for dileptons. Detailed comparisons between HSD and UrQMD for central Au+Au (Pb+Pb) collisions have been reported in Ref. [36] from AGS to top SPS energies with respect to a large experimental data set. Indeed, both hadronic approaches

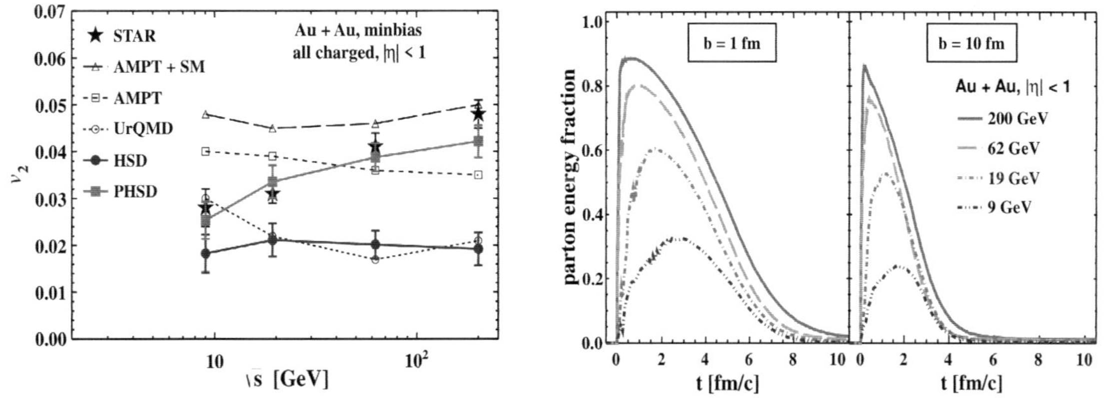

Figure 1. (l.h.s.) The average elliptic flow v_2 of charged particles at midrapidity for minimum bias collisions at $\sqrt{s_{NN}} = 9.2$, 19.6, 62.4 and 200 GeV (stars) is taken from the data compilation of Ref. [22]). The corresponding results from different models are compared to the data and explained in more detail in the text. (r.h.s.) Evolution of the parton fraction of the total energy density at midrapidity (from PHSD) for different collision energies at impact parameters $b = 1$ fm and 10 fm.

yield similar results on the level of 20-30% which is also the maximum deviation from the data sets. Accordingly, the HSD model also predicts an approximately energy-independent flow v_2 in quite close agreement with the UrQMD results. We may thus conclude that the rise of v_2 with bombarding energy is not do to hadronic interactions and models with partonic degrees-of-freedom have to be addressed.

The AMPT (A Multi Phase Transport model) [37, 38] uses initial conditions of a perturbative QCD (pQCD) inspired model which produces multiple minijet partons according to the number of binary initial nucleon-nucleon collisions. These (massless) minijet partons undergo scattering (without potentials) before they are allowed to fragment into hadrons. The string melting (SM) version of the AMPT model (labeled in Fig. 1 as AMPT-SM) is based on the idea that the existence of strings (or hadrons) is impossible for energy densities beyond a critical value of $\varepsilon \sim 1$ GeV/fm^3. Hence they melt the strings to (massless) partons. This is done by converting the mesons to a quark and anti-quark pair, baryons to three quarks *etc.* fulfilling energy-momentum conservation. The subsequent scattering of the quarks are based on a parton cascade with (adjustable) effective cross sections which are significantly larger than those from pQCD [37, 38]. Once the partonic interactions terminate, the partons hadronize through the mechanism of parton coalescence.

We find from Fig. 1 that the interactions between the minijet partons in the AMPT model indeed increase the elliptic flow significantly as compared to the hadronic models UrQMD and HSD. An additional inclusion of interactions between partons in the AMPT-SM model gives rise to another 20% of v_2 bringing it into agreement (for AMPT-SM) with the data at the maximal collision energy. So, both versions of the AMPT model indicate the importance of partonic contributions to the observed elliptic flow v_2 but do not reproduce its growth with $\sqrt{s_{NN}}$. The authors address this result to the partonic-equation-of state (EoS) employed which corresponds to a massless and noninteracting relativistic gas of particles. This EoS deviates severely from the results of lattice QCD calculations for temperatures below 2-3 T_c. Accordingly, the degrees-of-freedom are propagated without self-energies and a parton spectral function.

The PHSD approach incorporates the latter medium effects in line with a lQCD equation-of-

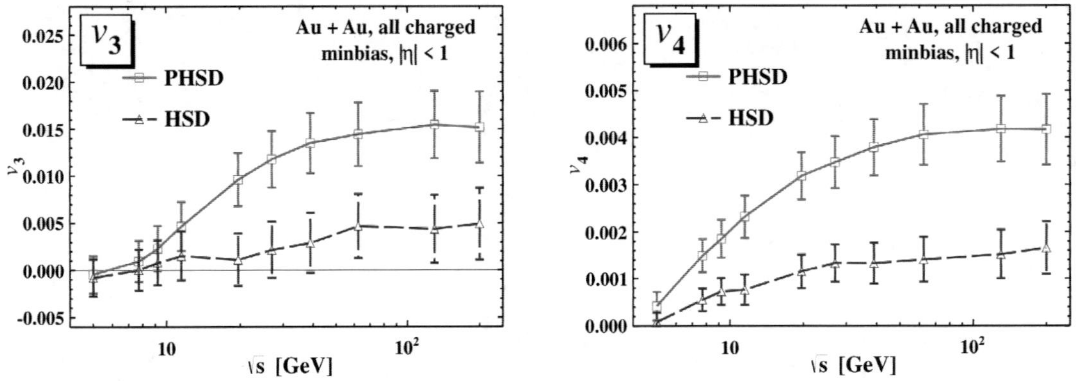

Figure 2. Average anisotropic flows v_3 (l.h.s.) and v_4 (r.h.s.) of charged particles at mid-pseudorapidity for minimum bias Au + Au collisions calculated within the PHSD (solid lines, red) and HSD (dashed lines, blue) models.

state and also includes a dynamical hadronization scheme based on covariant transition rates. As has been demonstrated in Refs. [23, 24] and explicitly shown in Fig. 1 (l.h.s.), the elliptic flow v_2 from PHSD (red line) agrees with the data from the STAR collaboration and clearly shows an increase with bombarding energy. Note that PHSD and AMPT-SM practically give the same elliptic flow at the top RHIC energy of $\sqrt{s_{NN}} = 200$ GeV.

An explanation for the increase in v_2 with collision energy is provided in Fig. 1 (r.h.s.). Here we show the partonic fraction of the energy density with respect to the total energy where the energy densities are calculated at mid-rapidity. As discussed above the main contribution to the elliptic flow is coming from an initial partonic stage at high $\sqrt{s}$. The fusion of partons to hadrons or, inversely, the melting of hadrons to partonic quasiparticles occurs when the local energy density is about $\varepsilon \approx 0.5$ GeV/fm^3. As follows from Fig. 1, the parton fraction of the total energy goes down substantially with decreasing bombarding energy while the duration of the partonic phase is roughly the same. The maximal fraction reached is the same in central and peripheral collisions but the parton evolution time is shorter in peripheral collisions. One should recall again the important role of the repulsive mean-field for partons in the PHSD model that leads to an increase of the flow v_2 with respect to HSD predictions (cf. also Ref. [39]). We point out in addition that the increase of v_2 in PHSD relative to HSD is also partly due to the higher interaction rates in the partonic medium because of a lower ratio of η/s for partonic degrees-of-freedom at energy densities above the critical energy density than for hadronic media below the critical energy density [40, 41]. The relative increase in v_3 and v_4 in PHSD essentially is due to the higher partonic interaction rate and thus to a lower ratio η/s in the partonic medium which is mandatory to convert initial spacial anisotropies to final anisotropies in momentum space [42].

2.2. Higher-order flow harmonics
Depending on the location of the participant nucleons in the nucleus at the time of the collision, the actual shape of the overlap area may vary: the orientation and eccentricity of the ellipse defined by the participants fluctuates from event to event. Note, however, that by averaging over many events an almond shape is regained for the same impact parameter.

Recent studies suggest that fluctuations in the initial state geometry can generate higher-order flow components [10, 16, 43, 44]. The azimuthal momentum distribution of the emitted

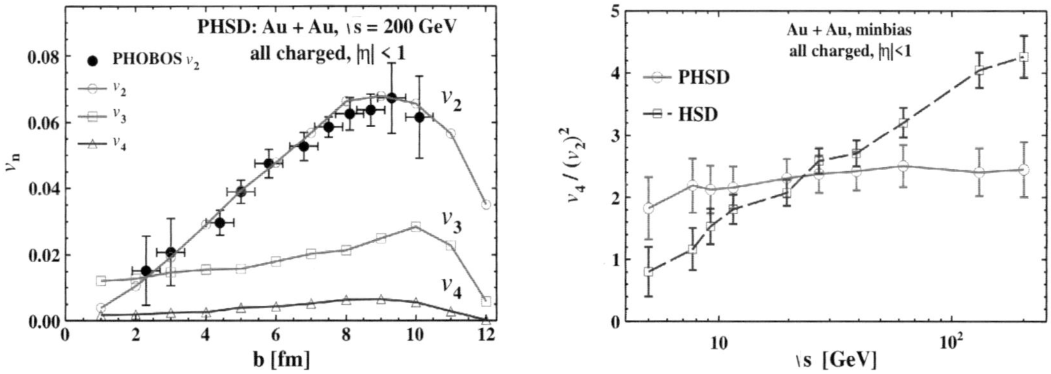

Figure 3. (l.h.s.) Impact parameter dependence of anisotropic flows of charged particles at mid-pseudorapidity for minimum bias collisions of Au+Au at $\sqrt{s_{NN}} = 200$ GeV. Experimental points are from Ref. [48]. (r.h.s.) Beam energy dependence of the ratio $v_4/(v_2)^2$ for Au+Au collisions. The solid and dashed curves are calculated within the PHSD and HSD models, respectively.

particles is commonly expressed in the form of Fourier series as

$$E\frac{d^3N}{d^3p} = \frac{d^2N}{2\pi p_T dp_T dy}\left(1 + \sum_{n=1}^{\infty} 2v_n(p_T)\cos(n(\psi - \Psi_n))\right), \tag{2}$$

where v_n is the magnitude of the n-th order harmonic term relative to the angle of the initial-state spatial plane of symmetry Ψ_n. The anisotropy in the azimuthal angle ψ is usually characterized by the even order Fourier coefficients with the reaction plane $\Psi_n = \Psi_{RP}$: $v_n = \langle \exp(\imath n(\psi - \Psi_{RP}))\rangle$ $(n = 2, 4, ...)$, since for a smooth angular profile the odd harmonics vanish. For the odd components, e.g. v_3, one should take into account event-by-event fluctuations with respect to the participant plane $\Psi_n = \Psi_{PP}$. We calculate the v_3 coefficients with respect to Ψ_3 as: $v_3\{\Psi_3\} = \langle\cos(3[\psi - \Psi_3])\rangle/Res(\Psi_3)$. The event plane angle Ψ_3 and its resolution $Res(\Psi_3)$ are calculated as described in Ref. [45] via the two-sub-events method [46, 47].

In Fig. 2 we display the PHSD and HSD results for the anisotropic flows v_3 and v_4 of charged particles at mid-pseudorapidity for Au+Au collisions as a function of $\sqrt{s_{NN}}$. The pure hadronic model HSD gives $v_3 \approx 0$ for all energies. Accordingly, the results from PHSD (dashed red line) are systematically larger than from HSD (dashed blue line). Unfortunately, our statistics are not good enough to allow for more precise conclusions. The hexadecupole flow v_4 stays almost constant in the energy range $\sqrt{s_{NN}} \geq 10$ GeV; at the same time the PHSD gives noticeably higher values than HSD which we attribute to the higher interaction rate in the partonic phase, i.e. a lower ratio of η/s in the partonic phase [40, 41].

Alongside with the integrated flow coefficients v_n the PHSD model reasonably describes their distribution over centrality or impact parameter b. A specific comparison at $\sqrt{s_{NN}} = 200$ GeV is shown in Fig. 3 for v_2, v_3 and v_4. While v_2 increases strongly with b up to peripheral collisions, v_3 and v_4 are only weakly sensitive to the impact parameter. The triangular flow is always somewhat higher than the hexadecupole flow in the whole range of impact parameters b.

2.3. Ratios of different harmonics

Different harmonics can be related to each other. In particular, hydrodynamics predicts that $v_4 \propto (v_2)^2$ [49]. The simplest prediction that $v_4 = 0.5(v_2)^2$ is given for a boosted thermal freeze-out distribution of an ideal fluid, Ref. [50]. In this work it was noted also that v_4 is largely

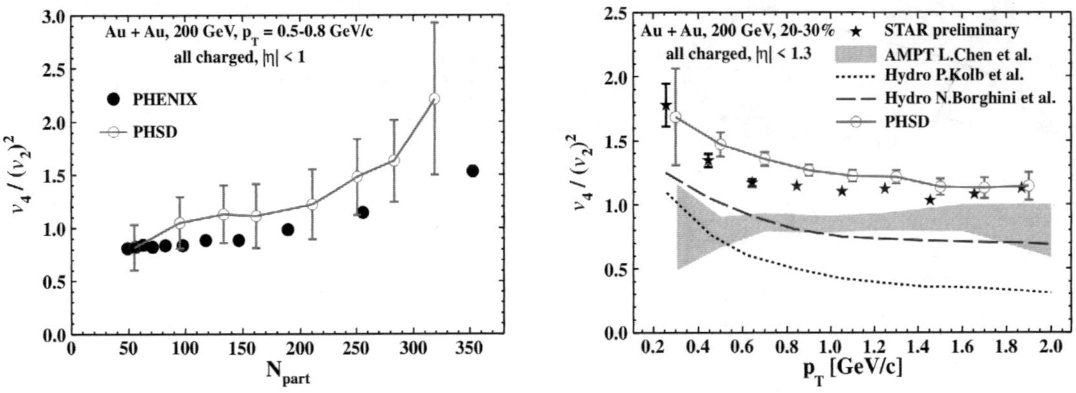

Figure 4. (l.h.s.) Participant number dependence of the $v_4/(v_2)^2$ ratio of charged particles for Au+Au ($\sqrt{s_{NN}}$ = 200 GeV) collisions. The experimental data points for 0.5< p_T <0.8 GeV/c are from Ref. [51].
(r.h.s.) Transverse momentum dependence of the ratio $v_4/(v_2)^2$ of charged particles for Au+Au (at $\sqrt{s_{NN}}$ = 200 GeV) collisions. The dashed and dot-dashed lines are calculated within the hydrodynamic approaches from Refs. [49] and [50], respectively. The shaded region corresponds to the results from the AMPT model [53]. The experimental data points are from the STAR Collaboration [52].

generated by an intrinsic elliptic flow (at least at high p_T) rather than the fourth order moment of the fluid flow. This is a motivation for studying the ratio $v_4/(v_2)^2$ rather than v_4 alone. As is seen in Fig. 4 (r.h.s.), indeed the ratio calculated within the PHSD model is practically constant in the whole range of $\sqrt{s_{NN}}$ considered but significantly deviates from the ideal fluid estimate of 0.5. In contrast, neglecting dynamical quark-gluon degrees-of-freedom in the HSD model, we obtain a monotonous growth of this ratio.

The dependence of the $v_4/(v_2)^2$ ratio versus the number of participants N_{part} is shown in Fig. 4 for charged particles produced in Au + Au collisions at $\sqrt{s_{NN}}$ = 200 GeV. The PHSD results are roughly in agreement with the experimental data points from Ref. [52] but overshoot them for $N_{part} \sim 250$.

As pointed out before, the ratio $v_4/(v_2)^2$ is sensitive to the microscopic dynamics. In this respect we show the transverse momentum dependence of the ratio $v_4/(v_2)^2$ in Fig. 4 for charged particles produced in Au+Au collisions at $\sqrt{s_{NN}}$ = 200 GeV (20-30% centrality). The PHSD results are quite close to the experimental data points from Ref. [52], however, overestimate the measurements by up to 20%. The hydrodynamic results – plotted in the same figure – significantly underestimate the experimental data and noticeably depend on viscosity. The partonic AMPT model [53] discussed above also predicts a slightly lower ratio than the measured one, however, being in agreement with both hydrodynamic models for p_T 0.8 GeV/c. Our interpretation of Fig. 4 (r.h.s.) is as follows: the data are not compatible with ideal hydrodynamics and a finite shear viscosity is mandatory (in viscous hydrodynamics) to come closer the experimental observations. The kinetic approaches AMPT and PHSD perform better but either overestimate (in AMPT) or slightly underestimate the scattering rate of soft particles (in PHSD). An explicit study of the centrality dependence of these ratios should provide further valuable information.

Journal of Physics: Conference Series **389** (2012) 012015 doi:10.1088/1742-6596/389/1/012015

3. Conclusions

In summary, relativistic collisions of Au+Au from $\sqrt{s_{NN}} = 5$ to 200 GeV have been studied within the PHSD approach which includes the dynamics of explicit partonic degrees-of-freedom as well as dynamical local transition rates from partons to hadrons and also the final hadronic scatterings. Whereas earlier studies have been carried out for longitudinal rapidity distributions of various hadrons, their transverse mass spectra and the elliptic flow v_2 as compared to available data at SPS and RHIC energies [25, 26], here we have focussed on the PHSD results for the collective flow coefficients v_2, v_3 and v_4 in comparison to recent experimental data in the large energy range from the RHIC Beam-Energy-Scan (BES) program as well as different theoretical approaches ranging from hadronic transport models to ideal and viscous hydrodynamics. We mention explicitly that the PHSD model from Ref. [26] has been used for all calculations performed in this study and no tuning (or change) of model parameters has been performed.

We have found that the anisotropic flows – elliptic v_2, triangular v_3, hexadecapole v_4 – are reasonably described within the PHSD model in the whole transient energy range naturally connecting the hadronic processes at lower energies with ultrarelativistic collisions where the quark-gluon degrees of freedom become dominant. The smooth growth of the elliptic flow v_2 with the collision energy demonstrates the increasing importance of partonic degrees of freedom. This feature is reproduced by neither hadron-string based kinetic models nor A Multi Phase Transport (AMPT) model treating the partonic phase in a simplified manner. Other signatures of the transverse collective flow, the higher-order harmonics of the transverse anisotropy v_3 and v_4 change only weakly from $\sqrt{s_{NN}} \sim 7$ GeV to the top RHIC energy of $\sqrt{s_{NN}} = 200$ GeV, roughly in agreement with experiment. As shown in this study, this success is related to a consistent treatment of the interacting partonic phase in PHSD whose fraction increases with the collision energy.

The analysis of correlations between particles emitted in ultrarelativistic heavy-ion collisions at large relative rapidity has revealed an azimuthal structure that can be interpreted as solely due to collective flow [54, 55, 56, 57]. This interesting new phenomenon, denoted as triangular flow, results from initial state fluctuations and a subsequent hydrodynamic-like evolution. Unlike the usual directed flow, this phenomenon has no correlation with the reaction plane and should depend weakly on rapidity. Event-by-event hydrodynamics [58] has been a natural framework for studying this triangular collective flow but it has been of interest also to investigate these correlations in terms of the PHSD model. We have found the third harmonics to increase steadily in PHSD with bombarding energy. The coefficient v_3 is compatible with zero for $\sqrt{s_{NN}} > 20$ GeV in case of the hadronic transport model HSD which does not develop 'ridge-like' correlations. In this energy range PHSD gives a positive v_3 due to dominant partonic interactions.

Different harmonics can be related to each other and in particular, hydrodynamics predicts that $v_4 \propto (v_2)^2$ [49]. In this work it was noted also that v_4 is largely generated by an intrinsic elliptic flow (at least at high p_T) rather than the fourth order moment of the fluid flow. Indeed, the ratio $v_4/(v_2)^2$ calculated within the PHSD model is approximately constant in the whole considered range of $\sqrt{s_{NN}}$ but significantly deviates from the ideal fluid estimate of 0.5. In contrast, neglecting dynamical quark-gluon degrees-of-freedom in the HSD model, we obtain a monotonous growth of this ratio.

The transverse momentum dependence of the ratio $v_4/(v_2)^2$ at the top RHIC energy has given further interesting information (*cf.* Fig. 4) by comparing the various model results to the data from STAR which are interpreted as follows: the STAR data are not compatible with ideal hydrodynamics and a finite shear viscosity is mandatory (in viscous hydrodynamics) to come closer the experimental ratio observed. The kinetic approaches AMPT and PHSD perform better but either overestimate (in AMPT) or slightly underestimate the scattering rate of soft particles (in PHSD). Our findings at LHC energies are close to those at top RHIC energy.

Journal of Physics: Conference Series **389** (2012) 012015 doi:10.1088/1742-6596/389/1/012015

References

[1] Ollitrault J Y 1992 *Phys. Rev.* D **46** 229
[2] Heinz U and Kolb P 2002 *Nucl. Phys.* A **702** 269
[3] Shuryak E V 2009 *Prog. Part. Nucl. Phys.* **62** 48
[4] Shuryak E V 2005 *Nucl. Phys.* A **750** 64
[5] Gyulassy M and McLerran L 2005 *Nucl. Phys.* A **750** 30
[6] Peshier A and Cassing W 2005 *Phys. Rev. Lett.* **94** 172301
[7] Aoki Y *et al.* 2006 *Nature* **443** 675
[8] Aoki Y *et al.* 2006 *Phys. Lett.* B **643** 46
[9] Borsanyi S *et al.* 2010 *JHEP* **1009** 073
[10] Adare A *et al.* 2007 *Phys. Rev. Lett.* **98** 172301
[11] Romatschke P and Romatschke U 2007 *Phys. Rev. Lett.* **99** 172301
[12] Xu Z, Greiner C and Stöcker H 2008 *Phys. Rev. Lett.* **101** 082302
[13] Bhalerao R S *et al.* 2005 *Phys. Lett.* B **627** 49
[14] Hama Y *et al.* 2008 *Phys. Atom. Nucl.* **71** 1558
[15] Alver B *et al.* 2007 *Phys. Rev. Lett.* **98** 242302
[16] Mishra A P *et al.* 2008 *Phys. Rev.* C **77** 064902
[17] Aamodt K *et al.* 2010 *Phys. Rev. Lett.* **105** 252302
[18] Taranenko A 2011 *arXiv:1101.5069*
[19] Abelev B I *et al.* 2010 *Phys. Rev.* C **81** 024911
[20] Lacey R A *et al.* 2007 *Phys. Rev. Lett.* **98** 092301
[21] Aggarwal M M *et al.* 2010 *arXiv:1007.2613*
[22] Nasim M, Kumar L, Netrakanti P K and Mohanty B 2010 *Phys. Rev.* C **82** 054908
[23] Konchakovski V P *et al.* 2012 *Phys. Rev.* C **85** 011902
[24] Konchakovski V P *et al.* 2012 *Phys. Rev.* C **85** 044922
[25] Cassing W and Bratkovskaya E L 2009 *Nucl. Phys.* A **831** 215
[26] Bratkovskaya E L, Cassing W, Konchakovski V P and Linnyk O 2011 *Nucl. Phys.* A **856** 162
[27] Cassing W 2007 *Nucl. Phys.* A **791** 365
[28] Cassing W 2007 *Nucl. Phys.* A **795** 70
[29] Cassing W 2009 *Eur. J. Phys.* **168** 3
[30] Ackermann K H *et al.* 2001 *Phys. Rev. Lett.* **86** 402
[31] Cassing W *et al.* 1990 *Phys. Rep.* **188** 363
[32] Bass S A *et al.* 1998 *Prog. Part. Nucl. Phys.* **41** 255
[33] Bratkovskaya E L, Cassing W and Stöcker H 2003 *Phys. Rev.* C **67** 054905
[34] Ehehalt W and Cassing W 1996 *Nucl. Phys.* A **602** 449
[35] Cassing W and Bratkovskaya E L 1999 *Phys. Rep.* **308** 65
[36] Bratkovskaya E L *et al.* 2004 *Phys. Rev.* C **69** 054907
[37] Lin Z W and Ko C M 2002 *Phys. Rev.* C **65** 034904
[38] Lin Z W, Ko C M, Li, B A Zhang B and Pal S 2005 *Phys. Rev.* C **72** 064901
[39] Cassing W and Bratkovskaya E L 2008 *Phys. Rev.* C **78** 034919
[40] Mattiello S and Cassing W 2010 *Eur. Phys. J.* C **70** 243
[41] Demir N and Bass S A 2009 *Phys. Rev. Lett.* **102** 172302
[42] Petersen H, Coleman-Smith C, Bass S A and Wolpert R 2011 *J. Phys.* G **38** 045102
[43] Petersen H and Bleicher M 2010 *Phys. Rev.* C **81** 044906
[44] Alver B and Roland G 2010 *Phys. Rev.* C **81** 054905
[45] Adare A *et al.* 2011 *Phys. Rev. Lett.* **107** 252301
[46] Poskanzer A M and Voloshin S A 1998 *Phys. Rev.* C **58** 1671
[47] Bilandzic A, Snellings R and Voloshin S A 2011 *Phys. Rev.* C **83** 044913
[48] Back B B *et al.* 2005 *Phys. Rev.* C **72** 051901
[49] Kolb P F 2003 *Phys. Rev.* C **68** 031902(R)
[50] Borghini N and Ollitrault J Y 2006 *Phys. Lett.* B **642** 227
[51] Gong X Y *et al.* 2011 *J. Phys.* G **38** 124146
[52] Bai Y *et al.* 2007 *J. Phys.* G **34** S903
[53] Chen L W, Ko C M and Lin Z W 2004 *Phys. Rev.* C **69** 031901
[54] Xu J and Ko C M 2011 *Phys. Rev.* C **84** 014903
[55] Schenke B, Jeon S and Gale C 2011 *Phys. Rev. Lett.* **106** 042301
[56] Teaney D and Yan L 2011 *Phys. Rev.* C **83** 064904
[57] Luzum M, Gombeaud C and Ollitrault J Y 2010 *Phys. Rev.* C **81** 054910
[58] Gardim F G, Grassi F, Hama Y, Luzum M and Ollitrault J Y 2011 *Phys. Rev.* C **83** 064901

Dilepton production from SIS to LHC energies

E L Bratkovskaya[1,2], O Linnyk[3], V P Konchakovski[3], W Cassing[3], V Ozvenchuk[2], J Manninen[1,2] and C M Ko[4]

[1] Institute for Theoretical Physics, University of Frankfurt, Frankfurt, Germany
[2] Frankfurt Institute for Advanced Studies, 60438 Frankfurt am Main
[3] Institute for Theoretical Physics, University of Giessen, Giessen, Germany
[4] Texas A& M Universtity, Texas, U.S.A.

E-mail: Elena.Bratkovskaya@th.physik.uni-frankfurt.de

Abstract. We study e^+e^- pair production in proton-proton and in nucleus-nucleus collisions from SIS to LHC energies within the parton-hadron-string dynamics (PHSD) approach which incorporates explicit partonic degrees-of-freedom in terms of strongly interacting quasiparticles (quarks and gluons) in line with an equation-of-state from lattice QCD as well as the dynamical hadronization and hadronic collision dynamics in the final reaction phase. We find a visible in-medium effect in the low mass dilepton sector from dynamical vector-meson spectral functions from SIS to SPS energies whereas at RHIC and LHC energies such medium effects become more moderate. In the intermediate mass regime from 1.1 to 3 GeV pronounced traces of the partonic degrees of freedom are found at SPS and RHIC energies which superseed the hadronic (multi-meson) channels as well as the correlated and uncorrelated semi-leptonic D-meson decays. The dilepton production from the strongly interacting quark gluon plasma (sQGP) becomes already visible at top SPS energies and more pronounced at RHIC and LHC energies.

1. Introduction

Dileptons, i.e. correlated electron and positron pairs, are one of the key observables in ultra-relativistic nuclear collisions experiments since dileptons are emitted during the whole collision evolution and thus one may probe various aspects at different stages of a relativistic nuclear collision by measuring differential dilepton spectra. Another important feature is that the produced leptons interact only electromagnetically and thus interact only very weakly with the strongly interacting partonic or hadronic medium created in the collisions. In another words, e.g. an initial state Drell-Yan-pair from the early stages of the collision is expected to survive the subsequent evolution of the fireball.

An important dilepton observable is the invariant mass spectrum. The mass spectrum can be roughly divided in 3 different regions, in each of which different physics dominates the radiation. In the low mass region ($M_{e^+e^-} < 1$ GeV) the radiation is dominated by the decays of light mesons (consisting of u, d and s (anti)quarks) and especially gives information about in-medium properties of the ρ^0 meson. In the intermediate mass region (1.1 GeV $< M_{e^+e^-} <$ 3 GeV) the dominant hadronic contribution to the invariant mass spectrum is obtained from the decays of open charm mesons, while above the J/ψ peak, first open beauty decays and later on initial state Drell-Yan radiation are expected to dominate the dilepton spectrum. On top of the previously mentioned sources, especially in the intermediate mass region, also radiation from the strongly interacting Quark-Gluon-Plasma (sQGP) can give a significant signal [1] as well as some other more exotic sources like simultaneous interactions of four pions [2, 3],[4, 5]. These partonic and hadronic channels have been studied in detail in Refs. [6, 7] at the top Super-Proton-Synchrotron (SPS) and Relativistic-Heavy-Ion-Collider (RHIC) energies and it has been found that the partonic channels clearly dominate over multi-pion sources in the intermediate dilepton mass regime. This contribution aims to summarize the perspectives of dilepton measurements from low SIS to high LHC energies based on the Parton-Hadron-String Dynamics (PHSD) transport model [8].

2. The PHSD approach

The dynamics of partons, hadrons and strings in relativistic nucleus-nucleus collisions is analyzed here within the Parton-Hadron-String Dynamics approach [8]. In this transport approach the partonic dynamics is based on Kadanoff-Baym equations for Green functions with self-energies from the Dynamical QuasiParticle Model (DQPM) [9] which describes QCD properties in terms of 'resummed' single-particle Green functions. In Ref. [10], the actual three DQPM parameters for the temperature-dependent effective coupling were fitted to the recent lattice QCD results of Ref. [11]. The latter lead to a critical temperature $T_c \approx 160$ MeV which corresponds to a critical energy density of $\epsilon_c \approx 0.5$ GeV/fm^3. In PHSD the parton spectral functions ρ_j $(j = q, \bar{q}, g)$ are no longer $\delta-$ functions in the invariant mass squared as in conventional cascade or transport models but depend on the parton mass and width parameters which were fixed by fitting the lattice QCD results from Ref. [11]. We recall that the DQPM allows one to extract a potential energy density V_p from the space-like part of the energy-momentum tensor as a function of the scalar parton density ρ_s. Derivatives of V_p w.r.t. ρ_s then define a scalar mean-field potential $U_s(\rho_s)$ which enters into the equation of motion for the dynamical partonic quasiparticles. Furthermore, a two-body interaction strength can be extracted from the DQPM as well from the quasiparticle width in line with Ref. [12]. The transition from partonic to hadronic d.o.f. (and vice versa) is described by covariant transition rates for the fusion of quark-antiquark pairs or three quarks (antiquarks), respectively, obeying flavor current-conservation, color neutrality as well as energy-momentum conservation [8, 10]. Since the dynamical quarks and antiquarks become very massive close to the phase transition, the formed resonant prehadronic color-dipole states ($q\bar{q}$ or qqq) are of high invariant mass, too, and sequentially decay to the groundstate meson and baryon octets increasing the total entropy.

On the hadronic side PHSD includes explicitly the baryon octet and decouplet, the 0^-- and 1^--meson nonets as well as selected higher resonances as in the Hadron-String-Dynamics (HSD) approach [13, 14]. Hadrons of higher masses (> 1.5 GeV in case of baryons and > 1.3 GeV for mesons) are treated as 'strings' (color-dipoles) that decay to the known (low-mass) hadrons, according to the JETSET algorithm [15]. Note that PHSD and HSD merge at low energy density, in particular below the critical energy density $\epsilon_c \approx 0.5$ GeV/fm^3. For more detailed descriptions of PHSD and its ingredients we refer the reader to Refs. [9, 10, 16, 17].

The PHSD approach was applied to nucleus-nucleus collisions from $s_{NN}^{1/2} \sim 5$ to 200 GeV in Refs. [8, 10] in order to explore the space-time regions of partonic matter. It was found that even central collisions at the top-SPS energy of $\sqrt{s_{NN}}$ =17.3 GeV show a large fraction of nonpartonic, *i.e.* hadronic or string-like matter, which can be viewed as a hadronic corona. This finding implies that neither hadronic nor only partonic models can be employed to extract physical conclusions in comparing model results with data.

3. Results for dilepton spectra in comparison to experimental data

We directly continue with the results from PHSD in comparison with the available experimental data on dilepton production from SIS to RHIC energies.

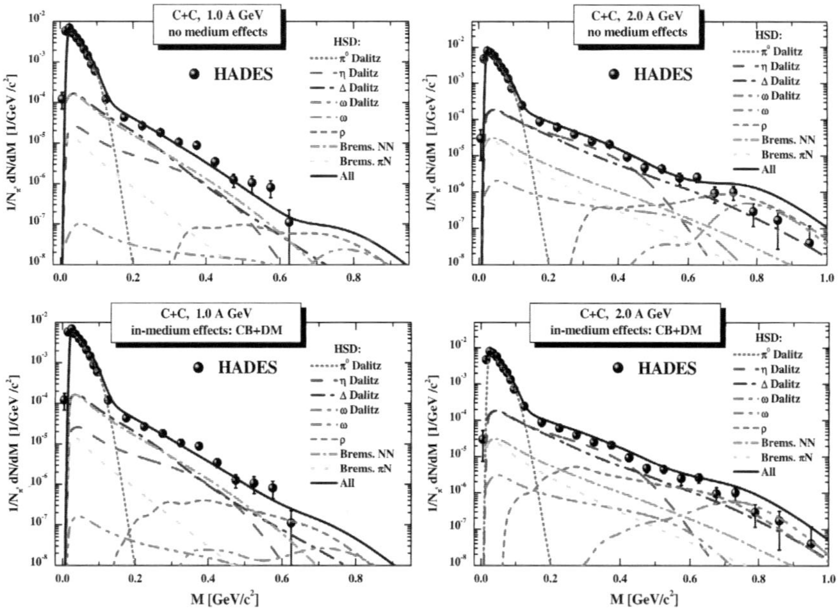

Figure 1. Results of the HSD transport calculation for the mass differential dilepton spectra - divided by the average number of π^0's - in case of $^{12}C + ^{12}C$ at 1.0 A GeV (left) and 2.0 A GeV (right) in comparison to the HADES data [40, 29]. The upper part shows the case of 'free' vector-meson spectral functions while the lower part gives the result for the 'dropping mass + collisional broadening' scenario. In both scenarios the HADES acceptance filter and mass resolution have been incorporated. The different color lines display individual channels in the transport calculation (see legend).

3.1. SIS energies

The dileptons produced in low energy heavy-ion collisions have been measured first by the DLS Collaboration at Berkeley [18, 19, 20, 21]. The observed dilepton yield [21] in the mass range from 0.2 to 0.5 GeV in C+C and Ca+Ca collisions at 1 A GeV was about of five times higher than the calculations by different transport models using the 'conventional' dilepton sources as bremsstrahlung, π^0, η, ω and Δ Dalitz decays and direct vector mesons (ρ, ω, ϕ) decays [22, 23, 24]. Even after including the different in-medium scenarios as collisional broadening and dropping mass for the ρ-meson spectral function did not solve the 'DLS puzzle' [25, 26, 27, 28].

The recent experimental data from the HADES Collaboration at GSI [29, 30, 31, 32, 33, 34], however, confirmed the measurement of the DLS Collaboration for C+C at 1.0 A GeV [30] as well as for the elementary reactions [35]. In the mean time also the theoretical transport approaches as well as effective models for the elementary NN reactions have been further developed.

A possible solution of the 'DLS puzzle' from the theoretical side has been suggested in Ref. [37] by incorporating stronger pn and pp bremsstrahlung contributions in line with the updated One-Boson-Exchange (OBE) model calculations from Ref. [36]. As shown in Ref. [37] the results of the HSD model (off-shell Hadron-String-Dynamics (HSD) transport approach) with 'enhanced' bremsstrahlung cross sections agree very well with the HADES data for C+C at 1 and 2 A GeV as well as with the DLS data for C + C and Ca + Ca at 1 A GeV, especially when including a collisional broadening in the vector-meson spectral functions. A similar finding has been obtained by other independent transport groups – IQMD [38] and Rossendorf BUU [39].

Fig. 1 shows a comparison of the HSD results to the HADES mass-differential dilepton spectra for C + C at 1.0 A GeV (left) and 2.0 A GeV (right) [40, 29] for the 'free' (upper

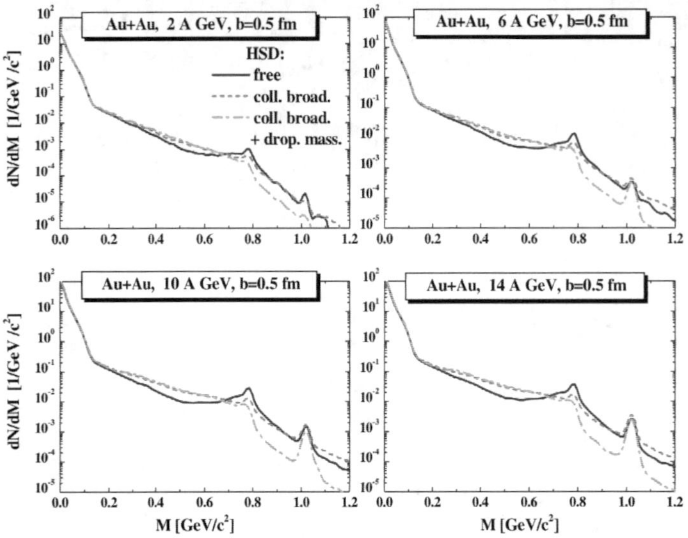

Figure 2. Results of the HSD transport calculation for the mass differential dilepton spectra for central Au+Au collisons from 2 to 14 A GeV calculated for different in-medium scenarios - collisional broadening and combined scenario (dropping mass + collisional broadening).

part) and the in-medium scenario (lower part). At the higher bombarding energy the η Dalitz decay provides the dominant contribution in the mass region from 0.2 to 0.5 GeV followed by Δ Dalitz decays and the combined bremsstrahlung channels. The mass region around 0.75 GeV is overestimated in the 'free' scenario, whereas including in-medium spectral functions for the vector mesons the description of the data is improved due to shifting of the strength from the vector-meson pole mass regime to lower invariant mass. However, the in-medium effects for the light C + C system are only very moderate. In order to observe a strong broadening of the ρ-meson spectral function one has to investigate a larger size system such as $Au + Au$. A corresponding measurement has been performed recently by the HADES Collaboration and the upcoming data will provide more accurate information on the in-medium effects.

In Fig. 2 we show the HSD predictions for the dilepton yields from central Au + Au collisions calculated for different energies from 2 to 14 A GeV applying the different in-medium scenarios: collisional broadening and combined approach (dropping mass + collisional broadening). One can see that both scenarios lead to an enhancement of the dilepton yield in the mass region $M = 0.3 - 0.8$ GeV by a factor of about 2. The largest in-medium effect is, however, attributed to the reduction of the dilepton yield between the ω and ϕ peaks due to the downward shift of the poles of the ρ and ω spectral functions. However, the latest scenario is not consistent with existing experimental data at higher energies, so one has to rely most likely on the relatively modest in-medium effects due to collisional broadening.

3.2. SPS energies

We step up in energy and compare our model results with experimental data for dileptons from In+In collisions at 160 A GeV measured by the NA60 Collaboration.

In Fig. 3 we present PHSD results for the dilepton excess over the known hadronic sources as produced in In+In reactions at 158 A GeV compared to the acceptance corrected data. We find here that the spectrum at invariant masses in the vicinity of the ρ peak is well reproduced by the ρ meson yield, if a broadening of the meson spectral function in the medium is assumed, while the partonic sources account for the yield at high masses. Our analysis shows that the

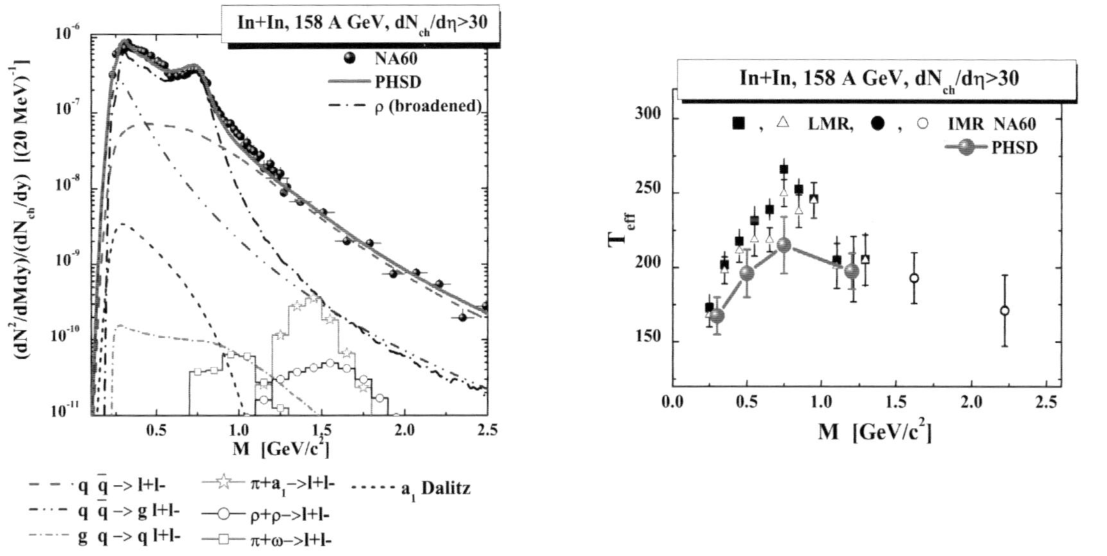

Figure 3. Left: Acceptance corrected mass spectra of excess dimuons from In+In at 158 A GeV integrated over p_T in $0.2 < p_T < 2.4$ GeV from PHSD compared to the data of NA60 [41]. The dash-dotted line shows the dilepton yield from the in-medium ρ with a broadened spectral function, the dashed line presents the yield from the $q + \bar{q}$ annihilation, the dash-dot-dot line gives the contribution of the gluon Bremsstrahlung process ($q\bar{q} \to gl^+l^-$), while the solid line is the sum of all contributions. For the description of the other lines, which correspond to the non-dominant channels, we refer to the figure legend. Right: The inverse slope parameter T_{eff} of the dimuon yield from In+In at 158 A GeV as a function of the dimuon invariant mass in PHSD compared to the data of the NA60 Collaboration [42, 41].

contributions of the '4π' processes (shown by the lines with symbols), first noted by the authors of Ref. [2], are very much supressed.

One concludes from Fig. 3 that the measured spectrum for $M > 1$ GeV is dominated by the *partonic* sources. Indeed, the domination of the radiation from the QGP over the hadronic sources in PHSD is related to a rather long – of the order or 3 fm/c – evolution in the partonic phase (in co-existence with the space-time separated hadronic phase) on one hand (cf. Fig. 10 of Ref. [43]) and the rather high initial energy densities created in the collision on the other hand (cf. Fig. 6 of Ref. [44]). In addition, we find from Fig. 3 that in PHSD the partonic sources also have a considerable contribution to the dilepton yield for $M < 0.6$ GeV. The yield from the two-to-two process $q+\bar{q} \to g+l^+l^-$ is especially important close to the threshold (≈ 0.211 GeV). This conclusion from the microscopic calculation is in qualitative agreement with the findings of an early (more schematic) investigation in Ref.[45].

The comparison of the mass dependence of the slope parameter evolution in PHSD and the data is shown explicitly in the right part of Fig. 3. Including partonic dilepton sources allows us to reproduce in PHSD the m_T-spectra as well as the finding of the NA60 Collaboration [41, 42] that the effective temperature of the dileptons (slope parameters) in the intermediate mass range is lower than that of the dileptons in the mass bin $0.6 < M < 1$ GeV, which is dominated by hadronic sources (cf. Fig. 3, right). The softening of the transverse mass spectrum with growing invariant mass implies that the partonic channels occur dominantly before the collective radial flow has developed. Also, the fact that the slope in the lowest mass bin and the highest one are approximately equal – both in the data and in PHSD – can be traced back to the two

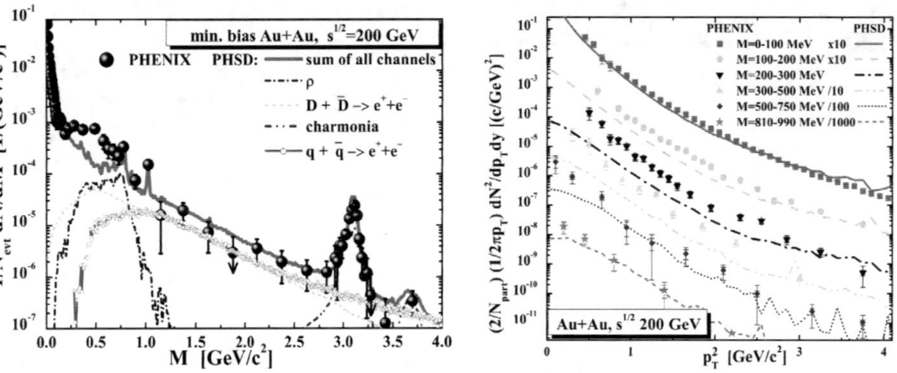

Figure 4. Left: The PHSD results for the invariant mass spectra of inclusive dileptons in Au+Au collisions at $\sqrt{s_{NN}}$ = 200 GeV within the PHENIX acceptance cuts in comparison to the data from the PHENIX Collaboration [46, 47] The different lines indicate the contributions from different channels as specified in the figure. Right: The PHSD results for the transverse-momentum spectra of dileptons from minimum bias Au+Au collisions at $\sqrt{s_{NN}}$ = 200 GeV in different mass bins compared to the data from the PHENIX Collaboration [46, 47].

windows of the mass spectrum that in our picture are influenced by the radiation from the sQGP: $M = 2M_\mu - 0.6$ GeV and $M > 1$ GeV. For more details we refer the reader to Ref. [7].

3.3. RHIC energies

Now we are coming to the top RHIC energy of $\sqrt{s_{NN}} = 200$ GeV and present the most important findings from the PHSD study in Ref. [6]. In the left part of Fig. 4 we show our results for the invariant mass spectra of inclusive dileptons in Au+Au collisions for the acceptance cuts on single electron transverse momenta p_{eT}, pseudorapidities η_e, azimutal angle ϕ_e, and dilepton pair rapidity y: $p_{eT} > 0.2$ GeV, $|\eta_e| < 0.35$, $-3\pi/16 < \phi_e < 5\pi/16$, $11\pi/16 < \phi_e < 19\pi/16$, $|y| < 0.35$.

In the low mass region $M = 0 - 1.2$ GeV, the dilepton yield in the PHSD is dominated by hadronic sources and essentially coincides with the earlier HSD result [48]. Note that the collisional broadening scenario for the modification of the ρ-meson was used in the calculations presented in Fig. 4 that underestimates the PHENIX data from 0.2 to 0.7 GeV substantially. In contrast, the partonic radiation as well as the yield from correlated D-meson decays are dominant in the mass region $M = 1 - 3$ GeV as seen in Fig. 4 (left), i.e. in the mass region between the ϕ and J/Ψ peaks. The dileptons generated by the quark-antiquark annihilation in the sQGP constitute about half of the observed yield in this intermediate-mass range. For $M > 2.5$ GeV the partonic yield even dominates over the D-meson contribution. Thus, the inclusion of the partonic radiation in the PHSD fills up the gap between the hadronic model results [48, 49] and the data of the PHENIX Collaboration for $M > 1$ GeV, however, the early expectation of a partonic signal in the low mass dilepton spectrum is not verified by the microscopic PHSD calculations.

In order to investigate the momentum dependance of the "missing low mass yield", we have calculated the p_T-spectra of dileptons in different bins of invariant mass M. In the right part of Fig. 4 we show the measured transverse momentum spectra of dileptons for minimum bias Au+Au collisions at $\sqrt{s_{NN}} = 200$ GeV (symbols) in comparison with the spectra from the PHSD (lines) for six mass bins as indicated in the figure. Whereas the PHSD can well describe the dilepton spectra in the mass intervals [0,100 MeV] and [810 MeV, 990 MeV], it underestimates the low p_T dileptons in the other mass bins, particularly in the mass bins [300 MeV, 500 MeV].

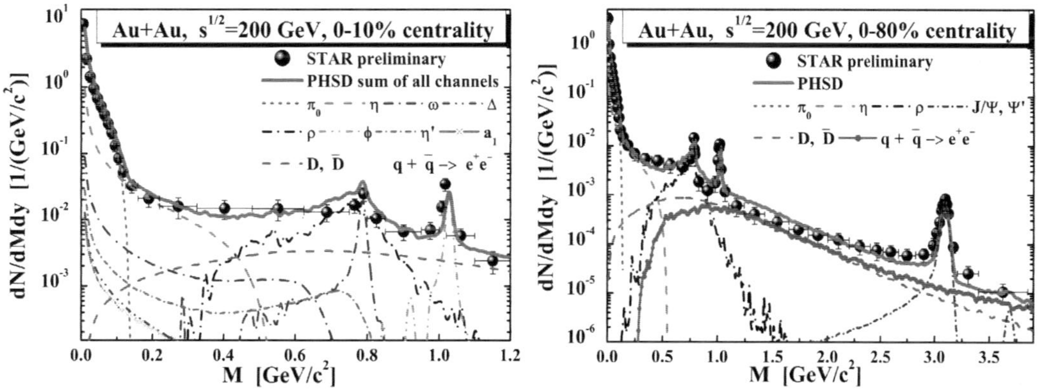

Figure 5. The PHSD results for the invariant mass spectra of dileptons in Au+Au collisions at $\sqrt{s_{NN}}$ = 200 GeV for M=0−1.2 GeV (left part) and for M=0 − 4 GeV (right part) and 0 - 10 % centrality within the cuts of the STAR experiment. The preliminary data from the STAR Collaboration are adopted from Ref. [50].

On the other hand, high p_T dileptons are reproduced quite well by the PHSD calculations. We conclude that the missing dilepton yield for masses from 0.15 to 0.6 GeV is essentially due to a severe underestimation of the data at low p_T by up to an order of magnitude. We recall that at top SPS energies the low p_T dilepton yield could be attributed to $\pi\pi$ annihilation channels, i.e. to the soft hadronic reactions in the expansion phase of the system. These channels are, however, insufficient to describe the very low slope of the p_T spectra at the top RHIC energy.

In order to shed some light on the 'PHENIX puzzle' we step to a comparison of the PHSD predictions with the preliminary STAR data measured for Au+Au collisions at $\sqrt{s_{NN}}$ = 200 GeV with the acceptance cuts on single electron transverse momenta p_{eT}, single electron pseudorapidities η_e and the dilepton pair rapidity y, i.e. $0.2 < p_{eT} < 5$ GeV, $|\eta_e| < 1$, $|y| < 1$. Our predictions for the dilepton yield within these cuts are shown in Fig. 5 for 0-80% centrality. One can observe generally a good agreement with the preliminary data from the STAR Collaboration [50] in the whole mass regime. Surprisingly, our calculations are also roughly in line with the low mass dilepton spectrum from STAR in case of central collisions whereas the PHSD results severely underestimate the PHENIX data for central collisions. The observed dilepton yield from STAR at masses below 1.2 GeV can be accounted for by the known hadronic sources, i.e. the decays of the π_0, η, η', ω, ρ, ϕ and a_1 mesons, of the Δ particle and the semileptonic decays of the D and $\bar{D}$ mesons, where the collisional broadening of the ρ meson is taken into account.

The discrepancy between the PHENIX and STAR data will have to be investigated closer by the experimental collaborations. Furthermore, the upgrade of the PHENIX experiment with a hadron blind detector should provide decisive information on the origin of the low mass dileptons produced in the heavy-ion collisions at $\sqrt{s_{NN}}$ = 200 GeV.

4. Conclusions

We close this contribution by noting that the dileptons are an interesting probe of the dynamical processes in heavy-ion collisions at all energy regimes. The low-mass dileptons provide information on the vector meson spectral function in the medium whereas the high mass part from 1.1 to 3 GeV can be attributed dominantly to the partonic annihilation in the QGP phase. At the higher RHIC and LHC energies the modifications of the low-mass sector are less pronounced than at SIS and SPS energies, however, the dilepton emissivity from the sQGP becomes substantial or even dominant relative to the background from *D*-meson decays.

The authors acknowledge the financial support through the "HIC for FAIR" framework of the "LOEWE" program and the Deutsche Forschungsgemeinschaft (DFG).

References

[1] Shuryak E V 1978 *Phys. Lett. B* **78** 150; 1978 *Sov. J. Nucl. Phys.* **28** 408; 1978 *Yad.Fiz.* **28** 796.
[2] Song C, Ko C M and Gale C 1994 *Phys. Rev. D* **50** 1827
[3] Li G Q and Gale C 1998 *Phys. Rev. C* **58** 2914 [arXiv:nucl-th/9807005].
[4] van Hees H and Rapp R 2006 *Phys. Rev. Lett.* **97** 102301
[5] van Hees H and Rapp R 2008 *Nucl. Phys. A* **806** 339
[6] Linnyk O, Cassing W, Manninen J, Bratkovskaya E L and Ko C M 2012 *Phys. Rev. C* **85** 024910
[7] Linnyk O, Bratkovskaya E L, Ozvenchuk V, Cassing W and Ko C M 2011 *Phys. Rev. C* **84** 054917
[8] Cassing W and Bratkovskaya E L 2009 *Nucl. Phys.* A **831** 215
[9] Cassing W 2007 *Nucl. Phys.* A **795** 70
[10] Bratkovskaya E L, Cassing W, Konchakovski V P and Linnyk O 2011 *Nucl. Phys.* A **856** 162
[11] Aoki Y *et al.* 2009 *JHEP* **0906** 088
[12] Peshier A and Cassing W 2005 *Phys. Rev. Lett.* **94** 172301
[13] Ehehalt W and Cassing W 1996 *Nucl. Phys.* A **602** 449
[14] Cassing W and Bratkovskaya E L 1999 *Phys. Rep.* **308** 65
[15] Bengtsson H U and Sjöstrand T 1987 *Comp. Phys. Commun.* **46** 43
[16] Cassing W 2007 *Nucl. Phys.* A **791** 365
[17] Cassing W 2009 *Eur. J. Phys.* **168** 3
[18] Matis H S *et al.*, DLS Collaboration 1995 *Nucl. Phys. A* **583** 617C
[19] Wilson W K *et al.* DLS Collaboration 1998 *Phys. Rev. C* **57** 1865
[20] Wilson W K *et al.* DLS Collaboration 1993 *Phys. Lett. B* **316** 245
[21] Porter R J *et al.* DLS Collaboration 1997 *Phys. Rev. Lett.* **79** 1229
[22] Wolf G, Cassing W and Mosel U 1993 *Prog. Part. Nucl. Phys.* **30** 273
[23] Bratkovskaya E L, Cassing W and Mosel U 1996 *Phys. Lett. B* **376** 12
[24] Xiong L, Wu Z G, Ko C M and Wu J Q 1990 *Nucl. Phys. A* **512** 772
[25] Ernst C, Bass S A, Belkacem M, Stoecker H and Greiner W 1998 *Phys. Rev. C* **58** 447
[26] Bratkovskaya E L, Cassing W, Rapp R, and Wambach J 1998 *Nucl. Phys. A* **634** 168
[27] Bratkovskaya E L and Ko C M *Phys. Lett. B* **445** 265
[28] Fuchs C, Faessler A, Cozma D, Martemyanov B V and Krivoruchenko M I 2005 *Nucl. Phys. A* **755** 499
[29] Agakishiev G *et al.*, HADES Collaboration 2008 *Phys. Lett. B* 663 43
[30] Pachmayer Y C *et al.*, HADES Collaboration 2008 *J. Phys. G* **35** 104159
[31] Sudol M *et al.*, HADES Collaboration 2009 *Eur. Phys. J. C* **62** 81
[32] Agakishiev G *et al.*, HADES Collaboration 2010 *Phys. Lett. B* **690** 118
[33] Lapidus K *et al.*, HADES Collaboration arXiv:0904.1128 [nucl-ex].
[34] Agakishiev G *et al.*, HADES Collaboration 2011 *Phys. Rev. C* **84** 014902
[35] Agakishiev G *et al.*, HADES Collaboration, 2012 *Phys. Rev. C*, in press, arXiv1203.2549 [nucl-ex]
[36] Kaptari L P and Kämpfer B 2006 *Nucl. Phys. A* **764** 338
[37] Bratkovskaya E L and Cassing W, 2008 *Nucl. Phys. A* **807** 214
[38] Thomere M, Hartnack C, Wolf G and Aichelin J 2007 *Phys. Rev. C* **75** 064902
[39] Barz H W, Kämpfer B, Wolf G and Zetenyi M arXiv:0910.1541 [nucl-th].
[40] Agakichiev G *et al.*, HADES Collaboration 2007 *Phys. Rev. Lett.* **98** 052302
[41] Arnaldi R *et al.*, NA60 Collaboration 2009 *Eur. Phys. C* **59** 607
[42] Arnaldi R *et al.*, NA60 Collaboration 2006 *Phys. Rev. Lett.* **96** 162302; Seixas J *et al.*, 2007 *J. Phys. G* **34** S1023; Damjanovic S *et al.*, 2007 *Nucl. Phys. A* **783**, 327c; Arnaldi R *et al.*, 2009 *Eur. Phys. J. C* **61** 711
[43] Cassing W and Bratkovskaya E L 2009 *Nucl. Phys. A* **831** 215
[44] Linnyk O, Bratkovskaya E L, and Cassing W 2008 *Int. J. Mod. Phys. E* **17** 1367
[45] Alam J, Hirano T, Nayak J K, and Sinha B arXiv:0902.0446.
[46] Toia A. *et al.*, PHENIX Collaboration, 2006 *Nucl. Phys. A* **774** 743
[47] Adare A. *et al.*, PHENIX Collaboration 2010 *Phys. Rev. C* **81** 034911
[48] Bratkovskaya E L, Cassing W and Linnyk O 2009 *Phys. Lett. B* **670** 428
[49] Manninen J, Bratkovskaya E L, Cassing W and Linnyk O 2011 *Eur. Phys. J. C* **71** 1615
[50] Zhao J *et al.*, STAR Collaboration arXiv:1106.6146[nucl-ex].

Journal of Physics: Conference Series **389** (2012) 012017 doi:10.1088/1742-6596/389/1/012017

Quark number susceptibilities and color screening at high temperatures

A. Bazavov, H.-T. Ding and P. Petreczky

Physics Department, Brookhaven National Laboratory, Upton, NY 11973, USA

Abstract. We discuss lattice calculations of quark number susceptibilities and the free energy of a static quark anti-quark pair in 2+1 flavor QCD at high temperatures using the HISQ action. We compare our lattice calculations with perturbative results.

1. Introduction

At high temperatures strongly interacting matter undergoes a deconfining transition to a new state, where the dominant degrees of freedom are quarks and gluons. At sufficiently high temperatures quarks and gluons are weakly interacting as a consequence of asymptotic freedom and the deconfined matter is called quark-gluon plasma in analogy with electromagnetic plasmas (for a recent review of this subject see Ref. [1]). The deconfinement transition is signaled by rapid increase in the energy density and the entropy density, indicating liberation of many degrees of freedom. Quark number susceptibilities are also a good probes of deconfinement as they show a rapid rise at the transition and at sufficiently high temperature are dominated by quark degrees of freedom [2, 3, 4, 5, 6, 7, 8, 9].

The deconfinement can also be characterized by onset of color electric screening that is analogous to the usual Debye screening[1]. Color screening can be studied in terms of the free energy of static quark anti-quark pair, which on the lattice is calculated in terms of the Polyakov loop correlators, or at asymptotically large distances in terms of the expectation value of the Polyakov loop [12].

In lattice QCD calculations at non-zero temperature discretization effects can be large. Therefore it is important to perform the calculations using improved actions at several lattice spacings. Because of relatively small computational costs staggered quark formulation is often used for these calculations. In this contribution we report on lattice calculations of the quark number susceptibilities and the free energy of static quark anti-quark pair using tree-level improved gauge action and highly improved staggered quark (HISQ) action on lattices with the temporal extent $N_\tau = 4,\ 6,\ 8,\ 10$ and 12. The choice of the actions and of the simulation parameters is described in Ref. [13]. We extended the calculations for $N_\tau = 6,\ 8$ and 12 to higher temperatures and also added calculations for $N_\tau = 4$ and 10.

2. Qurk number susceptibilities

Fluctuations and correlations of conserved charges are good probes of deconfinement because they are sensitive to the underlying degrees of freedom, i.e. they can tell whether the relevant

[1] QCD and non-Abelian plasmas in general also screen static magnetic fields, see e.g. Refs. [10, 11] for discussions.

degrees of freedom of the system at a given temperature are hadronic or partonic. Here we consider quadratic fluctuations of light and strange quark numbers, also known as quark number susceptibilities

$$\chi_i(T) = \frac{1}{T^3 V} \frac{\partial^2 \ln Z(T,\mu_i)}{\partial(\mu_i/T)^2}\bigg|_{\mu_i=0} . \quad (1)$$

(2)

Here $Z(T,\mu_i)$ is the partition function and indices $i = l$ and s stand for light and strange quark respectively. The numerical results for the strange and light quark number susceptibilities obatined with the HISQ action are shown in Fig. 1. The results for the strange quark number susceptibility for $T < 250$ MeV and $N_\tau = 6$, 8 and 12 are taken form Ref. [13]. We compare the strangeness fluctuations with the hadron resonance gas model including all resonances with masses smaller than 2.5GeV. We also compare our results with lattice calculations preformed with the stout action [9]. At low temperatures the lattice results agree with hadron resonance gas, while at high temperatures they approach a constant value which is 10% below the value for the ideal massless quark gas. Thus the temperature dependence of the strange quark number susceptibility shows the expected transition from hadronic degrees of freedom to quark degrees of freedom. For temperatures $T > 300$ MeV the difference between the strange and light quark number susceptibilities is quite small, i.e. less than 5%. For this reason the comparison with massless ideal gas is appropriate. In the right panel we show the light quark number susceptibility at high temperatures obtained with the HISQ action for $N_\tau = 6$, 8, 10 and 12. We also compare our results with the continuum-extrapolated stout data as well as with resummed perturbative calculations in next-to-leading log approximation (NLA) [14]. The $N_\tau = 8$ HISQ data are in good agreement with the continuum extrapolated stout data. On the other hand the $N_\tau = 10$ and 12 data are larger than the stout data. This may imply that for the HISQ action the continuum limit is approached from below and the continuum extrapolation for the stout action in Ref. [9] is not fully reliable. The $N_\tau = 12$ HISQ results agree well with the NLA perturbative results at the highest temperature. The $N_\tau = 6$ HISQ lattice data are significantly lower than other lattice results. This is due to cutoff effects which are large for $N_\tau = 6$. In fact, at high temperatures the cutoff effects in $N_\tau = 6$ data follow qualitatively the expectations based on the free field theory calculations [2, 4]. At quantitative level, however, the cutoff dependence is smaller than for the free quark gas as already pointed out in calculations with the p4 action [6]. It should be mentioned that the small deviations of quark number susceptibilities from the ideal gas value is indicative of the weakly interacting nature of the deconfined state. For strongly interacting plasma fluctuations of conserved charges are about half smaller than for the ideal gas [15].

3. Free energy of static quarks

In $SU(N)$ gauge theories there is a true deconfining phase transition related to $Z(N)$ symmetry. The order parameters of this phase transition are the expectation value of the Polyakov loop and the Polyakov loop correlator

$$L(T) = \langle \frac{1}{N_c} \mathrm{Tr} W(\vec{x}) \rangle, \; W(\vec{x}) = \prod_{\tau=0}^{N_\tau - 1} U_0(\tau, \vec{x}), \quad (3)$$

$$C_{PL}(r,T) = \frac{1}{N_c^2} \langle \mathrm{Tr} W^\dagger(r) \mathrm{Tr} W(0) \rangle. \quad (4)$$

The Polyakov loop transforms non-trivially under $Z(N)$ transformation and the expectation value of the Polyakov loop is zero in the confined phase and non-zero above the transition

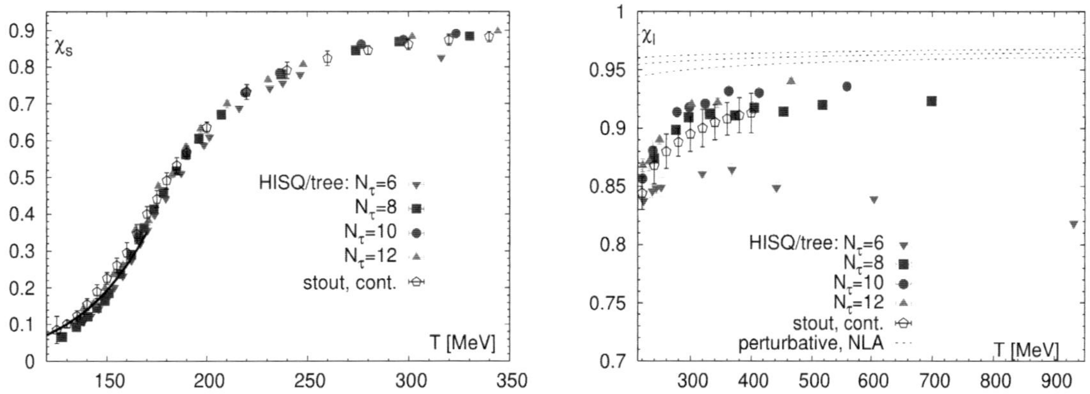

Figure 1. The strange (left) and light quark number (right) susceptibilities as function of the temperature calculated with the HISQ action. The solid black line in the left panel shows the results of the hadron resonance gas model. The band in the right panel shows the resummed perturbative results for the light quark number susceptibility in NLA [14]. We also show the continuum-extrapolated results for the stout action [8, 9].

temperature. The correlation function of the Polyakov loop is related to the free energy of a static quark anti-quark pair [12]. The qualitative change in the behavior of the Polyakov loop and its correlator above the phase transition temperature is related to color screening [12]. As it was pointed out already in Ref. [12] the Polyakov loop and the Polyakov loop correlator require renormalization to be interpreted as the free energy of an isolated static quark or the free energy of a static quark anti-quark, $(Q\bar{Q})$ pair. More precisely, they are related to the free energy difference of a system with static $Q\bar{Q}$ pair at some temperature and the system without $Q\bar{Q}$ pair at the same temperature. Since in the zero temperature limit the free energy of a static quark anti-quark pair should coincide with the static potential, the Polyakov loop renormalization is determined by the normalization constant of the static potential, namely

$$L_{ren}(T) = \exp(-c/(2T))L(T) = \exp(-F_\infty(T)/(2T)), \tag{5}$$
$$C_{PL}(r,T) = \exp(-F(r,T)/T + c/T), \quad F_\infty(T) = \lim_{r\to\infty} F(r,T), \tag{6}$$

where c is an additive normalization that ensures that the static potential has a certain value at a given distance. In the above expressions we made explicit that the free energy of an isolated static quark is half the free energy of $Q\bar{Q}$ pair at infinite separation. In QCD the $Z(3)$ symmetry is broken by light dynamical quarks and the expectation value of the Polyakov loop is always non-zero. Physically this corresponds to the fact that the static quark can be screened by light dynamical quarks resulting in string breaking, the free energy of a static $Q\bar{Q}$ pair saturates at some value. In Fig. 2 we show the free energy of static quark anti-quark pair as function of the distance at different temperatures. To facilitate the comparison with zero-temperature potential we subtracted a temperature dependent constant $T\ln 9$ arising from the normalization of the Polyakov loop correlator in Eq. (4). The normalization constant c has been previously calculated [13]. For the lowest temperature the free energy coincides with the zero-temperature potential calculated in Ref.[13] and is shown as the black line, up to distances where string breaking is expected to occur. At higher temperatures we see significant temperature dependence even at quite short distances. This feature of the $Q\bar{Q}$ free energy was also observed in $SU(3)$ gauge theory [16, 17]. To better understand the behavior of the free energy of a static $Q\bar{Q}$ pair it is worth to consider the so-called singlet free energy. The singlet free energy is useful for calculating

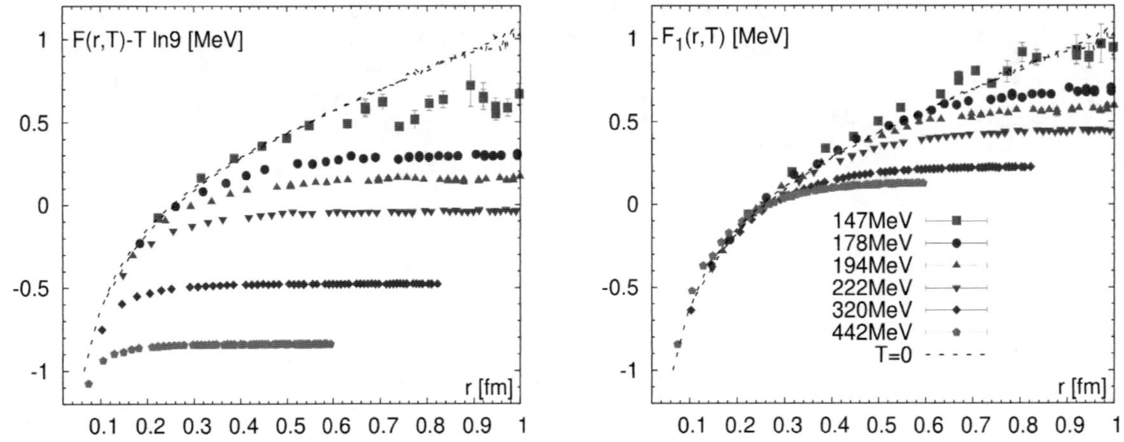

Figure 2. The free energy (left) and singlet free energy (right) of a static quark anti-quark pair as function of the separation r for different temperatures calculated with the HISQ action on $24^3 \times 6$ lattice. The black line shows the zero-temperature potential [13]. The same color (symbol) coding is used for different temperature values in the two panels.

quarkonium spectral functions at non-zero temperature in potential models [18, 19, 20], as direct lattice calculations of quarkonium spectral functions appear to be difficult [21, 22].

A static quark anti-quark pair could be either in a singlet or in an octet state, and thus (in a fixed gauge) one can define the so-called color singlet and octet free energy [23, 24]

$$\exp(-F_1(r,T)/T + c/T) = \frac{1}{3}\langle \mathrm{Tr}[W^\dagger(x)W(y)]\rangle, \tag{7}$$

$$\begin{aligned}\exp(-F_8(r,T)/T + c/T) &= \frac{1}{8}\langle \mathrm{Tr}W^\dagger(x)\mathrm{Tr}W(y)\rangle \\ &- \frac{1}{24}\langle \mathrm{Tr}\left[W^\dagger(x)W(y)\right]\rangle.\end{aligned} \tag{8}$$

Then the Polyakov loop correlator can be written as the thermal average over the singlet and the octet contributions [12, 23, 24]

$$C_{PL}(r,T) = \frac{1}{9}\exp(-F_1(r,T)/T + c/T) + \frac{8}{9}\exp(-F_8(r,T)/T + c/T). \tag{9}$$

At short distances, $rT \ll 1$, the singlet and octet free energies defined above coincide with the well-known perturbative singlet and octet potentials. The singlet and octet free energies can be also calculated at high temperature in the leading order HTL approximation and turn out to be gauge-independent [25]. However, it was not clear how to generalize the decomposition of the free energy into the singlet and octet contributions beyond the leading order. Recently using the effective theory approach, namely the potential non-relativistic QCD (pNRQCD) at finite temperature [26, 27], it was shown that this decomposition indeed holds at short distances [28]. The decomposition of the Polyakov loop correlator in Eq. (9) to large extent explains the strong temperature dependence of the free energy $F(r,T)$. Even if the singlet and octet free energies are temperature-independent the Polyakov loop correlator may have a significant temperature dependence, especially when the singlet and octet contributions become comparable. We calculated the singlet free energy using Coulomb gauge.The numerical results are shown in Fig. 2. At short distances the singlet free energy agrees with the zero-temperature potential as expected while at larger distances it approaches a constant value. The saturation of the free energy happens at smaller distances as temperature increases, as expected if color electric screening occurs. From the above discussions as well as from Fig. 2 it is also clear that the value of the singlet free energy at infinite separation is the same as for the free energy $F(r,T)$. The temperature dependence of the singlet free energy is much smaller than of the static $Q\bar{Q}$ free energy indicating that the octet contribution is significant at high temperatures. The qualitative features of the singlet free energy described above are not specific to the Coulomb gauge. In fact, the singlet free energy defined in terms of periodic Wilson loops [29] shows very similar behavior to the one calculated in the Coulomb gauge [30].

As discussed above the asymptotic value of the free energy (or singlet free energy) is related to the renormalized Polyakov loop. The numerical results for the renormalized Polyakov loop are shown in Fig. 3. Our calculations show that the renormalized Polyakov loop is remarkably insensitive to cutoff effects. We get a reliable estimate for the Polyakov loop expectation value already for $N_\tau = 4$. In the figure we compare our results with the calculations performed in $SU(3)$ gauge theory [31, 32]. The temperature dependence of the renormalized Polyakov loop in QCD and pure gauge theory is similar qualitatively but differences cannot be absorbed by scaling the temperature axis with the transition temperature. In particular in the transition region there are qualitative differences in the temperature dependence of the Polyakov loop. The increase in the expectation value of the Polyakov loop in the transition region is not as rapid in QCD as in pure gauge theory.

At high temperature the renormalized Polyakov loop becomes larger than one. This is expected from perturbation theory which predicts $F_\infty(T) = -4/3\alpha_s m_D$. Here m_D is the well-known leading order Debye mass $m_D = gT\sqrt(1+N_f/3)$ with N_f being the number of light quark flavors, $N_f = 3$ in our case. In Fig. 3 we compare our lattice results with NLO perturbative result of Refs. [33, 28]. As one can see, perturbation theory agrees with lattice results only for $T > 1$ GeV.

4. Conclusions

In this contribution we discussed the results on quark number susceptibilities and the free energy of static quark anti-quark pair obtained with the HISQ action on lattices with the temporal extent $N_\tau = 4$, 6, 8, 10 and 12. The lattice spacing dependence for quark number susceptibilities is small for $N_\tau \geq 8$, while for the free energy of static quark anti-quark pair it is small already for $N_\tau = 4$. This is presumably due to the fact that Polyakov loop is a gluonic quantity and cutoff effects are dominantly coming from the quark sector. The quark number susceptibilities show the expected change in degrees of freedom from hadronic to quark-like as the temperature increases. At high temperatures the lattice results on quark number susceptibilities agree well with resummed perturbative calculations. The increase in the renormalized Polyakov loop indicating the onset of color screening is slower than in quark number susceptibilities and agreement between the lattice results with perturbation theory is only found for $T > 1$ GeV.

Acknowledgements

This work has been supported by contract DE-AC02-98CH10886 with the U.S. Department of Energy. The numerical calculations have been performed using the USQCD resources at Fermilab and JLab, NERSC as well as the BlueGene/L at the New York Center for Computational Sciences (NYCCS).

[1] P. Petreczky, (2012), 1203.5320.
[2] MILC Collaboration, C. Bernard *et al.*, Phys. Rev. **D71**, 034504 (2005), hep-lat/0405029.
[3] Y. Aoki, Z. Fodor, S. Katz, and K. Szabo, Phys. Lett. **B643**, 46 (2006), hep-lat/0609068.
[4] M. Cheng *et al.*, Phys.Rev. **D79**, 074505 (2009), 0811.1006.
[5] Y. Aoki *et al.*, JHEP **0906**, 088 (2009), 0903.4155.

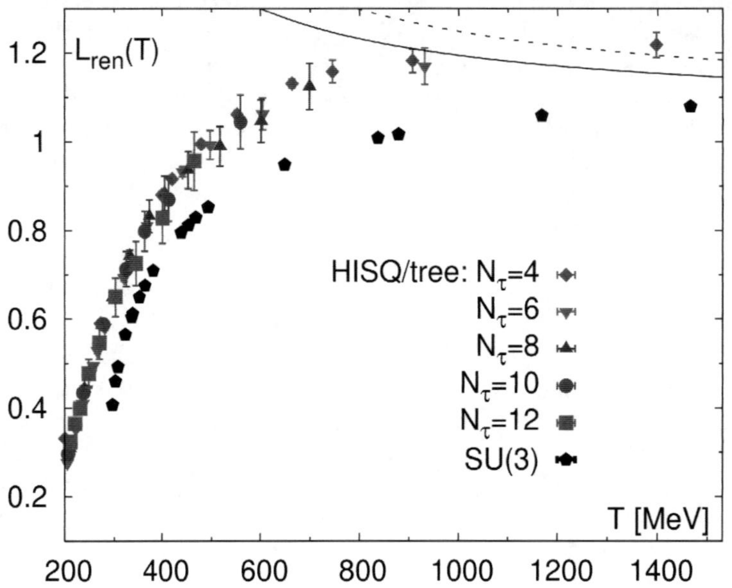

Figure 3. The renormalized Polyakov loop as function of the temperature calculated on different lattices with the HISQ action and compared with the results in pure glue theory. The dashed and solid lines show the next-to-leading perturbative results for 2+1 flavor QCD and $SU(3)$ gauge theory, respectively.

[6] RBC-Bielefeld Collaboration, P. Petreczky, P. Hegde, and A. Velytsky, PoS **LAT2009**, 159 (2009), 0911.0196.
[7] HotQCD collaboration, A. Bazavov and P. Petreczky, J.Phys.Conf.Ser. **230**, 012014 (2010), 1005.1131.
[8] Wuppertal-Budapest Collaboration, S. Borsanyi *et al.*, JHEP **1009**, 073 (2010), 1005.3508.
[9] S. Borsanyi *et al.*, JHEP **1201**, 138 (2012), 1112.4416.
[10] F. Karsch, M. Oevers, and P. Petreczky, Phys.Lett. **B442**, 291 (1998), hep-lat/9807035.
[11] A. Cucchieri, F. Karsch, and P. Petreczky, Phys.Lett. **B497**, 80 (2001), hep-lat/0004027.
[12] L. D. McLerran and B. Svetitsky, Phys. Rev. **D24**, 450 (1981).
[13] A. Bazavov *et al.*, Phys. Rev. **D85**, 054503 (2012), 1111.1710.
[14] A. Rebhan, (2003), hep-ph/0301130.
[15] P. Petreczky, Nucl.Phys. **A830**, 11C (2009), 0908.1917.
[16] O. Kaczmarek, F. Karsch, E. Laermann, and M. Lutgemeier, Phys. Rev. **D62**, 034021 (2000), hep-lat/9908010.
[17] P. Petreczky *et al.*, Nucl. Phys. **A698**, 400 (2002), hep-lat/0103034.
[18] F. Riek and R. Rapp, New J.Phys. **13**, 045007 (2011), 1012.0019.
[19] P. Petreczky, C. Miao, and A. Mocsy, Nucl.Phys. **A855**, 125 (2011), 1012.4433.
[20] A. Mocsy and P. Petreczky, Phys. Rev. **D73**, 074007 (2006), hep-ph/0512156.
[21] I. Wetzorke, F. Karsch, E. Laermann, P. Petreczky, and S. Stickan, Nucl.Phys.Proc.Suppl. **106**, 510 (2002), hep-lat/0110132.
[22] A. Jakovac, P. Petreczky, K. Petrov, and A. Velytsky, Phys. Rev. **D75**, 014506 (2007), hep-lat/0611017.
[23] S. Nadkarni, Phys. Rev. **D33**, 3738 (1986).
[24] S. Nadkarni, Phys. Rev. **D34**, 3904 (1986).
[25] P. Petreczky, Eur. Phys. J. **C43**, 51 (2005), hep-lat/0502008.
[26] N. Brambilla, A. Pineda, J. Soto, and A. Vairo, Nucl. Phys. **B566**, 275 (2000), hep-ph/9907240.
[27] N. Brambilla, J. Ghiglieri, A. Vairo, and P. Petreczky, Phys. Rev. **D78**, 014017 (2008), 0804.0993.
[28] N. Brambilla, J. Ghiglieri, P. Petreczky, and A. Vairo, Phys.Rev. **D82**, 074019 (2010), 1007.5172.
[29] A. Bazavov, P. Petreczky, and A. Velytsky, Phys. Rev. **D78**, 114026 (2008), 0809.2062.
[30] S. Digal, S. Fortunato, and P. Petreczky, Phys. Rev. **D68**, 034008 (2003), hep-lat/0304017.

[31] O. Kaczmarek, F. Karsch, P. Petreczky, and F. Zantow, Phys. Lett. **B543**, 41 (2002), hep-lat/0207002.
[32] M. Doring, K. Huebner, O. Kaczmarek, and F. Karsch, Phys.Rev. **D75**, 054504 (2007), hep-lat/0702009.
[33] Y. Burnier, M. Laine, and M. Vepsalainen, JHEP **1001**, 054 (2010), 0911.3480.

Dynamical freeze-out in event-by-event hydrodynamics

Hannu Holopainen[1] and Pasi Huovinen[2]

[1] Frankfurt Institute for Advanced Studies, Ruth-Moufang-Str. 1, D-60438 Frankfurt am Main, Germany,

[2] Institut für Theoretische Physik, Johann Wolfgang Goethe-Universität, Max-von-Laue-Str. 1, 60438 Frankfurt am Main, Germany

E-mail: holopainen@fias.uni-frankfurt.de

Abstract. In hydrodynamical modeling of the ultrarelativistic heavy-ion collisions the freeze-out is typically performed at a constant temperature or density. In this work we apply a dynamical freeze-out criterion, which compares the hydrodynamical expansion rate with the pion scattering rate. Recently many calculations have been done using event-by-event hydrodynamics where the initial density profile fluctuates from event to event. In these event-by-event calculations the expansion rate fluctuates strongly as well, and thus it is interesting to check how the dynamical freeze-out changes hadron distributions with respect to the constant temperature freeze-out. We present hadron spectra and elliptic flow calculated using (2+1)-dimensional ideal hydrodynamics, and show the differences between constant temperature and dynamical freeze-out criteria. We find that the differences caused by different freeze-out criteria are small in all studied cases.

1. Introduction

In hydrodynamical modeling of heavy-ion collisions the freeze-out is typically assumed to take place instantaneously on a very thin hypersurface of constant temperature or energy density, and the fluid is converted to particles using the Cooper-Frye formula. However, from microscopic point of view the system should freeze-out either when the mean free path of particles is larger than the system size, or when the expansion rate exceeds the scattering rate of particles [1]. Neither of these criterions is directly proportional to temperature nor density. The dynamical freeze-out criterion which directly compares scattering rate with the hydrodynamical expansion rate was applied to hydrodynamical modeling already some time ago [2], but it has not been widely used.

Earlier studies with smooth optical Glauber initial profiles have shown that while the freeze-out surface from the dynamical condition differ significantly from the constant temperature surface, the effect on observable particle spectra is small [3]. However, no studies of the effect on elliptic flow have been done yet. As well, event-by-event hydrodynamical calculations have become popular recently, and in such a case the flow develops more violently than when averaged initial state is used. Thus it is not obvious whether the difference between the two freeze-out conditions is small also when the initial density fluctuates event-by-event.

The problem of freeze-out is not fully resolved in the hybrid models either. In these models the hydrodynamical evolution is stopped while the hadrons are still interacting, the fluid is converted to individual hadrons, and the subsequent evolution of hadrons is described using a hadron cascade [4]. The switch from fluid to cascade is usually done on a constant temperature hypersurface, but the final results depend on the temperature where the switch is done [5, 6, 7]. It is possible that switching on a constant Knudsen number instead would lead to results which are less sensitive to the actual value of the switching parameter.

In this proceedings we compare the dynamical freeze-out criterion with the constant temperature one. We start by introducing our hydrodynamical model and the dynamical freeze-out criterion. We compare the properties of the freeze-out surfaces obtained using different freeze-out criteria and averaged initial state, and calculate the particle p_T-spectra and elliptic flow in these cases. Finally we apply the two different criteria to an event-by-event calculation, study the surfaces, and calculate the particle p_T-spectra and elliptic flow.

2. Hydrodynamical model

Here we use a slightly modified version of the ideal event-by-event hydrodynamics framework presented in [8]. We solve the ideal hydrodynamical equations

$$\partial_\mu T^{\mu\nu} = 0 \qquad \partial_\mu N^\mu = 0, \tag{1}$$

where $T^{\mu\nu}$ is the ideal energy-momentum tensor and N^μ is the net-baryon number current. We assume longitudinal boost invariance and thus the numerical problem reduces to (2+1)-dimensions. In order to close the set of equations we need an equation of state (EoS) and our choice here is s95p-v1 from Ref. [9].

Particle emission from the freeze-out surface is calculated using conventional Cooper-Frye formula

$$\frac{dN}{d^2p_T dy} = \int_\sigma f(x,p) p^\mu d\sigma_\mu, \tag{2}$$

where we integrate over the freeze-out surface σ, $f(x,p)$ is particle distribution and p^μ is the four momentum of the emitted particle. The freeze-out surface elements $d\sigma_\mu$ are found with `CORNELIUS++`[1] using the algorithm described in [7].

After the thermal particle spectra are calculated using Cooper-Frye, we sample particles using these spectra as a probability distributions. Sampling is done as described in [8]. Thus the number of each particle species is fixed, but the total energy of particles in each event can fluctuate. Then the strong and electromagnetic decays are done one particle at a time. In [8] decays were done using `PYTHIA`, which meant that all hadron resonances included in the EoS could not be used. Here we use a separate decay program, which makes it possible to include all hadrons present in the EoS.

In this contribution we consider Au+Au collisions at Relativistic Heavy Ion Collider (RHIC) with $\sqrt{s_{NN}} = 200$ GeV. Smooth initial conditions for the entropy density are obtained from the optical Glauber model using a mixture of binary (75%) and wounded nucleon (25%) profiles. For the fluctuating initial conditions a Monte Carlo Glauber model with same mixture is used. Monte Carlo Glauber model gives only the positions of the wounded nucleons and binary collisions and we need to distribute entropy around these positions before we can initialize hydrodynamical evolution. Our choice is a 2-dimensional Gaussian:

$$s(x,y) = \text{const.} \sum_{\text{wn,bc}} \frac{1}{2\pi\Sigma^2} \exp\Big[-\frac{(x-x_i)^2+(y-y_i)^2}{2\Sigma^2}\Big], \tag{3}$$

[1] `CORNELIUS` surface finding subroutine is available at the OSCAR code repository, `https://karman.physics.purdue.edu/OSCAR/`.

Journal of Physics: Conference Series **389** (2012) 012018 doi:10.1088/1742-6596/389/1/012018

where Σ is a free parameter controlling the width of the Gaussian. Typical values for this fluctuation size parameter is order of 0.5 fm, our choices here are 0.4 fm and 0.8 fm. For net-baryon density a pure wounded nucleon profile is used in both cases. We choose the initial time to be $\tau_0 = 0.6$ fm.

Because there are differences between the two Glauber models and the mixtures are not defined exactly in the same way, we cannot compare the results obtained with smooth and fluctuating initial conditions. Here we concentrate on the effects coming from the freeze-out condition and comparisons between smooth and fluctuating initial conditions with constant temperature freeze-out can be found *e.g.* from Ref. [8].

3. Dynamical freeze-out condition

For the system to maintain kinetic equilibrium, the scattering rate should be much larger than the expansion rate. In other words

$$K = \frac{\theta}{\Gamma} \ll 1, \tag{4}$$

where Γ is the scattering rate and θ is the hydrodynamical expansion rate. As a ratio of (an inverse of) a macroscopic length scale to (an inverse of) a microscopic length scale, the ratio K can be identified as a Knudsen number — which must be larger than one for hydrodynamics to be valid as well. Since we are interested in freeze-out, the region of last rescattering, we replace the vague requirement $K \ll 1$ by assuming that the system is reasonably close to equilibrium until $K = K_{\text{fo}} = 1$, and test the sensitivity of the spectra to the exact value of K_{fo} by using values $K_{\text{fo}} = 0.5 - 2$.[2] Our initial assumption $K_{\text{fo}} = 1$ is a posteriori justified by a good reproduction of the measured particle spectra. When we compare to results calculated using freeze-out at constant temperature, we use $T_{\text{fo}} = 140$ MeV for the same reason.

The expansion rate of a system is given by the hydrodynamical model. For a boost-invariant system it is evaluated as

$$\theta = \partial_\mu u^\mu = \partial_\tau u^\tau + \partial_x u^x + \partial_y u^y + u^\tau/\tau. \tag{5}$$

We evaluate the scattering rate of pions in kinetically (but not necessarily chemically) equilibrated hadron-resonance gas as

$$\Gamma = \frac{1}{n_\pi(T,\mu_\pi)} \sum_i \int d^3p_\pi d^3p_i f_\pi(T,\mu_\pi) f_i(T,\mu_i) \frac{\sqrt{(s-s_a)(s-s_b)}}{2E_\pi E_i} \sigma_{\pi i}(s), \tag{6}$$

where n_π is the density of pions, $f_i(T,\mu_i)$ is the thermal distribution function, $\sqrt{(s-s_a)(s-s_b)})/(2E_\pi E_i)$ is the relative velocity, $\sigma_{\pi i}$ is the cross section for scattering of pion with particle i, and the summation over i runs over all particle species included in the EoS. Cross sections are evaluated as in the UrQMD model [10], *i.e.* the main contribution comes from resonance formation which is evaluated using the relativistic Breit-Wigner formula. Details of this calculation will be published in a later paper.

Scattering rates are different for different particle species, and thus one could argue that freeze-out should take place when the scattering rate of a particular species is equal to the expansion rate, which would mean freeze-out at different times for different particle species. However, to be consistent, this approach would require us to modify fluid dynamics in such a way that when *e.g.* pions are decoupled, they are removed from the fluid. For the subsequent evolution of the remaining fluid we should then include the interactions with the non-equilibrated pion cloud, which is clearly beyond the capabilities of fluid dynamics. Thus we make the

[2] In Ref. [2] it was argued that $K_{\text{fo}} = 2$ would give the region of the last rescattering.

simplifying approximation that all the particle species decouple at the same time, and since the pions are the most abundant particles, we base our freeze-out description on the scattering rate of pions.

4. Results with smooth initial conditions

We start by studying the freeze-out systematics with smooth initial conditions. First in Fig. 1 we compare the freeze-out surfaces with both the constant temperature and the dynamical freeze-out criteria. We can see that with the dynamical condition the edges of the system decouple earlier and the center of the system lives little longer. This is because near the edges the pressure gradients are large and thus the expansion rate is large. In the center of the system the expansion is not so rapid and thus the center can maintain the kinetical equilibrium at lower temperatures, *i.e.* at smaller scattering rates.

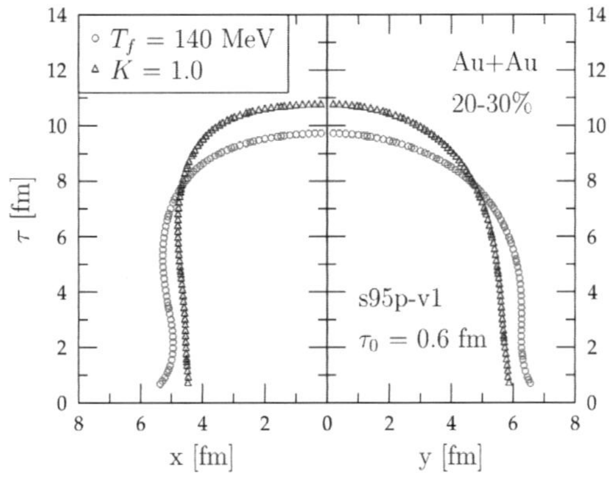

Figure 1. Freeze-out surface along positive x- and y-axis in 20-30% central Au+Au collisions at $\sqrt{s_{NN}} = 200$ GeV. Surfaces are shown with both dynamical and constant temperature criterion.

In Fig. 2 we plot the temperature on the freeze-out surfaces with both conditions. Naturally with the constant temperature criterion the curve is constant, but with the dynamical freeze-out the temperature can vary a lot. As already explained above, the edges decouple earlier with the dynamical conditions and thus the edges are hotter than in the constant temperature case. Accordingly, the center the system is little bit cooler with the dynamical freeze-out.

The radial velocity on the surface along the x-axis as a function of time is plotted in Fig. 3. Since with the dynamical condition the edges decouple earlier, the largest flow velocities are cut away as we can see in the figure. For the high-p_T ($\gtrsim 1$ GeV) emission the part of the surface where the flow velocity is largest is most important and one could expect that the two different freeze-out conditions could give different results. However, the effect of smaller transverse flow is countered by the larger temperature and thus it is difficult to say what happens based on these figures alone.

It is interesting to see how the freeze-out surface changes when we change the freeze-out condition, but the experiments cannot measure the properties of the freeze-out surface itself, but rather the particle spectra and its anisotropies. In Fig. 4 we have plotted the spectra of positively charged pions and protons. We see that results from different freeze-out conditions are basically on top of each other. This indicates that the increased temperature at the edges cancels the effect of reducing the maximum flow velocity.

We also plot the elliptic flow of charged hadrons in Fig. 5 with respect to the reaction plane (defined by impact parameter and beam axis). Also here the freeze-out criterion does not play a role and the result is the same with both conditions. From the particle spectra and elliptic flow we can conclude that the constant temperature freeze-out gives the same answers as dynamical freeze-out despite the fact that the surface itself shows some sensitivity to the criterion.

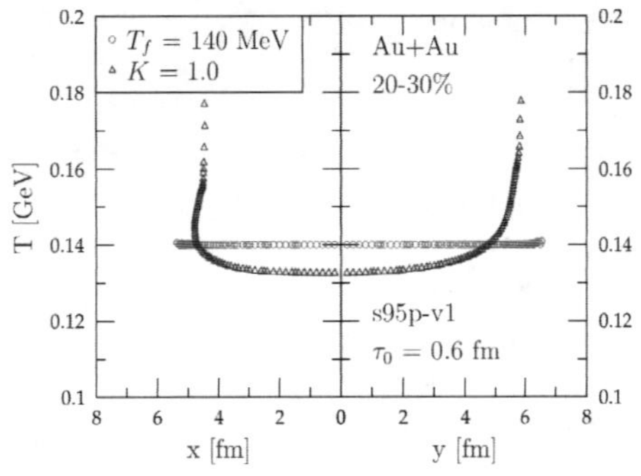

Figure 2. Temperature on the freeze-out surface along positive x- and y-axis in 20-30% central Au+Au collisions at $\sqrt{s_{NN}}$ = 200 GeV. Surfaces are shown with both dynamical and constant temperature criterion.

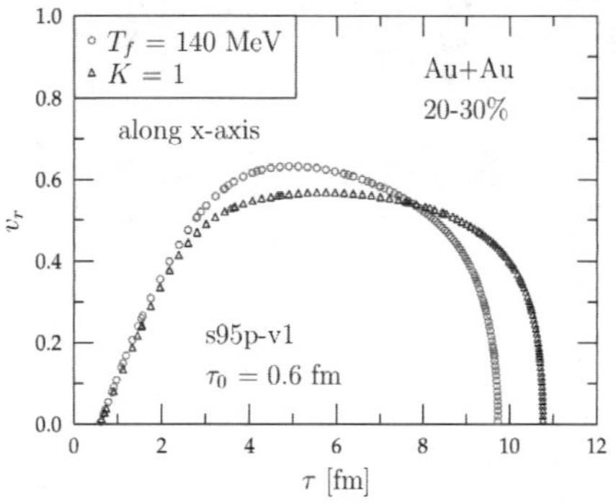

Figure 3. Radial velocity on the freeze-out surface along positive x- and y-axis in 20-30% central Au+Au collisions at $\sqrt{s_{NN}}$ = 200 GeV. Surfaces are shown with both dynamical and constant temperature criteria.

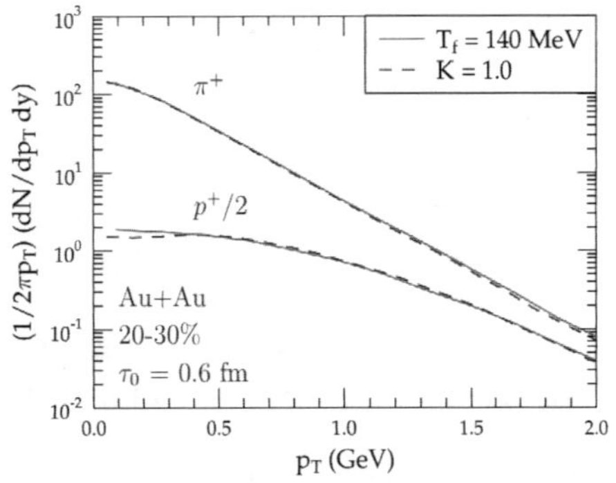

Figure 4. Spectra of positively charged pions and protons in 20-30% central Au+Au collisions at $\sqrt{s_{NN}}$ = 200 GeV. Results are shown with both dynamical and constant temperature criterion.

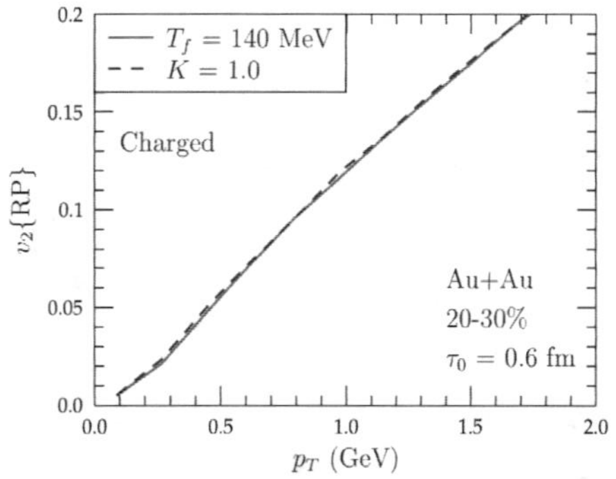

Figure 5. Elliptic flow of charged particles calculated with event plane method in 20-30% central Au+Au collisions at $\sqrt{s_{NN}}$ = 200 GeV. Results are shown with both dynamical and constant temperature criterion.

Finally, we show how the results depend on the freeze-out parameter $K_{\rm fo}$. In Fig. 6 we plot the freeze-out surfaces with $K_{\rm fo} = 0.5, 1.0$ and 2.0 and in Fig. 7 we show the p_T-spectra of pions and protons with these three values. We see that with larger value of $K_{\rm fo}$ the system lives longer. This naturally leads to the fact that more flow is developed and thus the spectra for pions and protons is flatter. Also because we have chemical equilibrium until freeze-out, the number of protons depends on the freeze-out value K since the average temperature on the surface depends on K. Introducing a separate chemical freeze-out would solve this problem and this will be done in the future.

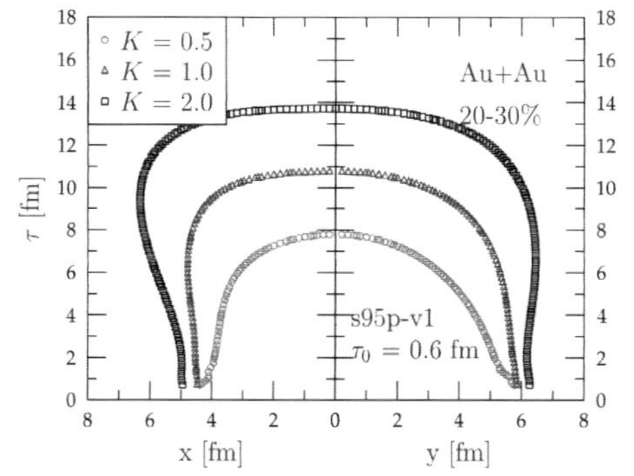

Figure 6. Freeze-out surface along positive x- and y-axis in 20-30% central Au+Au collisions at $\sqrt{s_{NN}} = 200$ GeV. Results are shown with three different freeze-out parameter values.

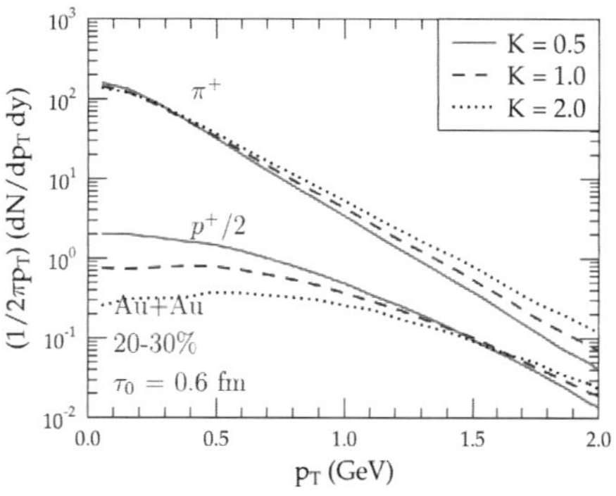

Figure 7. Spectra of positively charged pions and protons in 20-30% central Au+Au collisions at $\sqrt{s_{NN}} = 200$ GeV. Results are shown with three different freeze-out parameter values.

5. Results with fluctuating initial conditions

In the previous section we saw that the particle spectra and elliptic flow do not change when we switch from constant temperature freeze-out to the dynamical one. However, we used smooth initial conditions, where the expansion rate behaves smoothly. When we include event-by-event density fluctuations to the initial state, the flow develops more violently and the freeze-out criterion could affect the final spectra.

In the first calculations we used a quite small smearing parameter $\sigma = 0.4$ fm, which means that there is a lot of structure in the initial energy density distribution. Consequently the expansion rate can vary a lot during the evolution. An example of the freeze-out surface with dynamical condition is shown in Fig. 8. We can see that many long living small scale structures exist: Thin rod-like "horns" and narrow but wide "fins". Conceptually this kind of structures are difficult, since their size is smaller than the average mean free path of particles, and thus from microscopic point of view, they shouldn't exist.

The same event with double the smearing parameter is shown in Fig. 9, and we immediately notice that the surface is much smoother and only one fin remains. Thus before studying the effects of the fins and horns, one can use smoother initial states to check if the freeze-out condition is more important when the fluctuating initial state is used than when an averaged initial state is used. More advanced studies where the effect of the fins and horns are explored are left for future work.

The reason for the fin and horn formation is that with fluctuations there are some regions where the expansion rate goes to negative, *i.e.* there is compression. By definition the dynamical freeze-out condition does not allow those regions to freeze-out, no matter how long living or cold they are. It is obvious that the smaller the smearing parameter σ, the more pronounced the initial peaks and valleys are in the initial density profile, and the more common the regions are where the fluid compresses instead of expanding. Thus the freeze-out surface is much more complicated in Fig. 8 than in Fig. 9.

In Fig. 10 we plot the spectra of pions and protons and in Fig. 11 we plot the elliptic flow of charged hadrons. Here v_2 is calculated with event plane method in the same way as in Ref. [8]. We can see that also in this situation there are no differences between the two freeze-out conditions. The situation is similar with and without initial state fluctuations, although the

Figure 8. Freeze-out surface with fluctuating initial conditions with σ = 0.4 fm.

Figure 9. Freeze-out surface with fluctuating initial conditions with σ = 0.8 fm.

reason for this may be that we had to use very smooth event-by-event initial states in order to get rid off the multiple fins and horns, which we consider unphysical.

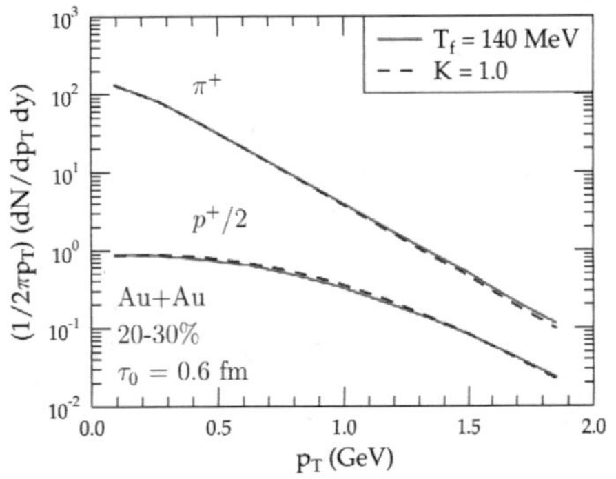

Figure 10. Spectra of positively charged pions and protons in 20-30% central Au+Au collisions at $\sqrt{s_{NN}}$ = 200 GeV. Results are shown with both dynamical and constant temperature criterion.

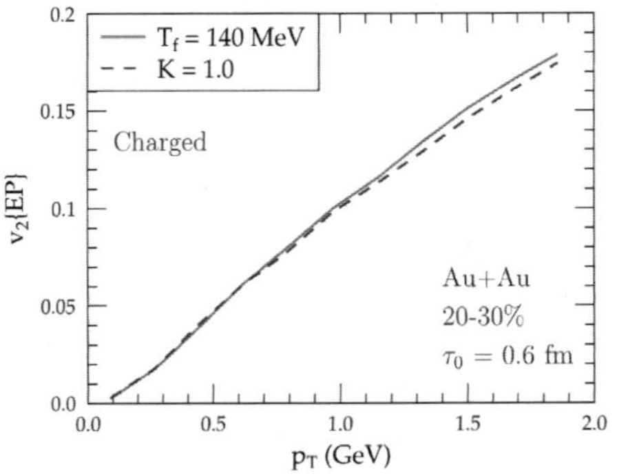

Figure 11. Elliptic flow of charged particles calculated with event plane method in 20-30% central Au+Au collisions at $\sqrt{s_{NN}}$ = 200 GeV. Results are shown with both dynamical and constant temperature criterion.

6. Conclusions

We have argued that the constant temperature freeze-out is an oversimplification, but the differences compared to the dynamical freeze-out criterion are small in particle spectra and elliptic flow. Freeze-out surfaces itself are different, but effects from the increased temperature and decreased radial flow at the edges in dynamical freeze-out cancel each other and finally there are no visible effects in the particle spectra and elliptic flow.

However, here our event-by-event studies were limited to the case where initial fluctuation smearing parameter was large. Because of that our initial states are very smooth and we can expect that the behavior is very similar to the optical Glauber model case. With smaller smearing parameters the freeze-out surface had many small scale structures (which we called horns and fins) which are most probably unphysical. Thus we did not evaluate the particle distributions yet in that case. Ultimately one needs to evaluate the spectra in this case too, but before that we have to devise a way to distinguish the effect of these small scale structures on the spectra to know when the change in the spectra might be unphysical.

On the other hand it has been shown that finite viscosity smoothens the density distribution at the end of the evolution considerably [11]. Thus the small scale structures may disappear altogether in a viscous calculation, and it would be interesting and easier to apply dynamical freeze-out criterion in the context of viscous hydrodynamics.

The work of H.H. was supported by the Extreme Matter Institute (EMMI) and the work of P.H. by BMBF under contract no. 06FY9092.

References

[1] Bondorf J P, Garpman S I A and Zimanyi J 1978 *Nucl. Phys.* A **296** 320
[2] Hung C M and Shuryak E V 1998 *Phys. Rev.* C **57** 1891
[3] Eskola K J, Niemi H and Ruuskanen P V 2008 *Phys. Rev.* C **77** 044907
[4] Petersen H, Steinheimer J, Burau G, Bleicher M and Stocker H 2008 *Phys. Rev.* C **78** 044901
[5] Song H, Bass S A and Heinz U 2011 *Phys. Rev.* C **83** 024912
[6] Hirano T, Huovinen P, Murase K and Nara Y 2012 Integrated Dynamical Approach to Relativistic Heavy Ion Collisions *Preprint* arXiv:1204.5814 [nucl-th]
[7] Huovinen P and Petersen H 2012 Particlization in hybrid models *Preprint* arXiv:1206.3371 [nucl-th]
[8] Holopainen H, Niemi H and Eskola K J 2011 *Phys. Rev.* C **83** 034901
[9] Huovinen P and Petreczky P 2010 *Nucl. Phys.* A **837** 26
[10] Bass S A *et al.* 1998 *Prog. Part. Nucl. Phys.* **41** 255
[11] Schenke B, Jeon S and Gale C 2011 *Phys. Rev. Lett.* **106** 042301

RHIC and LHC phenomena with an unified parton transport

Ioannis Bouras[a], Andrej El[a], Oliver Fochler[a], Felix Reining[a], Florian Senzel[a], Jan Uphoff[a], Christian Wesp[a], Zhe Xu[b] and Carsten Greiner[a]

[a]Institut für Theoretische Physik, Johann Wolfgang Goethe-Universität
Max-von-Laue-Str. 1, D-60438 Frankfurt am Main, Germany
[b]Department of Physics, Tsinghua University
Beijing 100084, China

E-mail: `carsten.greiner@th.physik.uni-frankfurt.de`

Abstract. We discuss recent applications of the partonic pQCD based cascade model BAMPS with focus on heavy-ion phenomeneology in hard and soft momentum range. The nuclear modification factor as well as elliptic flow are calculated in BAMPS for RHIC end LHC energies. These observables are also discussed within the same framework for charm and bottom quarks. Contributing to the recent jet-quenching investigations we present first preliminary results on application of jet reconstruction algorithms in BAMPS. Finally, collective effects induced by jets are investigated: we demonstrate the development of Mach cones in ideal matter as well in the highly viscous regime.

1. Introduction

In collisions of heavy ions at ultrarelativistic energies at the Relativistic Heavy-Ion Collider (RHIC) and the Large Hadron Collider (LHC) a new state of matter, the quark-gluon plasma (QGP), has been created. Although the QGP is not available for direct observation, its properties can be deduced from the measurement of the produced hadrons in the final state.

The large values of the measured hadronic elliptic flow v_2 [1, 2, 3], which is the second coefficient of the Fourier series of the azimuthal particle multiplicity, suggests that equilibration of quarks and gluons occurs on a very short time scale ≤ 1 fm/c. This also suggests that the shear viscosity over entropy density ratio η/s of the QGP is very small, which means that the QGP behaves like a nearly perfect fluid. All these conclusions can be drawn from comparison of experimental results with hydrodynamic calculations. However, an understanding of the mechanism of fast thermalization can not be achieved in the scope of hydrodynamic models. The early pre-equilibrium dynamics of the QGP must be studied in the scope of the kinetic theory.

In contrast to the hydrodynamic approach, kinetic transport theory is a microscopic theory and thus allows to study processes of soft and hard processes simultaneously. This is in particular important for detailed understanding of further properties of the quark-gluonic medium, such as the suppression of jets and heavy-quarks. Suppression of jets, also known as jet quenching, is quantified by comparing the hadron multiplicities measured in heavy-ion collisions with appropriately scaled multiplicities from $p+p$ collisions [4, 5, 6]. In addition, very exciting jet-associated particle correlations were observed [7], which might be the result of a conical

Published under licence by IOP Publishing Ltd

emission off propagating shock waves in form of Mach Cones. These Mach Cones might be induced by high-energy partons traversing the expanding medium [8]. Observations of these effects is consistent with the picture of a nearly perfect fluidity of the QGP.

The kinetic transport model BAMPS (Boltzmann Approach to Multiparton Scatterings) [9] has been developed to provide a unified description of dynamics of the early QGP stage of heavy-ion collisions (HIC) including perturbative QCD based elastic and inelastic processes. BAMPS has been applied to provide explanation of fast thermalization on a very short time scale ≤ 1 fm/c [10] as well as a small value of $\eta/s \approx 0.08 - 0.2$ for $\alpha_s = 0.6 - 0.3$ [11, 12]. Furthermore recent calculations with BAMPS provide results on elliptic flow [13, 14] and jet quenching [15] at RHIC energies, which is for the first time done in a consistent and fully pQCD–based microscopic transport model. In addition, BAMPS has been used in certain works as a reference for hydrodynamic calculations. This opens the possibility to study hydrodynamic phenomena for arbitrary viscosity.

In these proceedings we discuss application of BAMPS to describe a number of phenomena observed in the recent heavy-ion experiments. In Sec. 2 we introduce calculations of the nuclear modification factor $R_{AA}(p_T)$ for RHIC and LHC conditions. In Sec. 3 BAMPS results on elliptic flow and suppression of charm and bottom quarks for RHIC and LHC energies are introduced. In Sec. 4 for the first time in the framework of BAMPS we introduce preliminary results on application of jet reconstruction algorithms. Finally in Sec. 5 the formation and propagation of shock waves in form of Mach Cones are discussed for a wide range of viscosity to entropy density ratio η/s.

2. Nuclear modification factor and elliptic flow from partonic transport simulations

As established in [13, 14, 15] the partonic medium in BAMPS simulations of ultra relativistic heavy ion collisions features a small ratio of the shear viscosity to the entropy, η/s, and develops a strong collectivity with an integrated v_2 that is in good agreement with experimental results over a large centrality range for a fixed coupling of $\alpha_s = 0.3$ and a kinetic freeze-out energy density $\varepsilon_c = 0.6\,\mathrm{GeV/fm^3}$. These parameters are used for all calculations that are presented in this section.

Fig. 1 shows the integrated partonic v_2 as a function of centrality from simulations of Au + Au at 2.76 A TeVand of Au + Au at 200 A GeVcompared to experimental data from the ALICE experiment [16] at LHC and from the STAR experiment [17] at RHIC. Using the same set of parameters ($\alpha_s = 0.3$, $\varepsilon_c = 0.6\,\mathrm{GeV/fm^3}$) that have been fixed to the RHIC data [13, 14] also the integrated elliptic flow at LHC can be described over a large range in centrality. Accordingly the simulated differential v_2 of Pb + Pb collisions at LHC energies shows no significant deviation from the Au + Au results at 200 A GeVin the low to intermediate p_T region which is also in agreement with experimental findings [18].

One of the main virtues of the transport model BAMPS is that it allows for the investigation of different observables within a consistent framework and consequently also the nuclear modification factor, R_{AA}, is studied using the same parameters that provide a description of the elliptic flow ($\alpha_s = 0.3$, $\varepsilon_c = 0.6\,\mathrm{GeV/fm^3}$). Fig. 2 shows the nuclear modification factor R_{AA} obtained from BAMPS simulations of central, 0-5%, Pb + Pb collisions at 2.76 A TeV. The results are both shown on the partonic level for gluons and light quarks and on the hadronic level for neutral pions based on AKK fragmentation functions [19]. The suppression of high-p_T particles in simulations with BAMPS is distinctly stronger than the experimentally observed suppression, which is also observed in simulations of R_{AA} at RHIC energies [20]. Additionally the rise towards larger transverse momenta that is present in the LHC data is not reproduced.

The strong quenching observed in BAMPS calculations is due to the energy loss in $2 \rightarrow 3$ interactions that include an effective implementation of the LPM effect [21] via a mean free path-

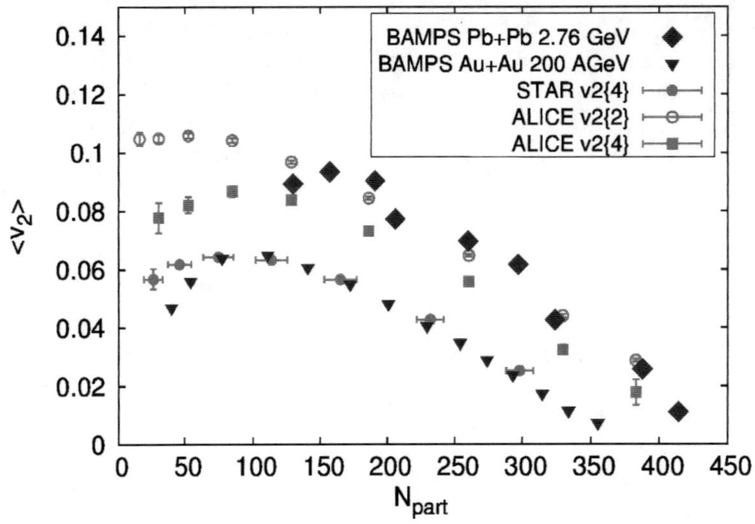

Figure 1. Integrated partonic v_2 from BAMPS as a function of N_{part} for Pb + Pb at 2.76 A TeV($|y| < 0.8$) and Au + Au at 200 A GeV($|y| < 0.5$) compared to the measured v_2 of charged particles from ALICE [16] and from STAR [17].

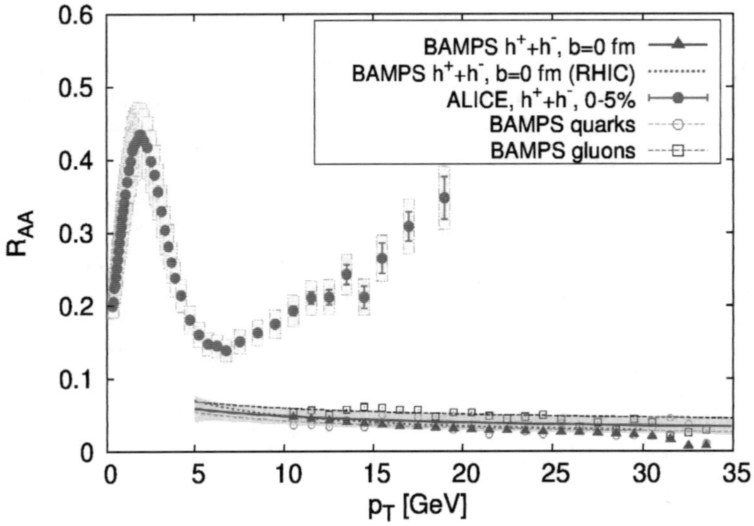

Figure 2. Nuclear modification factor R_{AA} of charged hadrons, gluons and quarks from BAMPS simulations of Pb + Pb at $b = 0$ fm compared to results from ALICE for 0-5% central Pb + Pb collisions [18]. Lines indicate R_{AA} computed from fits to the simulated parton spectra, while symbols indicate R_{AA} computed directly from the parton spectra as obtained from BAMPS. For comparison the R_{AA} of charged hadrons from simulations of Au + Au at 200 A GeVand $b = 0$ fm is also shown.

dependent cutoff [9, 15]. The strong quenching is then caused by [20] *a*) a strong energy loss that is caused by a complex interplay of the Gunion-Bertsch matrix element and the effective

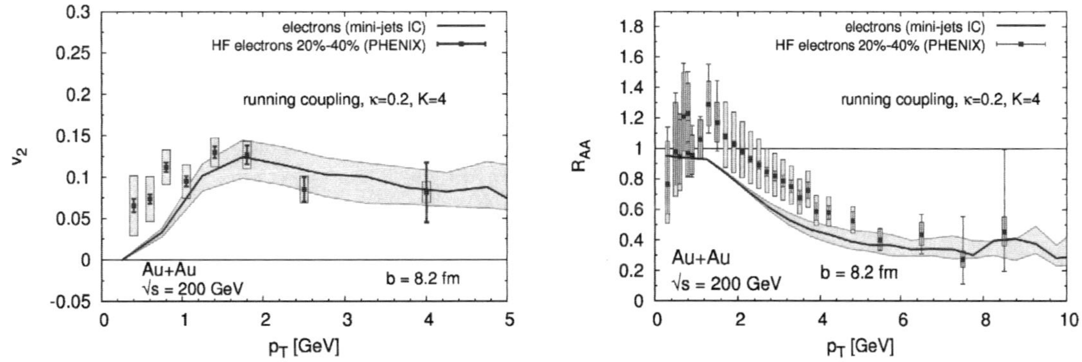

Figure 3. Elliptic flow v_2 (left) and nuclear modification factor R_{AA} (right) of heavy flavor electrons for Au+Au collisions at RHIC with an impact parameter of $b = 8.2$ fm together with data [22]. The elastic cross section of $gQ \to gQ$ is multiplied with the factor $K = 4$ to mimic the influence of radiative processes.

implementation of the LPM effect [20], *b*) a conversion of quark into gluon jets in $2 \to 3$ interactions and *c*) a small difference in the energy loss of gluons and quarks caused by the iterative computation of interaction rates required by the inclusion of the LPM cutoff.

Thus, while the collectivity of the medium can be well described within the current approach, the quenching of high-p_T particles is overestimated. Future studies will therefore focus on the implementation of a running coupling for light quarks and gluons and also systematically explore the modeling of the LPM effect. These modifications are qualitatively expected to bring the results for the nuclear modification factors into better agreement with experimental data.

3. Elliptic flow and suppression of heavy quarks

Heavy quarks are a good probe to study the properties of the QGP. They are well calibrated in a sense that they are produced entirely in the early stage of the heavy ion collision due to their large mass [23] and are also tagged during hadronization due to flavor conservation. Whereas heavy quarks at RHIC can only be measured indirectly via heavy flavor electrons, at LHC for the first time it is possible to reconstruct D mesons and, therefore, receive information only about charm quarks.

The heavy flavor electron data from RHIC [26, 27, 22] and the heavy flavor electron, muon and D meson data from LHC [28, 24] show that the suppression of heavy quarks is on the same order as for light quarks. From the theory perspective it was thought that radiative processes involving heavy quarks are suppressed due to the dead cone effect [29, 30], which means that gluon radiation at small angles is suppressed and, therefore, the energy loss is smaller compared to light partons. Elliptic flow v_2 measurements of particles associated with open heavy flavor also show that heavy quarks interact strongly with the other particles of the medium. Whether these observations can be explained by collisional or radiative energy loss or other effects is currently in debate.

The elliptic flow v_2 and the nuclear modification factor R_{AA} are important observables for heavy quarks. Although those particles are rare probes, both observables are experimentally accessible for fragmentation and decay products of heavy quarks such as D mesons or heavy flavor electrons. The R_{AA} reflects how much energy heavy quarks lose in the QGP. The v_2 is large if heavy quarks interact often with the medium and pick up its collective flow.

All the calculations for heavy quarks in this section are done with a running coupling and an

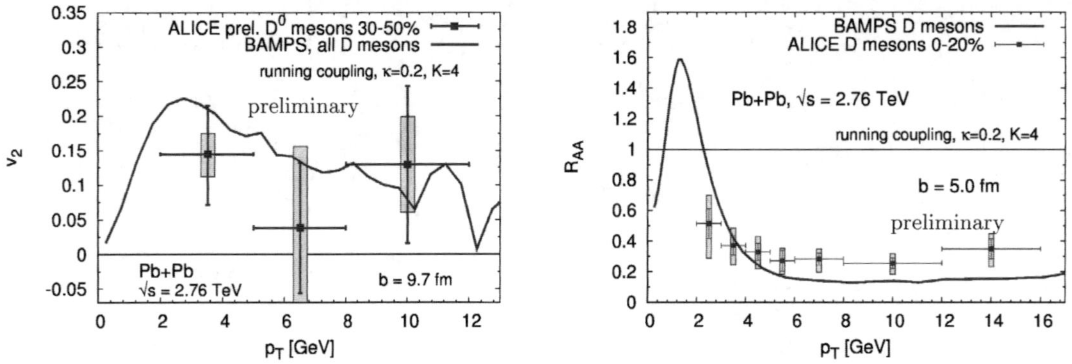

Figure 4. Preliminary BAMPS results on elliptic flow v_2 (left) and nuclear modification factor R_{AA} (right) of D mesons at Pb+Pb collisions at LHC with an impact parameter b together with data [24, 25]. The cross section of $gQ \to gQ$ is multiplied with the factor $K = 4$.

improved Debye screening. The latter means that the screening mass of the t channel of elastic scatterings is determined such that the energy loss matches the energy loss of a heavy quark calculated within the hard thermal loop approach. More information how this matching is done can be found in Ref. [31, 32, 33, 34].

Detailed studies in BAMPS [33, 35, 36, 37, 38] show that elastic energy loss of heavy quarks alone is not compatible with the experimental data at RHIC and LHC. However, elastic energy loss explains a significant portion of the overall suppression. If we employ a running coupling and improved Debye screening the experimental data for both v_2 and R_{AA} for both RHIC and LHC can be explained if the elastic cross section is multiplied with the artificial factor $K = 4$. This indicates that radiative energy loss should be three times larger than the elastic energy loss. However, this must be checked in a forthcoming study. First results on implementing radiative energy loss of heavy quarks in BAMPS look promising [35].

Figure 3 compares our results of the v_2 and R_{AA} at RHIC to the heavy flavor electron data from Ref. [22]. The agreement with the experimental data is very good for both observables if one employs a factor $K = 4$ for the elastic cross section to mimic the effect of radiative energy loss. At LHC for the first time it is possible to reconstruct D mesons and, therefore, distinguish between charm and bottom quarks. In Fig. 4 our results on D mesons is compared to data from ALICE. For the same parameters, that describe the RHIC data, a good agreement is also found at LHC. The suppression of D mesons at LHC is slightly larger than the data. This can be due to a different relation between collisional and radiative processes at LHC compared to RHIC or due to the fact that we represent the rather large centrality class $0 - 20$ % by only one impact parameter. We note that muon data from charm and bottom quarks at forward rapidity is also well described for the same parameters [35].

4. Jet reconstruction within Bamps

Another observable to determine the parton energy loss inside a heavy-ion medium is the reconstruction of full dijets. The initial hard scattering processes of the approaching nucleons lead to back-to-back parton pairs, which gain a high amount of virtuality during these scattering processes. In the subsequent evolution of partons, they try to decrease their virtuality by splitting processes like q $\to$ qg or g $\to$ gg, which can be described by the DGLAP evolution equation [39, 40, 41]. These fragmentation processes lead to particle showers with a broad angle and momentum distribution. In order to provide a description of the energy loss mechanism

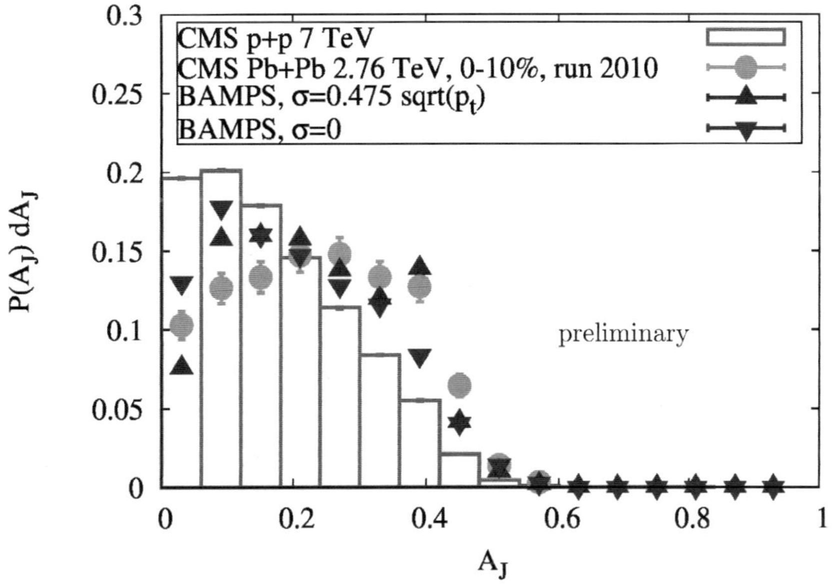

Figure 5. Momentum imbalance A_J in central Pb + Pbcollisions at the LHC with and without gaussian smearing for $\alpha_s = 0.3$ [45].

inside the created medium, jet reconstruction methods [42, 43, 44] are used. They combine single shower particles to a common "full jet" based on their distance $\Delta R = \sqrt{(\Delta y)^2 + (\Delta\phi)^2}$ to the jet axis.

In p + p collisions, where no medium creation is expected, these splitting processes already lead to an imbalance in the momenta of the reconstructed jets with the two highest transverse momenta. These jets are associated with the initial back-to-back parton pair and momentum asymmetry is caused by stochastic distributed vacuum splitting processes out of the considered jet cone. Experimental results [46, 45, 47] in $\sqrt{s_{NN}} = 2.76\,\text{TeV}$ Pb + Pb collisions at the LHC showed an enhancement of this momentum imbalance in central HIC in comparison to p + p-collisions. As a measure of this enhancement the momentum imbalance A_J,

$$A_J = \frac{p_{\text{t;Leading}} - p_{\text{t;Subleading}}}{p_{\text{t;Leading}} + p_{\text{t;Subleading}}}, \qquad (1)$$

is defined, where $p_{\text{t;Leading}}$ ($p_{\text{t;Subleading}}$) is the reconstructed transverse momentum of the jet with the highest (second highest) transverse momentum. The additional suppression of balanced events in HIC are supposed to be the result of different in-medium energy loss of the two partons within the created bulk medium, which is a consequence of a non-central spatial production point of the initial dijet pair.

In this section we present our first preliminary results on momentum imbalance simulated within the transport model BAMPS. For the initial momentum spectra of the partons we use a distribution sampled according to a parametrized parton distribution [48], starting at $p_{t;0} = 100\,\text{GeV}$. Because BAMPS describes only scattering processes of particles on the mass-shell, it is necessary to model the initial splitting processes of the virtual partons properly for reproducing the findings in p + p collisions. Therefore the shower routines of PYTHIA [49] are used to model the virtual splitting processes. Because the medium modification of the created

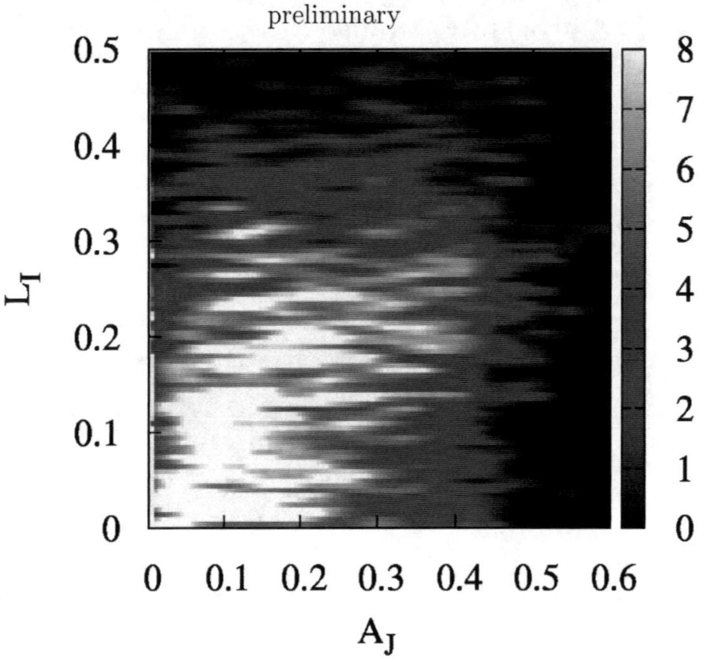

Figure 6. Distribution of length imbalance L_i depending on A_J.

parton showers is to be evaluated within the BAMPS framework, it is necessary to switch off hadronization processes and terminate the splitting processes within PYTHIA prematurely. Therefore, the standard PYTHIA global termination criterion in the virtuality $Q_0 = 1\,\text{GeV}$ is replaced by an energy-dependent minimum virtuality scale $Q_0 = \sqrt{\frac{E_{\text{parton}}}{\tau}}$ depending on the individual parton energy and a global shower time τ. Throughout this section the shower time is assumed as $\tau = 0.2\,\text{fm}$. Calculations within a static medium showed that the energy loss of the reconstructed jets is, for realistic values of τ, nearly independent of the used shower time. The initial spatial production points of the parton pairs are determined by a Glauber modelling of the initial nucleus-nucleus collisions based on a Woods-Saxon density profile.

The so created parton showers are evolved within an offline recorded BAMPS background event. At every timestep the shower particles can interact with medium particles which then become shower particle by their own. With this procedure it is possible to clearly discriminate between shower and background particles.

In the following we compare our simulations with the experimental data measured by CMS. All event trigger conditions by CMS ($p_{\text{t;Leading}} > 120\,\text{GeV}$, $p_{\text{t;Subleading}} > 50\,\text{GeV}$, $\Delta\phi > \frac{2\pi}{3}$ and $|\eta_{\text{jet}}| < 2$) and an effective handling of the detector response and background fluctuations were used. For that an independent Gaussian smearing of the reconstructed jet momenta is applied. The width σ is chosen in such a way that the smeared hadronic PYTHIA events without shower termination can reproduce the measured p + p data by CMS [45].

Figure 5 shows the calculated A_J distribution for central $\sqrt{s_{NN}} = 2.76\,\text{TeV}$ Pb + Pb collisions (0-10%, which corresponds to a mean impact parameter $b = 3.4\,\text{fm}$) with and without smearing of the reconstructed jet momenta. As one can see, already the "true" jet momenta lead to an increase in the momentum asymmetry, though it is insufficient to reproduce the measured experimental data at $A_J > 0.3$. Therefore it can be assumed that the background fluctuations of the medium and the detector response play significant roles in explaining the strong imbalance in dijet momenta.

One advantage of simulations within a full 3+1D transport model is the availability of microscopic particle informations like space and momentum coordinates at every timestep. With this information it is possible to further investigate the processes leading to the observed momentum imbalance. The imbalance of the in-medium path lengths of the parton pair is studied introducing the length imbalance observable L_i

$$L_i = \frac{L_{\text{long}} - L_{\text{short}}}{L_{\text{long}} + L_{\text{short}}} \,. \tag{2}$$

Lower values of L_i correspond to equal paths of the partons inside the medium and thus more central production of the partons. The in-medium path lengths of the initial partons ($L_{\text{long}}/L_{\text{short}}$) are determined by their spatial production point, their initial transverse momentum direction and their distance to the Wood-Saxon surface. Figure 6 shows the distribution of the lenght imbalance in bins of A_J. We observe that the length imbalance seems to be correlated to the underlying momentum asymmetry. One can state that the different transverse momenta of the reconstructed jets are mainly caused by the different in-medium path length of the two initial partons and hence a different energy loss. This suggests that there are events in which the parton pairs are produced in a more peripheral region so that one parton has to travel a longer distance through the medium than the other one, before leaving the collision zone.

We showed that the observation of a momentum imbalance by CMS is in agreement with simulated BAMPS events. Therefore we showed that the consideration of background fluctuations and detector responses plays significant role. This momentum imbalance is caused by a different in-medium path length of the two initial partons. Recent experimental results by CMS [47] with a lower cone radius $R = 0.3$ and lower subleading jet trigger $p_{\text{t;Subleading Jet}} > 30\,\text{GeV}$ show a broader and flatter A_J distribution which can only be explained within BAMPS by usage of a higher cone radius ($R = 0.5$). This implies further investigations of the influence of bulk particles on the momentum of the reconstructed dijets. In addition, to understand the excess on energy loss of single hadrons simulated within BAMPS, it is highly necessary to study the relation between the momentum imbalance A_J and the nuclear modification factor R_{AA}.

5. Transition from ideal to viscous Mach cones in BAMPS

Highly energetic partons propagating through the hot and dense QGP rapidly lose their energy and momentum as the energy is deposited in the medium. Measurements of two- and three-particle correlations in heavy-ion collisions show a complete suppression of the away-side jet, whereas for lower p_T a double peak structure is observed in the two-particle correlation function [7]. One possible and promising origin of these structures is assumed to be the interaction of fast partons with the soft matter which generates collective motion of the medium in form of Mach cones. [8, 50].

For this purpose we investigate the propagation and formation of Mach cones in the microscopic transport model BAMPS (Boltzmann Approach of MultiParton Scatterings) [9] in the limit of vanishing mass and very small shear viscosity over entropy density ratio η/s of the matter. Two different scenarios for the jet are used. In addition, by adjusting η/s, the influence of the viscosity on the profile of the Mach cone and the corresponding two-particle correlation is explored for the first time. The results presented are based on a recent publication [51].

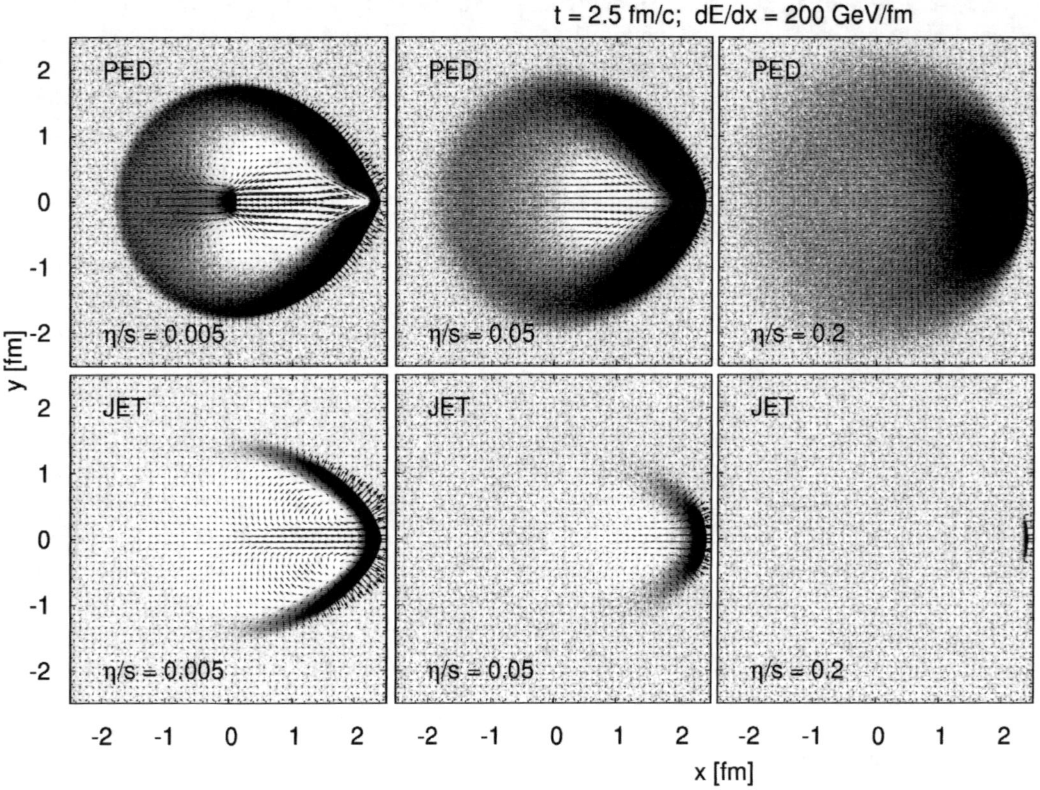

Figure 7. (Color online) Transition from ideal to viscous Mach cones. Shape of a Mach cone shown for different jet scenarios and different viscosity over entropy density ratios, $\eta/s = 0.005$, 0.05 and 0.5. The energy deposition is $dE/dx = 200$ GeV/fm. The upper panel shows the pure energy deposition scenario (PED); the lower panel shows the propagation of a highly energetic jet (JET) depositing energy and momentum in x-direction. Depicted are the LRF energy density within a specific range; as an overlay we show the velocity profile with a scaled arrow length. The results are a snapshot of the evolution at $t = 2.5$ fm/c.

Shock waves are phenomena which have their origin in the collective motion of matter. In a simplified one-dimensional setup shock waves have already been studied within the framework of BAMPS for the perfect fluid limit [52, 53]. Furthermore BAMPS calculations have demonstrated that the shock profile is smeared out when viscosity is large. It was also found that a clear observation of the shock within the short time available in HIC requires a small viscosity.

In the following we study the evolution of "Mach cone"-like structures with different scenarios of the jet-medium interaction by using the parton cascade BAMPS. We focus on investigation of Mach cone evolution in absence of any other effects - i.e. we neglect such effects as initial fluctuations or expansion, which are however relevant in HIC. We use a static box with $T_{\text{med}} = 400$ MeV and binary collisions with an isotropic cross section. Furthermore, we keep the mean free path λ_{mfp} of the medium particles constant in all spatial cells by adjusting the cross section according to $\sigma = 1/(n\lambda_{\text{mfp}})$, where n is the particle density. The related shear viscosity for isotropic binary collisions is given by $\eta = 0.4\, e\, \lambda_{\text{mfp}}$ [54].

The Mach Cones studied here are induced by two different sources. The first of them we refer to as the pure energy deposition scenario (PED) [55]. This is simulated by a moving

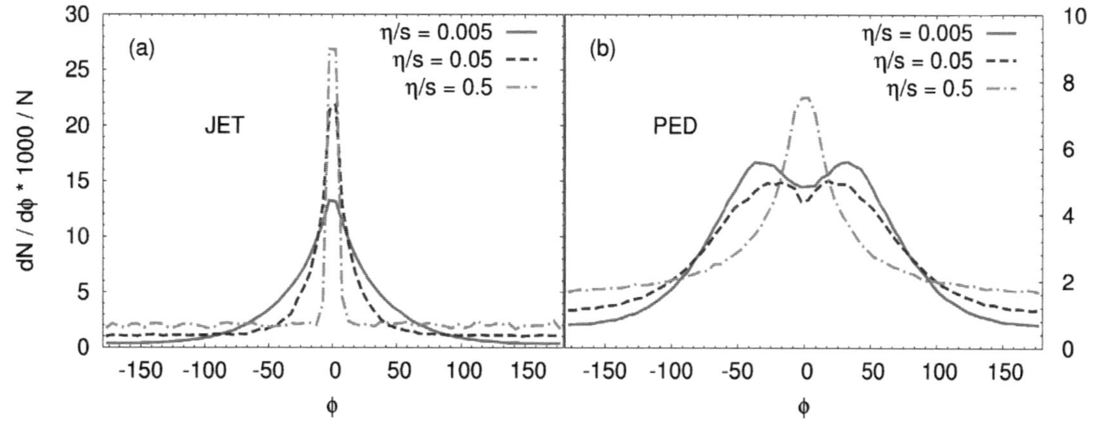

Figure 8. (Color online) Two-particle correlations $dN/(Nd\phi)$ for different viscosities extracted from calculations shown in Fig. 7. The results are shown in the for the JET (a) and PED (b) scenario for $dE/dx = 200$ GeV/fm.

source depositing momentum end energy isotropically according to the thermal distribution $f(x,p) = exp(-E/T)$. The second source we refer to as JET. This is simulated by a highly massless particle (jet) which has only momentum in x-direction, i.e. $p_{\mathrm{x}} = E_{\mathrm{jet}}$. After each timestep the energy of the jet particle is reset to its initial value. For both scenarios the sources are initialized at $t = 0$ fm/c at the position $x = -0.1$ fm and propagate in x-direction with $v_{\mathrm{source}} = 1$, i.e. with the speed of light.

In Fig. 7 we show the Mach Cone structure for both PED scenario (upper panel) and JET scenario (lower panel) with $\eta/s = 0.005$, 0.05 and 0.5 from left to right, respectively. We show a snapshot at $t = 2.5$fm/c. The energy deposition rate is fixed to $dE/dx = 200$ GeV/fm. In both scenarios, PED and JET, for $\eta/s = 0.005$ (left panel), we observe a conical structure, but with obvious differences. The PED case with the isotropic energy deposition induces a spherical shock into back region; this structure is missing in the JET scenario because of the high forward peaked momentum deposition. Another difference is that in the JET scenario a clearly visible head shock appears. This in turn is missing in the PED scenario. Furthermore a (anti)-diffusion wake is induced by the JET (PED) scenario.

Adjusting the shear viscosity over entropy density ratio $\eta/s = 0.05-0.5$ we observe a smearing out of the Mach cone structure. For a sufficient high $\eta/s = 0.5$ the conical structure in both scenarios disappears. This is true for shock fronts as well as for the (anti-) diffusion wake. The difference between the PED and the JET case is that as η/s increases, in the PED scenario the resulting "Mach cone" solution covers approximately the same spatial region regardless of a value of η/s, while in the JET case the structure is concentrated more and more near the projectile as the viscosity increases.

In Fig. 8 we show the two-particle correlations extracted from BAMPS calculations of the Mach Cones shown in Fig. 7. For the JET scenario (a) and sufficiently small $\eta/s = 0.005$ we observe only a peak in direction of the jet. The typical double peak structure, which has been proposed as a possible signature of the Mach cone in HIC, can only be observed for the PED scenario (b) and small η/s. However, the PED scenario has no correspondence in heavy-ion physics. We conclude that Mach cones can not be connected to double peak structures by any realistic picture of jets in HIC. In addition, finite values of the η/s destroy any kind of Mach cone signatures.

Acknowledgements
The authors are grateful to the Center for the Scientific Computing (CSC) at Frankfurt for the computing resources. This work was supported by the Helmholtz International Center for FAIR within the framework of the LOEWE program launched by the State of Hesse.

[1] Stephen Scott Adler et al. Elliptic flow of identified hadrons in Au + Au collisions at s(NN)**(1/2) = 200-GeV. *Phys. Rev. Lett.*, 91:182301, 2003.
[2] John Adams et al. Particle dependence of azimuthal anisotropy and nuclear modification of particle production at moderate p(T) in Au + Au collisions at s(NN)**(1/2) = 200-GeV. *Phys. Rev. Lett.*, 92:052302, 2004.
[3] B. B. Back et al. Centrality and pseudorapidity dependence of elliptic flow for charged hadrons in Au + Au collisions at s(NN)**(1/2) = 200-GeV. *Phys. Rev.*, C72:051901, 2005.
[4] John Adams et al. Transverse momentum and collision energy dependence of high p(T) hadron suppression in Au + Au collisions at ultrarelativistic energies. *Phys. Rev. Lett.*, 91:172302, 2003.
[5] C. Adler et al. Centrality dependence of high p(T) hadron suppression in Au + Au collisions at s(NN)**(1/2) = 130-GeV. *Phys. Rev. Lett.*, 89:202301, 2002.
[6] K. Adcox et al. Suppression of hadrons with large transverse momentum in central Au + Au collisions at s**(1/2)(N N) = 130-GeV. *Phys. Rev. Lett.*, 88:022301, 2002.
[7] Fuqiang Wang. Measurement of jet modification at RHIC. *J. Phys.*, G30:S1299–S1304, 2004.
[8] Horst Stoecker. Collective Flow signals the Quark Gluon Plasma. *Nucl. Phys.*, A750:121–147, 2005.
[9] Zhe Xu and Carsten Greiner. Thermalization of gluons in ultrarelativistic heavy ion collisions by including three-body interactions in a parton cascade. *Phys. Rev.*, C71:064901, 2005.
[10] Andrej El, Zhe Xu, and Carsten Greiner. Thermalization of a color glass condensate and review of the 'Bottom-Up' scenario. *Nucl. Phys.*, A806:287–304, 2008.
[11] Zhe Xu and Carsten Greiner. Shear viscosity in a gluon gas. *Phys. Rev. Lett.*, 100:172301, 2008.
[12] Andrej El, Azwinndini Muronga, Zhe Xu, and Carsten Greiner. Shear viscosity and out of equilibrium dissipative hydrodynamics. *Phys. Rev.*, C79:044914, 2009.
[13] Zhe Xu, Carsten Greiner, and Horst Stocker. PQCD calculations of elliptic flow and shear viscosity at RHIC. *Phys. Rev. Lett.*, 101:082302, 2008.
[14] Zhe Xu and Carsten Greiner. Elliptic flow of gluon matter in ultrarelativistic heavy- ion collisions. *Phys. Rev.*, C79:014904, 2009.
[15] Oliver Fochler, Zhe Xu, and Carsten Greiner. Towards a unified understanding of jet-quenching and elliptic flow within perturbative QCD parton transport. *Phys. Rev. Lett.*, 102:202301, 2009.
[16] K. Aamodt et al. Elliptic Flow of Charged Particles in Pb+Pb Collisions at $\sqrt{s_{NN}} = 2.76$ TeV. *Phys. Rev. Lett.*, 105:252302, 2010.
[17] John Adams et al. Azimuthal anisotropy in Au + Au collisions at s(NN)**(1/2) = 200-GeV. *Phys. Rev.*, C72:014904, 2005.
[18] K. Aamodt et al. Suppression of charged particle production at large transverse momentum in central Pb+Pb collisions at $\sqrt{s_{NN}} = 2.76$ TeV. *Phys.Lett.*, B696:30–39, 2011.
[19] S. Albino, B.A. Kniehl, and G. Kramer. AKK update: Improvements from new theoretical input and experimental data. *Nucl.Phys.*, B803:42–104, 2008.
[20] Oliver Fochler, Zhe Xu, and Carsten Greiner. Energy loss in a partonic transport model including bremsstrahlung processes. 2010.
[21] L. D. Landau and I. Pomeranchuk. Limits of applicability of the theory of bremsstrahlung electrons and pair production at high energies. *Dokl. Akad. Nauk Ser. Fiz.*, 92:535–536, 1953.
[22] A. Adare et al. Heavy quark production in $p+p$ and energy loss and flow of heavy quarks in au+au collisions at $\sqrt{s_N N} = 200$ gev. *Phys.Rev.*, C84:044905, 2011.
[23] Jan Uphoff, Oliver Fochler, Zhe Xu, and Carsten Greiner. Heavy quark production at RHIC and LHC within a partonic transport model. *Phys. Rev.*, C82:044906, 2010.
[24] B. Abelev et al. Suppression of high transverse momentum D mesons in central Pb–Pb collisions at $\sqrt(s_{NN}) = 2.76$ TeV. 2012.
[25] Chiara Bianchin. Measurement of D^0 v_2 in Pb-Pb collisions at $\sqrt(s_{NN}) = 2.76$ TeV with ALICE at the LHC. 2011.
[26] B. I. Abelev et al. Transverse momentum and centrality dependence of high-pt non-photonic electron suppression in Au+Au collisions at $\sqrt{s_{NN}} = 200$ GeV. *Phys. Rev. Lett.*, 98:192301, 2007.
[27] A. Adare et al. Energy Loss and Flow of Heavy Quarks in Au+Au Collisions at $\sqrt{s_{NN}} = 200$ GeV. *Phys. Rev. Lett.*, 98:172301, 2007.

[28] S. Masciocchi. Investigation of charm and beauty production via semileptonic decays of heavy-flavour hadrons in pp at 7 TeV and Pb–Pb at 2.76 TeV with ALICE. *J.Phys.G*, G38:124069, 2011.
[29] Yuri L. Dokshitzer and D. E. Kharzeev. Heavy quark colorimetry of QCD matter. *Phys. Lett.*, B519:199–206, 2001.
[30] Raktim Abir, Carsten Greiner, Mauricio Martinez, Munshi G. Mustafa, and Jan Uphoff. Soft gluon emission off a heavy quark revisited. *Phys.Rev.*, D85:054012, 2012.
[31] P. B. Gossiaux and J. Aichelin. Towards an understanding of the RHIC single electron data. *Phys. Rev.*, C78:014904, 2008.
[32] Andre Peshier. Turning on the Charm. 2008.
[33] Jan Uphoff, Oliver Fochler, Zhe Xu, and Carsten Greiner. Elliptic flow and energy loss of heavy quarks in ultra-relativistic heavy ion collisions. *Phys.Rev.*, C84:024908, 2011.
[34] Alex Meistrenko, Andre Peshier, Jan Uphoff, and Carsten Greiner. Collisional energy loss of heavy quarks. 2012. 7 pages, 8 figures, 1 table, submitted to Physics Letters B.
[35] Jan Uphoff, Oliver Fochler, Zhe Xu, and Carsten Greiner. Open heavy flavor at RHIC and LHC in a partonic transport model. 2011.
[36] Oliver Fochler, Jan Uphoff, Zhe Xu, and Carsten Greiner. Jet quenching and elliptic flow at RHIC and LHC within a pQCD-based partonic transport model. *J.Phys.G*, G38:124152, 2011.
[37] Jan Uphoff, Oliver Fochler, Zhe Xu, and Carsten Greiner. Heavy quarks at RHIC and LHC within a partonic transport model. *Nucl.Phys.*, A855:444–447, 2011.
[38] Jan Uphoff, Oliver Fochler, Zhe Xu, and Carsten Greiner. Production, elliptic flow and energy loss of heavy quarks in the quark-gluon plasma. 2010.
[39] V. N. Gribov and L. N. Lipatov. Deep inelastic e p scattering in perturbation theory. *Sov. J. Nucl. Phys.*, 15:438–450, 1972.
[40] Yuri L. Dokshitzer. Calculation of the Structure Functions for Deep Inelastic Scattering and e+ e- Annihilation by Perturbation Theory in Quantum Chromodynamics. *Sov. Phys. JETP*, 46:641–653, 1977.
[41] Guido Altarelli and G. Parisi. Asymptotic Freedom in Parton Language. *Nucl. Phys.*, B126:298, 1977.
[42] Gavin P Salam. Towards jetography. *The European Physical Journal C*, 67(3-4):637–686, 2010.
[43] Matteo Cacciari, Gavin P Salam, and Gregory Soyez. The anti-k_t jet clustering algorithm. *Journal of High Energy Physics*, (04), 2008.
[44] Matteo Cacciari, Juan Rojo, Gavin P Salam, and Gregory Soyez. Jet Reconstruction in Heavy Ion Collisions. October 2010.
[45] S. Chatrchyan, V. Khachatryan, A. Sirunyan, A. Tumasyan, W. Adam, et al. Observation and studies of jet quenching in PbPb collisions at $\sqrt{s}_{NN} = 2.76$ TeV. *Physical Review C*, 84(2), 2011.
[46] G. Aad et al. Observation of a Centrality-Dependent Dijet Asymmetry in Lead-Lead Collisions at $\sqrt{s}_{NN} = 2.76$ TeV with the ATLAS Detector at the LHC. *Physical Review Letters*, 105(25):1–18, 2010.
[47] CMS-Collaboration. Jet momentum dependence of jet quenching in PbPb collisions at sqrt(sNN)=2.76 TeV. February 2012.
[48] M. Glück, E. Reya, and A. Vogt. Dynamical parton distributions of the proton and small x physics. *Z. Phys.*, C67:433–448, 1995.
[49] Torbjrn Sjstrand, Stephen Mrenna, and Peter Skands. PYTHIA 6.4 physics and manual. *Journal of High Energy Physics*, (05):026–026, 2006.
[50] Ioannis Bouras et al. Relativistic Shock Waves and Mach Cones in Viscous Gluon Matter. *J. Phys. Conf. Ser.*, 230:012045, 2010.
[51] I. Bouras, A. El, O. Fochler, H. Niemi, Z. Xu, et al. Transition From Ideal To Viscous Mach Cones In A Kinetic Transport Approach. *Phys.Lett.*, B710:641–646, 2012.
[52] I. Bouras et al. Relativistic shock waves in viscous gluon matter. *Phys. Rev. Lett.*, 103:032301, 2009.
[53] I. Bouras et al. Investigation of shock waves in the relativistic Riemann problem: A comparison of viscous fluid dynamics to kinetic theory. *Phys. Rev.*, C82:024910, 2010.
[54] S. R. de Groot et al. *Relativistic Kinetic Theory: Principles and Applications.* North Holland, Amsterdam, 1980.
[55] Barbara Betz, Jorge Noronha, Giorgio Torrieri, Miklos Gyulassy, Igor Mishustin, et al. Universality of the Diffusion Wake from Stopped and Punch-Through Jets in Heavy-Ion Collisions. *Phys.Rev.*, C79:034902, 2009.

Direct Photons: Flow, Thermal Yield and High p_T R_{AA} from the PHENIX experiment at RHIC

Takao Sakaguchi, for the PHENIX Collaboration
Brookhaven National Laboratory, Physics Department, Upton, NY 11973-5000, USA
E-mail: takao@bnl.gov

Abstract. Electromagnetic radiation has been of interest in heavy ion collisions because it sheds light on early stages of the collisions where hadronic probes do not provide direct information since hadronization and hadronic interactions occur later. The latest results on direct photon measurements in Au+Au collisions together with ones in d+Au collisions from the PHENIX experiment at RHIC provide thermodynamic properties of the matter produced in the heavy ion collisions. An unexpectedly large positive elliptic flow measured for direct photons are hard to be explained by many models.

1. Introduction

Many intriguing phenomena have been observed at RHIC since it started running in 2000. The high transverse momentum (p_T) hadron production from the initial hard scattering was observed, and the large suppression of their yields suggested that the matter is sufficiently dense to cause parton energy loss prior to hadronization [1]. The large elliptic flow of particles and its scaling in terms of particle kinetic energy suggests that the system is locally in equilibrium as early as $0.3\,\mathrm{fm}/c$, and the flow occurs already on the partonic level.

Because photons interact with the medium and other particles only electromagnetically and are largely unaffected by final state interactions, they serve as a direct and penetrating probe of the early stages at high temperature and high density [2]. At leading order, the production processes of photons are annihilation ($q\bar{q} \to \gamma g$) and Compton scattering ($qg \to \gamma q$) (Figure 1). Their yields are proportional to $\alpha\alpha_s$, where α is the electromagnetic coupling constant and α_s is the strong coupling constant, and are ~40 times lower than those of hadrons from strong interactions ($\propto \alpha_s^2$). A calculation predicts that the photon contribution from the QGP state is predominant in the p_T range of $1< p_T <3\,\mathrm{GeV}/c$ [3]. For $p_T >3\,\mathrm{GeV}/c$, the signal is dominated by a contribution from initial hard scattering, and $p_T <1\,\mathrm{GeV}$, the signal is from hadron gas through processes of $\pi\pi(\rho) \to \gamma\rho(\pi)$, $\pi K^* \to K\gamma$ and etc. Figure 2 shows a landscape of photon sources as a function of the time they are produced. The vertical axis corresponds to transverse momenta of photons. We have one additional degree of freedom, virtual mass, in photon measurement, which will be explained in detail in a later section. These photons can be measured after a huge amount of background photons coming from hadron decays (π^0, η, η' and ω, etc.) are subtracted off from inclusive photon distributions. The typical signal to background ratio is ~1 % at 2 GeV, and ~10 % at 5 GeV in case of p+p collisions. The signal from QGP is predicted to be ~10 % of the inclusive photons. For Au+Au collisions, thanks to a large suppression of high p_T hadrons, the ratio is enhanced by the same degree. PHENIX [4]

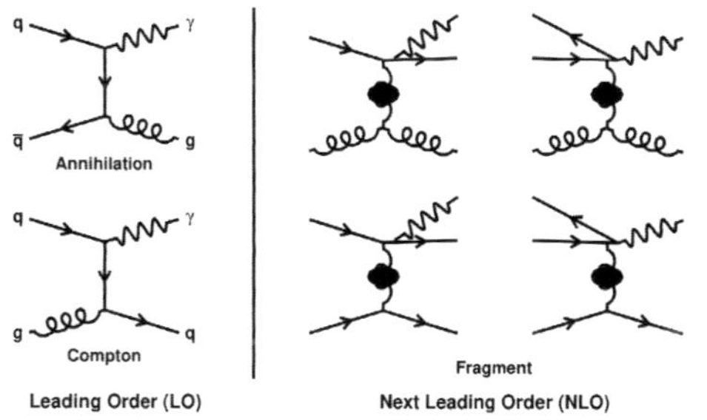

Figure 1. Photon production process.

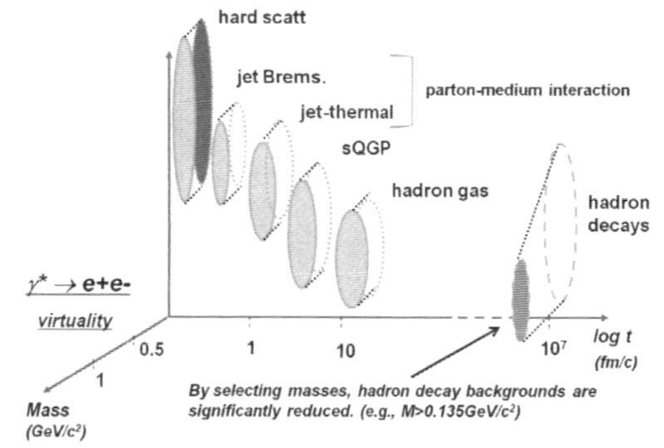

Figure 2. Sources of photons from various stages of collisions.

has measured photons throughout the first decade of RHIC operations. We present here the latest results on direct photons.

2. Measurement of direct photons in p+p collisions

One of the big successes by now in electromagnetic radiation measurements is the observation of high p_T direct photons that are produced in initial hard scattering [5] in relativistic heavy ion collisions. Figure 3 shows the nuclear modification factors (R_{AA}) for direct photons, π^0 and η for 0-10 % central Au+Au collisions at the same center-of-mass (cms) energy. R_{AA} is defined as the ratio of the yield in nucleus-nucleus collisions divided by that in p+p collisions scaled by T_{AA}. The high p_T hadron suppression is interpreted as a consequence of an energy loss of hard-scattered partons in the hot and dense medium. It was strongly supported by the fact that the high p_T direct photons are not suppressed. The small suppression seen in the highest p_T is likely due to the fact that the ratio of the yields in Au+Au to p+p was computed without taking the isospin dependence of direct photon yields into account [6]. Figure 4 shows the same direct photon R_{AA} as Fig. 3, with a theory calculation with isospin effect [6]. The large uncertainty on the measurement at high p_T partly comes from the statistical errors in p+p measurement.

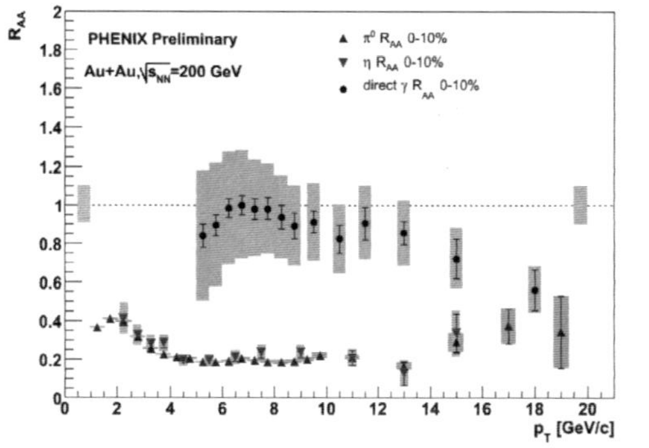

Figure 3. Nuclear modification factors (R_{AA} for photons, π^0 and η in 10 % central Au+Au collisions at $\sqrt{s_{NN}}$=200 GeV.

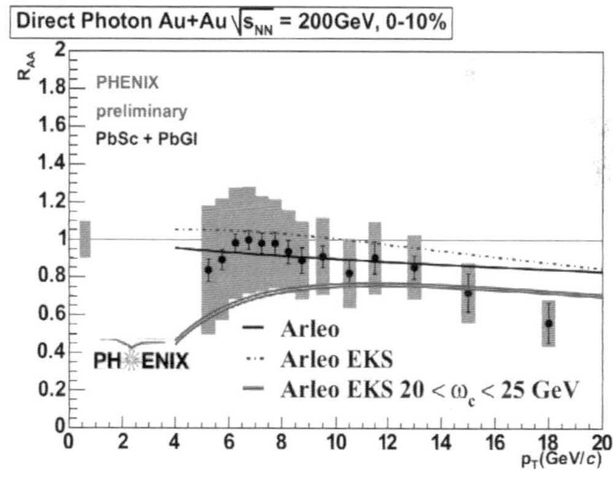

Figure 4. R_{AA} for direct photons in 10 % central Au+Au collisions with thoery calculations.

We have measured the direct photons in p+p collisions with RHIC Year-6 data which has larger statistics compared to the one used for calculating R_{AA} in Figs. 3 and 4, as shown in Fig. 5. A

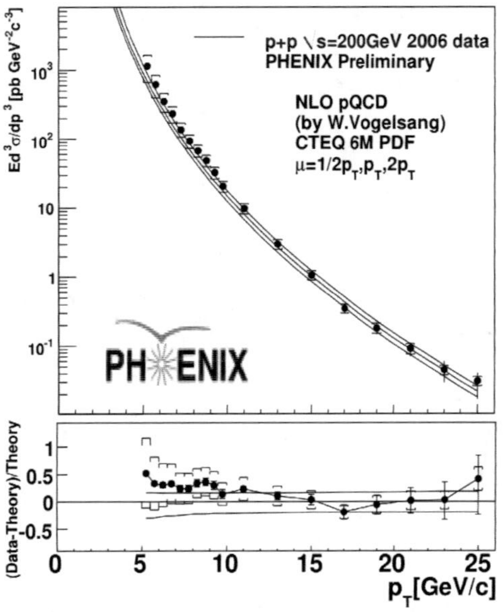

Figure 5. Direct photon spectra in p+p collisions at $\sqrt{s}$=200 GeV from RHIC Year-6 run.

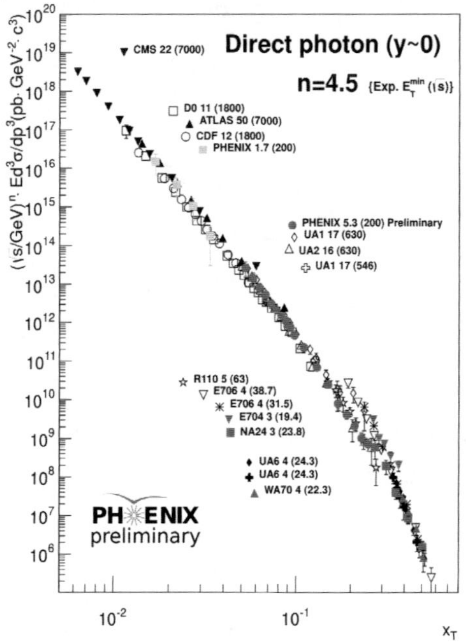

Figure 6. x_T-scaling of direct photons from various experiments. The power used here is n=4.5.

NLO pQCD calculation [7] describes the data well given the quoted systematic uncertainties in p+p measurement. In the pQCD picture, the cross-section should scale with x_T ($\equiv 2p_T/\sqrt{s}$), and can be expressed as the following formula:

$$E\frac{d^3\sigma}{dp^3} = \frac{1}{\sqrt{s}^{n_{eff}(x_T,\sqrt{s})}}G(x_T)$$

where n_{eff} is constant and is 4 for the LO case. We plotted this new data in x_T with the data from other experiments whose center of mass energies vary over two order of magnitudes (from 20 GeV to 7 TeV). We also included the low p_T photons obtained by the virtual photon channel as described in the following section. We found that n_{eff} = 4.5 gives an universal scaling of the cross-section for all the data collected. This is consistent with the other study shown in a literature [8]. This fact implies that the direct photon cross-section is explained very well with the pQCD framework over wide center of mass energies. The calculation of R_{AA} of direct photons in Au+Au collisions using this p+p data is now in progress for publication.

3. Measurement of direct photons through its internal conversion

There is a huge background arising from π^0 decaying into two photons, which makes it very difficult to look at the direct photon signal at low p_T, where thermal photons from QGP manifest, with traditional calorimetry of (real) photons. However, if we look at photons with a small mass

Journal of Physics: Conference Series **389** (2012) 012020 doi:10.1088/1742-6596/389/1/012020

(virtual photons) instead, we can select the mass region where π^0 contribution ceases (Fig 7). For the case of $p_T >> M$, the yield of virtual photons is expected to be dominated by internal conversion of real photons [9, 10]. For obtaining direct photon yield, we fit the measured invariant mass distribution with the function:

$$F = (1-r)f_c + rf_d,$$

where f_c is the cocktail calculation (photons from various hadron decays), f_d is the mass distribution for direct photons, and r is the free parameter in the fit. Next, using the Kroll-Wada formula [11] to account for the Dalitz decays of π^0, η and direct photons, r is defined as the ratio of direct photons to inclusive photons:

$$r = \frac{\gamma^*_{\text{dir}}(m_{ee} > 0.15)}{\gamma^*_{\text{inc}}(m_{ee} > 0.15)} \propto \frac{\gamma^*_{\text{dir}}(m_{ee} \approx 0)}{\gamma^*_{\text{inc}}(m_{ee} \approx 0)} = \frac{\gamma_{\text{dir}}}{\gamma_{\text{inc}}} \equiv r_\gamma$$

Then, the invariant yield of direct photons is calculated as $\gamma_{\text{inc}} \times r_\gamma$. As described in [10], the procedure is demonstrated in Fig 7 for 1.0< p_T <1.5 GeV/c. The dotted lines show the

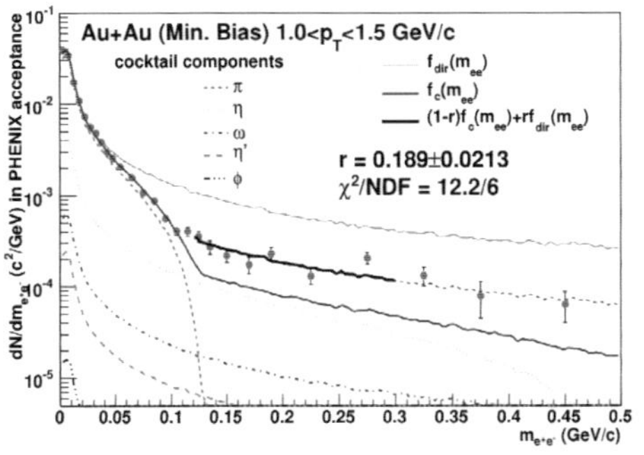

Figure 7. Invariant mass distributions of electron-pairs and comparison with possible hadron sources of electron-pairs.

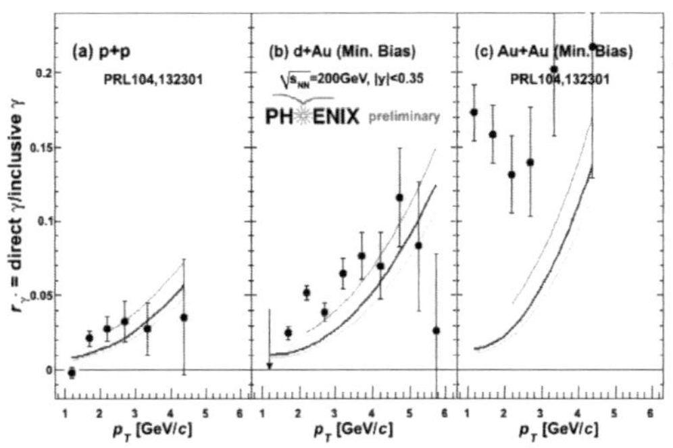

Figure 8. Ratio of direct photons to background photons (r values).

contributions from various hadrons, the solid blue is the sum of these contributions, and the solid red line shows the distribution from direct photons converted internally. The r value is determined by the fit to the data. The error of the fit corresponds to the statistical error. We applied the procedure as a function of p_T for various centrality selections in p+p and Au+Au collisions. Figure 8 shows the r value for p+p, d+Au minimum bias and Au+Au minimum bias collisions at $\sqrt{s_{NN}}$ =200 GeV. The lines are NLO pQCD expectations scaled by corresponding thickness functions. We then obtained the p_T spectra as shown in Fig 9. The distributions are for 0–20 %, 20–40 % centrality and minimum bias events for Au+Au collisions. For p_T <2.5 GeV/c the Au+Au yield is visibly higher than the scaled p+p yield. The distributions were fitted with the p+p fit plus exponential function to obtain slopes and dN/dy for three centralities. The slopes are estimated to be ~220 MeV. The lines show the theoretical expectation from a literature [3]. One may question whether or not the excess arises from a source that exists only in Au+Au collisions. For example, an effect that could increase the yield is cold-nuclear-matter (CNM) effect such as k_T broadening (Cronin effect). To quantify the contribution we analyzed 2008 d+Au data with the same procedure [12]. Figure 10 shows the Au+Au yield compared to the d+Au yield scaled by N_{coll}. It clearly shows that there is an enhancement over CNM effects in Au+Au collisions.

Journal of Physics: Conference Series **389** (2012) 012020 doi:10.1088/1742-6596/389/1/012020

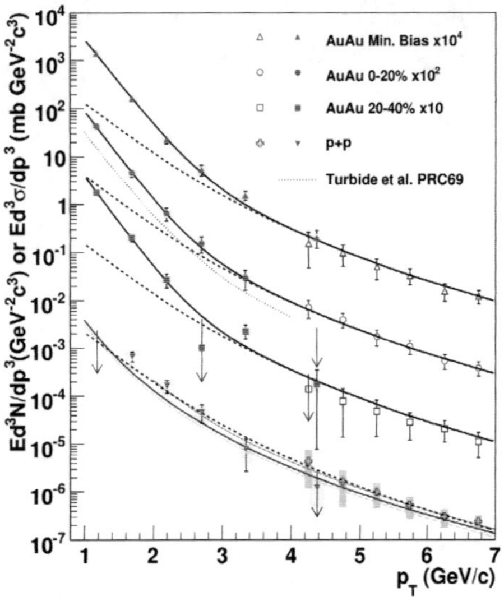

Figure 9. Direct photon spectra obtained from the measurement of internal conversion of photons in Au+Au collisions.

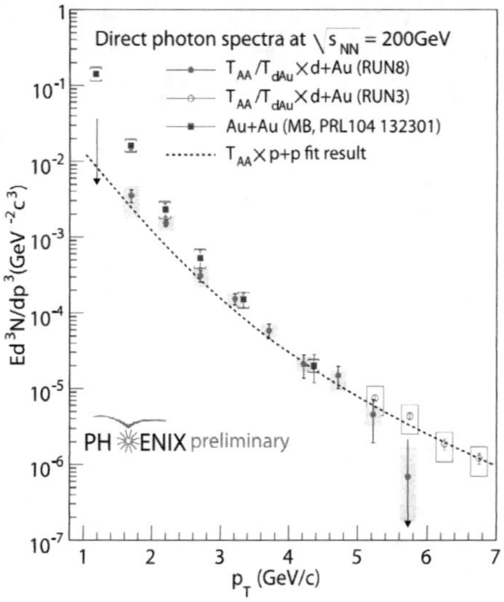

Figure 10. Direct photon yield in Au+Au and d+Au collisions scaled by the difference of N_{coll}.

4. Probing nuclear effect using direct photons and π^0

As mentioned above, the direct photons in d+Au collisions give an insight to the initial state of the collisions, since they are emitted from the initial stage directly, penetrate the medium and provide the kinematic information. Figure 11 shows the R_{dA} of the direct photons obtained from the spectra in Fig. 10 and direct real photons measured using RHIC Year-3 d+Au data. The R_{dA} is consistent with 1 within quoted errors. In Fig. 12, we compared the data with calculations including Cronin enhancement (k_T broadening), isospin effect, anti-shadowing and energy loss of partons in the cold nuclear matter [13]. All the calculations are consistent with the data given the quoted errors. The R_{dA} of direct photons provide a test of the recombination model by

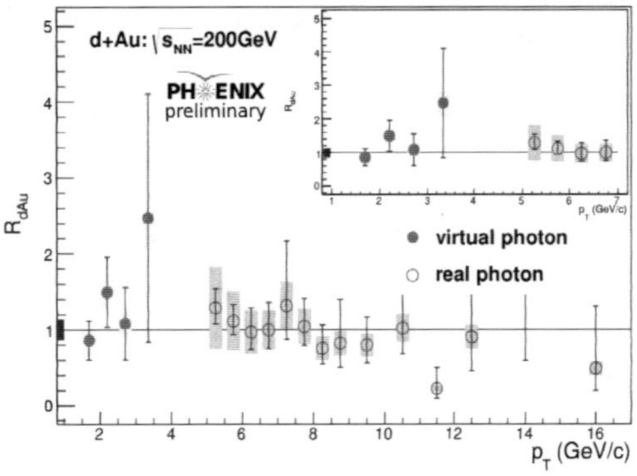

Figure 11. R_{dA} of direct photons in d+Au collisions.

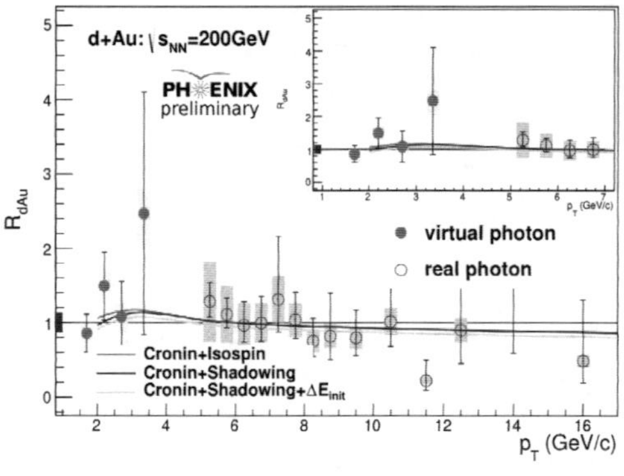

Figure 12. R_{dA} of direct photons in d+Au collisions with theory calculations from Vitev [13].

comparing with the R_{dA} of π^0's [14]. The recombination model predicts that the R_{dA} for direct photons is smaller than that of π^0's, since the k_T broadening is doubly effective in creating π^0's. Figure 13 shows the π^0 R_{dA} in a previous publication [15]. Within the current statistical and

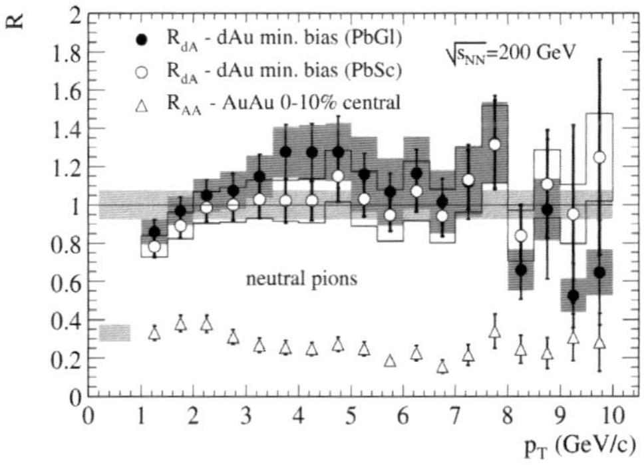

Figure 13. R_{dA} of π^0's published in a previous publication [15].

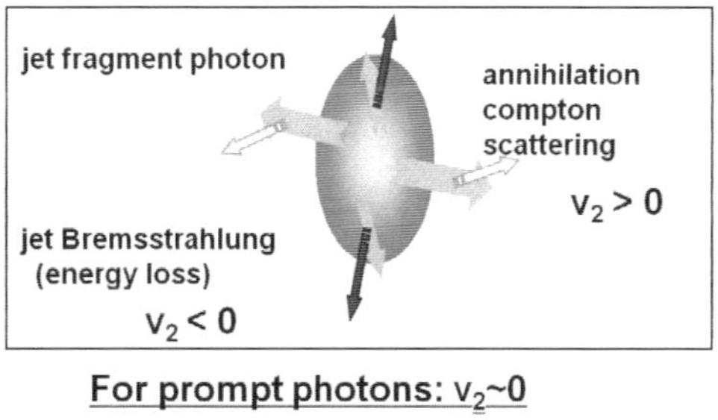

Figure 14. Source dependence of elliptic flow (v_2) of direct photons.

systematic uncertainties, it is difficult to conclude whether or not the recombination mechanism plays a significant role in π^0 production.

5. Elliptic flow of direct photons –A photon source detector–

On exploring the matter produced, one wants to explore a new degree of freedom of the observables. The angular dependence of the photon yield with respect to the plane defined by impact parameter (event plane) is one of the degrees that can be investigated. It is predicted that the second order of the Fourier transfer coefficient (v_2, elliptic flow) of angular distributions of photons show the different sign and/or magnitude, depending on the production processes [16] (Fig. 14). The observable is powerful to disentangle the contributions from various photon sources in the p_T region where they intermix. The photons from hadron-gas interaction and thermal radiation may follow the collective expansion of a system, and give a positive v_2. The amount of photons produced by jet-photon conversion or in-medium bremsstrahlung increases as the medium to traverse increases. Therefore these photons show a negative v_2. The fragmentation photons will give positive v_2 since larger energy loss of jets is expected orthogonal to the event plane.

PHENIX has measured the v_2 of direct photons by subtracting the v_2 of hadron decay photons off from that of the inclusive photons, following the formula below:

$$v_2{}^{dir.} = (R \times v_2{}^{incl.} - v_2{}^{bkgd.})/(R-1), \quad R = (\gamma/\pi^0)_{meas}/(\gamma/\pi^0)_{bkgd}$$

The centrality dependence of elliptic flows of π^0 and inclusive photons are shown in Fig. 15(a), (c) and (e), and the ones for direct photons are shown in Fig. 15(b), (d) and (f) [17]. The v_2 of direct photons is large and positive, and comparable to the flow of hadrons for p_T <3 GeV/c. It is also seen that the magnitude increases as going from central to mid-central collisions. This large positive flow is hard to be explained by many models. Several models qualitatively predicted the positive flow of the photons assuming the photons are boosted with hydrodynamic expansion of the system, but the amount is significantly lower than the measurement [18]. There is one model that gives relatively large flow by including hadron-gas interaction shown as Fig. 16 [19]. The model changed the relative ratio of the various sources of photons since the previous publication [20]. As a result, the photons from hadron gas interaction dominate for

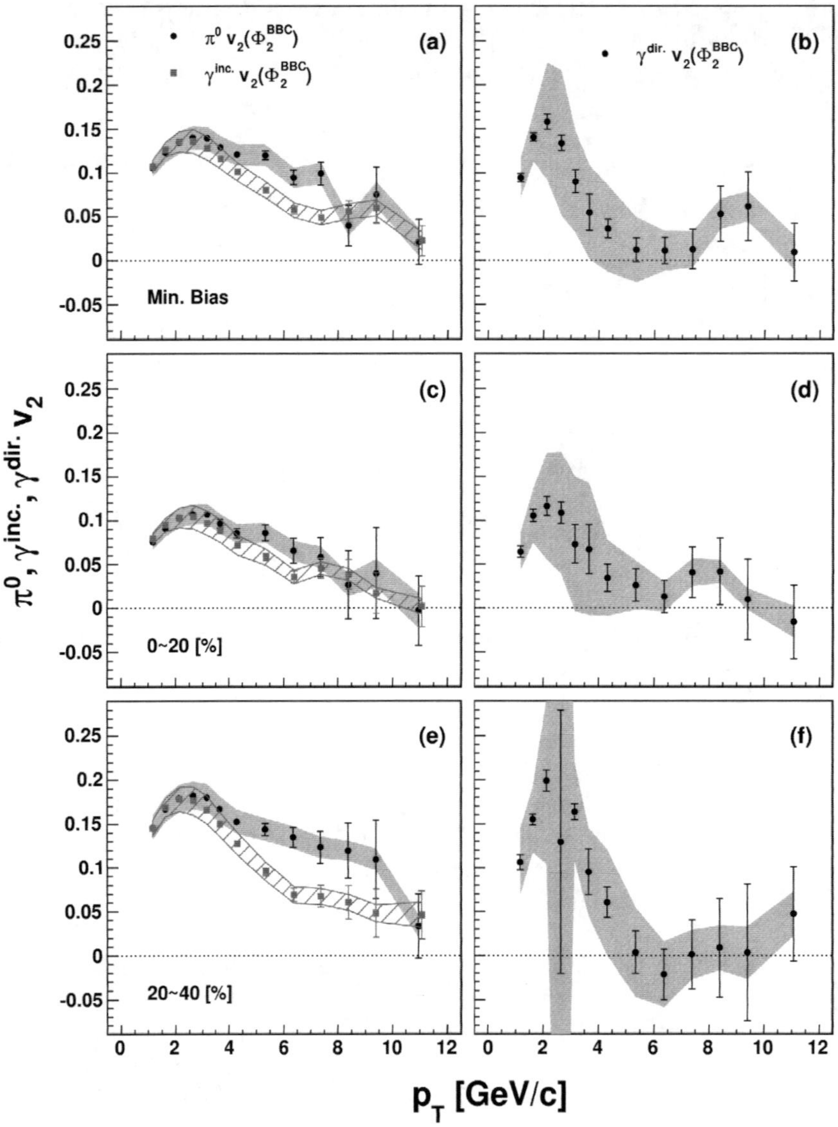

Figure 15. Elliptic flow of (a, c and e, left) π^0 and inclusive photons and (b, d, and f, right) direct photons in Au+Au collisions.

p_T <2.5 GeV/c, as shown in Fig. 17. The model describes both v_2 and the yield for the 0-20 % centrality data rather well, but less in 20-40 % centrality.

6. Summary

Direct photons are a powerful tool to investigate the collision dynamics. PHENIX has measured direct photons over wide p_T ranges, including hard scattering and thermal photons, and extracted quantities, such as slope parameters, that reflect thermodynamic properties of the matter. The measurement in d+Au proved that the enhancement of the yield in low p_T in Au+Au collisions is not from the cold nuclear matter effect. The d+Au direct photon yields

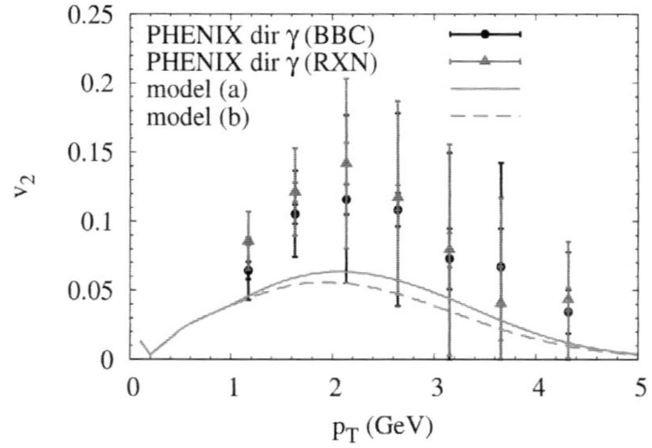

Figure 16. A model calculation for v_2 for direct photons in 0-20 % central Au+Au collisions [19].

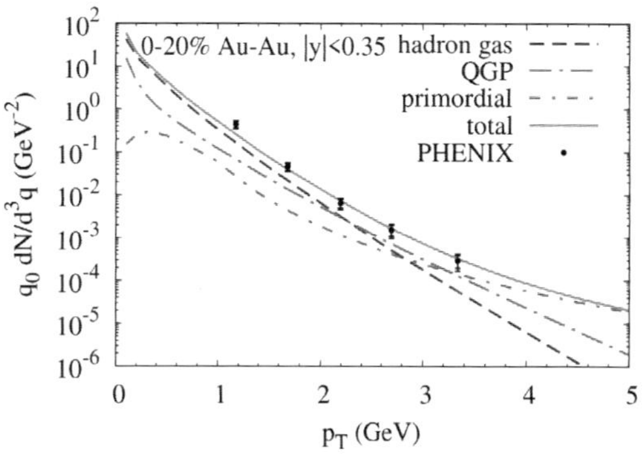

Figure 17. Direct photon invariant yield from the same model that explained the v_2 in Fig. 16.

are compared with a model including k_T broadening, isospin, anti-shadowing and parton energy loss in the cold nuclear matter. Within quoted errors, the model calculations are consistent with data. A comparison of the R_{dA} of π^0's and direct photons is a test of the recombination model, but with the current errors, it is hard to make a statement on the validity of the model. An unexpectedly large positive elliptic measured for direct photons are hard to be explained by many models except for the one assuming photons from hadron-gas interaction is dominant for $p_T < 2.5\,\mathrm{GeV}/c$.

References

[1] K. Adcox, et al. (PHENIX Collaboration), Nucl. Phys. A757 (2005) 184.
[2] P. Stankus, Ann. Rev. Nucl. Part. Sci. 55 (2005) 517.
[3] S. Turbide, R. Rapp, C. Gale, Phys. Rev. C 69 (2004) 014903.
[4] K. Adcox, et al., Nucl. Instrum. Meth. A499 (2003) 469.
[5] S. S. Adler *et al.* (PHENIX Collaboration), Phys. Rev. Lett. 94, 232301 (2005).
[6] F. Arleo, JHEP **0609**, 015 (2006).
[7] L. E. Gordon, W. Vogelsang, Phys. Rev. D 48 (1993) 3136.
[8] D. d'Enterria and J. Rojo, Nucl. Phys. B **860**, 311 (2012).
[9] A. Adare, et al. (PHENIX Collaboration), Phys. Rev. Lett. 104 (2010) 132301.
[10] A. Adare, et al. (PHENIX Collaboration), Phys. Rev. C 81 (2010) 034911.
[11] N. M. Kroll, W. Wada, Phys. Rev. 98 (1955) 1355.
[12] T. Sakaguchi, Nucl. Phys. **A855**, 141 (2011).
[13] I. Vitev and B. -W. Zhang, Phys. Lett. B **669**, 337 (2008).
[14] R. C. Hwa, Eur. Phys. J. C **43**, 233 (2005).
[15] S. S. Adler *et al.* [PHENIX Collaboration], Phys. Rev. Lett. **96**, 202301 (2006).
[16] S. Turbide, C, Gale and R.J. Fries, Phys. Rev. Lett. 96, 032303 (2006); R. Chatterjee et al., Phys. Rev. Lett. 96, 202302 (2006)
[17] A. Adare *et al.* [PHENIX Collaboration], arXiv:1105.4126 [nucl-ex].
[18] R. Chatterjee, D. Srivastava, Phys. Rev. C 79 (2009) 021901.
[19] H. van Hees, C. Gale and R. Rapp, Phys. Rev. C **84**, 054906 (2011).
[20] S. Turbide, C. Gale, E. Frodermann and U. Heinz, Phys. Rev. C **77**, 024909 (2008).

28th Winter Workshop on Nuclear Dynamics 2012 IOP Publishing
Journal of Physics: Conference Series **389** (2012) 012021 doi:10.1088/1742-6596/389/1/012021

Direct photon-hadron and Di-hadron correlations measured in PHENIX

Justin Frantz and Dmitri Kotchetkov

Department of Physics and Astronomy, Ohio University, Athens, OH 45701, U.S.A.

Abstract. We review results of PHENIX two-particle correlations for both di-hadrons and direct photon hadron pairs. These measurements can reveal details of QGP induced jet energy loss in heavy ion collisions. The spectra of per-trigger yields on the awayside from trigger photons in the photon-hadron correlations or fragmentation function for direct photons of greater than 5 GeV/c in transverse momentum appear to show some modification compared to the same measurement in p+p collisions, including a suppression at high z (> ~0.4), and much less suppression, possibly even enhancement at low z. A framework for calculating parton or jet level observables from the PHENIX 2-particle yields results that can be compared to recent LHC jet reconstruction observables. Under certain assumptions, one of these observables, the mean fractional energy loss, appears to be larger for lower jet energies (~10 GeV) that the PHENIX data is sensitive to, than for much higher energy jets (~100 GeV) at the LHC. However the current PHENIX di-hadron measurements for this comparison which may have inherent biases need to be followed up by higher statistics direct photon-hadron measurements before definite statements could be made.

1. Fragmentation function

Over the course of the Relativistic Heavy Ion Collider (RHIC) physics program, two-particle correlations proved to be an effective probe to characterize the medium created in nucleus-nucleus collisions. Particularly, two-particle correlations can be used to study jet energy loss due to the medium. One can assign a certain threshold for a transverse momentum p_T and select outgoing trigger particles that satisfy that threshold. In every event, a difference in measured azimuths $\Delta\phi$ between trigger particles and all other, so-called "associated", particles describe how the trigger and associated particles correlate with each other. The spectrum of $\Delta\phi$ in unmodified hard scattering di-jet events has two pronounced peaks, one at about $\Delta\phi = 0$ and the second at about $\Delta\phi = \pi$. The first peak represents correlations between the triggers and those associated particles that belong to the same jet as the trigger particle. The second peak shows how the triggers correlate with the associated particles, members of the jet that opposes the trigger. In a heavy ion collision a suppression of the distribution around $\Delta\phi = \pi$, in regard to the similar distribution measured in p+p collisions, indicate energy loss of the jets that pass through the created medium.

The fragmentation function *D(z)* describes the probability for outgoing parton in hard scattering collision to form a hadron with momentum fraction $z = p^h/p^{jet}$ (p^h is the momentum of the hadron and p^{jet} is the momentum of the jet; here it is implied that the momentum of the jet approximately equal to the momentum of the parton). PHENIX, as a collider detector with a modest geometrical acceptance, successfully studies hard scattering events with z approximated by its transverse component $z_T = |p_T^h|/|p_T^{jet}|$. As it is defined, the fragmentation function *D(z)* is a characteristic of the final state of the hard scattering process; thus *D(z)* is supposed to be independent of the colliding species. But various factors in other stages of the collision process might affect *D(z)*. For example, patrons can traverse through the dense medium and lose their energy. That would result in the lower transverse momentum at the start of the fragmentation. Thus one can quantify effects of the medium by comparing *D(z)* measured in heavy ion collisions with *D(z)* measured in p+p collisions.

2. Direct photons as a "tomographic" tool

Direct photons can be produced in hard scattering events through processes of quark-antiquark annihilation and Compton-like scattering. Also, direct photons can be produced through fragmentation of outgoing partons. Photons do not participate in strong interactions, so once produced in heavy ion collisions, they pass through the medium unaffected. Conservation of momentum requires a photon and an opposing jet to have equal transverse momenta. Thus, *D(z)* can be directly measured with z_T defined as $z_T = |p_T^h|/|p_T^{\gamma_{dir}}|$. Contrary to γ-hadron correlations, hadron-hadron correlations can yield improved *D(z)* measurements when the hard scattering events occur on or near the surface of the medium so one of the outgoing partons, presumably, do not suffer from a sizable loss of energy. It is evident that certain bias is inherent in *D(z)* measurements from hadron-hadron correlations.

A method of using γ-hadron correlations to measure fragmentation functions is naturally very attractive. However, in practice such a task is very challenging; the rates of direct photon production processes are about a thousand times lower than the rates of two-jet events. PHENIX has been successful in measuring γ-hadron correlation yields for Au+Au and p+p collisions using a statistical subtraction method in which the γ-hadron yield is defined as

$$Y_{direct\text{-}h} = (R_\gamma Y_{inclusive\text{-}h} - Y_{decay\text{-}h})/(R_\gamma - 1) \qquad (1)$$

where

R_γ is the ratio of the rate of the inclusive photons to the rate of the decay photons (a sample of the inclusive photons is a sum of a sample of the direct photons and a sample of the decay photons),

$Y_{inclusive\text{-}h}$ is the γ-hadron yield, in which photons are inclusive,

$Y_{decay\text{-}h}$ is the γ-hadron yield, in which photons are all decay.

3. Medium effects on the fragmentation function measured from photon-hadron correlations

The PHENIX published results [1, 2] presented measurements from Au+Au data recorded in 2004 and from p+p data recorded in 2005 and 2006. Those results do not include p_T-dependent measurements of the fragmentation functions *D(z)* in Au+Au collisions. During 2007 PHENIX detector accumulated Au+Au data sample as large as 4 times compared to that from 2004. Increased statistics allowed to complete *D(z)* measurements for the Au+Au events.

Figure 1 shows the fragmentation functions $D(z) = (\mathrm{d}N_{trig})^{-1}\mathrm{d}N/\mathrm{d}z_T$ for p+p events (2005 and 2006 data) and for 0-20% most central Au+Au events (2007 data). The functions are computed in regard to the partons that oppose direct photons (i.e. in the away side jet) for different p_T ranges of the trigger direct photons. The yield of the γ-hadron pairs d*N* is extracted for the so-called "head" region that spans $|\Delta\phi - \pi| < \pi/5$. Data from p+p collisions were analyzed with transverse momenta of trigger photons equal to $p_T = 5 \div 15$ GeV. Associated hadrons were chosen with $p_T = 1 \div 10$ GeV. The fragmentation function for Au+Au data was extracted considering the trigger photons with $p_T = 5 \div 15$ GeV and the associated hadrons with $p_T = 1 \div 7$ GeV.

It is seen that trigger p_T scales approximately with z_T. With the fit function $\mathrm{d}N/\mathrm{d}z_T = Ne^{-bz_T}$ applied for both collision systems the slope parameter *b* was measured to be $b = 6.9 \pm 0.6$ for p+p

Journal of Physics: Conference Series **389** (2012) 012021 doi:10.1088/1742-6596/389/1/012021

collisions and $b = 9.5 \pm 1.4$ for Au+Au collisions. A steeper slope b in case of the Au+Au data is in agreement with a hypothesis that the medium created in the heavy ion collision causes constant fractional energy loss of the away-side parton. For the case of p+p collisions, the value $b = 6.9$ is more consistent with the slope $b = 8$ of the quark fragmentation function than with the slope $b = 11$ of the gluon fragmentation function [2].

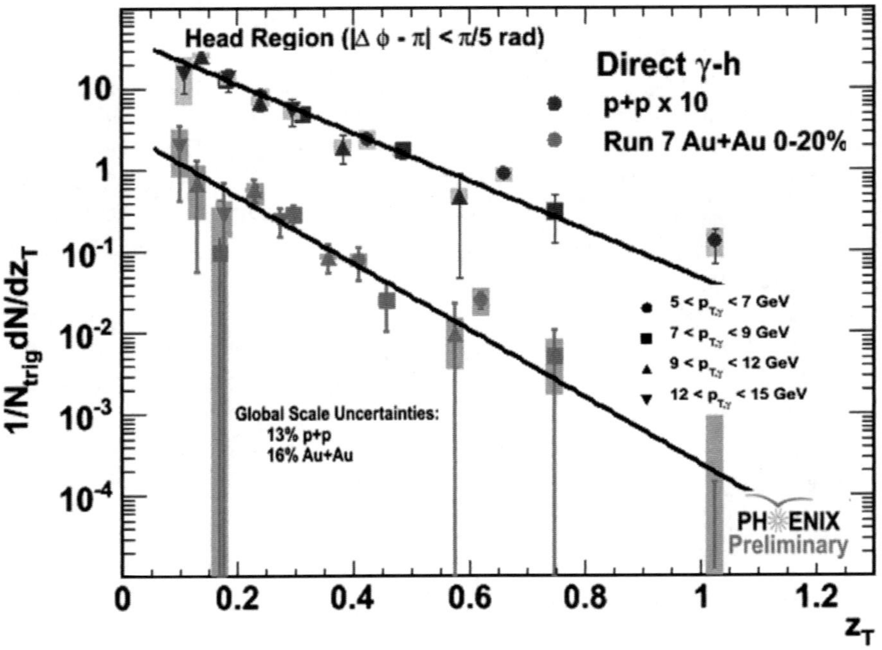

Figure 1. Fragmentation functions measured for γ-hadron correlated pairs in p+p and 0-20% most central Au+Au collsisions. For each collision system results are fitted with an exponential function.

It is reasonable to anticipate more manifested energy loss effects when z_T is small. That is why the fragmentation functions can be studied in more detail when z_T is replaced by an alternative variable $\xi = -ln(z_T)$. In Figure 2 the fragmentation function $D(\xi) = (dN_{trig})^{-1}dN/d\xi$ extracted for p+p events is plotted together with the fit function to the data from TASSO measurements [3]. Those measurements were done with e^++e^- collisions. The fit function is arbitrarily scaled by a factor of 10 to match the PHENIX p+p data. Similarity of $D(\xi)$ shapes between PHENIX and TASSO results hints on the fact that the $D(\xi)$ measured in p+p collisions is consistent with the quark fragmentation mechanism. Data from Au+Au collisions have an expanded range of $p_T = 0.5 \div 7$ GeV for the associated hadrons. At lower ξ the fragmentation function is suppressed in regard to the p+p data, as energy loss in the medium is more pronounced for the hadrons with higher transverse momenta. The Au+Au data points are fitted well with the prediction (red line) from the model [4]. That model explains that in collisions of heavy ions the fragmentation function enhances with increasing ξ, because the energy, lost by the partons in the medium, is spent on soft particle production. Figure 3 shows the ratio of the fragmentation functions, which are plotted in Figure 2 (the ratio I_{AA} is computed as Au+Au result over p+p result with a reduced number of ξ bins). The ratio I_{AA} manifests suppression of associated hadrons in the medium created in Au+Au collision, except when $\xi > 1.6$. However even starting at $\xi > 1.0$ there appears to be a rise in the I_{AA}. This behavior seems qualitatively similar to measurements by the STAR collaboration [6] of jet-hadron correlations using fully reconstructed jets. The statistical strength of the statement made by our direct photon hadron data should be improved very soon with the use of a new larger 2010 dataset.

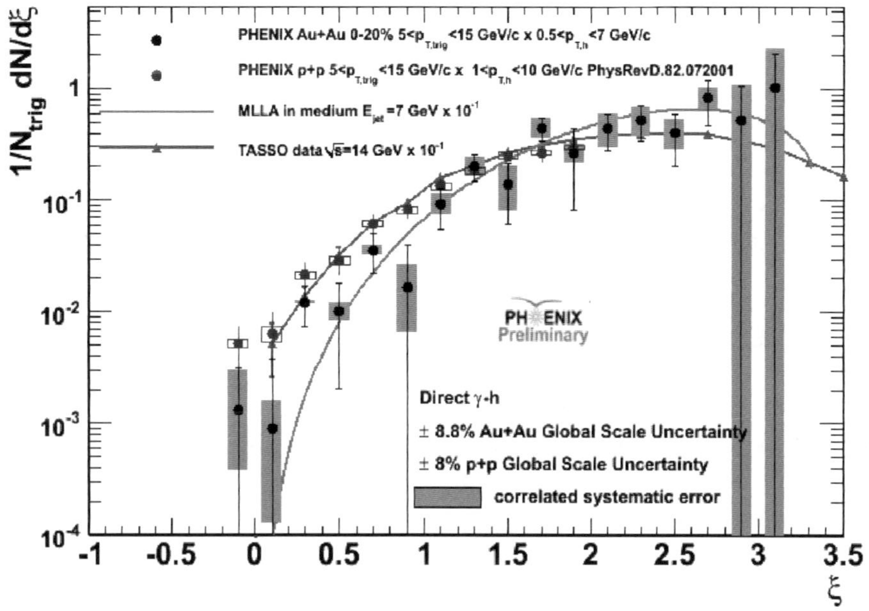

Figure 2. Fragmentation functions measured for γ-hadron correlated pairs in p+p and 0-20% most central Au+Au collsisions; $\xi = -ln(z_T)$ is used as a functional variable instead of z_T. The green line represents a scaled fit to TASSO data [3]. The red line is a prediction from the model [4].

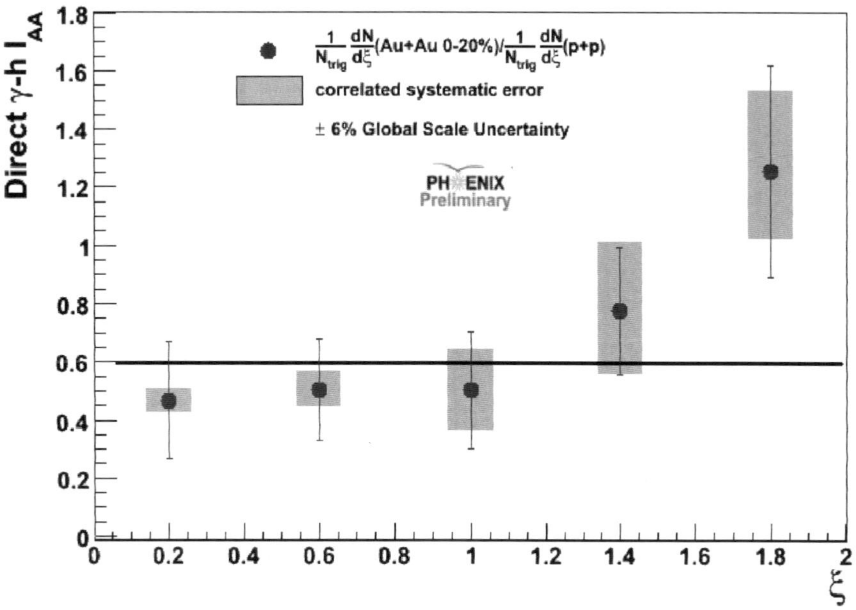

Figure 3. Ratio of the γ-hadron fragmentation function measured in 0-20% most central Au+Au events to the γ-hadron fragmentation function measured in p+p collisions.

4. PHENIX measurements in the LHC era

The Large Hadron Collider (LHC) provides nuclear physics community with unparallel research opportunities. At the LHC the energy of Pb+Pb collisions far exceeds the RHIC energy of Au+Au collisions. Two LHC general purpose detectors, ATLAS and CMS, both have very large acceptances and arrays of versatile detectors which allow in-depth studies of jets. PHENIX, in contrast, can detect hadrons and photons in a narrow pseudorapidity range of $\eta < |0.35|$ and in a modest azimuth range of $\phi = 2 \times 90^0$. Lack of a hadronic calorimeter further limits PHENIX jet physics program. However, recent LHC data already showed that the LHC and RHIC can successfully complement each other.

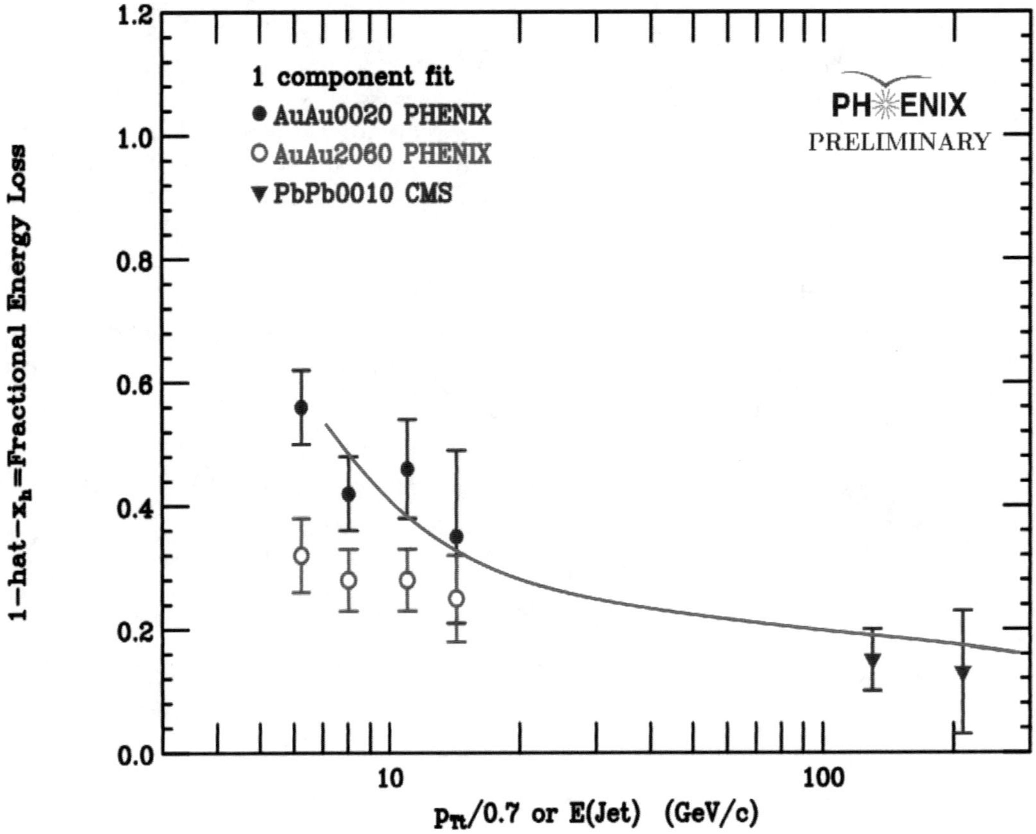

Figure 4. Fractional energy loss for PHENIX data calculated with two-particle hadron-hadron correlations for partons compared with the same quantities calculated for reconstructed jets measured by the CMS collaboration directly. For the CMS data, the partner jet is also restricted to have energy > 50 GeV.

The LHC energy loss data has been mostly made use of full jet reconstruction, and the RHIC data we are concerned with are mostly measured with 2-particle correlations. Ignoring details of the influence of the actual experimental definitions of jets via various jet finding algorithms, one should be able to make a connection to the underlying "parton" parent of the jets via either type of measurement. Fortunately PHENIX has developed a calculation framework [7] that can calculate averages of various jet (parton) level observables from two particle observables such as the shape of the spectral yield of associated particles conditional on the trigger. The quantities include for example, mean <z> of trigger jets, mean di-jet intrinsic momentum imbalance $<k_T>$ between trigger and second jet in a di-jet pair, and the mean fraction of transverse momenta of the awayside jet over the trigger jet $<\hat{x}_h> = \langle \frac{E\ assoc\ jet}{E\ trigger\ jet} \rangle$ or equivalently 1-$<\hat{x}_h>$ the mean fractional energy loss of the awayside jet. Since in PHENIX the rapidity range is limited so that $p_{Tjet} \approx E_{jet}$, these latter quantities can be related to recent LHC jet data, measured both in the jet asymmetry A_J [8,9] which is defined as $|E_1 - E_2|/(E_1+E_2) = (1-\hat{x}_h)/(1+\hat{x}_h)$ or in direct measurements of itself of $\hat{x}_h$ itself [9,10]. This allows one to compare the mean fractional energy loss between jets of different energy as is done in Figure 4, if one assumes that the medium effects are not strongly dependent on the center of mass energy of the collisions, e.g. if the sQGP exists at the phase boundaries in both cases for the dominant times of the evolution. This could test energy loss models ability to reproduce a "Bethe-Bloch formula"-like dependence. On the other hand if such a quantity can be measured for at both LHC and RHIC for the same energy jets, and this quantity is different, it could elucidate differences in the medium for the two center of mass energy ranges. As Fig.4 shows, the first assumption of similar QGP effects, it appears that the lower energy jets suffer a larger fractional energy loss. More details of this analysis can be found in [10]. The RHIC data is derived from hadron-hadron correlations, and therefore for the RHIC data, a scale factor of $1/<z_{leading\ particle}>$ taken approximately as 1/0.7 is used to scale the leading trigger π^0 momentum to its parent jet energy. Deriving the same data from direct photon –hadron correlations can be cleaner since, to leading order, no such scale factor is needed, and the photons avoid possible biases (e.g. geometric) coming from the source distribution of the trigger hadrons, since their parent jet distributions could already be modified. However improvement of the statistical precision of the direct photon data is necessary before such comparisons can be made. The new 2010 dataset which will be analyzed soon could also offer such improved statistics.

5. Conclusions

Measurements of two particle correlations, both hadron-hadron and especially direct photon-hadron correlations reveal details of QGP induced jet energy loss in heavy ion collisions. The fragmentation function for relatively low p_T direct photons (> 5 GeV/c) appear to show some modification, including a suppression at large z, and much less suppression, possibly even enhancement. The statistical precision of the result needs improved to make a more definite statement, something that may be forthcoming with the availability of a new higher statistics 2010 dataset.

A framework for calculating parton or jet level observables from the PHENIX 2-particle yields results that can be compared to recent LHC jet reconstruction observables. Under certain assumptions, the fractional energy loss for lower jet energies that the PHENIX data is sensitive to, appears to be larger than for much higher energy jets at the LHC. This statement needs many caveats, many of which may be able to be removed in the future with the availability of higher statistics direct photon – hadron measurements and possibly jet reconstruction analyses at RHIC.

6. References

[1] Adare A *et al.* (PHENIX Collaboration) 2009 *Phys. Rev.* C **80** 024908
[2] Adare A *et al.* (PHENIX Collaboration) 2010 *Phys. Rev.* D **82** 072001
[3] Braunschweig W *et al.* (TASSO Collaboration) 1990 *Z. Phys.* C **47** 187
[4] Borghini N and Wiedemann U A arXiv:hep-ph/0506218
[5] Sickles A M (for the PHENIX Collaboration) 2011 *J. Phys. G: Nucl. Part. Phys.* **38** 124129
[6] Ohlson Alice (for the STAR Collaboration) 2011 *J.Phys.Conf. Ser.* **316** 012015
[7] Adler, S.S. *et al.* (PHENIX Collaboration) 2006 *Phys. Rev. D* **74** 072002

[8] Aad, G. et. al. (ATLAS Collaboration) 2010 *Phys.Rev.Lett.* **105** 252303
[9] Chatrchyan, S. *et al.* (CMS Collaboration) 2011 *Phys. Rev. C* **84**, 024906
[10] Tannenbaum, M.J. (for the PHENIX Collaboration) 2012 *AIP Conf.Proc.* **1441** 862-864

Anti- and Hypermatter Research at the Facility for Antiproton and Ion Research FAIR

J. Steinheimer[1], Z. Xu[2], K. Gudima[4,5], A. Botvina[4,6], I. Mishustin[4,7], M. Bleicher[4] and H. Stöcker[4,8]

[1] Lawrence Berkeley National Laboratory, Berkeley, CA 94720, USA
[2] Physics Department, Brookhaven National [3] Laboratory, Upton, NY 11973, USA
[4] FIAS, Johann Wolfgang Goethe University, Frankfurt am Main, Germany
[5] Institute of Applied Physics, Academy of Sciences of Moldova, MD-2028 Kishinev, Moldova
[6] Institute for Nuclear Research, Russian Academy of Sciences, 117312 Moscow, Russia
[7] Kurchatov Institute, Russian Research Center, 123182 Moscow, Russia
[8] GSI Helmholtzzentrum für Schwerionenforschung GmbH, Planckstr. 1, D-64291 Darmstadt, Germany

E-mail: `jsfroschauer@lbl.gov`

Abstract. Within the next six years, the Facility for Antiproton and Ion Research (FAIR) is built adjacent to the existing accelerator complex of the GSI Helmholtz Center for Heavy Ion Research at Darmstadt, Germany. Thus, the current research goals and the technical possibilities are substantially expanded. With its worldwide unique accelerator and experimental facilities, FAIR will provide a wide range of unprecedented fore-front research in the fields of hadron, nuclear, atomic, plasma physics and applied sciences which are summarized in this article. As an example this article presents research efforts on strangeness at FAIR using heavy ion collisions, exotic nuclei from fragmentation and antiprotons to tackle various topics in this area. In particular, the creation of hypernuclei and antimatter is investigated.

1. The FAIR Project

The Facility for Antiproton and Ion Research, FAIR [1, 2, 3], will provide an extensive range of particle beams from protons and their antimatter partners, antiprotons, to ion beams of all chemical elements up to the heaviest one, uranium, with in many respects world record intensities. As a joint effort of 16 countries the new facility builds, and substantially expands, on the present accelerator system at GSI, both in its research goals and its technical possibilities. Compared to the present GSI facility, an increase of a factor of 100 in primary beam intensities, and up to a factor of 10000 in secondary radioactive beam intensities, will be a technical property of the new facility.

The start version of FAIR, the so-called *M*odularized Start Version [4, 5], includes a basic accelerator SIS100 (module 0) as well as three experimental modules (module 1-3). The superconducting synchrotron SIS100 with a circumference of 1100 meters and a magnetic rigidity of 100 Tm is at the heart of the FAIR accelerator facility. Following an upgrade for high intensities, the existing GSI accelerators UNILAC and SIS18 will serve as an injector. Adjacent to the SIS100 synchrotron are two storage-cooler rings and experiment stations, including a superconducting nuclear fragment separator (Super-FRS) and an antiproton production target.

The Modularized Start Version secures a swift start of FAIR with outstanding science potential for all scientific pillars of FAIR within the current funding commitments. Moreover, after the start phase and as additional funds become available the facility will be upgraded by experimental storage rings enhancing capabilities of secondary beams and upgraded by SIS300 providing particle energies 20-fold higher compared to those achieved so far at GSI.

2. The Experimental Program of FAIR

The main thrust of FAIR research focuses on the structure and evolution of matter on both a microscopic and on a cosmic scale. The approved FAIR research program embraces 14 experiments, which form the four scientific pillars of FAIR and offers a large variety of unprecedented forefront research in hadron, nuclear, atomic and plasma physics as well as applied sciences. Already today, over two 2500 scientists and engineers are involved in the design and preparation of the FAIR experiments. They are organized in the experimental collaborations APPA, CBM, NuSTAR, and PANDA.

2.1. APPA – Atomic Physics, Plasma Physics and Applications

Atomic physics with highly charged ions [6] will concentrate on two central research themes: a) the correlated electron dynamics in strong, ultra-short electromagnetic fields including the production of electron-positron pairs and b) fundamental interactions between electrons and heavy nuclei - in particular the interactions described by Quantum Electrodynamics, QED. Here bound-state QED in critical and supercritical fields is the focus of the research program. In addition, atomic physics techniques will be used to determine properties of stable and unstable nuclei and to perform tests of predictions of fundamental theories besides QED.

For Plasma physics the availability of high-energy, high-intensity ion-beams enables the investigation of High Energy Density Matter in regimes of temperature, density and pressure not accessible so far [7]. It will allow probing new areas in the phase diagram and long-standing open questions of basic equation of state (EoS) research can be addressed. The biological effectiveness of high energy and high intensity beams was never studied in the past. It will afford to investigate the radiation damage induced by cosmic rays and protection issues for the Moon and Mars missions. Furthermore, the intense ion-matter interactions with projectiles of energies above 1 GeV/u will endorse systematic studies of material modifications.

2.2. CBM/HADES – Compressed Baryonic Matter

Violent collisions between heavy nuclei promise insight into an unusual state in nature, that of highly compressed nuclear matter. Results from lattice QCD indicate that the transition from confined to deconfined matter at vanishing net baryon density is a smooth crossover, whereas in the region of high baryon densities, accessible with heavy-ion reactions at lower beam energies, a first-order phase transition is expected [8]. Its experimental confirmation would be a substantial progress in the understanding of the properties of strongly interacting matter.

The CBM experiment [9, 10] as well as HADES [11, 12] at SIS100/300 will explore the QCD phase diagram in the region of very high baryon densities and moderate temperatures by investigating heavy-ion collision in the beam energy range 2–35 AGeV. This approach includes the study of the nuclear matter equation-of-state, the search for new forms of matter, the search for the predicted first order phase transition to the deconfinement phase at high baryon densities, the QCD critical endpoint, and the chiral phase transition, which is related to the origin of hadron masses. The CBM experiment at FAIR is being designed to perform this search with a large range of observables, including very rare probes like charmed hadrons. Ratios of hadrons containing charm quarks as a function of the available energy may provide direct evidence for a deconfinement phase.

Journal of Physics: Conference Series **389** (2012) 012022 doi:10.1088/1742-6596/389/1/012022

The properties of hadrons are expected to be modified in a dense hadronic environment which is eventually linked to the onset of chiral symmetry restoration at high baryon densities and/or high temperatures. The dileptonic decays of the light vector mesons (ρ,ω,ϕ) provide the tool to study such modifications since the lepton daughters do not undergo strong interactions and can therefore leave the dense hadronic medium essentially undistorted by final-state interaction. As a detector system dedicated to high-precision di-electron spectroscopy at beam energies of 1–2 AGeV, the modified HADES detector at SIS100 will measure e^+e^- decay channels as well as hadrons [13, 14] up to 10 AGeV beam energy. Complementarity, the CBM experiment will cover the complete FAIR energy range by measuring both the e^+e^- and the $\mu^+\mu^-$ decay channels.

Most of the rare probes like lepton pairs, multi-strange hyperons and charm will be measured for the first time in the FAIR energy range. The goal of the CBM experiment as well as HADES is to study rare and bulk particles including their phase-space distributions, correlations and fluctuations with unprecedented precision and statistics. These measurements will be performed in nucleus–nucleus, proton–nucleus, and proton–proton collisions at various beam energies. The unprecedented beam intensities will allow studying extremely rare probes with high precision.

2.3. NuSTAR – Nuclear Structure, Astrophysics and Reactions

The main scientific thrusts in the study of nuclei far from stability are aimed at three areas of research: (i) the structure of nuclei, the quantal many-body systems built by protons and neutrons and governed by the strong force, toward the limits of stability, where nuclei become unbound, (ii) nuclear astrophysics delineating the detailed paths of element formation in stars and explosive nucleosynthesis that involve short-lived nuclei, (iii) and the study of fundamental interactions and symmetries exploiting the properties of specific radioactive nuclei.

The central part of the NuSTAR program at FAIR [15, 16] is the high acceptance Super-FRS with its multi-stage separation that will provide high intensity mono-isotopic radioactive ion beams of bare and highly-ionized exotic nuclei at and close to the driplines. This separator, in conjunction with high intensity primary beams with energies up to $1.5A$ GeV, is the keystone for a competitive NuSTAR physics program. This opens the unique opportunity to study the evolution of nuclear structure into the yet unexplored territory of the nuclear chart and to determine the properties of many short-lived nuclei which are produced in explosive astrophysical events and crucially influence their dynamics and associated nucleosynthesis processes.

2.4. PANDA – AntiProton ANnihilation in Darmstadt

The big challenge in hadron physics is to achieve a quantitative understanding of strongly interacting complex systems at the level of quarks and gluons. In $p\bar{p}$-annihilation, particles with gluonic degrees of freedom as well as particle-antiparticle pairs are copiously produced, allowing spectroscopic studies with unprecedented statistics and precision. The PANDA experiment at FAIR [17, 18, 19] will bring new fundamental knowledge in hadron physics by pushing the precision barrier toward new limits. The charmonium ($c\bar{c}$) spectroscopy will take advantage by precision measurements of mass, width, decay branches of all charmonium states. Particular emphasis is placed on mesons with open and hidden charm, which extends ongoing studies in the light quark sector to heavy quarks, and adds information on contributions of the gluon dynamics to hadron masses. The search for exotic hadronic matter such as hybrid mesons or heavy glueballs gains enormously by precise scanning of resonance curves of narrow states as well. Recently, this field has attracted much attention with the surprise observation at electron-positron colliders of the new X, Y and Z states with masses around 4 GeV. These heavy particles show very unusual properties, whose theoretical interpretation is entirely open. Additionally the precision gamma-ray spectroscopy of single and double hypernuclei will allow extracting information on their structure and on the hyperon-nucleon and hyperon-hyperon interaction.

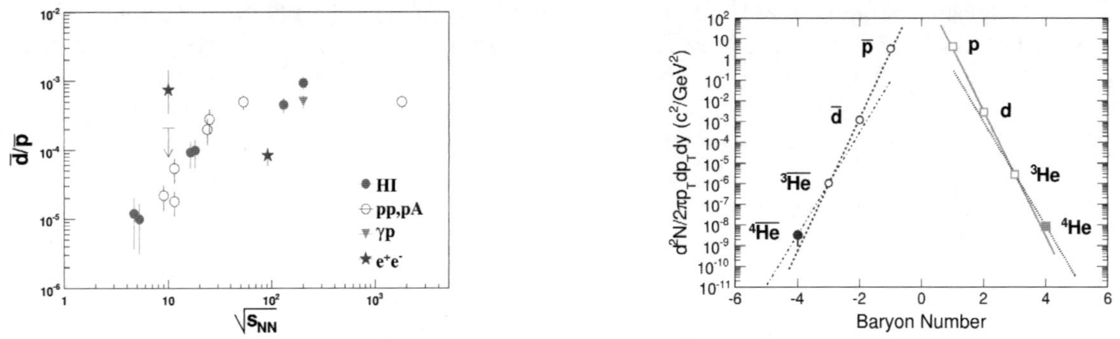

Figure 1. (Color online) Left: Antideuteron and antiproton yield ratio as a function of $\sqrt{S_{NN}}$ of the colliding target and projectile. Right: Matter and antimatter invariant differential yields as a function of the baryon number. The solid and dashed lines are a fit to experimental data. The dotted and dashed dotted lines are extrapolations from yields of $|A| = 3$ and $|A| = 4$.

3. The creation of antimatter

In 1928, Dirac predicted the existence of negative energy states of electrons based on the application of symmetry principles to quantum mechanics. After the discoveries of antiprotons and antineutrons, one of the important questions was whether the building blocks in the antimatter world have the same force to glue together the antinucleons into nuclei and eventually anti-atoms by adding positrons. Nuclei are abundant in the universe, but antinuclei with $|A| \geq 2$ have not been found in nature. Relativistic heavy ion collisions, simulating the condition at the early universe, provide an environment with abundant antinucleons and antihyperons and produce antinuclei and antihypernuclei by coalescing them together [20]. This offers the first opportunity for discovery of antihypernuclei [21] and heavier antinuclei [22] having atomic mass numbers (or baryon numbers) $|A| > 2$. The production of antimatter nuclei can be explained by coalescence of antiprotons and antineutrons close in position and momentum. Figure 1 compiles all the antideuteron production in e^+e^-, γp, pp, pA and AA collisions [23]. The results are shown for $\bar{d}/\bar{p}$ ratio as a function of beam energy. One can see that this ratio increases from 10^{-5} at low energy to 10^{-3} at high energy. Each additional antinucleon into the heavier antimatter decreases its production rate by that same penalty factor. At a center of mass energy of 100 GeV and above, this factor is relatively flat at slightly below 10^{-3}. It is interesting to note that this effective measure of antibaryon density shows no difference among pp, pA and AA collisions. In heavy ion collisions, more antiprotons are produced in each collision than in pp collisions. However, if more pp collisions are collected to match the amount of antiproton yields in heavy-ion collisions, one can essentially produce the same amount of heavy antimatter in pp and heavy-ion collisions. Now we understand that there are two deciding facts that RHIC discovered the last two heavy antimatters: sufficient energy to provide the highest antibaryon density for antinuclear production, and high luminosity heavy-ion collisions for effective data collection and particle identification.

Figure 1 shows the matter and antimatter yields as a function of baryon numbers as measured by the STAR Collaboration at RHIC [22]. The fit lines yield the production reduction rate by a factor of 1.6×10^{-3} (1.1×10^{-3}) for matter (antimatter) for each additional nucleon (antinucleon). The sensitivity of current and planned space based charged particle detectors is below what would be needed to observe antihelium produced by nuclear interactions in the cosmos. This implies that any observation of antihelium or even heavier antinuclei in space would indicate the existence of a large amount of antimatter elsewhere in the universe. In particular, finding

Journal of Physics: Conference Series **389** (2012) 012022 doi:10.1088/1742-6596/389/1/012022

antimatter ^{4}He in the cosmos is one of the major motivations for space detectors such as the Alpha Magnetic Spectrometer [24]. We have shown that antimatter ^{4}He exists and provided a measure of the background rate in nuclear collisions for possible future observations in cosmic radiation.

The next stable antimatter nucleus would be $A = 6$ (^{6}He; ^{6}Li). However, the penalty factor on the production rate for an additional antinucleon is about 1500 as shown in Fig. 1. This means that the $A = 6$ antinuclei are produced at a rate 2×106 lower than that of an $A = 4$ antialpha particle. Unless production mechanisms or collider technology change dramatically, it is unlikely that $A = 6$ antinuclei can be produced in collider or fixed-target experiments (STAR 2011). On the other hand, the ratio of the $^4\mathrm{He}/^3\mathrm{He} = 3.1 \times 10^{-3}$ and $^4\overline{\mathrm{He}}/^3\overline{\mathrm{He}} = 2.4 \times 10^{-3}$. There is a factor of 2 higher yield of $|A| = 4$ over $|A| = 3$ than the extrapolation from the fit. The excess is visible even in a log-scale plot of 13 orders of magnitude. This ratio is also much higher than that shown in Fig. 1 for the $|A| = 2$ over $|A| = 1$. The indicative enhancement of higher antialpha yields suggests that even higher enhanced yields of heavier antimatter. Besides the possible high yields of $|A| = 6$ antimatter, the heaviest antimatter that can be produced and detected with a tracking detector in high-energy accelerators are likely to be $A = 4$ or 5 unstable antinuclei: $^4\mathrm{He}^* \to \mathrm{t} + \mathrm{p}$, $^4\mathrm{Li} \to ^3\mathrm{He} + \mathrm{p}$, and $^5\mathrm{Li} \to ^4\mathrm{He} + \mathrm{p}$. New trigger scheme and high data acquisition rate have been proposed to improve the effective data taking rate by two orders of magnitude in STAR during the heavy ion collisions [25]. This should confirm if the enhancement indeed exists and provide a possible path for discovering even heavier antimatter. In addition, as mentioned in the previous section, the antimatter yield reduction factor is similar in $p+p$ and AA collisions. One expects that the penalty factor to persist for antimatter heavier than antideuteron in $p + p$ collisions. A comparison between the antimatter yields as shown in Fig. 1 in $p + p$ and $A + A$ collisions will provide a reference for whether the enhancement seen in antialpha production in AA collisions is due to new production mechanism. Both RHIC and LHC have sufficient luminosity in $p + p$ collisions to produce antialpha. The only experimental issue is how to trigger and identify those particles. STAR has proposed a new trigger and TPC readout schemes for heavy antimatter search by using the Electromagnetic Calorimeter (EMC) for charged hadrons and only readout small sector of TPC associated with that struck EMC.

4. Hypermatter

Relativistic heavy ion collisions are an abundant source of strangeness [26]. Exotic forms of deeply bound objects with strangeness have been proposed [27] as states of matter, either consisting of baryons or quarks. The H di-baryon was predicted by Jaffe [28] and later, many more bound di-baryon states with strangeness were proposed using quark potentials [29, 30] or the Skyrme model [31]. However, the non-observation of multi-quark bags, e.g. strangelets is still one of the open problems of intermediate and high energy physics. Hypernuclei however are known to exist and be produced in heavy ion collisions already for a long time [32, 33, 34, 35]. The interest in the field of hypernuclear physics was fueled by the recent discoveries of the first anti-hypertriton [36] and anti-α [37] (the largest antiparticle cluster ever reported).

It is of great interest to estimate the yields for hypernuclei, and comparably the proton and Λ yields, from a consistent model for heavy ion collisions. In such models the cluster is formed at, or shortly after, the (chemical-)freeze out of the system. A general assumption is, that these clusters are then formed through coalescence of different newly produced hadrons. To estimate the production yield we can employ two distinct approaches which allow us to estimate the theoretical uncertainties associated with different treatment of the process. First we use a hadronic transport model (DCM model) to provide us with the phase space information of all hadrons produced in a heavy ion collision. This information then serves as an input for a coalescence prescription. Secondly we assume thermal production of clusters from a fluid dynamical description to heavy ion collisions.

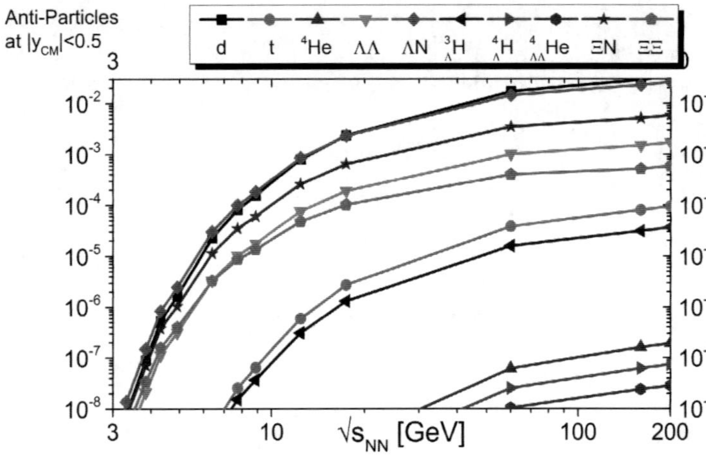

Figure 2. (Color online) Yields of antiparticle clusters with $|y| < 0.5$ of most central collisions of Pb+Pb/Au+Au as a function of $\sqrt{s_{NN}}$. Shown are only the results from the thermal production in the UrQMD hybrid model (lines with symbols).

Though thermal production differs significantly in its assumptions from a coalescence approach one would expect to obtain different results, depending on the method used. However it can be shown that both approaches can lead to very similar results [38]. More detailed information on the calculations performed for the results in this section can be found in [38, 39].

When the beam energy of the collisions is increased, the system created becomes almost net-baryon free. This means that the probability to create an antiparticle cluster approaches that of the particle cluster. Figure 2 shows the results for antiparticle cluster production at mid-rapidity ($|y| < 0.5$) in collisions of Pb+Pb/Au+Au at center of mass energies of $\sqrt{s_{NN}} = 3$ - 200 GeV. We show only results for the UrQMD hybrid model because the DCM calculations are restricted to energies up to $E_{lab} = 160A$ GeV where the statistics needed for a meaningful estimate are quite significant. The yields of the antiparticle clusters show a monotonous increase with beam energy. They show that, at the highest RHIC energy (and at the LHC) the reconstruction of $^4_\Lambda$He might be a feasible task.

As another promising experimental observable the double ratio R_H defined as:

$$R_H = ^3_\Lambda H/^3He \cdot p/\Lambda \tag{1}$$

for collisions of Pb+Pb/Au+Au and a wide range of beam energies. This ratio is especially interesting, as it has been proposed to be sensitive on the local correlation of strangeness and baryon number, therefore being a measure of the baryon-strangeness correlation c_{BS} [40].

Our results for R_H are shown in figure 3 as an excitation function of the beam energy $\sqrt{s_{NN}}$. R_H is evaluated for the mid rapidity region of most central ($b < 3.4$ fm) heavy ion collisions. The lines depict results from the UrQMD-hybrid model and the symbols denote DCM coalescence results. Experimental data are depicted as green symbols with error bars. For the hadrons the feed down from resonances is taken into account as well as the feed down to the ^3He from the hypertriton. Because experiments usually cannot distinguish between Λ's and Σ^0's, we show R_H in the cases where the Λ yield includes Σ^0 (black solid line and squares) and where the yield is corrected for the Σ^0 (red dashed line and circles). This is in fact important as there is no experimental indication for a bound $^3_{\Sigma^0}H$ hypernucleus.

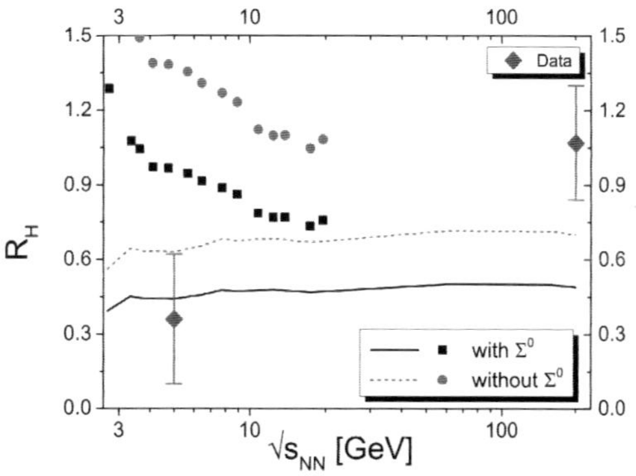

Figure 3. The Strangeness Population Factor $R_H = (^3_\Lambda H/^3He) \cdot (p/\Lambda)$ as a function of $\sqrt{s_{NN}}$ for most central collisions of Pb+Pb/Au+Au. We compare results from the thermal production in the UrQMD hybrid model (lines) with coalescence results with the DCM model (symbols). The red line and symbols denote values of R_H where the Λ yield has been corrected for the Σ^0 contribution.

The double ratio R_H from the hybrid model turns out to be almost energy independent. The same behavior has been observed in previous thermal calculations [41]. On the other hand, the coalescence result increases with decreasing beam energy and is in general larger than the thermal result. This observation leads to the conclusion that the information on correlations of baryon number and strangeness is lost in the thermal calculation because here R_H essentially only depends on the temperature. On the other hand, in the microscopic treatment the correlation information survives and R_H captures the trend of c_{BS}.

5. Summary

After about ten years of negotiations, R&D and writing reports, on 4th of October nine countries finally signed the international agreement on the construction of the Facility for Antiproton and Ion Research (FAIR). Construction of the first FAIR buildings will start in 2012 and first beams will be delivered in 2018. The initial version of FAIR, the so-called *M*odularized Start Version, includes the superconducting synchrotron SIS100 and three experimental modules to perform experiments for all research pillars. It will allow to carry out an outstanding and world-leading research program in hadron, nuclear, atomic and plasma physics as well as applied sciences. Due to the high luminosity, exceeding current facilities by orders of magnitude, it will be possible to conduct experiments that could not be done anywhere yet. FAIR will expand the knowledge in various scientific fields beyond current frontiers. Moreover, the exploitation of exiting strong cross-topical synergies promise novel insights.

6. Acknowledgments

This work was supported by HGS-HIRe and the Hessian LOEWE initiative through the Helmholtz International center for FAIR (HIC for FAIR). Computational resources were provided by Frankfurt Center for Scientific Computing (CSC). J. S. acknowledges a Feodor Lynen fellowship of the Alexander von Humboldt foundation. This work was supported by the Office of Nuclear Physics in the US Department of Energy's Office of Science under Contract No. DE-AC02-05CH11231. I.M. acknowledges partial support from grant NS-215.2012.2 (Russia).

[1] H.H. Gutbrod *et al.* (Eds.), *FAIR Baseline Technical Report*, (2006).

Journal of Physics: Conference Series **389** (2012) 012022 doi:10.1088/1742-6596/389/1/012022

[2] W. F. Henning, Nucl. Phys. A **805**, 502 (2008).
[3] H. Stöcker, Proc. of 11th Europ. Part. Acc. Conf., Genoa, Italy (2008).
[4] C. Sturm, B. Sharkov and H. Stöcker, Nucl. Phys. A **834**, 682c (2010).
[5] Green Paper of FAIR: *The Modularized Start Version*, www.gsi.de/documents/DOC-2009-Nov-124-1.pdf (2009)
[6] Th. Stöhlker *et al.*, Nucl. Instrum. Meth. B **261**, 234 (2007).
[7] I.V. Lomonosov and N.A. Tahir, Nucl. Phys. News 16 **1**, 29 (2006).
[8] B. Friman, C. Höhne, J. Knoll, S. Leupold, J. Randrup, R. Rapp, P. Senger (Editors), Lect. Notes Phys. **814**, 1 (2011)
[9] P. Senger *et al.* [CBM Collab.], Phys. Part. Nucl. **39**, 1055 (2008).
[10] P. Senger *et al.* [CBM Collab.], Prog. Part. Nucl. Phys. **62**, 375 (2009).
[11] G. Agakishiev *et al.* [HADES Collab.], Eur. Phys. J. A **41**, 243 (2009).
[12] J. Stroth *et al.* [HADES Collab.], Prog. Part. Nucl. Phys. **62**, 481 (2009).
[13] I. Fröhlich [HADES Collab.], arXiv:0906.0091 [nucl-ex].
[14] P. Tlusty *et al.* [HADES Collab.], AIP Conf. Proc. **1322**, 116 (2010).
[15] R. Krücken *et al.* [NuSTAR Collab.], J. Phys. G **31**, S1807 (2005).
[16] B. Rubio and T. Nilsson [NuSTAR Collab.], Nucl. Phys. News **16**, 9 (2006).
[17] Physics Performance Report for PANDA, arXiv:0903.3905v1 [hep-ex].
[18] K. Fohl *et al.* [PANDA Collab.], Eur. Phys. J. ST **162**, 213 (2008).
[19] J. S. Lange [PANDA Collab.], Int. J. Mod. Phys. A **24**, 369 (2009).
[20] J. Adams *et al.* [STAR Collab.], Nucl. Phys. A **757**, 102 (2005).
[21] B. I. Abelev [STAR Collab.], Science **328** 58 (2010).
[22] H. Agakishiev *et al.* [STAR Collab.], Nature **473**, 353 (2011).
[23] H. Liu and Z. Xu, arXiv:nucl-ex/0610035.
[24] S. P. Ahlen, V. M. Balebanov, *et al.*, Nucl. Instrum. Meth. A **350**, 351 (1994).
[25] STAR decadal plan 2011, www.bnl.gov/npp/docs/STAR_decadal_Plan_Final.pdf.
[26] P. Koch, B. Müller and J. Rafelski, Phys. Rept. **142**, 167 (1986).
[27] A. R. Bodmer, Phys. Rev. D **4**, 1601 (1971).
[28] R. L. Jaffe, Phys. Rev. Lett. **38**, 195 (1977).
[29] J. T. Goldman *et al.*, Phys. Rev. Lett. **59**, 627 (1987).
[30] J. T. Goldman *et al.*, Mod. Phys. Lett. A **13**, 59 (1998).
[31] B. Schwesinger, F. G. Scholtz and H. B. Geyer, Phys. Rev. D **51**, 1228 (1995).
[32] P. Braun-Munzinger and J. Stachel, J. Phys. G **21**, L17 (1995).
[33] J. K. Ahn *et al.*, Phys. Rev. Lett. **87**, 132504 (2001).
[34] H. Takahashi *et al.*, Phys. Rev. Lett. **87**, 212502 (2001).
[35] A. Andronic, P. Braun-Munzinger, J. Stachel and H. Stöcker, Phys. Lett. B **697**, 203 (2011).
[36] STAR Collab., Science **328**, (2010) 58.
[37] STAR Collab., Nature **473** (2011) 353.
[38] J. Steinheimer, K. Gudima, A. Botvina, I. Mishustin, M. Bleicher and H. Stocker, Phys. Lett. B **714**, 85 (2012)
[39] J. Steinheimer *et al.*, Phys. Lett. B **676**, 126 (2009).
[40] S. Zhang, J. H. Chen, H. Crawford, D. Keane, Y. G. Ma and Z. B. Xu, Phys. Lett. B **684**, 224 (2010)
[41] A. Andronic, P. Braun-Munzinger, J. Stachel and H. Stocker, Phys. Lett. B **697**, 203 (2011)

Journal of Physics: Conference Series **389** (2012) 012023 doi:10.1088/1742-6596/389/1/012023

Open heavy-flavour measurements in Pb-Pb collisions at $\sqrt{s_{NN}}$=2.76 TeV with ALICE

Raphaelle Bailhache on behalf of the ALICE collaboration

Goethe University Frankfurt, Institut fuer Kernphysik, Max-von-Laue-Str.1 , 60438 Frankfurt am Main, GE

E-mail: rbailhache@ikf.uni-frankfurt.de

Abstract. In ultra-relativistic heavy-ion collisions, heavy quarks, i.e charm and beauty, are produced in the early stage of the reaction. Therefore, they are uniquely suited to probe the Quark-Gluon-Plasma (QGP), which is formed in such reactions, and to study the parton-medium interactions. In 2010 and 2011, Pb-Pb collisions at $\sqrt{s_{NN}}$=2.76 TeV have been recorded by ALICE at the LHC. We present results of open heavy-flavour production using hadronic D meson decays, as well as semi-electronic and semi-muonic decays of D and B mesons. The corresponding nuclear modification factors are measured as a function of the transverse momentum and the collision centrality in the ALICE central-barrel at mid-rapidity and in the forward muon detector. First elliptic flow measurements are also shown.

1. Introduction

Ultra-relativistic nuclear collisions allow to study nuclear matter under extreme conditions. In such collisions a deconfined state of quarks and gluons, the Quark-Gluon Plasma (QGP), is expected to be formed. Among the possible probes of the QGP, heavy quarks are of particular interest since they are produced at the beginning of the collision and interact strongly with the surrounding medium. Therefore the measurement of open heavy-flavour production allows to test models of in-medium parton energy-loss and provides essential information on the properties of the strongly-interacting system formed in the early stages of heavy-ion collisions. In addition, the elliptic flow measurements of D mesons probe the level of thermalization of heavy quarks in the medium at low momentum, and the path length dependence of the parton energy loss at higher momentum.

In November 2010 and 2011, ALICE (A Large Ion Collider Experiment) collected data during the first Pb-Pb collisions at a center-of-mass energy per nucleon pair $\sqrt{s_{NN}}$=2.76 TeV delivered by the LHC (Large Hadron Collider) at CERN. A detailed description of the experiment can be found in [1]. The data readout was triggered by a minimum-bias interaction trigger based on the trigger signals from two forward scintillator hodoscopes (VZERO-A and VZERO-C) and two innermost layers of Silicon Pixel Detectors (SPD). In total 17 million minimum-bias and 16 million semi-central (15-50 %) Pb-Pb collisions, recorded in 2011 and 2011 respectively, were used for the analysis. The summed amplitudes in the VZERO scintillator tiles were used to determine the centrality of the Pb-Pb collision, together with a model of particle production based on a Glauber description of the nuclear collision geometry.

Journal of Physics: Conference Series **389** (2012) 012023 doi:10.1088/1742-6596/389/1/012023

2. D meson production at mid-rapidity ($|y|<0.5$)

The D^0, D^+ and D^{*+} mesons and their charge conjugates are reconstructed from their decays into charged hadrons, $D^0 \to K^-\pi^+$ (with branching ratio, BR of 3.87±0.05 %), $D^+ \to K^-\pi^+\pi^+$ (BR of 9.13±0.19 %), and $D^{*+} \to D^0\pi^+$ (BR of 67.7±0.5 %), in the central barrel of ALICE. Due to their large lifetime ($c\tau$=123 μm, 312 μm for D^0 and $D^\pm$ respectively), the D mesons decay further away from the primary vertex. The tracking capabilities of the Inner Tracking System (ITS) and Time Projection Chamber (TPC) are used to reconstruct the displaced secondary vertices. The TPC and Time Of Flight (TOF) detector provide moreover the possibility to identify $\pi^\pm$ and $K^\pm$.

Figure. 1 shows the invariant mass distributions for D^0 candidates and their charge conjugates for different D^0 p_t intervals in 0-20 % central Pb-Pb collisions. In the case of the D^{*+} decay, the D^0 candidate is first reconstructed via its decay topology and then combined with low momentum $\pi^\pm$. The D^0, D^+ and D^{*+} yields are extracted by fitting the invariant mass distributions. The p_t-differential production yields in the p_t range $2<p_t<16$ GeV/c at mid-rapidity ($|y|<0.5$) are determined after efficiency and feed-down correction for B decays in 0-20 % and 40-80 % central Pb-Pb collisions [2]. The fraction of prompt D mesons, f_{prompt}, depends on the D meson species, p_t, the analysis selection criteria, the parameters used in the FONLL B prediction [4], and the hypothesis on the B suppression in Pb-Pb collisions. The values of f_{prompt} range from about 0.95 in the p_t range $2<p_t<3$ GeV/c to about 0.85 at high p_t.

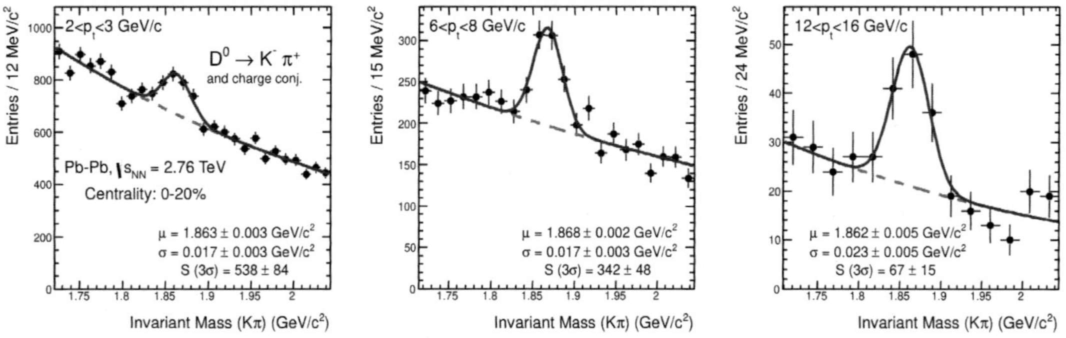

Figure 1. Invariant mass distributions for D^0 candidates and their charge conjugates in selected p_t intervals for 3.2×10^6 0-20 % central Pb-Pb collisions at $\sqrt{s_{NN}}$=2.76 TeV. The uncertainties on the signal yields reported in the figures are statistical only [2].

The D meson production in Pb-Pb collisions can be compared to pp collisions at the same $\sqrt{s_{NN}}$ energy with the nuclear modification factor R_{AA}:

$$R_{AA}(p_t) = \frac{1}{\langle T_{AA}\rangle} \frac{dN_{AA}/dp_t}{d\sigma_{pp}/dp_t} \tag{1}$$

where $\langle T_{AA}\rangle$ is the average nuclear overlap function for the given centrality class proportional to the number of nucleon-nucleon collisions, dN_{AA}/dp_t is the measured yield for this centrality class in Pb-Pb collisions, and $d\sigma_{pp}/dp_t$ is the corresponding cross section in pp collisions. A suppression of D meson production in central Pb-Pb collisions compared to pp collisions implies a nuclear modification factor smaller than 1.0.

The D^0, D^+ and D^{*+}, and their charge conjugates are measured in pp collisions at $\sqrt{s}$=2.76 TeV in the transverse momentum range $1<p_t<12$ GeV/c [3]. The analysis was performed on an event sample collected in 2011 with a minimum-bias trigger, corresponding

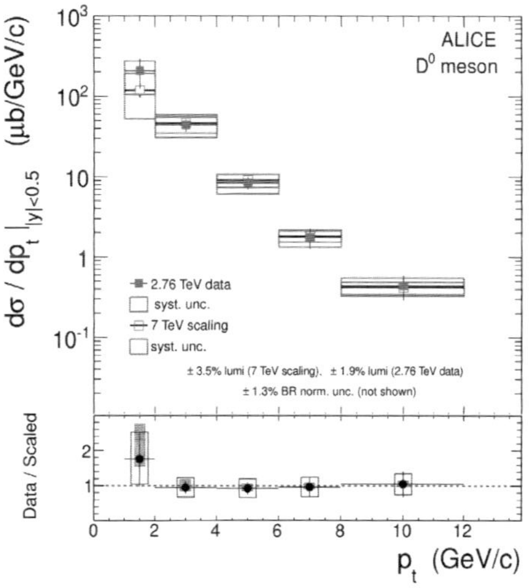

Figure 2. The p_t-differential cross section for prompt D^0 mesons in pp collisions at $\sqrt{s}$=2.76 TeV compared to the scaled ALICE measurement at $\sqrt{s}$=7 TeV (top), together with their ratio (bottom). The filled boxes represent the scaling uncertainties, whereas the empty boxes the measurement systematics [3].

to an integrated luminosity of 1.35 nb^{-1}. Since the statistics is limited and does not allow a comparison with the Pb-Pb measurements for every p_t interval, the reference used for the calculation of the D meson R_{AA} is obtained from a pQCD-based (FONLL [4]) energy scaling of the 7 TeV p_t differential cross sections to 2.76 TeV. Figure. 2 shows the 2.76 TeV measured p_t differential cross section, together with the 7 TeV results scaled down to 2.76 TeV. The agreement is very good.

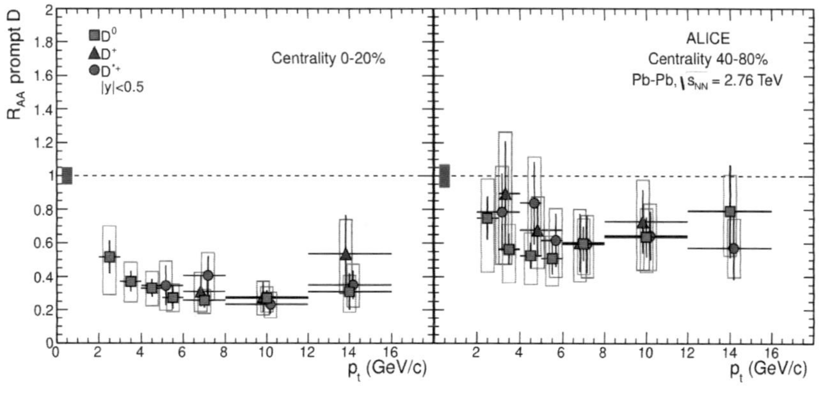

Figure 3. R_{AA} for prompt D^0, D^+ and D^{*+} in 0-20 % (left) and 40-80 % (right) central Pb-Pb collisions at $\sqrt{s_{NN}}$=2.76 TeV. Statistical (bars), systematic (empty boxes), and normalization (full box) uncertainties are shown [2].

The nuclear modification factor of prompt D^0, D^+ and D^{*+} is shown for central (0-20 %) and semi-peripheral (40-80 %) Pb-Pb collisions in Fig. 3. The central values of R_{AA} are calculated under the assumption that the nuclear modification factors for feed-down from B mesons and prompt D mesons are equal ($R_{AA}^{feed-down}=R_{AA}^{prompt}$). The systematic uncertainties are estimated by varying the hypothesis in the range $1/3<R_{AA}^{feed-down}/R_{AA}^{prompt}<3$. The D mesons are suppressed by a factor 3-4 above 5 GeV/c in 0-20 % Pb-Pb collisions. The suppression is reduced when going to more peripheral collisions.

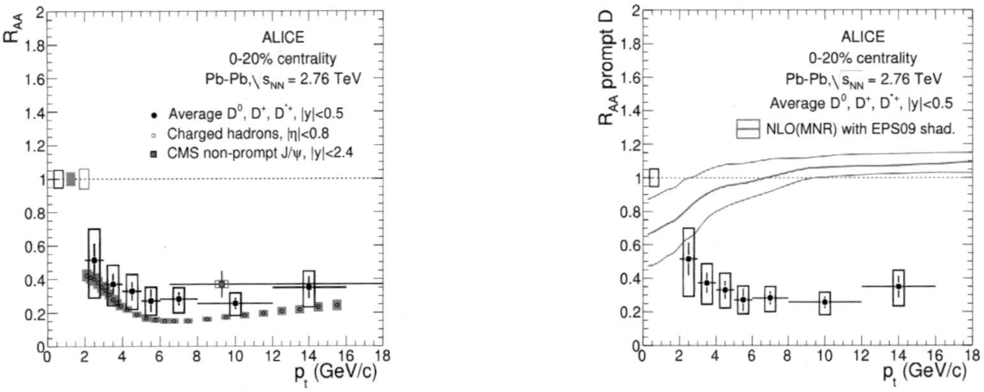

Figure 4. Average R_{AA} of D mesons in 0-20 % central Pb-Pb collisions at $\sqrt{s_{NN}}$=2.76 TeV compared to: left, the nuclear modification factors of charged hadrons [5] and non-prompt J/ψ from B decays [6]: right, the expectation from NLO pQCD [8] with nuclear shadowing [7].

On the left panel of Fig. 4, the nuclear modification of prompt D mesons, charged hadrons [5] and non-prompt J/ψ from B decays [6] are compared in 0-20 % Pb-Pb collisions. There is an indication for $R_{AA}^{D}>R_{AA}^{charged}$, whereas the suppression of non-prompt J/ψ, although the y range is wider, is clearly weaker than that of charged hadrons.

On the right panel of Fig. 4, the nuclear modification of prompt D mesons in 0-20 % Pb-Pb collisions is compared to the expectation from NLO pQCD [8] with nuclear shadowing [7]. The effect of shadowing on the R_{AA} is of the order of ± 15 % for p_t>6 GeV/c. This suggests that the strong suppression observed in the data is dominated by final-state effects. The upcoming pPb collisions at 4 ZTeV, which will be delivered by the LHC in 2012, will allow to measure directly the initial state effects.

The nuclear modification factor of prompt D mesons in 0-20 % central Pb-Pb collisions at $\sqrt{s_{NN}}$=2.76 TeV are compared to several theoretical models in Fig. 5, along with the comparison to the charged-hadron R_{AA}, see Ref. [2] for more details and references to the models. The model based on radiative energy loss with in-medium D meson dissociation (I) and the ones based on the radiative plus collisional energy loss (II,VII) describe reasonably well at the same time the charm and charged-hadron suppression. A model based on AdS/CFT drag coefficients underestimates significantly the charm R_{AA} and has large errors for the light-flavour R_{AA}.

3. Measurements of open heavy-flavour production via semi-leptonic decays of D and B mesons

Open heavy-flavour hadrons can be measured indirectely via their semi-eletronic decays at mid-rapidity ($|y|<0.8$) in the central barrel of ALICE and via their semi-muonic decays at forward rapidity ($2.5<y<4.0$) in the ALICE muon spectrometer.

Electrons are tracked with the ITS and TPC, and identified using the energy loss in the TPC gas and the measured time of flight in TOF. In Fig. 6, the p_t differential electron spectrum at mid-rapidity, fully corrected for acceptance and detector efficiency, is compared to a cocktail of known

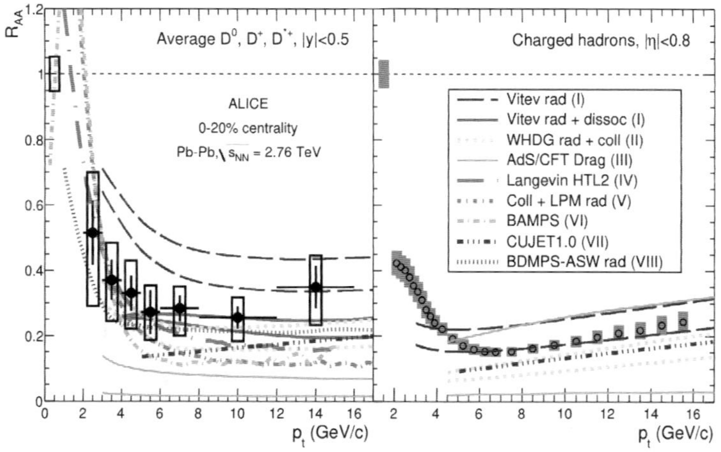

Figure 5. Average R_{AA} of D mesons (left) and charged hadrons (right) in 0-20 % central Pb-Pb collisions at $\sqrt{s_{NN}}$=2.76 TeV compared to model calculations [2].

background electrons for central Pb-Pb collisions (0-10 %). At low p_t, the main contributions from the estimated background come from electrons from Dalitz decays of π^0 and from gamma conversion in the detector material. At high p_t, the signal to background ratio increases. The $\pi^\pm$ spectra measured in ALICE, are used as input for the cocktail in the different centrality classes, together with the decay kinematics from PYTHIA, and the knowledge of the material budget within 6 % uncertainty. The contribution from direct photons is estimated based on Next-to-Leading order calculations [10] assuming a scaling with the number of nucleon-nucleon collisions.

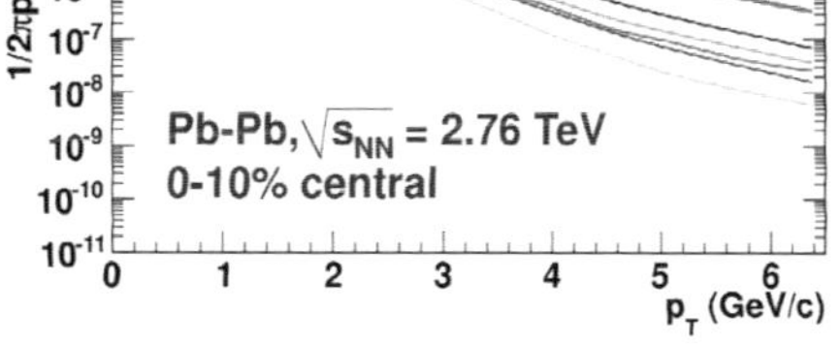

Figure 6. Inclusive electron spectrum in 0-10 % central Pb-Pb collisions at $\sqrt{s_{NN}}$=2.76 TeV at mid-rapidity, compared to a cocktail of known background electrons.

At low p_t (p_t<3.5 GeV/c), an excess is observed in central Pb-Pb collisions, which could point

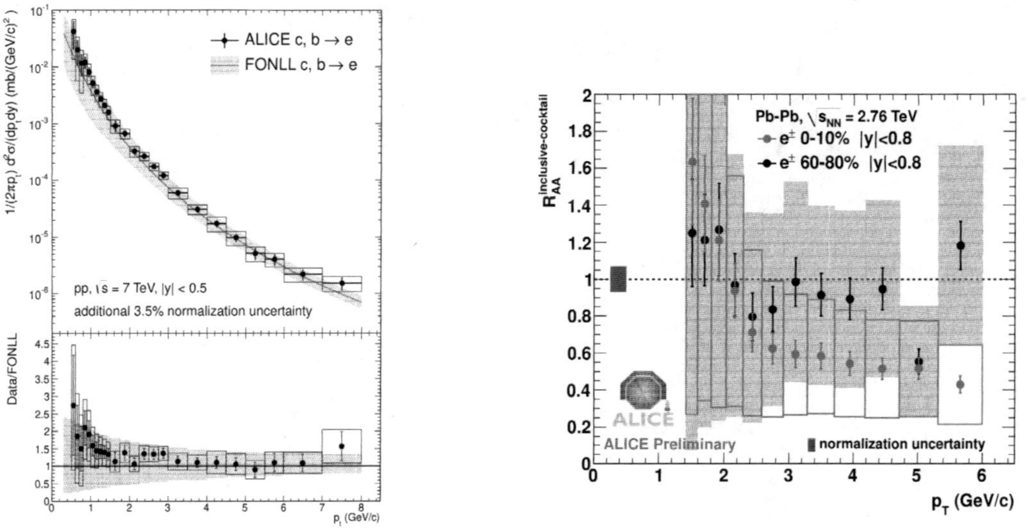

Figure 7. Left: The p_t-differential cross section at mid-rapidity of electrons from charm and beauty hadron decays in pp collisions at 7 TeV, compared to FONLL calculations [9]. Right: Nuclear modification factors of cocktail-subtracted electrons in central (0-10 %) and peripheral (60-80 %) Pb-Pb collisions at 2.76 TeV

to missing background sources in the cocktail, e.g. thermal radiation measured by PHENIX in central collisions in central Au-Au collisions at $\sqrt{s_{NN}}$=200 GeV [11].

Electrons from heavy-flavour hadron decays are measured at mid-rapidity in pp collisions at 7 TeV. The analysis was performed with data collected in 2010, using two different approaches for the electron identication: the TPC, TOF and Transition Radiation Detector (TRD) in the p_t range $0.5<p_t<8$ GeV/c, and the TPC and ElectroMagnetic Calorimeter in the p_t range $3<p_t<6$ GeV/c. The left panel of Fig. 7 shows the comparison of the p_t-differential cross section of electrons from heavy-flavour hadron decays with FONLL calculations [9]. The theory reproduces well the data. The 7 TeV data were scaled in the same way as the D meson measurements to 2.76 TeV to obtain the reference for the computation of the nuclear modification factor of heavy-flavour electrons.

On the right panel of Fig. 7, R_{AA} of the cocktail-subtracted electrons is shown for central (0-10 %) and peripheral (60-80 %) Pb-Pb collisions. Above 3.5 GeV/c, heavy-flavour electrons are expected to dominate. The suppression factor is of the order of 1.5 to 4 for central Pb-Pb collisions, whereas it is consistent with 1.0 for peripheral collisions.

Muons with $p_t>0.5$ GeV/c are reconstructed and identified in the muon spectrometer ($2.5<y<4.0$). From the inclusive muon spectrum, the muon background component, mainly muons from primary $\pi^{\pm}$ and $K^{\pm}$ decays, must be subtracted. The p_t-distributions of $\pi^{\pm}$ and $K^{\pm}$ measured at mid-rapidity in ALICE are used as input to estimate the muon background at foward rapidity. The pp reference used for the determination of the R_{AA} of heavy-favour decay muons is obtained from the analysis of muon-triggered events collected during the pp run at $\sqrt{s}$=2.76 TeV in 2011 [12]. The corresponding integrated luminosity is 19 nb^{-1}. Figure. 8 shows on the left panel the measured p_t differential cross section of muons from heavy-flavour decays in the p_t range $2<p_t<10$ GeV/c in pp collisions at $\sqrt{s}$=2.76 TeV, together with FONLL calculations. The ratio between data and FONLL calculations is shown in the bottom panel. The measurements are well reproduced by the model, although at the upper limit of the calculations.

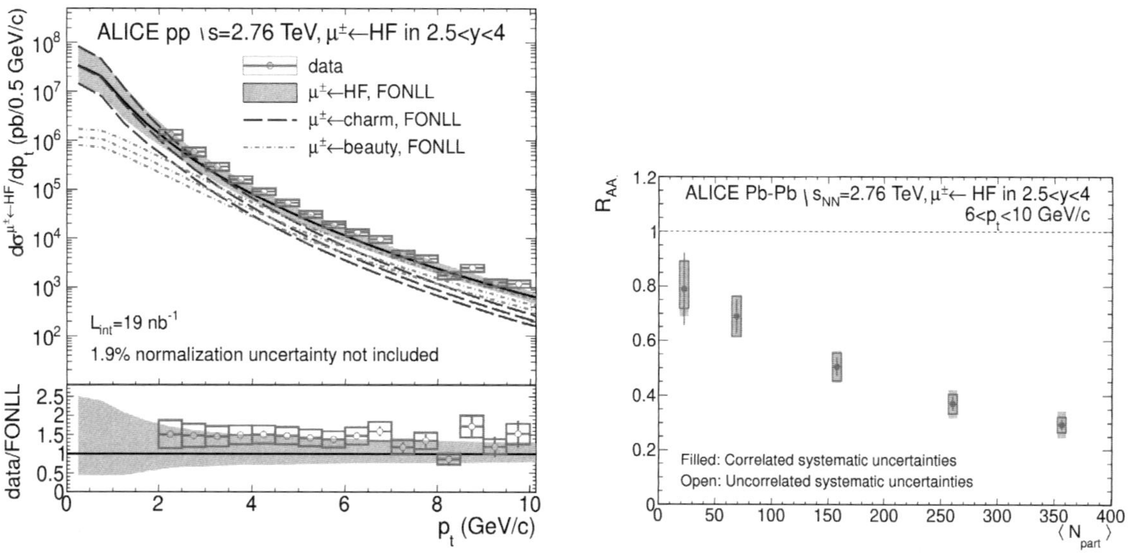

Figure 8. Left: p_t-differential inclusive cross section of muons from heavy-flavour decays in the rapidity range 2.5<y<4, in pp collisions at $\sqrt{s}$=2.76 TeV, compared to FONLL calculations. Error bars (open boxes) are statistical (systematics) uncertainties. The lower panel shows the ratio between data and FONLL calculations. Right: R_{AA} of muons from heavy-flavour decays as a function of the mean number of participating nucleons, in $2.5<y<4$ and $6<p_t<10$ GeV/c. The horizontal bars indicate the uncertainty on $\langle N_{part}\rangle$ [12].

The nuclear modification of heavy-flavour muons with $6<p_t<10$ GeV/c is presented on the right panel of Fig. 8 as function of the mean number of participants $\langle N_{part}\rangle$ in Pb-Pb collisions [12], determined with the Glauber model. In this p_t range, the contribution of muons from B decays becomes dominant in pp collisions according to the central value of the FONLL calculations. The heavy-flavour muons exhibit a strong suppression increasing with $\langle N_{part}\rangle$, reaching a factor of about 3-4 in the 10 % most central collisions. The suppression of D mesons at mid-rapidity is similar to that of muons from heavy flavour decays at foward rapidity.

4. Elliptic flow of D mesons at mid-rapidity ($|y|<0.5$)

In heavy-ion collisions the spatial anisotropy in non-central collisions is converted into a momentum anisotropy, if the medium mean free path allows for sufficient rescattering. The azimuthal distribution of the final particles reflects then the initial anisotropy and the medium characteristics, and can be quantified via the coefficients of a Fourier expansion, in particular the second coefficient, called the elliptic flow v_2.

The elliptic flow of D^0 and D^+ mesons, measured with the event plane method, is shown in Fig. 9 in 30-50 % central Pb-Pb collisions at $\sqrt{s_{NN}}$=2.76 TeV. For comparison the v_2 of charged particles is also plotted. There is an indication for a non-zero elliptic flow of D^0 and D^+ mesons qualitatively near the charged hadron result.

5. Summary

We presented the nuclear modification factor R_{AA} of D mesons in central (0-20 %) and semi-peripheral (40-80 %) Pb-Pb collisions at $\sqrt{s_{NN}}$=2.76 TeV at mid-rapidity in the transverse momentum range $2<p_t<16$ GeV/c. The D mesons are suppressed by a factor 3-4 above 5 GeV/c in 0-20 % Pb-Pb collisions, increasing with centrality. Compared to charged hadrons,

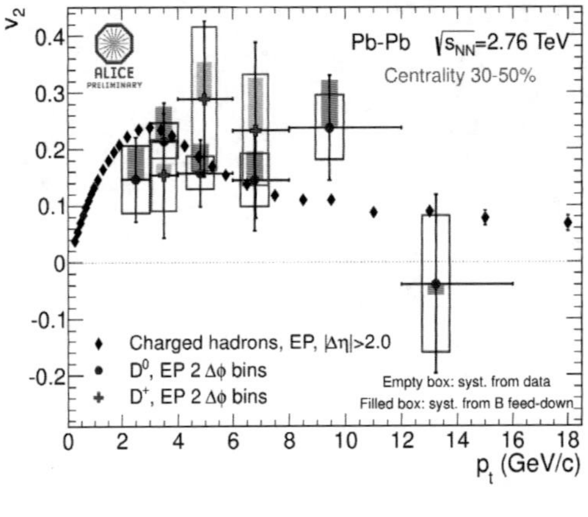

Figure 9. Elliptic flow of D^0, D^+ mesons and charged particles in 30-50 % Pb-Pb collisions at $\sqrt{s_{NN}}$=2.76 TeV.

there is a hint for $R^D_{AA}>R^{charged}_{AA}$. Effects due to shadowing can not explain such a large suppression. Theoretical models with radiative energy loss describe reasonably well at the same time the charm and the charged-hadron suppression. We showed the nuclear modification factor of electrons from heavy-flavour hadron decays at mid-rapidity for central (0-10 %) and peripheral (60-80 %) Pb-Pb collisions up to 6 GeV/c. The suppression factor is of the order of 1.5 to 4 for central Pb-Pb collisions. The nuclear modification factor of muons from heavy-flavour hadron decays was measured at forward rapidity ($2.5<y<4$) in the transverse momentum range $6<p_t<10$ GeV/c as a function of the centrality of the Pb-Pb collisions. The suppression increases with centrality and reaches a factor of about 3-4 in the 10 % most central collisions, similar to the D mesons. Finally the measured elliptic flow of D^0 and D^+ mesons in 30-50 % Pb-Pb collisions shows an indication for a non-zero elliptic flow.

6. References

[1] ALICE Collaboration, JINST, 3:S08002, 2008
[2] ALICE Collaboration, arXiv:1203.2160 nucl-ex
[3] ALICE Collaboration, arXiv:1205.4007 nucl-ex
[4] M. Cacciari, M. Greco and P. Nason, JHEP 9805 007 (1998)
[5] ALICE Collaboration, paper in preparation
[6] CMS Collaboration, arXiv:1201.5069 nucl-ex, 2012
[7] K. J. Eskola, H.Paukkunen and C. A. Salgado, JHEP 0904 (2009) 065
[8] M.Mangano, P.Nason and G.Ridolfi, Nucl. Phys. B373 (1992) 295
[9] ALICE Collaboration, arXiv:1205.5423 nucl-ex
[10] Private communication with V.Vogelsang
[11] Adare A et al, PHENIX Collaboration, Phys. Rev. Lett. 104 132301 (2010)
[12] ALICE Collaboration, arXiv:1205.6443 nucl-ex

Measurements of Open Heavy Flavor Production in STAR

David Tlustý for STAR collaboration

Nuclear Physics Institute, Academy of Sciences Czech Republic, Na Truhlářce 39/64, 180 86 Praha 8, Czech Republic

Czech Technical University in Prague, Faculty of Nuclear Sciences and Physical Engineering, Břehová 7, 11519, Prague 1, Czech Republic

E-mail: tlustdav@fjfi.cvut.cz

Abstract. In this article, we report the measurements of D^0, D^* in $p+p$ at 0.6 GeV/$c < p_T <$ 6 GeV/c and D^0 in Au+Au collisions at 0.2 GeV/$c < p_T <$ 5 GeV/c via hadronic decays $D^0 \rightarrow K^-\pi^+$, $D^{*+} \rightarrow D^0\pi^+ \rightarrow K^-\pi^+\pi^+$ and non-photonic electrons spectra at 3 GeV/$c < p_T <$ 10 GeV/c reconstruction in $\sqrt{s_{NN}} = 200$ GeV $p+p$ collisions at mid-rapidity.

1. INTRODUCTION

Heavy quark production is dominated by initial gluon fusion and can be described by perturbative QCD (pQCD) due to their large mass [1]. The measurement of heavy quark production total cross section in $p+p$ and Au+Au collisions is important to understand both open heavy flavor and heavy quarkonia production mechanisms in the nuclear matter. The study of heavy quark production in relativistic nuclear collisions follows two different approaches: (i) the direct reconstruction of heavy flavor mesons and (ii) the identification of electrons from semi-leptonic decays of such mesons.

The data presented in this paper were collected with the Solenoidal Tracker at RHIC (STAR) detector [2] (Figure 1). Main detector subsystems used for these analyses are the large cylindrical time projection chamber (TPC), which is able to track charged particles in the pseudo-rapidity range $|\eta| <1.8$ with full azimuthal coverage [3], time of flight (TOF) significantly improving charged hadrons identification [4] and barrel electromagnetic calorimeter (BEMC) being able to trigger on high-p_T particles and improving electron-hadron separation [5]. Both TOF and BEMC subsystems provide full azimuthal coverage as TPC, but slightly reduces pseudo-rapidity range $|\eta| <1.0$. A uniform magnetic field of 0.5 T is applied along the beamline by the surrounding solenoidal coils, allowing the p_T of charged particles to be determined .

2. DATA ANALYSES

2.1. Direct reconstruction

Direct reconstruction of open charm mesons is being performed by the STAR collaboration using decay channels $D^0 \rightarrow K^-\pi^+$ with branching ratio of 3.83%, $D^* \rightarrow D^0\pi^+$ with branching ratio of 67.7% in p+p collisions, and former in Au+Au

Published under licence by IOP Publishing Ltd

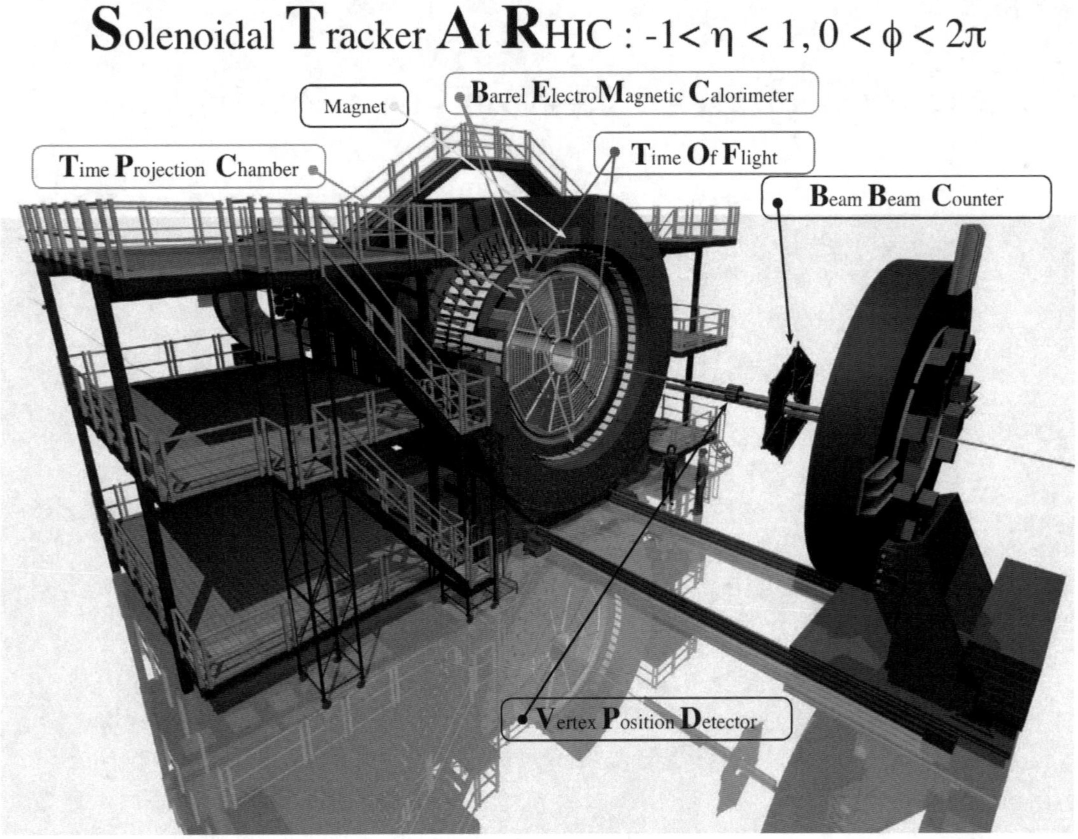

Figure 1. The STAR detector. TPC (Time Projection Chamber) is main detector for tracking and PID (provides $\mathrm{d}E/\mathrm{d}x, \vec{p}$), TOF (Time Of Flight) is used for PID improvement and pileup tracks removal, BEMC (Barrel Electromagnetic Calorimeter) used for electron identification improvement and pile-up tracks removal.

collisions at $\sqrt{s_{NN}}$ = 200 GeV. Because of the small branching ratio and the lack of dedicated detector determining decay vertices, the direct reconstruction of D-mesons requires the analysis of a large amount of data. The available statistics limits the study of such mesons to the p_T region ($0.6 < p_T < 6$ GeV and $0.2 < p_T < 8$ GeV). Kaons and pions are identified using the TPC $\mathrm{d}E/\mathrm{d}x$ and TOF $1/\beta$. The resulting invariant mass spectrum of kaon-pion pairs contains a substantial amount of background from random combinatorics that can be subtracted using various independent methods:

(i) **Mixed-Event**: Events are categorized according to the event multiplicity. Pions from one event are paired with kaons from other random events from an event pool with similar global features.

(ii) **Track-Rotation**: Each π is paired with K with reversed 3-momenta (within same event).

(iii) **Same-Sign**: pions are paired with same charged kaons (within same event). The geometric mean for positive N_{++} and negative N_{--} pair is calculated as $2\sqrt{N_{++}N_{--}}$.

We used Mixed-Event for Au+Au collisions and an average of Track-Rotation and Same-Sign for $p+p$ collision where difference between raw signals obtained by subtraction of background was the main source of systematic uncertainties. Figure 2 left and center panels show raw D^0 signals after combinatorial background reconstruction. A second-order polynomial function is used to fit the residual background. The background fitting with a first-order polynomial function gives negligible difference. A Gaussian function is used to fit the signal. The raw yield of the D^0 is obtained by fitting the data (blue solid circles) with a fit function representing the sum of signal and background (red dashed curve) in the mass region of $1.72 < M_{K\pi} < 2.05$ GeV/c^2 . The signal after the residual background subtraction is shown as the red open circles. The Gaussian function used to describe the signal is shown as the blue dashed curve.

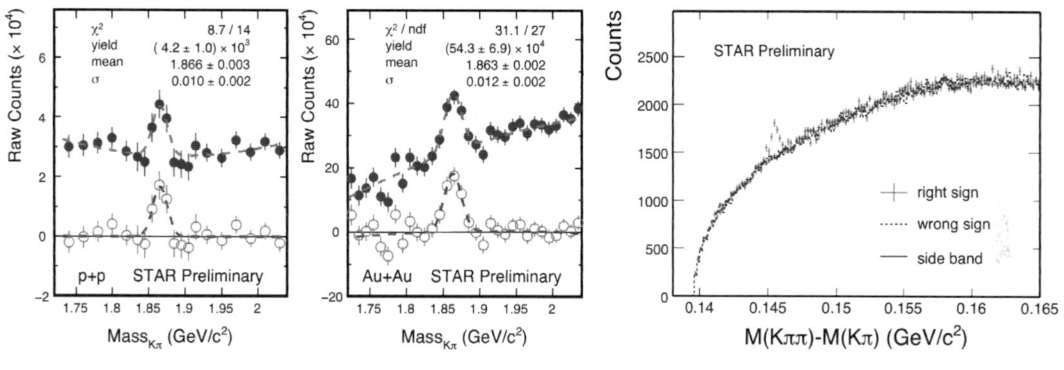

Figure 2. Left and center panels: D^0 signals in p+p and Au+Au 0-80% minbias collisions after same-sign and mix-event background subtraction, respectively. Right panel: D^* signal in p+p collisions. Combinatorial background is reproduced by the distributions from the wrong-sign (black dotted) and side-band (blue solid) methods.

In the case of D^*, we followed the same analysis technique as described in [6]. Compared to the cuts used in [6], the p_T threshold cut for the π^+ (from D^* decays), denoted as π_s^+, was lowered to 0.15 GeV/c. The ratio, r, of transverse momenta from the D^0 and π_s^+ was required to be $7 < r < 20$.These two changes were implemented to improve the statistics near the lower bound in p_T. The invariant mass difference $\Delta M = M(K\pi\pi) - M(K\pi)$ was calculated in reconstructing the D^* signal to take advantage of the partial cancellation in the detector resolution in measured mass distributions. The ΔM distributions are shown in the right panel of Fig. 2. The right-sign combinations $K^{\mp}\pi^{\pm}\pi_s^{\pm}$ were used to select the $D^{*\pm}$ candidates. Two independent methods wrong-sign combinations (opposite sign of π_s) and D^0 side-band combinations - were used for combinatorial background reconstruction. The plot illustrates that both methods reproduce the combinatorial background very well.

2.2. Identification of electrons from semi-leptonic decays

The analysis of non-photonic electrons consists of three main steps: selection of a clean electron sample; subtraction of electron background arising from interactions in material and decays; and residual corrections of the signal yield. The inclusive electron identification was done using TPC $\mathrm{d}E/\mathrm{d}x$ + BEMC information (matching of track momentum and electromagnetic energy) + Shower Max Detector (SMD) information (the shape of the electro magnetic shower, that is much wider for electrons). The analysis details and a discussion of the sources of uncertainty can be found in [9]. The main background in this analysis is the substantial flux of photonic electrons from photon conversion in the detector material and Dalitz decay of π^0 and η mesons. These contributions need to be subtracted in order to extract the non-photonic electron yield according to formula

$$N(\mathrm{npe}) = N(\mathrm{inc}) \cdot \varepsilon_{\mathrm{purity}} - \frac{N(\mathrm{pho})}{\varepsilon_{\mathrm{pho}}}, \tag{1}$$

where $N(\mathrm{npe})$ is the non-photonic electron yield, $N(\mathrm{inc})$ is the inclusive electron yield, $N(\mathrm{pho})$ is the photonic electron yield, $\varepsilon_{\mathrm{pho}}$ is the photonic electron reconstruction efficiency defined as the fraction of the photonic electrons identified through invariant mass reconstruction, and $\varepsilon_{\mathrm{purity}}$ is the purity reflecting hadron contamination in the inclusive electron sample.

3. RESULTS

3.1. Direct reconstruction

Raw Yields $N_{\mathrm{raw}}^{D\,\mathrm{mesons}}$ were calculated in 7 p_T bins (3 for D^0, 4 for D^*) for p+p data and 5 p_T bins for Au+Au data. Then the invariant charm cross section $\mathrm{d}\sigma_{\mathrm{p+p}}^{c\bar{c}}/(2\pi p_T \mathrm{d}p_T \mathrm{d}y)$ was calculated by formula

$$\frac{\mathrm{d}\sigma^{c\bar{c}}}{2\pi p_T \mathrm{d}p_T \mathrm{d}y} = \frac{N_{\mathrm{raw}}^{D\,\mathrm{mesons}} \sigma^{\mathrm{NSD}}}{2\pi p_T \Delta p_T \Delta y\,\mathrm{BR}} \frac{\epsilon_{\mathrm{T}}}{N f_{\mathrm{frag.}}} \tag{2}$$

in each p_T bin (σ^{NSD} is non-single diffractive p+p inelastic cross section, $f_{\mathrm{frag.}}$ is the ratio of charm quarks hadronized to open charm mesons and ϵ_{T} represents the trigger bias correction including the vertex reconstruction efficiency in the measurement). The charm cross section at mid rapidity $\mathrm{d}\sigma^{c\bar{c}}/\mathrm{d}y$ was obtained from power-law function fit to $\mathrm{d}\sigma^{c\bar{c}}/(2\pi p_T \mathrm{d}p_T \mathrm{d}y)$ points (see Fig. 3) as $202 \pm 56(\mathrm{stat.}) \pm 40(\mathrm{sys.}) \pm 20(\mathrm{norm.})$ μb. In Au+Au collisions we calculate invariant yield $\mathrm{d}^2N/(N_{\mathrm{ev}} p_T \mathrm{d}p_T \mathrm{d}y)$.

The $\mathrm{d}\sigma^{c\bar{c}}/\mathrm{d}y$ at mid-rapidity in Au+Au collisions was extracted, from the average of a power-law(dot-dashed curve) and a blast-wave (dashed curve) fit (Fig. 4) as $186 \pm 22(\mathrm{stat.}) \pm 30(\mathrm{sys.}) \pm 18(\mathrm{norm.})$ μb assuming that the $f_{\mathrm{frag.}}$ does not change from p+p to Au+Au collisions. The power-law fit to p+p scaled by the average number of binary collisions (N_{bin}) is shown as solid curve. The charm cross section for three centrality bins, 0-20%, 20-50% and 50-80%, is obtained according to the integrated yields. The charm production cross section per nucleon-nucleon collision at mid-rapidity as a function of N_{bin} is shown in the right panel of Fig. 3. Within errors, the results are in agreement and follow the number of binary collisions scaling, which indicates that charm quark is produced via initial hard scatterings at early stage of the collisions at RHIC. The FONLL(orange band)and NLO [7] (light-blue band) uncertainties are also shown here for comparison.

Journal of Physics: Conference Series **389** (2012) 012024 doi:10.1088/1742-6596/389/1/012024

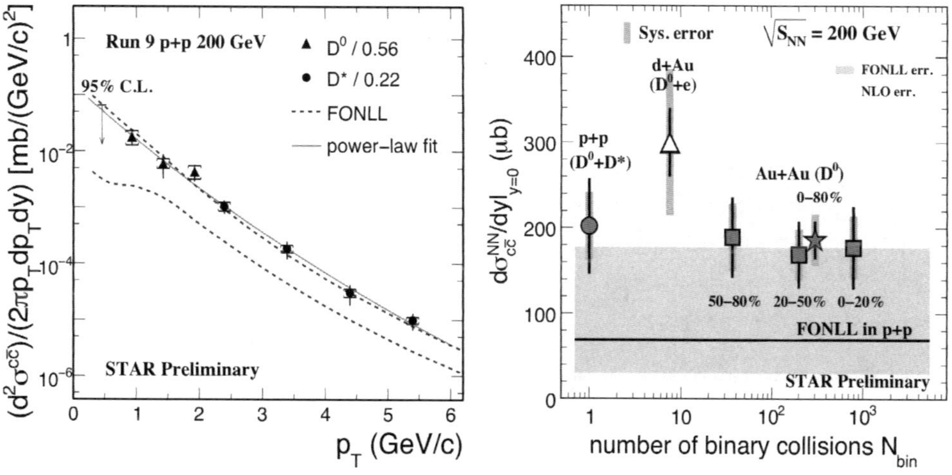

Figure 3. Left Panel: $c\bar{c}$ pair production cross sections (symbols) as a function of p_T in 200 GeV p+p collisions. Right Panel: The charm production cross section per nucleon-nucleon collision at mid-rapidity as a function of $N_{\rm bin}$.

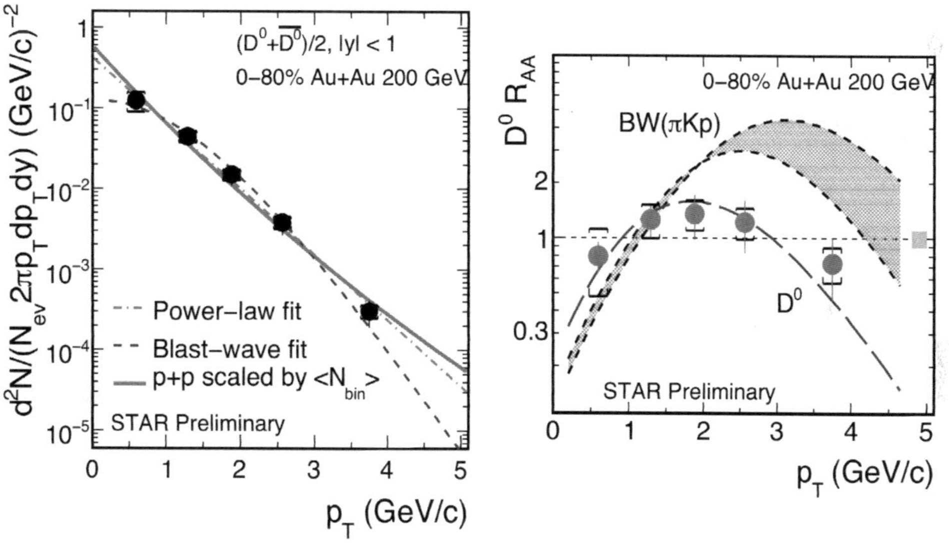

Figure 4. Left Panel: D^0 p_T spectrum in 200 GeV 0-80% Au+Au collisions. Right Panel: D^0 nuclear modification factor $R_{\rm AA}$ as a function of p.

The $R_{\rm AA}^{D^0}$ was obtained via dividing D^0 yields in Au+Au 0-80% minbias collisions by the power-law fit to p+p yields scaled by $N_{\rm bin}$, shown in right panel of Fig. 4. The uncertainty of the p+p power-law shape is taken into account as systematic error. No suppression is observed at $p_T < 3$ GeV/c. The dashed curve shows the blast-wave fit. The shaded band is the predicted $R_{\rm AA}^{D^0}$ blast-wave parameters from light-quark hadrons [8], which is different from data. This might indicate that D^0 mesons freeze out earlier than light flavor hadrons or has less radial flow.

3.2. Identification of electrons from semi-leptonic decays

The invariant cross section for non-photonic electron production is calculated according to

$$E\frac{d^3\sigma}{dp^3} = \frac{1}{\mathcal{L}}\frac{1}{2\pi p_T \Delta p_T \delta y}\frac{N_{\rm npe}}{\varepsilon_{\rm rec}\varepsilon_{\rm trig}\varepsilon_{\rm eid}\varepsilon_{\rm BBC}}, \quad (3)$$

where $N_{\rm npe}$ is the nonphotonic electron raw yield, $\varepsilon_{\rm rec}$ is the product of the single electron reconstruction efficiency and the correction factor for momentum resolution and finite spectrum bin width, $\varepsilon_{\rm trig}$ is the high-tower trigger efficiency, $\varepsilon_{\rm eid}$ is the electron identification efficiency, $\mathcal{L}$ is the integrated luminosity with the z-position of vertex cuts, and $\varepsilon_{\rm BBC} = 0.87 \pm 0.08$ is the BBC trigger efficiency.

Figure 5(a) shows the non-photonic electron $\left(\frac{e^+ + e^-}{2}\right)$ invariant cross section obtained by combining the Run2008 and the Run2005 results using the "best linear unbiased estimate" [10]. The corrected result of our early published measurement using year 2003 data [11] is shown in the plot as well. Fig. 5(b) shows the ratio of each individual measurement, including PHENIX results, to the FONLL calculation. One can see that all measurements at RHIC on non-photonic electron production in $p+p$ collisions are now consistent with each other. The corrected run 2003 data points have large uncertainties because of the small integrated luminosity (100nb^{-1}) in that run. FONLL is able to describe the RHIC measurements within its theoretical uncertainties.

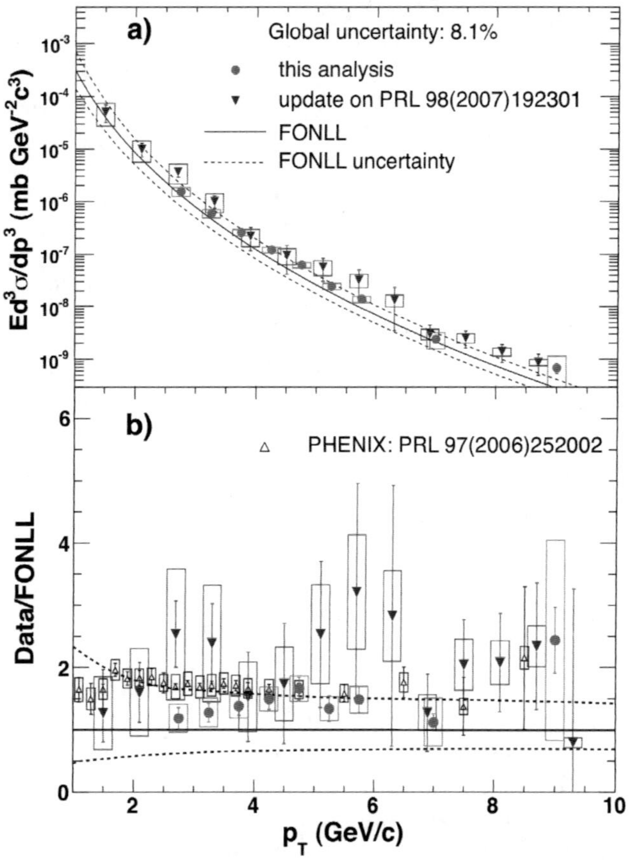

Figure 5. (a) Invariant cross section of non-photonic electron production $\left(\frac{e^+ + e^-}{2}\right)$ in $p + p$ collisions from [9] analysis (closed circles) after combining results from Run2005 and Run2008. The published STAR result [11] (closed triangles) is also shown. (b) Ratio of data over FONLL [1] from all measurements at RHIC including PHENIX results [12] (open triangles).

Electrons from bottom and charm meson decays are the two dominant components of the non-photonic electrons. Mostly due to the decay kinematics, the azimuthal

Journal of Physics: Conference Series **389** (2012) 012024 doi:10.1088/1742-6596/389/1/012024

correlations between the daughter electron and daughter hadron are different for bottom meson decays and charm meson decays. A study of these azimuthal correlations has been carried out on STAR data and is compared with a PYTHIA simulation to obtain the ratio of the bottom electron yield to the heavy-flavor decay electron yield $\frac{e_b}{e_b+e_c}$ [13], where PYTHIA was tuned to reproduce STAR measurements of D mesons p_T spectra [14]. Using the measured $e_b/(e_b + e_c)$ together with the measured non-photonic electron cross section with the electrons from J/Ψ, Υ decay and Drell-Yan processes subtracted, we are able to disentangle these two components. Figure 6 shows the invariant cross section of electrons $\left(\frac{e^+ + e^-}{2}\right)$ from bottom (upper left) and charm (upper right) mesons as a function of p_T and the corresponding FONLL predictions, along with the ratio of each measurement to the FONLL calculations (lower panels).

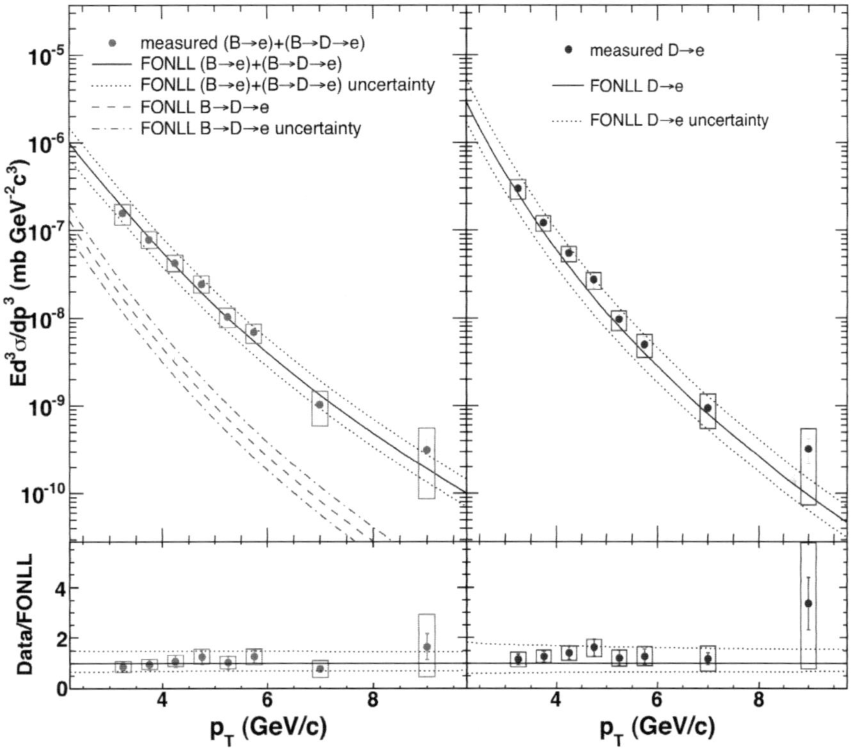

Figure 6. Invariant cross section of electrons $\left(\frac{e_b}{e_b+e_c}\right)$ from bottom (upper left) and charm meson (upper right) decay, together with the ratio of the corresponding measurements to the FONLL predictions for bottom (lower left) and charm electrons (lower right). The solid circles are experimental measurements. The error bars and the boxes are, respectively, the statistical and systematic uncertainties. The solid and dotted curves are the FONLL predictions and their uncertainties. The dashed and dot-dashed curves are the FONLL prediction for $B \to D \to e$

From the measured spectrum, we determine the integrated cross section of electrons $\left(\frac{e^+ + e^-}{2}\right)$ at 3 GeV/c < p_T < 10 GeV/c from bottom and charm meson decays to be, respectively,

$$\left.\frac{\mathrm{d}\sigma_{(B\to e)+(B\to D\to e)}}{\mathrm{d}y}\right|_{y_e=0} = 4.0 \pm 0.5(\mathrm{stat}) \pm 1.1(\mathrm{syst})\ \mathrm{nb}$$

$$\left.\frac{\mathrm{d}\sigma_{D\to e}}{\mathrm{d}y}\right|_{y_e=0} = 6.2 \pm 0.7(\mathrm{stat}) \pm 1.5(\mathrm{syst})\ \mathrm{nb}$$

4. CONCLUSIONS

Open charm mesons (D^0, D^{*+}) are measured in p+p and Au+Au collisions at $\sqrt{s_{\mathrm{NN}}}$ = 200 GeV at STAR. Charm cross sections per nucleon-nucleon collision at mid-rapidity follow the number of binary collisions scaling. The charm pair production cross sections per nucleon-nucleon collision at mid rapidity are measured to be $\mathrm{d}\sigma^{c\bar{c}}/(2\pi p_T \mathrm{d}p_T \mathrm{d}y)$ = $202 \pm 56(\mathrm{stat.}) \pm 40(\mathrm{sys.}) \pm 20(\mathrm{norm.})$ μb in p+p and $186 \pm 22(\mathrm{stat.}) \pm 30(\mathrm{sys.}) \pm 18(\mathrm{norm.})$ μb in Au+Au minimum bias collisions. Blast-wave predictions with light-quark hadron parameters are different from data, which might indicate that D^0 mesons freeze out earlier than light flavor hadrons or have less radial flow.

STAR measurements of high p_T non-photonic electron production in $p+p$ collisions at $\sqrt{s}$ = 200 GeV using data from Run2005 and Run2008 and PHENIX result are consistent with each other. We are able to disentangle the electrons from bottom and charm meson decays in the non-photonic electron spectrum using the measured ratio of $e_B/(e_B + e_D)$ and the measured non-photonic cross section.

In the near future the STAR Heavy Flavor Tracker [15] will provide the necessary resolution to reconstruct secondary vertices of charm mesons, which will increase the precision of charm measurements.

REFERENCES

[1] M. Cacciari, P. Nason and R. Vogt, *Phys. Rev. Lett.* **95**, 122001 (2005).
[2] K. H. Ackermann *et al.* (STAR Collaboration), Nucl. Instrum. Methods A **499**, 624 (2003).
[3] M. Anderson *et al.*, Nucl. Instrum. Methods A **499**, 659 (2003).
[4] M. Shao *et al.*, Nucl. Instrum. Methods A **558**, 419 (2006).
[5] M. Beddo *et al.*, Nucl. Instrum. Methods A **499**, 725 (2003).
[6] B.I. Abelev *et al.* [STAR Collaboration], Phys. Rev. D **79**, 112006 (2009).
[7] R. Vogt, Eur.Phys.J.ST **155** 213, (2008).
[8] B. I. Abelev *et al.*, Phys. Rev. C **79**, 34909 (2009).
[9] B.I. Agakishiev *et al.* [STAR Collaboration], Phys. Rev. D **83**, 052006 (2011).
[10] A. Valassi, Nucl. Instrum. Methods Phys. Res., Sect. A **500**, 391 (2003).
[11] B. I. Abelev *et al.* (STAR Collaboration), Phys. Rev. Lett. **98**, 192301 (2007); B.I. Abelev *et al.* (STAR Collaboration), arXiv:nucl-ex/0607012v3
[12] A. Adare *et al.* (PHENIX Collaboration), Phys. Rev. Lett. **97**, 252002 (2006).
[13] M. M. Aggarwal *et al.* (STAR Collaboration), Phys. Rev. Lett. **105**, 202301 (2010).
[14] X. Lin, arXiv:hep-ph/0602067.
[15] Z. Xu *et al.*, J. Phys. G **32**, S571(2006).

This work was supported by grant INGO LA09013 of the Ministry of Education, Youth and Sports of the Czech Republic, and by the Grant Agency of the Czech Technical University in Prague, grant No. SGS10/292/OHK4/3T/14.

Measurements of Υ Production and Nuclear Modification Factor at STAR

Anthony Kesich for the STAR Collaboration
One Shields Avenue, Davis, CA, 95616, USA
E-mail: akesich@ucdavis.edu

Abstract. Thermal suppression of quarkonium production in heavy-ion collisions, due to Debye screening of the quark-antiquark potential, has been proposed as a clear signature of quark-gluon plasma (QGP) formation. At RHIC energies, the Υ meson is a clean probe of the early system thanks to negligible levels of enhancement from $b\bar{b}$ recombination and non-thermal suppression from co-mover absorption. We report on our measurement of the $\Upsilon \rightarrow e^+e^-$ cross section in Au+Au collisions at $\sqrt{S_{NN}}$=200 GeV. We compute the Nuclear Modification Factor by comparing these results to p+p collisions. In order to have a complete assessment of both hot and cold nuclear matter effects on Upsilon production we also report on results from d+Au collisions.

1. Introduction

The key goal of the RHIC program for the past decade has been to characterize the Quark-Gluon Plasma. By measuring its properties and evolution, we have a unique tool to study the underlying process governing it all: the strong nuclear force. Heavy quarkonia are unique probes into the dynamics of QGP evolution. Being massive, heavy quarkonia are produced almost exclusively via initial hard scatterings. They thereby allow us to extract information about the initial conditions of the QGP as well as observe effects due to evolution and cooling. Lattice-based calculations predict that the formation of a QGP will screen the inter-quark potential, thereby decreasing the meson's binding energy and allowing thermal excitation to dissociate the quark-antiquark pair. By measuring the extent of this suppression and its dependence on system size, we can extract information about the initial temperature of the produced plasma [1, 2].

2. Analysis Techniques

The analyses shown in this paper were done using three different datasets: p+p from 2009, d+Au from 2008, and Au+Au from 2010. All systems were at end energy of $\sqrt{s_{NN}} = 200$ GeV. In all cases, for an event to be recorded, it first had to satisfy a High Tower trigger ($E_T \gtrsim 4.3$ GeV). Thanks to increased bandwidth in 2010 due to our DAQ1000 upgrade, we were able to record this trigger without any prescaling, so there were no additional requirements in the Au+Au system.

However, for the p+p and d+Au systems, we had an additional Upsilon trigger that ran on all events which fired the High Tower trigger. This trigger looks for a second energetic tower no closer than 90° to the tower which fired the first trigger. It then calculates an invariant mass

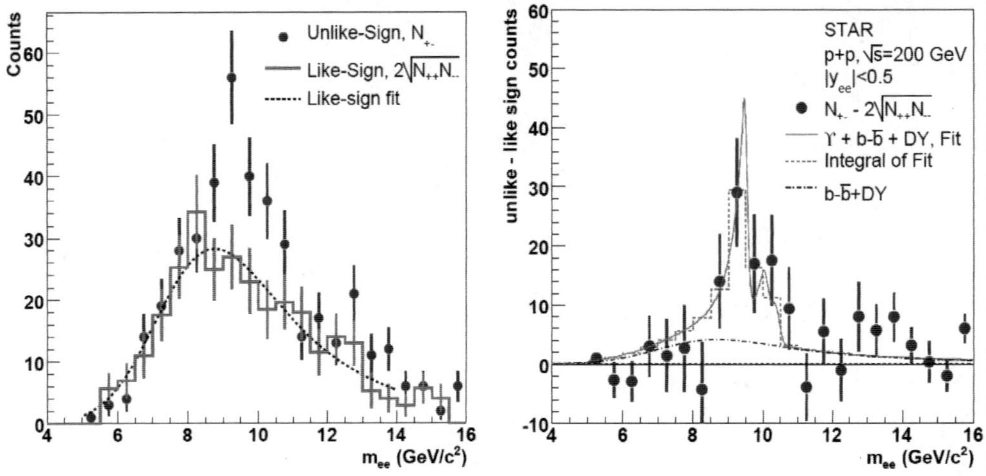

Figure 1. *Left:* Invariant mass spectrum of identified electrons. The red curve is the like-sign spectrum. The blue curve is the unlike-sign spectrum. The peak in the background is due to kinematic constraints imposed on our candidates by the Upsilon trigger. *Right:* To remove combinatorial pairs, we subtract the like-sign spectrum from the unlike-sign. The red curve represents our Υ signal. The dotted black curve is the combination of Drell-Yan and $b\bar{b}$ recombination.

assuming the energy in the towers came from electrons. If that energy is near the Υ mass, we record the event.

We relied on two main detectors in these analyses: the Time Projection Chamber (TPC) and the Barrel Electromagnetic Calorimeter (BEMC). The TPC is a gas-ionization chamber used to track charged particles. By looking at track curvature, we can measure the momentum and charge of produced particles. Also, by looking at the energy lost to ionization, we can identify particles based on their characteristic energy loss as a function of momentum. The BEMC is a lead-scintillator sandwich-style calorimeter used to measure energy carried by electromagnetic particles. By matching tracks from the TPC to deposited energy in the BEMC, we can reject BEMC signals from photons. Furthermore, we can identify electrons by comparing the energy deposited in the BEMC to the momentum measured in the TPC. Finally, the BEMC is a fast detector and serves as the basis for both triggers used in these analyses.

To reconstruct Upsilon candidates, we look at the invariant mass spectrum of unlike-sign (oppositely charged) identified electrons. For mass reconstruction, the momentum as measured by the TPC is used. In order to remove and contributions from random, combinatorial pairs, we subtract the like-sign spectrum from the unlike-sign spectrum. The resulting subtracted spectrum contains three signals: Drell-Yan production, $b\bar{b}$ recombination, and the desired $\Upsilon(1S+2S+3S) \rightarrow e^+e^-$ signal. The shape of the Drell-Yan signal is obtained from PYTHIA simulations. The $b\bar{b}$ background is obtained from pQCD calculation by R. Vogt. The shape of our Upsilon signal comes mainly from detector effects such as bremsstrahlung radiation and momentum resolution. It is obtained via embedding PYTHIA simulations into our detector model and real data.

3. Upsilon Production in p+p Collisions

In the left plot of Fig. 1, we show the invariant mass spectrum for both like-sign (red) and unlike-sign (blue) identified electron pairs. The peak in both signals around 9 GeV is a kinematic

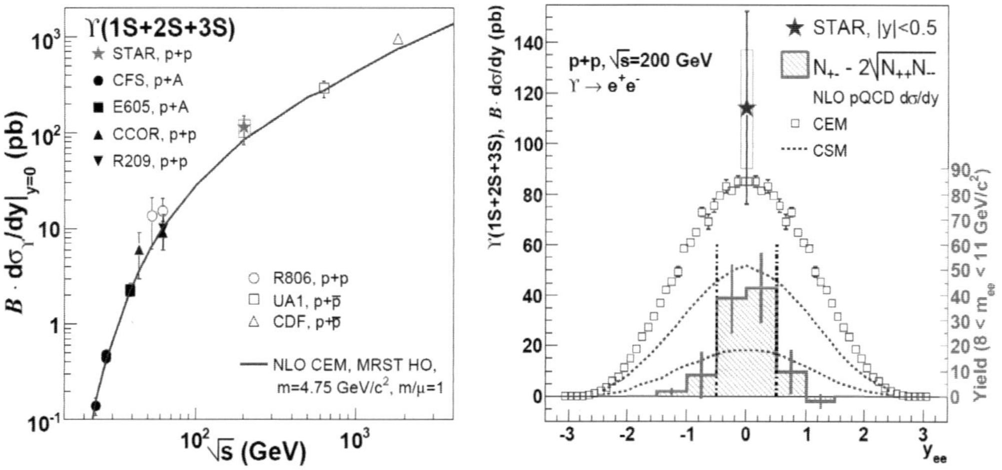

Figure 2. *Left:* Our measurement (red star) compared to the world data trend. *Right:* Comparison of our measurement to theory. The dotted blue line is a Color Evaporation Model calculation [3]. The blue squares are calculated using a Color Singlet Model [4].

constraint imposed by the condition of the Upsilon trigger. We use the shape of the like-sign spectrum to measure the kinematic conditions imposed by the trigger for use in our later fits. Furthermore, we subtract the like-sign spectrum from the unlike-sign to account for the combinatorial background. The results are show in the right plot of Fig. 1. We fit the subtracted spectrum to the integral of three signals: Drell-Yan, $b\bar{b}$ recombination, and the Upsilon signal. The Drell-Yan and $b\bar{b}$ signals are shown as a black dotted line. To obtain our yield of Υ mesons, we subtract the fit to Drell-Yan and $b\bar{b}$ from the subtracted spectrum. This helps remove any dependence we have on our choice of line-shape. We measure a cross section at $|y| < 0.5$ of $114 \pm 38^{+23}_{-24}$ pb.

The left plot of Fig. 2 shows the comparison between our measurement and the world data trend. The blue curve comes from a pQCD calculation by R. Vogt using the Color Evaporation Model. The right plot of Fig. 2 compares our measurement to two NLO pQCD calculations. The Color Singlet Model is shown as a blue dotted line [4]. The Color Evaporation Model is down as blue boxes [3]. Our measurement is much more consistent with the CEM calculation.

4. Upsilon Production in d+Au Collisions

The d+Au dataset was collected using the Upsilon trigger, imposing the same kinematic peak in the signal as was seen in the p+p analysis. We used the same technique as in p+p to isolate the signal. We report a combined Upsilon, Drell-Yan and $b\bar{b}$ cross section for $|y| < 0.5$ in the mass window 7 GeV<m<11 GeV of $35 \pm 4 \pm 5$ nb.

5. Upsilon Production in Au+Au Collisions

As a result of the installation of DAQ1000, which allows us to record Au+Au events at a rate of 1 KHz, we no longer needed the Upsilon trigger. As a result, we no longer had a kinematic peak imposed on the background (see Fig. 4, left). Furthermore, once we subtract the like-sign spectrum from the unlike-sign spectrum, our Drell-Yan and $b\bar{b}$ signal also do not have a kinematic peak (Fig. 4, center).

Since we do not know a priori what the ratio between the 1S, 2S, and 3S states should be, we

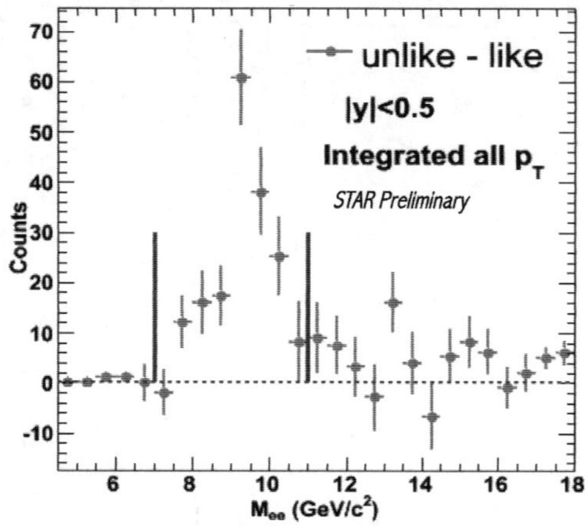

Figure 3. Invariant mass spectrum of unlike-sign minus like-sign pairs of identified electrons. This contains the Upsilon signal, Drell-Yan production, and b$\bar{b}$ recombination. We see an 8σ signal in the mass window of 7-11 GeV.

fit our signal to lineshapes representing different scenarios (such as the excited states existing in the proportion to the ground state as in p+p, no excited states, half of p+p, etc.). We also still subtract the fit to the Drell-Yan and b$\bar{b}$ background from the subtracted spectrum to reduce further any dependence on our lineshapes.

By selecting on the number of charged tracks produced, we can separate our measurement into three centrality bins: 0% to 10%, 10% to 30%, and 30% to 60%. Using our p+p collisions as a baseline, we compute a nuclear modification factor, R_{AA}, as a function of the average number of participants (obtained via Glauber simulations). The results are shown in the right plot of Fig. 4. We compare our results to calculations done using lattice-based QCD results with a hydrodynamical model incorporating expansion, cooling, and temperature gradients [2]. The model is run using two different inter-quark potentials: one based on the internal energy of the quark-antiquark pair, the other on the free energy. Our results are more consistent with the internal energy model. However, more statistics are needed to reach any definitive conclusions.

6. Conclusions

Thanks to the large acceptance of the STAR detector and our dedicated Upsilon trigger, we were able to measure Υ mesons in p+p, d+Au, and Au+Au collisions at $\sqrt{s_{NN}} = 200$ GeV. Measured p+p production is consistent with NLO calculations. Our results shows increasing suppression with the number of participants. Furthermore, the results are consistent with hydrodynamical suppression models for the QGP.

This material is based upon work supported by the National Science Foundation under Grant No. 0645773.

References

[1] T. Matsui, H. Satz, Physics Letters B 178 (4) (1986) 416 – 422.
[2] M. Strickland, D. BazowarXiv:1112.2761v4, [link].
URL `http://arxiv.org/abs/1112.2761v4`
[3] S. Brodsky, J.-P. Lansberg, Physical Review D 81 (5) (2010) 051502.
[4] A. Frawley, T. Ullrich, R. Vogt, Physics Reports 462 (4–6) (2008) 125 – 175.

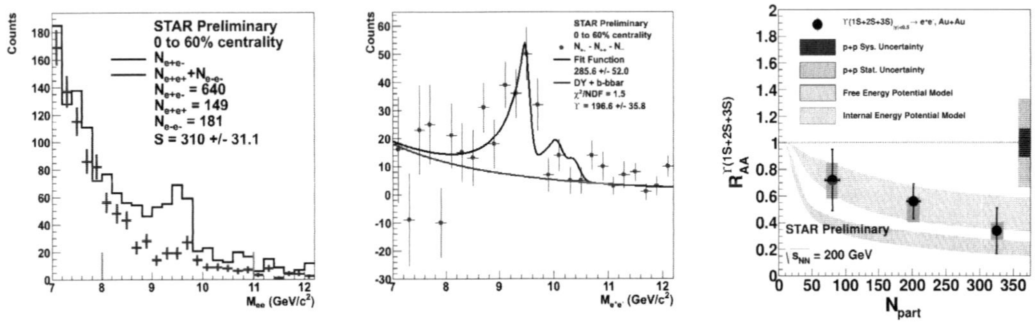

Figure 4. *Left:* Invariant mass spectrum of identified electrons in Au+Au system in the 0% to 60% centrality range. Like-sign pairs are in blue; unlike-sign in black. Note that the spectrum no longer has a kinematic peak imposed by the Upsilon trigger. *Center:* Unlike-sign signal subtracted from like-sign signal, 0%-60% centrality. The blue curve is the combination of Drell-Yan and $b\bar{b}$. The black curve includes the Upsilon signal also. *Right:* $R_{AA}$for 2010 Au+Au compared to 2009 p+p. The magenta and orange curves are theoretical predictions using a combination of lattice-based QCD and hydrodynamical expansion and cooling [2].

Υ, χ_c and ψ' production and nuclear modifications from PHENIX

Cesar Luiz da Silva for the PHENIX Collaboration
LANL, Po Box 1663.H846, Los Alamos NM 87544-0600, USA
E-mail: slash@bnl.gov

Abstract. We report on the most recent measurements of bottomonia and excited states of charmonium made by PHENIX in mid- and forward rapidities at $\sqrt{s}$ = 200 GeV. We also discuss the prospects for future measurements and concerns in using the color screening of different quarkonia states in heavy ion collisions as a quantitative observable for the temperature reached at the strong interacting quark gluon plasma.

1. Introduction

For a long time the suppression of quarkonia yields in heavy ion collisions relative to binary scaling and cold nuclear effects has been proposed as a signature for the formation of a quark gluon plasma [1]. Indeed anomalous suppression of J/ψ observed in SPS experiments was the first verification of a quark-gluon-plasma behavior in heavy ion collisions [2]. A similar suppression was observed at RHIC [3] and different levels of suppression were observed in LHC in different kinematic regions [ALICE,ATLAS and CDF in this volume]. All these measurements have concerns on how cold nuclear matter effects can be accounted for in heavy ion collisions [4, 5]. At RHIC and LHC charm coalescence and recombination can also contribute to the nuclear modifications [6].

A series of theoretical studies of the collor screening effect on quarkonia states is available in the literature. A nice collection of references on these studies is presented in [7] which includes studies using lattice QCD, QCD sum rules, Ads/QCD, resummed perturbation theory, effective field theories and potential models. Figure 1 shows the range of quark-gluon plasma (QGP) temperatures relative to the critical temperature T_c where each heavy quarkonia state is dissociated. The ranges are from variations from the several estimations and also cover the transition between unmodified and totaly dissociated quarkonia peak in the calculations. For comparison, the figure also shows the expected temperature reached at RHIC collisions calculated by hydrodynamics models fitted to PHENIX thermal photon data [8]. The experimental verification of these dissociations would represents a strong constraint to the temperature of the medium created in A+A collisions as well as a test for the color screening models.

One important caveat in use quarkonia dissociation as a thermometer is how the different quarkonia states cross the nucleus. The different states of charmonium and bottomonium are produced with the same initial state but evolve to their final states as color singlet or color octet state with some additional time for its color neutralization. If quarkonia cross the nucleus as a pre-resonant object, different final states will appear with the same suppression. Figure 2 shows

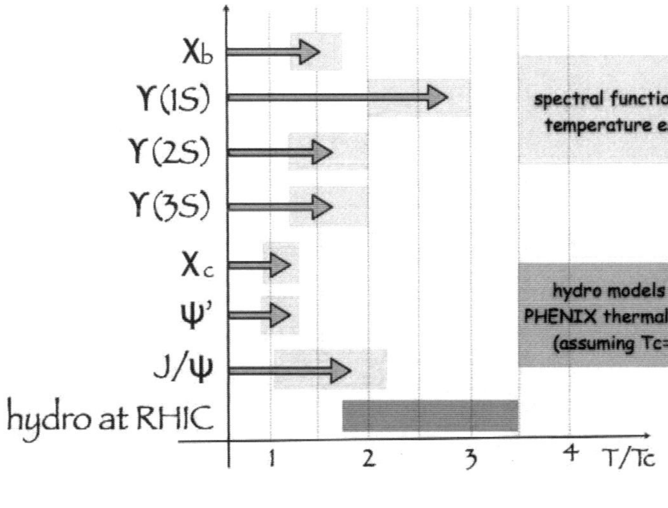

Figure 1. Ranges of medium temperatures where quarkonia states are dissociated. Ranges are from variation in the theoretical calculations therein [7] and the transition between unmodified and totally dissociated quarkonia peak.

how ψ' , produced as a color octet alike J/ψ , and χ_c , produced as a color singlet, would be suppressed in cold nuclear matter relative to J/ψ one. In fact, J/ψ and ψ' data taken in p+A collisions by E866/NuSea [9] reveal a different nuclear absorption between J/ψ and ψ' for small x_F and a pre-resonant like behavior for $x_F > 0.2$, where the magnitude of the absorption of both states are undistinguished. However, the lifetime for the charmonium octet state calculated using FERMILAB and NA50 [10] data was found to be too small demanding the emission of unphisically large energy gluons [11].

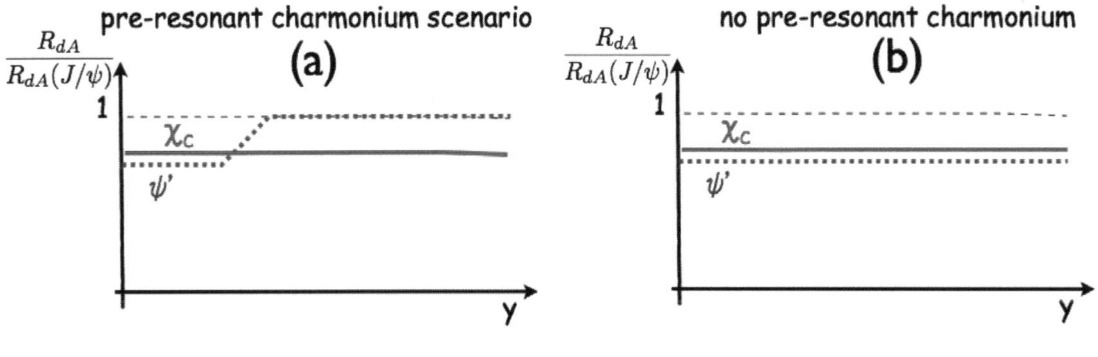

Figure 2. Rapidity dependence of the ψ' and χ_c suppression relative to direct J/ψ in the case, or not, of the existence of pre-resonant state for ψ' and J/ψ .

PHENIX detector has an extensive program in quarkonia measurements which includes di-lepton decays of J/ψ , ψ' and Υ family in the central arms ($|y| < 0.35$) and muon arms ($1.2 < |y| < 2.2$) as well as radiative decays of χ_c in central arms. The excited state charmonia and the Υ measurements are described in the following sections as well as their implications in understanding the behavior of different quarkonia states in QGP.

2. Excited state charmonia

The ground state charmonium (J/ψ) has been studied in p+p [12], d+Au [4, 5], Cu+Cu [13] and Au + Au collisions at $\sqrt{s_{NN}} = 200$ GeV [3, 14] as well as a low energy study which has started with $\sqrt{s_{NN}}$ of 39 GeV and 62 GeV [15]. The focus of this section is the measurement of ψ' and χ_c states.

2.1. ψ'

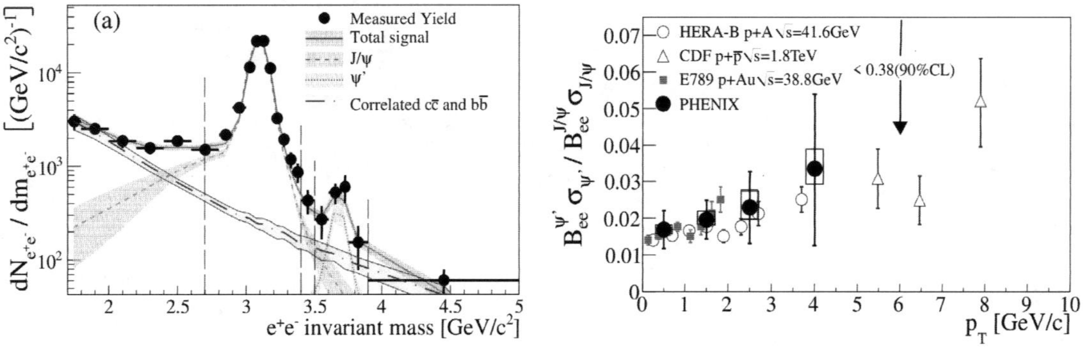

Figure 3. Di-electron signal in the J/ψ and ψ' mass range and the relative ψ' / (J/ψ) yield obtained by PHENIX compared to lower energy experiments [12].

The first measurements of ψ' was made in p+p collisions at mid-rapidity [12]. Correlated heavy flavor decays in electrons can contribute with up to 40% of the di-electron signal in the ψ' mass region. A careful estimation of the heavy flavor contribution was based on PYTHIA [16] using Next-to-Leading-Order (NLO) subprocesses and also randomized opening angle distribution. The yield of di-electron ψ' decays relative to J/ψ one is of about 2% and follows a rising trend when increasing p_T observed in lower energy experimental results. The feed-down fraction of ψ' in the J/ψ yields is

$$F_{\psi'}^{J/\psi} = B\left(\psi' \rightarrow J/\psi + X\right) \frac{\sigma_{\psi'}}{\sigma_{J/\psi}} = (9.4 \pm 2.4)\%, \tag{1}$$

which also agrees with previous measurements.

Preliminary result of ψ' R_{dA} will be available very soon.

2.2. χ_c

The channel we used to measured χ_c is $\chi_c \rightarrow J/\psi + \gamma \rightarrow e^+e^- + \gamma$. It is a challenging measurement given that the energy of the γ decayed is in average 600 MeV but can extends to lower energies where calorimetry measurement turns to be difficult. In addition, the correlated background in the $J/\psi + \gamma$ invariant mass spectrum used to count χ_c decays is significant. Radiative decays of J/ψ , electron bremsstrahlung, jets, ψ' decays in neutral particles, NNLO $c\bar{c}$ contributions and bottom decays all can contribute to the correlated background. After consider all these sources and side band e^+e^- mass ranges the χ_c feed-down fraction to J/ψ can be directly determined. Figure 4 shows the $e^+e^-\gamma$ mass distributions measured in p+p collisions [12] and in d+Au collisions after the subtraction of the combinatorial and correlated background. The feed-down fraction seems to hold in the presence of cold nuclear matter but large statistical uncertainties still forbids a firm conclusion.

Journal of Physics: Conference Series **389** (2012) 012026 doi:10.1088/1742-6596/389/1/012026

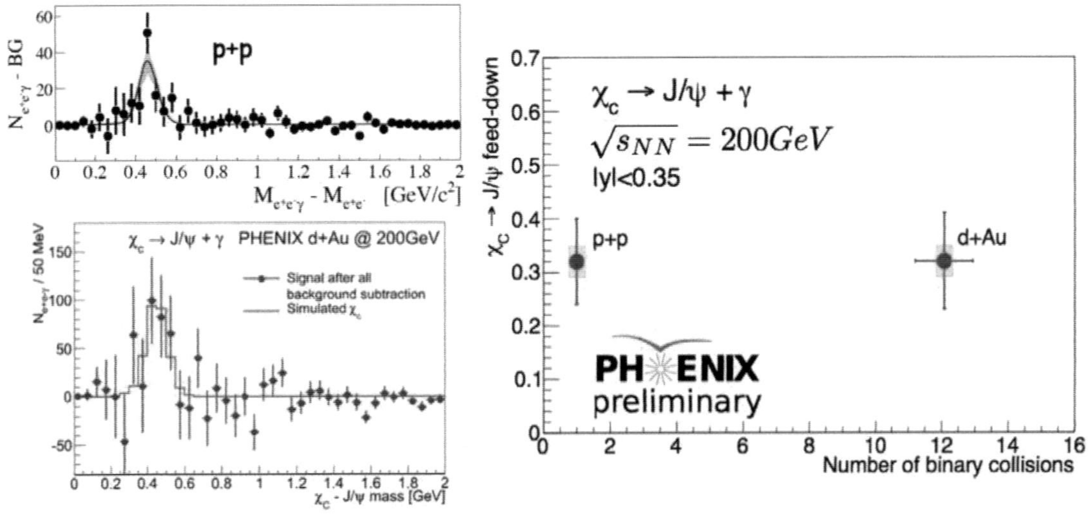

Figure 4. χ_c signal observed in p+p [12] and d+Au collisions after subtraction of the combinatorial and correlated backgrond and corresponding feed-down fractions to J/ψ .

2.3. Outcome of ψ' and χ_c feed-down to J/ψ in Au + Au collisions.

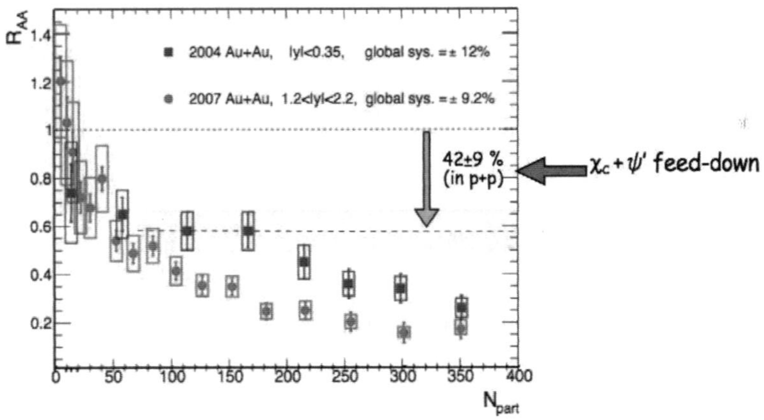

Figure 5. Nuclear modification factor of J/ψ in Au + Au collisions [14] and the modification if ψ' and χ_c states are totally dissociated in QGP.

There is no ψ' and χ_c measurements in Au + Au collisions at RHIC up to date. However, their nuclear modifications should influence 42±9 % of the J/ψ yield, this is the fraction of J/ψ which come from ψ' and χ_c decays. In the view of the estimated temperature ranges where ψ' and χ_c are totally melted in QGP and the medium temperature produced at RHIC as obtained in hydro calculations (Figure 1), we should expect these charmonium states are not present in central Au + Au collisions making R_{AA} of the J/ψ decrease. Even this scenario cannot describe all suppression observed for J/ψ , which indicates the presence of cold nuclear matter effects and perhaps some dissociation of the J/ψ as well.

3. Υ (1S+2S+3S)

PHENIX has observed Υ counts in central and muon arms in $p+p$, d+Au and Au + Au collisions. Given the limited mass resolution obtained in both PHENIX spectrometers and the small distance in mass between the Υ 1S, 2S and 3S, the measurement is made for the sum of these states. According to results from $p+\bar{p}$ collisions in CDF [17] and $p+p$ collisions in LHCb [18], 1S, 2S and 3S states contribute with the total Υ yield in ~73%, 17% and 10% respectively. Decays of the 2S, 3S and χ_b(1P+2P) states corresponds to 11%, 1% and 37% respectively of the 1S yield in $p+p$ and $p+\bar{p}$ collisions.

Di-lepton invariant mass distributions measured in the Υ mass range also contain correlated bottom decays and principally Drell-Yan. These contributions are accounted for in $p+p$ and d+Au collisions by using PYTHIA simulations, NLO calculations for Drell-Yan [19] and also including cold nuclear matter effects in heavy ion collisions [20].

Figure 6 shows the Υ peaks in $p+p$ and Au + Au collisions and the nuclear modification factor R_{AA} at mid-rapidity. PHENIX result is in agreement with STAR result at the same energy and rapidity [21]. For simplicity lets ignore cold nuclear matter effects for now, if Υ (2S+3S) are totally dissociated in Au + Au collisions, R_{AA} of the measured Υ would be around 0.64. If χ_b also disappear in QGP, R_{AA} would be around 0.37. The most central STAR result supports that only the 1S state survive, however, parton distribution modifications and breakup in hadrons can also be important in the strong suppression observed in Au + Au collisions. Upcoming measurements of the rapidity dependence of Υ nuclear modification factors in d+Au collisions can provide a more quantitative measurement of the cold nuclear matter effects.

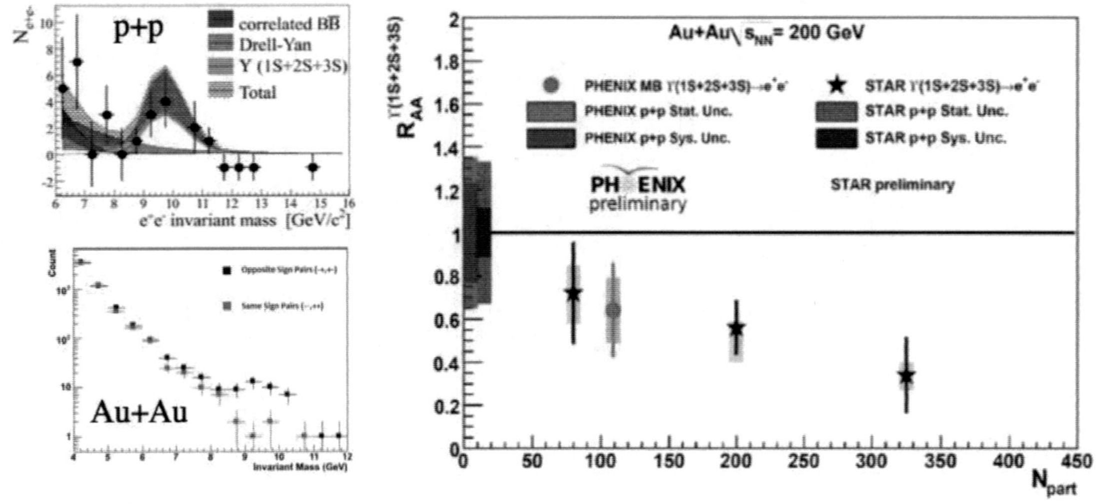

Figure 6. Υ (1S+2S+3S) yields in $p+p$ and Au + Au collisions and nuclear modification factor measured in Au + Au collisions compared to STAR results [21].

4. Future Measurements

PHENIX has now entered in the vertex detectors era opening the possibility to explore many more aspects of quarkonia production at RHIC. The two vertex detectors are already in operation in the central and (VTX) muon arms (FVTX). The VTX can improve significantly the signal/background of the di-electron measurements. A direct measurement of the B-meson contribution to the high p_T J/ψ and ψ' yields can now be done. Finally, VTX can be used as

Journal of Physics: Conference Series **389** (2012) 012026 doi:10.1088/1742-6596/389/1/012026

a photon converter to low-mass di-electrons suitable for the measurement of low energy γ in radiative χ_c decays.

Currently, di-muon invariant mass is measured by using the muon tracker in front of thick hadron absorbers. Multiple scattering in these absorbers limits the measurement of the opening angle of di-muons. The FVTX is installed between the collision point and the absorbers allowing it to improve the measurement of the di-muon opening angle. With this improvement we expect to discriminate ψ' from J/ψ peaks in the di-muon invariant mass spectrum. Rapidity dependent ψ' measurements can help to understand the role of the pre-resonant state in our results.

Future high luminosity runs will provide a better statistical precision in the Υ measurements at RHIC. The current PHENIX apparatus is going to be replaced by a solenoidal detector (sPHENIX) with 2T magnetic field covering $|y| < 1.1$ [22] which can resolve the Υ states. Detector studies to resolve Υ states are also being carried on in the forward rapidity ($1.1 < y < 4$). These upgrades will opem a new field of study in quarkonia at RHIC.

References

[1] Matsui T and Satz H 1986 *Phys. Lett. B* **178** 416
[2] Abreu M *et al.* (NA50) 1997 *Physics Letters B* **410** 337 – 343 ISSN 0370-2693
[3] Adare A *et al.* (PHENIX Collaboration) 2007 *Phys. Rev. Lett.* **98** 232301
[4] Adare A *et al.* (PHENIX Collaboration) 2011 *Phys. Rev. Lett.* **107**(14) 142301
[5] Adare A *et al.* (PHENIX Collaboration) 2012 (*Preprint* 1204.0777)
[6] Thews R L, Schroedter M and Rafelski J 2001 *Phys. Rev. C* **63**(5) 054905
[7] Suzuki K, Gubler P, Morita K and Oka M 2012 (*Preprint* 1204.1173)
[8] Adare A *et al.* (PHENIX Collaboration) 2010 *Phys. Rev. C* **81** 034911
[9] Leitch M J *et al.* ((FNAL E866/NuSea Collaboration)) 2000 *Phys. Rev. Lett.* **84**(15) 3256–3260
[10] Alessandro B *et al.* (NA50) 2006 *Eur. Phys. J. C* **48** 329 (*Preprint* nucl-ex/0612012)
[11] Arleo F, Gossiaux P B, Gousset T and Aichelin J 2000 *Phys. Rev. C* **61**(5) 054906
[12] Adare A *et al.* (PHENIX Collaboration) 2012 *Phys. Rev. D* **85**(9) 092004
[13] Adare A *et al.* (PHENIX Collaboration) 2008 *Phys. Rev. Lett.* **101**(12) 122301
[14] Adare A *et al.* (PHENIX Collaboration) 2011 *Phys. Rev. C* **84**(5) 054912
[15] da Silva C L (PHENIX Collaboration) 2011 *J.Phys.G* **G38** 124031
[16] Sjostrand T, Mrenna S and Skands P 2006 *JHEP* **05** 026
[17] Papadimitriou V (CDF Collaboration) 1994 (*Preprint* hep-ex/9501010)
[18] Aaij R *et al.* (LHCb Collaboration) 2012 *Eur.Phys.J.* **C72** 2025 (*Preprint* 1202.6579)
[19] Kubar J, Le Bellac M, Meunier J and Plaut G 1980 *Nucl.Phys.* **B175** 251
[20] Neufeld R, Vitev I and Zhang B W 2011 *Phys.Lett.* **B704** 590–595 (*Preprint* 1010.3708)
[21] Kesich A (STAR Collaboration) 2012 (*Preprint* 1207.7166)
[22] Aidala C *et al.* 2012 (*Preprint* 1207.6378)

28th Winter Workshop on Nuclear Dynamics 2012 IOP Publishing
Journal of Physics: Conference Series **389** (2012) 012027 doi:10.1088/1742-6596/389/1/012027

Quarkonium production in 2.76 TeV PbPb collisions in CMS

Guillermo Breto Rangel, on the behalf of the CMS collaboration

guillermo.breto.rangel@cern.ch
University of California, Davis
One Shields Ave, Davis CA 95616-8677, USA

E-mail: guillermo.breto.rangel@cern.ch

Abstract. The Compact Muon Solenoid (CMS) is fully equipped to measure hard probes in the di-muon decay channel in the high multiplicity environment of nucleus-nucleus collisions. Such probes are especially relevant for studying the quark-gluon plasma since they are produced at early times and propagate through the medium, mapping its evolution. CMS has measured the nuclear modification factors of non-prompt J/Ψ (from b-hadron decays), prompt J/Ψ and Υ(1S) in PbPb collisions at $\sqrt{s_{NN}} = 2.76$ TeV. Suppression of the excited Υ-states has also been measured with respect to the Υ(1S). Results from the 2010 data taking period are reported and an outlook on the 2011 data analysis will be given.

1. Introduction

Quarkonia have been singled out as a key signature for the study of the quark gluon plasma (QGP) since not only are they produced early and propagate through the medium mapping its evolution but their higher masses allow a non relativistic quantum treatment of their properties. Due to the small relative velocity of the quarks in the quarkonium states a hierarchy of scales is set in the order of m, mv and mv^2 where potential Effective Field Theories can be applied. In other words, the behavior of quarkonia is characterized by different energy and momentum scales related to non-relativistic bound states with heavy quark velocity v and mass m. The scale related to the inverse distance between the Q and $\overline{Q}$ is mv while mv^2 is the scale related to the binding energy of the state. It's worth noting that this scale hierarchy feature is only found in heavy Quarkonia making their study of prime importance. In particular, the J/Ψ in heavy-ion collisions was suggested to be a promising probe of the QGP video as the deconfined medium should screen the two quarks leading to a suppression of their production[1]. The suppression is predicted to occur above a critical temperature of the medium, and sequentially, in the order of the $Q\overline{Q}$ binding energy, making Quarkonia a perfect thermometer due to their different radii. The full spectroscopy of quarkonium states as a possible thermometer for the QGP has been suggested since the early 1990's [7]. For the J/Ψ, studies have been performed at different energies and with different collision systems without yet giving a fully understood complete description [2, 3, 4, 5]. Measuring the charmonium production at the LHC energies in PbPb collisions will help constrain predictions, in particular those with a large recombination probability for prompt J/Ψs. Indeed, the abundance of charm quarks in the medium could lead to a strong production enhancement at LHC energies [6]. In addition to charmonium precision

Journal of Physics: Conference Series **389** (2012) 012027 doi:10.1088/1742-6596/389/1/012027

studies, the LHC center-of-mass energy allows copious Υ production in PbPb collisions. Detailed measurements of bottomonia will help characterize the dense matter produced in heavy-ion collisions complementing the measurements accessible at RHIC energies. Also, since bottom $Q\overline{Q}$ quarks and anti-quarks are relatively rare within the plasma, the probability for regeneration of bottomonium states through recombination is much smaller than for charm quarks. As well as the fact that the effects of initial state nuclear suppression are expected to be reduced.

This paper reviews CMS prompt and non-prompt J/Ψ as well as Υ measurements in PbPb collisions at $\sqrt{s_{NN}} = 2.76$ TeV, as well as in pp collisions at the same energy which serves as a reference for the observation of hot nuclear effects in PbPb. Due to the high granularity of the tracker which translates into excellent momentum resolution and precise extrapolation of charged particle trajectories, CMS is able to differentiate between the primary and secondary (displaced) vertices making it feasible to distinguish non-prompt J/Ψ from prompt J/Ψ in both pp and PbPb collisions. The nuclear modification factor (R_{AA}) of prompt, non-prompt J/Ψ, and Υ(1S) in PbPb is measured as a function of transverse momentum (p_T), rapidity (y) and number of nucleons participating (N_{part}) in the collision [8]. Finally, the relative suppression of the excited states compared to the ground state is quantified in the form of a double ratio [9].

2. CMS Detector, Muon reconstruction, and Datasets

A detailed description of the CMS detector can be found elsewhere [10]. Its central feature is a superconducting solenoid of 6 m internal diameter, providing a magnetic field of 3.8 T. Within the field volume are the silicon pixel and strip tracker, the crystal electromagnetic calorimeter, and the brass/scintillator hadron calorimeter. Muons are measured in gas-ionisation detectors embedded in the steel return yoke. In addition, CMS has extensive forward calorimetry, in particular two steel/quartz-fiber Čerenkov hadron forward (HF) calorimeters, which cover the pseudorapidity range $2.9 < |\eta| < 5.2$.

In this paper, quarkonia are identified through their dimuon decay. The silicon pixel and strip tracker measures charged-particle trajectories for the range $|\eta| < 2.5$. The tracker consists of 66M pixel and 10M strip detector channels, providing a vertex resolution of ~ 15 μm in the transverse plane. Muons are detected for the $|\eta| < 2.4$ range, with detection planes based on three technologies: drift tubes, cathode strip chambers, and resistive plate chambers. Due to the strong magnetic field and the fine granularity of the silicon tracker, the muon transverse momentum measurement based on information from the silicon tracker alone has a resolution between 1 and 2% for a typical muon in this analysis. CMS is therefore very well suited to measure dimuons. In pp, the mass resolution obtained measuring J/Ψ in $|y| < 0.5$ is 20 MeV/c^2, and 67 MeV/c^2 for the Υ(1S) in $|\eta^{\mu}| < 1$.

In March 2011, CMS recorded a little more than $\mathcal{L}_{int} = 220$ nb^{-1} pp events at $\sqrt{s} = 2.76$ TeV. These data is used as a reference for the PbPb measurement. In November 2010, CMS recorded $\mathcal{L}_{int} = 7.28$ μb^{-1} of PbPb events, leading to about the same amount of quarkonia statistics as the reference pp run at 2.76 TeV. Both data sets have been analyzed following similar conditions [8, 9]: (1) events are selected by the CMS two-level trigger keeping any dimuon activity in the muon chambers, (2) offline muon reconstruction is seeded with $\simeq 99\%$ efficient tracks in the muon detectors, which are then matched to tracks reconstructed in the silicon tracker by means of an algorithm optimized for the heavy-ion environment [13, 14], (3) the same analysis procedure is followed for the offline selection with very loose criteria. Signal extraction is based on the procedures in CMS 7 TeV publications for the signal extraction [11, 12].

Using both data sets, the production measured in PbPb collisions is compared to expectations from an independent superposition of nucleon-nucleon collisions typically expressed in terms of the nuclear modification factor:

$$R_{AA} = \frac{\mathcal{L}_{pp}}{N_{MB}} \frac{N_{PbPb}(Q\overline{Q})}{N_{pp}(Q\overline{Q})} \cdot \frac{\varepsilon_{pp}}{\varepsilon_{PbPb}}. \quad (1)$$

Here is the nuclear overlap function[1], $\mathcal{L}_{pp}$ is the pp luminosity, N_{MB} is the measured number of equivalent minimum bias events in PbPb , $\frac{N_{PbPb}(Q\overline{Q})}{N_{pp}(Q\overline{Q})}$ is the raw yield ratio, and $\frac{\varepsilon_{pp}}{\varepsilon_{PbPb}}$ the multiplicity dependent fraction of the efficiency ($\frac{\varepsilon_{pp}}{\varepsilon_{PbPb}} \sim 1.17$ for the most central bin). This paper explores the R_{AA} for three different Quarkonium states.

3. Lower Mass Quarkonia

For the lower mass Quarkonia (J/Ψs) we can further subdivide them into two types depending on whether they come from the primary vertex (prompt J/Ψs) or are produced from decays of B mesons (non-prompt J/Ψs), in other words from the secondary vertex. Prompt J/Ψs contain both direct J/Ψ production and J/Ψs coming from the feed-down of higher excited states such as ψ' and χ_c. Due to the long lifetime of the B meson, non-prompt J/Ψs are produced at a distance L_{xy} from the primary vertex and can therefore be separated from the prompt contribution given the good secondary vertex resolution. This disentanglement is done in CMS by reconstructing the $\mu^+\mu^-$ vertices and performing a 2-dimensional unbinned simultaneous fit of the invariant mass distribution and the pseudo-proper decay length, $l = L_{xy}\frac{m}{p_T}$, leaving the B fraction as a free parameter(see [11] for details). Note that we need to use the pseudo-proper decay length, as we do not exactly know the kinematics of the B meson but only the kinematics of the J/Ψ instead (after we reconstruct the dimuon candidate).

Despite the busier heavy-ion environment the CMS detector performs extremely well, such that the good momentum resolution can be used to separate non-prompt from prompt J/Ψ just as in the pp case, making use of the distance between the non-prompt vertex and the primary vertex. An example of the 2-dimensional fit in PbPb collisions is shown on Fig. 1 for J/Ψ with $p_T > 6.5$ GeV/c. For more details see [8, 20, 21].

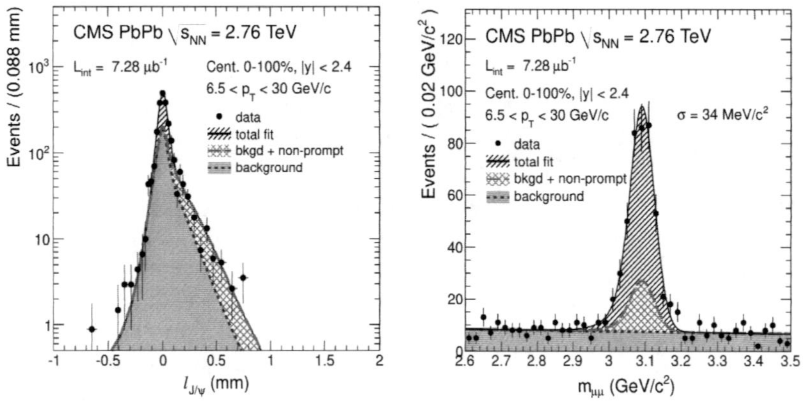

Figure 1. Pseudo-proper decay length (left) and invariant mass distribution (right) for J/Ψ with $p_T > 6.5$ GeV/c in PbPb collisions.The black dots are the data, the dotted line filled in blue the background, the dashed line filled in red the background and non-prompt contribution, and the black straight line the total fit.

For the first time, the secondary J/Ψ R_{AA} is measured in heavy-ion collisions. Fig. 2 (left) illustrates B-meson suppression via their J/Ψ decays through the R_{AA} as a function of N_{part}: $R_{AA} = 0.36 \pm 0.08(stat) \pm 0.03(syst)$ in the 20% most central collisions. This suppression could

[1] Ratio of the number of binary nucleon-nucleon collisions N_{coll} calculated from a Glauber model of the nuclear collision geometry [15, 16] and the inelastic nucleon-nucleon cross section $\sigma_{inel}^{NN} = (64 \pm 5)$ mb at $\sqrt{s} = 2.76$ TeV [17]

be a hint of b-quark energy loss. The level of suppression is of the same order of magnitude as charged hadrons as observed on Fig. 2 (right) where the non-prompt J/Ψ R_{AA} is plotted as a function of p_T for 0–20% while the bosons and charged hadrons are presented as a function of the transverse mass for 0–10% [19].

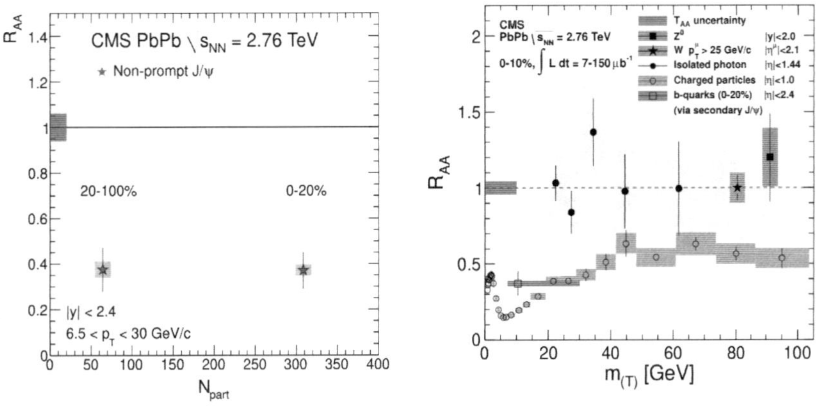

Figure 2. Left: Non-prompt J/Ψ R_{AA} in two centrality bins 0–20% and 20–100% in closed symbols. Right: R_{AA} vs. m_T for Z (squares), isolated photons (closed circles) and charged hadrons (open circles) compared to the secondary J/Ψ measurement as a function of p_T (open squares).

Fig. 3 shows the prompt J/ψ R_{AA} (filled squares) as a function of p_T, y and N_{part}. A factor three suppression is observed for the two p_T bins. CMS points are compared to measurements at $\sqrt{s_{NN}} = 200$ GeV from PHENIX [22] at mid- (open squares) and forward (open circles) rapidity for lower p_Ts, and from STAR [23] up to $p_T = 8$ GeV/c[2]. The tendency of high-p_T J/Ψ's to survive at RHIC is not seen at the LHC. Furthermore, CMS measures less suppression at forward rapidity for high-p_T J/Ψ. One should remember that the x probed with $\langle p_T^{J/\Psi} \rangle = 10$ GeV/c by CMS over $|y| < 2.4$ are $x_1 \sim 0.02$ and $x_2 \sim 5 \cdot 10^{-4}$. Therefore, anti-shadowing could play a role in the suppression observed and could contribute to seeing an opposite trend than PHENIX as a function of y, or an increase of the R_{AA} when going to low p_T and more forward regions as for ALICE measurements [24]. For $p_T > 3$ GeV/c and $1.6 < y < 2.4$, CMS measures $R_{AA} = 0.39 \pm 0.06(stat.) \pm 0.03(syst.)$. Finally, in the 10% most central collisions, CMS observes a factor five suppression, much greater than measured by STAR.

4. ϒ Family

CMS is able to disentangle the ϒ(1S) contribution from the higher states in PbPb as in pp collisions. Fig. 4 compares the ϒ invariant mass distribution at $\sqrt{s} = 2.76$ TeV in pp (left) and PbPb (right) collisions, for $p_T^{\mu} > 4$ GeV/c. The higher state contribution relative to the ground state is strikingly smaller in PbPb collisions. In order to quantify this suppression, an extended unbinned maximum likelihood simultaneous fit to the pp and PbPb mass spectra is performed, following the method described in [9], using the parameters detailed in [25]. The ratio of $\Upsilon(2S+3S)/\Upsilon(1S)$ in PbPb over the same ratio in pp benefits from an almost complete cancellation of possible acceptance and/or efficiency differences among the

[2] PHENIX and STAR measurements are inclusive measurements but the contamination from secondary J/Ψ is expected to be small at RHIC energies.

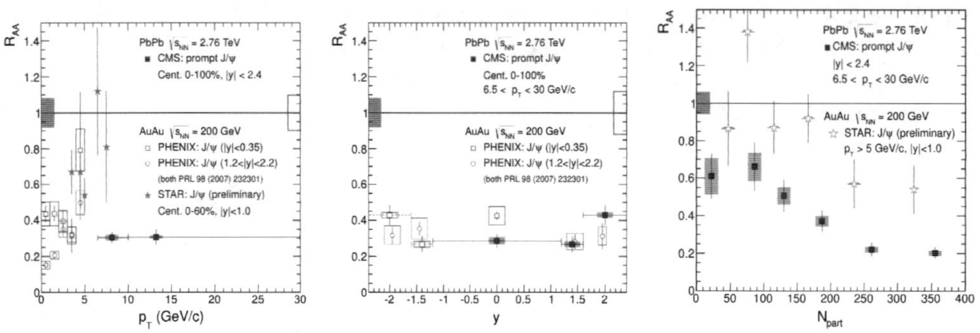

Figure 3. CMS prompt J/Ψ R_{AA} measurement (filled squares) compared to PHENIX mid (open squares) and forward (open circles) rapidity measurement and STAR higher p_T measurement (stars) as a function of p_T (left), y (center), and N_{part} (right).

reconstructed resonances. The double ratio obtained is

$$\frac{\Upsilon(2S+3S)/\Upsilon(1S)|_{\text{PbPb}}}{\Upsilon(2S+3S)/\Upsilon(1S)|_{\text{pp}}} = 0.31^{+0.19}_{-0.15}\,(\text{stat.}) \pm 0.03\,(\text{syst.}), \tag{2}$$

where the systematic uncertainty (10%) arises from varying the lineshape in the simultaneous fit, thus taking into account partial cancellations of systematic effects. Finally, using an ensemble of one million pseudo-experiments generated with the signal lineshape obtained from the pp data, Fig. 4 (left), the background lineshapes from both data sets, and a double ratio (Eq. 2) equal to unity within statistical and systematic uncertainties (absence of a suppression), the probability of finding the measured value of 0.31 or a downward fluctuation is estimated to be 0.9%., corresponding to 2.4 sigma in a one-tailed integral of a Gaussian distribution.

The Υ(1S) suppression has a been studied as a function of p_T, y and centrality as shown on Fig. 5. A suppression by a factor ~ 2.3 is observed for low p_T. This seems to disappear for $p_T > 6.5$. The rapidity dependence indicates a slightly smaller suppression at forward rapidity. In both cases however, the statistical uncertainties are too large for any strong conclusions. In addition, Υ(1S) are suppressed by a factor two in 0–10% central collisions. The CMS measurement is compared to STAR Υ(1S+2S+3S) preliminary result in AuAu collisions at $\sqrt{s_{NN}} = 200$ GeV [26] showing a suppression of the same order of magnitude but with large uncertainty.

5. Conclusions

In summary, this paper presented highlights of CMS quarkonia measurements in pp and PbPb collisions at $\sqrt{s} = 2.76$ GeV recorded in 2010 and 2011 (for the pp case). The high statistics accumulated at the LHC allows the performance of detailed studies that further constrain production mechanisms. CMS has performed the measurements of the prompt and non-prompt , as well as of the $\Upsilon(1S)$ and $\Upsilon(2S+3S)$ mesons via their decay into pairs in PbPb and pp collisions at 2.76 TeV. Prompt has been separated from non-prompt for the first time in heavy-ion collisions. A strong suppression of prompt J/Ψ with $p_T > 6.5$ GeV/c is measured in central collisions, and already in peripheral collisions, showing a clear dependence with centrality. Non-prompt , though strongly suppressed, show no strong centrality dependence within uncertainties. This is the first hint of b-quark energy loss in the hot medium. Furthermore, Υ(1S) are suppressed by 40% in the 20% most central collisions. The comparison of the ratios of $\Upsilon(nS)$-states in pp and PbPb collisions, taken at the same center-of-mass energy, is consistent with the

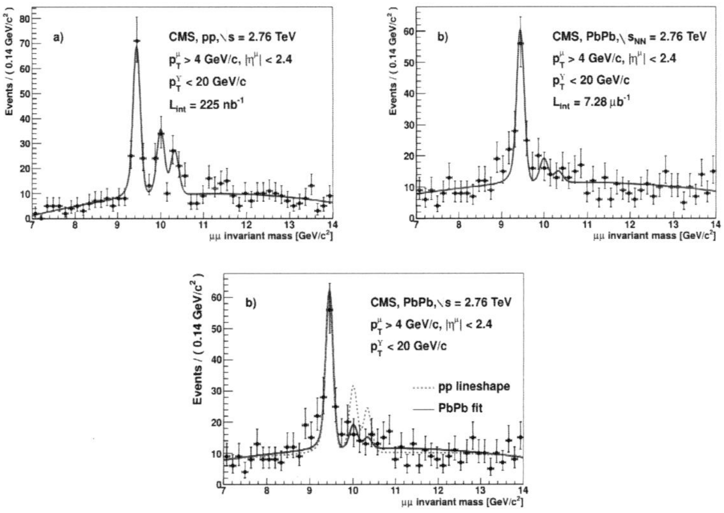

Figure 4. Invariant mass distribution measured in pp (top left) and PbPb (top right) collisions at 2.76 TeV for $p_T^\mu > 4$ GeV/c. Below overlaying pp and PbPb fit results: the dashed-red line shows the fit results obtained by imposing the Y(1S) as obtained in the nominal fit (continuous-blue line; same as plot (b) above), and the 2S/1S and 3S/1S ratios to the values obtained from the fit to the pp data (plot (a) above).

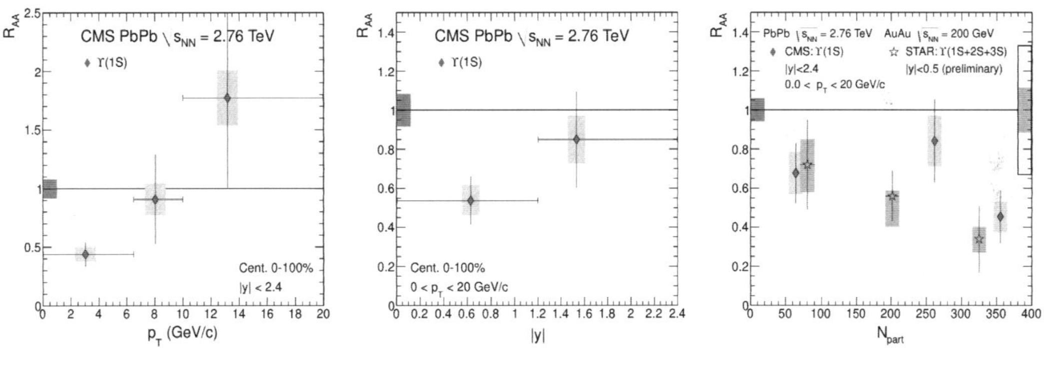

Figure 5. Υ(1S) R_{AA} as a function of p_T (left), y (middle), and N_{part} (right), compared to STAR inclusive preliminary measurement (stars) for the latter.

partial disappearance of the higher states with respect to the ground state in the PbPb collisions. Those two observations could indicate that the Υ(1S) suppression is due to the melting of the excited states only in PbPb collisions. Measuring the amount of suppression caused by shadowing through pA collisions together with more precise measurements will be crucial for interpreting what the melting is due to. As an outlook, in 2011 CMS recorded around 15 times more statistics, based on a PbPb dataset of 150 μb^{-1}, we measured the double ratio Y(2S)/Y(1S) in PbPb relative to pp collisions to be $0.21 \pm 0.07\,(stat.) \pm 0.02\,(syst.)$. No noticeable dependencies are observed on the dimuon rapidity or transverse momentum, within the current statistical precision. No clear centrality dependence of the double ratios can be inferred from the data, while such a dependence is observed when inspecting the individual $\Upsilon(1S)$ and $\Upsilon(2S)$ nuclear modification factors, R_{AA}. The centrality-integrated R_{AA} values for the $\Upsilon(1S)$ and $\Upsilon(2S)$ states are $0.56 \pm 0.08\,(stat.) \pm 0.07\,(syst.)$ and $0.12 \pm 0.04\,(stat.) \pm 0.02\,(syst.)$, respectively. The 3S state is not observed prominently in the data, and upper limits on its suppression ratios are set.

References

[1] T. Matsui and H. Satz, *Phys. Lett.* **B178** (1986) 416, http://dx.doi.org/10.1016/0370-2693(86)91404-8doi:10.1016/0370-2693(86)91404-8.
[2] NA38 Collaboration, *Phys. Lett.* **B345** (1995) 617, http://dx.doi.org/10.1016/0370-2693(94)01614-Idoi:10.1016/0370-2693(94)01614-I.
[3] NA50 Collaboration, *Eur. Phys. J.* **C49** (2007) 559, http://www.arXiv.org/abs/nucl-ex/0612013arXiv:nucl-ex/0612013.
[4] NA60 Collaboration, *Phys. Rev. Lett.* **99** (2007) 132302, http://dx.doi.org/10.1103/PhysRevLett.99.132302.
[5] PHENIX Collaboration, http://arxiv.org/abs/1103.6269arXiv:nucl-ex/1103.6269.
[6] X. Zhao, R. Rapp, http://arxiv.org/abs/1102.2194arXiv:1102.2194.
[7] C. Miao *et. al.*, *Nucl. Phys.* **A855** (2011) 125, http://dx.doi.org/10.1016/j.nuclphysa.2011.02.028.
[8] CMS Collaboration, *"Suppression of non-prompt J/Ψ, prompt J/Ψ, and $\Upsilon(1S)$ in PbPb collisions at $\sqrt{s_{NN}}$ = 2.76 TeV, JHEP 1205 (2012) 063* http://www.springerlink.com/content/131423084582p16g/ **arXiv:1201.5069.** (2012).
[9] CMS Collaboration, *"Indications of suppression of excited Υ states in PbPb collisions at $\sqrt{s_{NN}} = 2.76$ TeV", Phys. Rev. Lett. 107 (2011) 052302,*http://doi:10.1103/PhysRevLett.107.052302arXiv:1105.4894.
[10] CMS Collaboration, *JINST* **0803** (2008) S08004, http://dx.doi.org/10.1088/1748-0221/3/08/S08004 doi:10.1088/1748-0221/3/08/S08004.
[11] CMS Collaboration, *Eur. Phys. J.* **C71** (2011) 1575, http://dx.doi.org/10.1140/epjc/s10052-011-1575-8doi:10.1140/epjc/s10052-011-1575-8
[12] CMS Collaboration, *"Upsilon production cross section in pp collisions at $\sqrt{s_{NN}} = 7$ TeV", Phys.Rev. D83 (2011) 112004,* ,http://doi:10.1103/PhysRevD.83.112004arXiv:1012.5545
[13] C. Roland, for the CMS Collaboration, *NIM* **A566** (2006) 123, http://dx.doi.org/10.1016/j.nima.2006.05.023 doi:j.nima.2006.05.023.
[14] CMS Collaboration, *J. Phys.* **G34** (2007) 2307, http://dx.doi.org/10.1088/0954-3899/34/11/008 doi:10.1088/0954-3899/34/11/008.
[15] CMS Collaboration, *"Observation and studies of jet quenching using dijet momentum imbalance in PbPb collisions at $\sqrt{s_{NN}}$ = 2.76 TeV with the CMS detector", Phys.Rev.C84:024906,(2011).*,http://arxiv.org/abs/1102.1957arXiv:1102.1957
[16] B. Alver *et. al.*, *Phys. Rev.* **C77** (2008) 014906, http://dx.doi.org/10.1103/PhysRevC.77.014906.
[17] PDG Collaboration, *J. Phys.* **G37** (2010) 07502, ttp://dx.doi.org/10.1088/0954-3899/37/7A/07502110.1088/0954-3899/37/7A/075021.
[18] I. P. Lokhtin *et. al.*, *Eur. Phys. J.* **C45** (2006) 211–217, http://dx.doi.org/10.1140/epjc/s2005-02426-3 doi:10.1140/epjc/s2005-02426-3.
[19] Y-J. Lee, for the CMS Collaboration, *These proceedings.*
[20] T. Dahms, for the CMS Collaboration, *These proceedings.*
[21] M. Jo, for the CMS Collaboration, *These proceedings.*
[22] PHENIX Collaboration, *Phys. Rev. Lett.* **98** (2007) 232201, http://dx.doi.org/10.1103/PhysRevLett.98.232301.
[23] Z. Tang, for the STAR Collaboration, *QM2011.*
[24] P. Pillot , for the ALICE Collaboration, *These proceedings.*
[25] Z. Hu, for the CMS Collaboration, *These proceedings.*
[26] R. Reed, for the STAR Collaboration, *These proceedings.*

28th Winter Workshop on Nuclear Dynamics 2012 IOP Publishing
Journal of Physics: Conference Series **389** (2012) 012028 doi:10.1088/1742-6596/389/1/012028

J/ψ Measurements with the ALICE Experiment at the LHC

Julian Book (for the ALICE Collaboration)

Institut für Kernphysik, Goethe Universität, Frankfurt am Main, Germany

E-mail: Julian.Book@cern.ch

Abstract. The ALICE detector is the dedicated heavy-ion experiment at the Large Hadron Collider (LHC) at CERN. It provides excellent capabilities to study quarkonium production in proton-proton (pp) and heavy-ion (A-A) collisions in the kinematic range of $|y| < 0.9$ and $2.5 < y < 4.0$ down to a transverse momentum of zero. Quarkonia, bound states of heavy (charm or bottom) quark anti-quark pairs such as the J/ψ, are expected to be produced by hard processes. Thus they will provide insights into the earliest and hottest stages of A-A collisions where the formation of a Quark-Gluon Plasma is expected. Results of the inclusive and prompt J/psi production cross section in pp collisions at $\sqrt{s}$=2.76 and 7 TeV will be presented. A polarization measurement in the forward region and the dependence of the J/ψ yield on the charged particle multiplicity in pp collisions will be discussed. The nuclear modification factor (R_{AA}) at forward rapidity in Pb-Pb collisions at $\sqrt{s_{\mathrm{NN}}}$=2.76 TeV will be shown and its implications discussed.

1. Introduction

J/ψ suppression was proposed as a signature for a deconfined state of strongly interacting matter, called the Quark-Gluon-Plasma (QGP), that is expected to be created in realtivistic heavy-ion collisions [1]. There are several medium induced effects that can play a role on the production yield of J/ψ under the exterme conditions of such a hot and dense phase. The melting of the $c\bar{c}$ due to Debye-screening, known from QED, can lead to a reduced production rate, while there might be contributions enhancing the yields such as random combinations of uncorrelated charm and anti-charm quark at high charm quark densities or even a thermal production of $c\bar{c}$ at LHC energies.

In the past, many experiments [2, 3, 4] have measured the J/ψ production in heavy-ion, proton-proton and proton-ion collisions to understand the production mechanisms and the behaviour of quarkonia under different conditions. The NA50 experiment at the CERN SPS observed a suppression of J/ψ and ψ' in central Pb-Pb and in p-A collisions, while the NA60 collaboration measured p-A and In-In collisions with the outcome that a large fraction (70-80%) of the suppression in most central Pb-Pb collisions, seen by NA50, is due to cold nuclear matter effects (CNM). Some years later at RHIC the PHENIX experiment observed in Au-Au and d-Au collsions at ten times higher centre-of-mass energies than at SPS a similar suppression at midrapidty but larger suppression at forward rapidities, see Ref. [5]. Finally, the STAR collaboration has reported a smaller suppression at high transverse momentum ($p_{\mathrm{T}} \geq 5$ GeV/c) at midrapity.

2. Experimental setup

ALICE [6] is the dedicated heavy-ion experiment at the LHC. Quarkonia can be measured in two different parts, the central barrel with a pseudo rapidity coverage of $|\eta| < 0.9$ and the muon spectrometer located at $-4.0 < \eta < -2.5$. J/ψ can be reconstructed down to pt=0 in both decay channels, J/$\psi \rightarrow e^+e^-$ at midrapidity and J/$\psi \rightarrow \mu^+\mu^-$ at forward rapidity.
The main detectors used for the J/ψ reconstruction are the Inner Tracking System (ITS) and the Time Projection Chamber (TPC) in the dielectron analyses and will be complemented by the ElectroMagnetic CALorimeter (EMCAL) and the Transition Radiation Detector (TRD) in future analyses. The large-volume TPC serves as the main tracking device and provides Particle IDentification (PID) via the specific energy-loss of the particles in the detector gas. The ITS, consisting of six layers of silicon (two layers of pixel, strip and drift), provides excellent vertex reconstruction and improves the momentum resolution by its tracking capabilities. The muon spectrometer, used in the studies of J/ψ decaying into two muons, consists of a tracking system behind a ten interaction length thick absorber and is completed by a triggering system behind a 1.2 m thick iron wall.

3. Inclusive J/ψ cross sections in proton-proton collisions at 2.76 and 7 TeV

A detailed measurement of J/ψ production in pp collisions is performed as a function of the centre-of-mass energy, transverse momentum, rapidity and the charged particle density. In addition, preliminary results on the fraction of J/ψ coming from B-hadron decays are obtained. Furthermore, a polarization measurement of J/$\psi \rightarrow \mu^+\mu^-$ has been performed in two different reference frames. Data sets with an integrated luminosity of 15.6 (19.9) nb^{-1} at $\sqrt{s}$=7 (2.76) TeV in the dimuon decay channel and 5.6 (1.1) nb^{-1} in the dielectron analysis have been collected in 2010(11). The J/ψ signal of the dimuon and dielectron analysis have been divided into several rapidity (y) and transverse momentum (p_T) bins at both energies. Good agreement is found with the results by the LHCb collaboration at forward rapidity and with the other LHC experiments (CMS and ATLAS) at midrapidity, see Fig. 1. Furthermore ALICE has the unique capability to measure the inclusive J/ψ production in a large rapidity range down to p_T=0 (see Fig. 2).

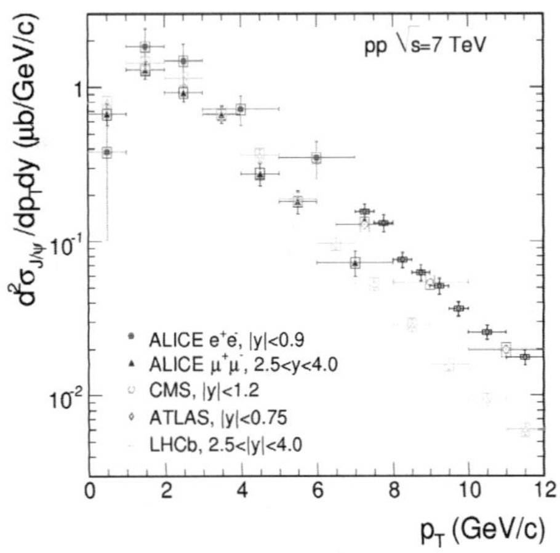

Figure 1. The differential inclusive J/ψ cross section $d^2\sigma_{J/\psi}/dp_Tdy$ at $\sqrt{s}$=7 TeV compared to results from other LHC experiments with similar rapidity ranges [7, 8, 9, 10].

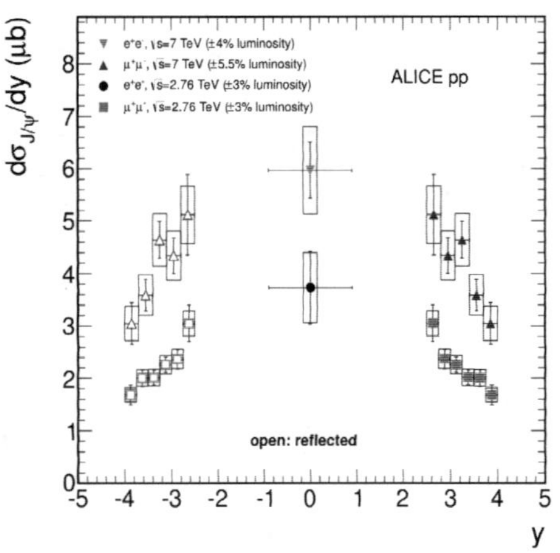

Figure 2. $d\sigma_{J/\psi}/dy$ for centre-of-mass energies of 2.76 and 7 TeV with a transverse momentum reach down to 0 [11].

A completely new study of quarkonia was performed, where the inclusive J/ψ yield was measured in dependence of the charged particle density ($\mathrm{d}N_{\mathrm{ch}}/\mathrm{d}\eta$) in proton-proton collisions at $\sqrt{s}$=7 TeV. The charged particle densities were obtained by the ITS silicon pixel detector in the pseudo rapidity range $|\eta|< 1.0$. An approximately linear increase with the midrapidity $\mathrm{d}N_{\mathrm{ch}}/\mathrm{d}\eta$ was seen in both rapidity regions (see Fig. 3). Even with a rapidity gap of 3.25 the correlation stays, which is a very intriguing observation. An enhancement relative to the minimum bias yield up to a factor 5 (8) at forward (midrapidity) was found. Details of this analysis can be found in [12].

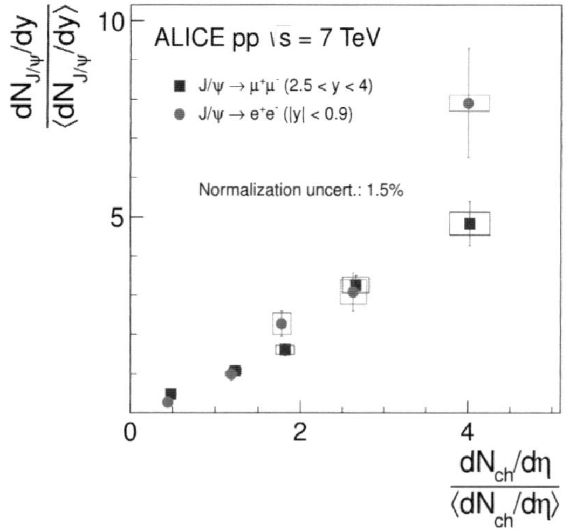

Figure 3. The midrapidity charged particle density dependence of the inclusive J/ψ yield in $|y| < 0.9$ and $2.5 < y < 4.0$ in pp collisions at $\sqrt{s}$= 7 TeV. The vertical (horizontal) axis shows values relative to the minimum bias yield (charged particle density).

ALICE is the first experiment at LHC, that has measured the polarisation of J/ψ decaying into two muons. This analysis was performed in pp collisions at $\sqrt{s}$=7 TeV[13]. The distribution of the J/ψ decay products can be expressed as,

$$W(\Theta,\phi) = \frac{1}{3+\lambda_\Theta}(1+\lambda_\Theta\cos^2\Theta + \lambda_\phi\sin^2\Theta\cos 2\phi + \lambda_{\Theta\phi}\sin 2\Theta\cos\phi) \qquad (1)$$

where $\Theta(\phi)$ are the polar (azimuthal) angles in a given reference frame (Collins-Soper (CS) or helicity (HE)) and $\lambda_\Theta, \lambda_\phi, \lambda_{\Theta\phi}$ are the polarization parameters. It was found that the polarization of the J/ψ in the forward region is consistent with very little or no polarisation and shows a weak p_{T} dependence in the HE frame, see Fig. 4.
With the excellent resolution of the ITS it was possible to study the fraction of J/ψ coming from B-hadrons in the central part of the experiment. The fraction was obtained to be 13.7%$\pm$5.4(stat.)$^{+2.5}_{-1.8}$(syst.)$^{+2.2}_{-2.7}$(syst.polar.) and agrees with other LHC results [8, 9, 10].

4. Nuclear modification factor (R_{AA}) in Pb-Pb collisions at $\sqrt{s_{\mathrm{NN}}}$=2.76 TeV

The inclusive J/ψ production was studied in Pb-Pb collisions at $\sqrt{s_{\mathrm{NN}}}$=2.76 TeV with the muon spectrometer [14]. The data sample collected end of 2010 belongs to an integrated luminosity of 2.88 μb^{-1}. 2350$\pm$139(stat.)$\pm$189(syst.) J/ψ have been reconstructed in the region $2.5 < y < 4.0$

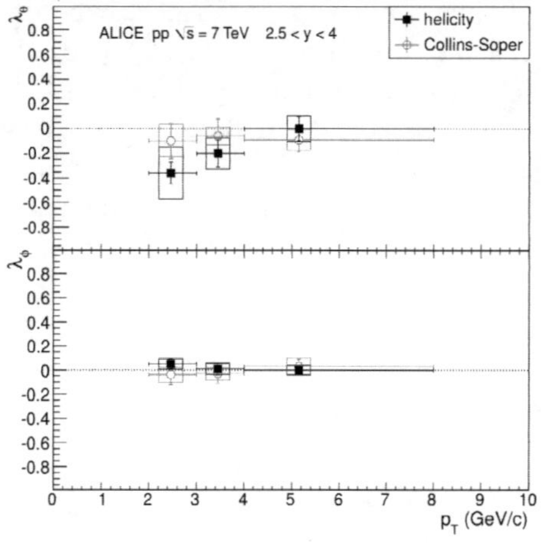

Figure 4. The polarization parameter λ_Θ and λ_ϕ as a function of p_T for inclusive J/ψ in $2.5 < y < 4.0$ [13].

with a invariant mass resolution of 75 MeV/c^2. The pp results shown in Fig. 2 at the same energy ($\sqrt{s}$=2.76 TeV) serve as a reference for the determination of the inclusive J/ψ nuclear modification factor $R_{AA} = Y_{Pb-Pb}/\langle N_{coll}\rangle Y_{pp}$. $\langle N_{coll}\rangle$ corresponds to the average number of binary nucleon-nucleon collisions in a given centrality class. Y_{Pb-Pb} and Y_{pp} are the corrected yields with equal kinematics at the given system. Figure 5 compares our results to several theoretical predictions [15, 16, 17] that include a J/ψ regeneration component from deconfined charm quarks in the medium or generation at the QCD phase boundary (hadronization). The data compared to PHENIX results at central and forward rapidity, see Fig. 6, show a smaller suppression than at RHIC. Of course, cold nuclear matter effects are expected to be different at the two energies and have to be taken into account.

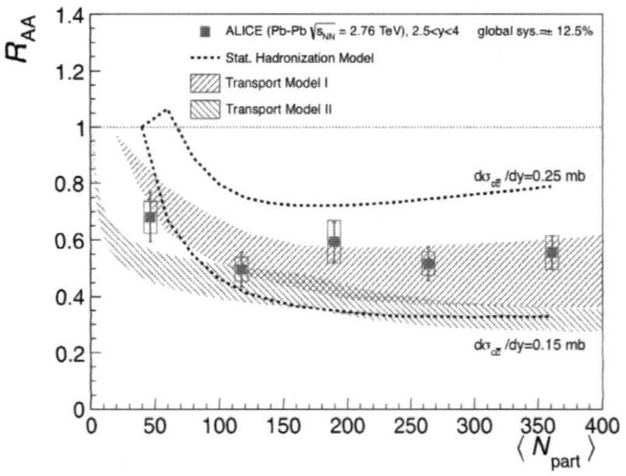

Figure 5. The inclusive J/ψ nuclear modification factor as a function of the average number of participating nucleons measured in $\sqrt{s_{NN}}$=2.76 TeV Pb-Pb collisions and compared to theoretical predictions.

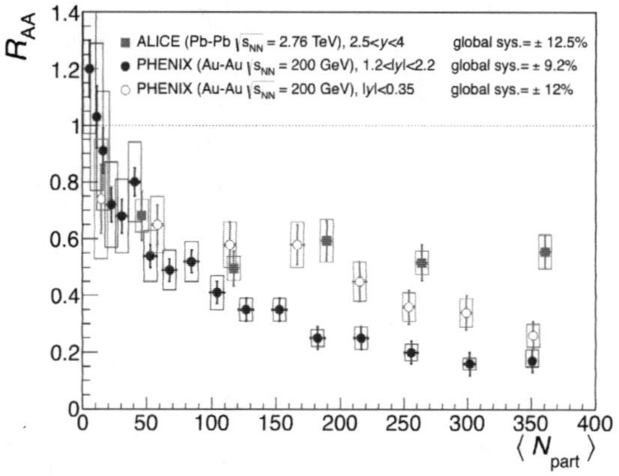

Figure 6. Comparision of the inclusive J/ψ R_{AA} to results at central and forward rapidity obtained by the PHENIX collaboration for Au-Au collisions at $\sqrt{s_{NN}}$=200 GeV [5, 18].

5. Conclusions

The ALICE J/ψ measurements in pp at $\sqrt{s}$=2.76 and 7 TeV and in Pb-Pb at $\sqrt{s_{\rm NN}}$=2.76 TeV have been presented. The results cover a large rapidity and transverse momentum range down to $p_{\rm T}$=0. In pp collisions the J/ψ yield in $|y| < 0.9$ and $2.5 < y < 4.0$ was found to increase approximately linear with the charged particle multiplicity in the range of $|\eta| < 1.0$. For the first time at the LHC the J/ψ polarization was measured and is compatible with 0 in the measured $p_{\rm T}$ range ($2 < p_{\rm T} < 8$ GeV/c). The inclusive J/ψ nuclear modification factor in Pb-Pb collisions is larger than that observed at RHIC energies in central Au-Au collisions. Theoretical (re)generation models are able the describe the observed shape and amount of suppression within their uncertanties. This might be a hint for a J/ψ regeneration in hot matter or at the chemical freeze-out at LHC energies. However, effects related to cold nuclear matter remain unkown at LHC energies and need to be studied with the data of the upcoming p-Pb run.

Acknowledgments

The author thanks C. Blume for his support and the possibilty to present these results. This work was supported by the Helmholtz Research School for Quark Matter Studies.

References

[1] Matsui T and Satz H 1986 *Phys.Lett.* **B178** 416
[2] Alessandro B *et al.* (NA50 Collaboration) 2006 *Eur.Phys.J.* **C48** 329 (*Preprint* nucl-ex/0612012)
[3] Alessandro B *et al.* (NA50 Collaboration) 2007 *Eur.Phys.J.* **C49** 559–567 (*Preprint* nucl-ex/0612013)
[4] Arnaldi R (NA60 Collaboration) 2009 *Nucl.Phys.* **A830** 345C–352C (*Preprint* 0907.5004)
[5] Adare A *et al.* (PHENIX Collaboration) 2011 *Phys.Rev.* **C84** 054912 (*Preprint* 1103.6269)
[6] Aamodt K *et al.* (ALICE Collaboration) 2008 *JINST* **3** S08002
[7] Aamodt K *et al.* (ALICE Collaboration) 2011 *Phys.Lett.* **B704** 442–455 (*Preprint* 1105.0380)
[8] Aad G *et al.* (ATLAS Collaboration) 2011 *Nucl.Phys.* **B850** 387–444 (*Preprint* 1104.3038)
[9] Khachatryan V *et al.* (CMS Collaboration) 2011 *Eur.Phys.J.* **C71** 1575 (*Preprint* 1011.4193)
[10] Aaij R *et al.* (LHCb Collaboration) 2011 *Eur.Phys.J.* **C71** 1645 (*Preprint* 1103.0423)
[11] Abelev B *et al.* (ALICE Collaboration) 2012 (*Preprint* 1203.3641)
[12] Abelev B *et al.* (ALICE Collaboration) 2012 *Phys.Lett.* **B712** 165–175 (*Preprint* 1202.2816)
[13] Abelev B *et al.* (ALICE Collaboration) 2012 *Phys. Rev. Lett.* **108**(8) 082001
[14] Abelev B *et al.* (ALICE Collaboration) 2012 (*Preprint* 1202.1383)
[15] Liu Y p, Qu Z, Xu N and Zhuang P f 2009 *Phys.Lett.* **B678** 72–76 (*Preprint* 0901.2757)
[16] Zhao X and Rapp R 2011 *Nucl.Phys.* **A859** 114–125 (*Preprint* 1102.2194)
[17] Andronic A, Braun-Munzinger P, Redlich K and Stachel J 2011 *J.Phys.G* **G38** 124081 (*Preprint* 1106.6321)
[18] Adare A *et al.* (PHENIX Collaboration) 2007 *Phys.Rev.Lett.* **98** 232301 (*Preprint* nucl-ex/0611020)

Journal of Physics: Conference Series **389** (2012) 012029 doi:10.1088/1742-6596/389/1/012029

Results on high transverse momentum quarkonium production and dissociation in heavy ion collisions

Ivan Vitev

Los Alamos National Laboratory, Theoretical Division, Los Alamos, NM 87545, USA

E-mail: ivitev@lanl.gov

Abstract. We calculate the yields of quarkonia in heavy ion collisions at RHIC and the LHC as a function of the transverse momentum. Based upon NRQCD, these results include both color-singlet and color-octet contributions and feed-down effects from excited states. In reactions with ultra-relativistic nuclei, we focus on the consistent implementation of dynamically calculated nuclear matter effects, such as coherent power corrections, cold nuclear matter energy loss and the Cronin effect, in the initial state and collisional dissociation of quarkonia in the final state, as they traverse through the QGP. Theoretical results are presented for J/ψ and Υ and compared to experimental data where applicable. At RHIC, a good description of the high-p_T J/ψ modification observed in central Cu+Cu and Au+Au collisions can be achieved within the model uncertainties. We find that J/ψ measurements in proton-nucleus reactions are needed to constrain the magnitude of cold nuclear matter effects. At the LHC, a good description of the experimental data can be achieved only in mid-central and peripheral Pb+Pb collisions.

1. Introduction

Melting of heavy quarkonium states, like the J/ψ and the Υ, due to color screening in a deconfined quark-gluon plasma (QGP) [1] has been proposed as one of the principal signatures for its formation. An expected experimental consequence of this melting in the thermal medium created in heavy ion collisions (HIC) is a suppression of the yields of heavy mesons, when compared to their yields in nucleon-nucleon (NN) collisions scaled with the number of binary interactions.

Description of quarkonia is provided by non-relativistic quantum chromodynamics (NRQCD) [2]. In this picture, the $Q\bar{Q}$ state in the color-singlet combination is the lowest order Fock space component of the full quarkonium wavefunction. Higher Fock components in the wavefunction have additional partons and are suppressed by positive powers of the small parameter $v = p/M$ (velocity). In our calculation we will take into account the Fock components that allow for both color-singlet and color-octet $Q\bar{Q}$ contributions. The NRQCD formalism has been used to calculate differential yields of heavy mesons as a function of the transverse momentum p_T in p+p collisions. Our goal is to combine the NRQCD formalism with cold nuclear matter (CNM) effects and the effects of propagation of quarkonia through a hot QGP to calculate the final yields in HIC [3]. We place special emphasis on the application of the collisional meson dissociation mechanism [4, 5], which has provided good description of D and

B meson suppression, to quarkonia. We present results for the nuclear modification ratio,

$$R_{AB}(p_T) = \frac{d\sigma_{AB}^{Q\bar{Q}}/dydp_T}{N_{AB}^{\rm col.} d\sigma_{pp}^{Q\bar{Q}}/dydp_T}, \tag{1}$$

in reactions with heavy nuclei. We discuss Au+Au and Cu+Cu collisions at RHIC at $\sqrt{S} = 0.2$ TeV per nucleon pair and Pb+Pb collisions at LHC at $\sqrt{S} = 2.76$ TeV per nucleon pair.

2. Quarkonium production in p+p collisions

In this section we describe the production of quarkonia at high transverse momentum in p+p collisions. It provides a baseline for the calculation of the nuclear modification factor $R_{AB}(p_T)$ defined above in Eq. 1. It also gives the initial unquenched spectrum of "proto-quarkonia" in the heavy ion collision, excluding CNM effects and effects of the propagation of the quarkonium states through the QGP medium.

The dominant processes in evaluating the differential yields of heavy mesons as a function of p_T are the $2 \to 2$ processes of the kind $g+q \to H+q$, $q+q \to H+g$ and $g+g \to H+g$, where H refers to the heavy meson. We label the process generically as $a+b \to c+d$, where a and b are light incident partons, c refers to H and d is a light final-state parton. Given the matrix element for the process, $\mathcal{M}_{ab\to cd}$, the cross section has the form

$$\frac{d\sigma}{dp_T dy} = \int dx_a \phi_a(x_a,\mu_f)\phi_b(x_b,\mu_f)\frac{2p_T}{x_a - \frac{m_T}{\sqrt{S}}e^y} x_a x_b \frac{d\sigma}{d\hat{t}}(ab\to cd) \;, \tag{2}$$

where ϕ_a (ϕ_b) is the distribution function of parton a (b) in the incident hadron traveling in the $+z$ ($-z$) direction. We denote by x_a (x_b) the large lightcone momentum fraction carried by the parton. In Eq. (2) $\sqrt{S}$ is the center-of-mass energy of the incident hadrons and $\hat{s}$, $\hat{t}$ and $\hat{u}$ are the parton level Mandelstam variables. Momentum-energy conservation fix

$$x_b = \frac{1}{\sqrt{S}}\frac{x_a\sqrt{S}m_T e^{-y} - m_H^2}{x_a\sqrt{S} - m_T e^y}. \tag{3}$$

We take the factorization and renormalization scales μ_f, μ_R to be $m_T = \sqrt{p_T^2 + m_H^2}$. The invariant cross section is given by

$$\frac{d\sigma}{d\hat{t}} = \frac{|\mathcal{M}|^2}{16\pi\hat{s}^2} \;. \tag{4}$$

We use NRQCD [2] results to calculate the production of quarkonia in p+p collisions. NRQCD provides a systematic procedure to compute any quantity as an expansion in the relative velocity v of the heavy quarks in the meson. For example, the wavefunction of the J/ψ meson (analogous expressions hold for the $\psi(2S)$, Υ, $\Upsilon(2S)$ and $\Upsilon(3S)$) is written as

$$\begin{aligned} |J/\psi\rangle \;=\; & |Q\bar{Q}([^3S_1]_1)\rangle + \mathcal{O}(v)|Q\bar{Q}([^1S_0]_8 g)\rangle + \mathcal{O}(v^2)|Q\bar{Q}([^3S_1]_8 gg)\rangle \\ & + \mathcal{O}(v^1)|Q\bar{Q}([^3P_0]_8 g)\rangle + \mathcal{O}(v^1)|Q\bar{Q}([^3P_1]_8 g)\rangle + \mathcal{O}(v^1)|Q\bar{Q}([^3P_2]_8 g)\rangle + \cdots \end{aligned} \tag{5}$$

The differential cross section for the prompt (as opposed to inclusive which includes contributions from $B-$hadron decay) and direct (as opposed to the indirect from the decay of heavier charmed mesons) production of J/ψ can also be calculated in NRQCD. It can be written as the sum of the contributions,

$$\begin{aligned} d\sigma(J/\psi) = {} & d\sigma(Q\bar{Q}([^3S_1]_1))\langle\mathcal{O}(Q\bar{Q}([^3S_1]_1) \to J/\psi)\rangle + d\sigma(Q\bar{Q}([^1S_0]_8))\langle\mathcal{O}(Q\bar{Q}([^1S_0]_8) \to J/\psi)\rangle \\ & + d\sigma(Q\bar{Q}([^3S_1]_8))\langle\mathcal{O}(Q\bar{Q}([^3S_1]_8) \to J/\psi)\rangle + d\sigma(Q\bar{Q}([^3P_0]_8))\langle\mathcal{O}(Q\bar{Q}([^3P_0]_8) \to J/\psi)\rangle \\ & + d\sigma(Q\bar{Q}([^3P_1]_8))\langle\mathcal{O}(Q\bar{Q}([^3P_1]_8) \to J/\psi)\rangle + d\sigma(Q\bar{Q}([^3P_2]_8))\langle\mathcal{O}(Q\bar{Q}([^3P_2]_8) \to J/\psi)\rangle + \cdots , \end{aligned} \tag{6}$$

Journal of Physics: Conference Series **389** (2012) 012029 doi:10.1088/1742-6596/389/1/012029

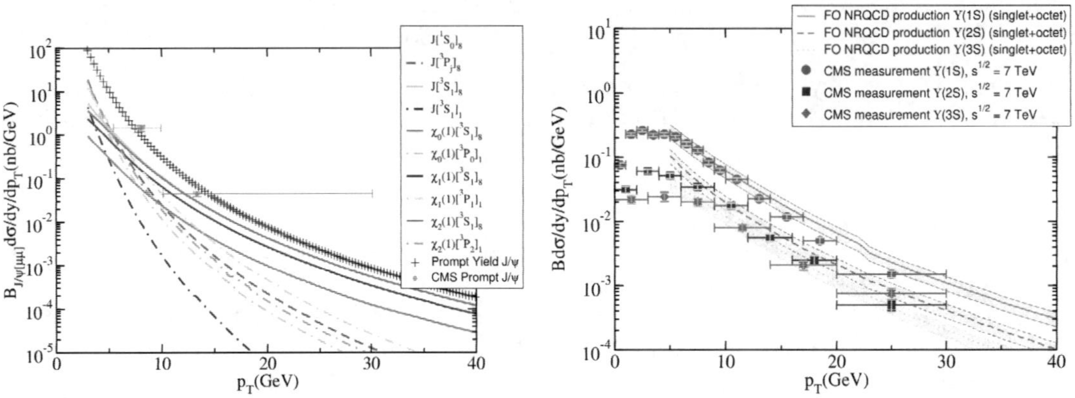

Figure 1. Left panel: The cocktail of contributions that gives the p+p yield for J/ψ production at the LHC. The symbol + represents the net yield and gives the p+p baseline. The data is from CMS [6]. Right panel: Calculation of the cross section of Υ states compared to data from CMS at 7 TeV [7]. The solid line (brown) is for $\Upsilon(1S)$, the dashed (red) for $\Upsilon(2S)$ and the dotted (gray) for $\Upsilon(3S)$

where the quantity in the brackets [ ] represents the angular momentum quantum numbers of the $Q\bar{Q}$ pair in the Fock expansion. The subscript on [ ] refers to the color structure of the $Q\bar{Q}$ pair, 1 being the color-singlet and 8 being the color-octet. The dots represent terms which contribute at higher powers of v. The short distance cross sections $d\sigma(Q\bar{Q})$ correspond to the production of a $Q\bar{Q}$ pair in a particular color and spin configuration, while the long distance matrix element $\langle \mathcal{O}(Q\bar{Q}) \to J/\psi \rangle$ corresponds to the probability of the $Q\bar{Q}$ state to convert to the quarkonium wavefunction. This probability includes any necessary prompt emission of soft gluons to prepare a color neutral system that matches onto the corresponding Fock component of the quarkonium wavefunction. Expressions similar to the ones shown in Eqs. (5), (6) hold for other quarkonium states (such as Υ, χ_c, χ_b, $\cdots$). The parton level cross sections and non-perturbative probabilities are available in the literature, see a list of references in [3].

An illustration of the lowest-order (LO) NRQCD calculation of quarkonium production is given in Fig. 1. The left panel shows the cocktail of contributions that give the baseline production yield for J/ψ at LHC at $\sqrt{S} = 2.76$ GeV in the NRQCD formalism. For the direct production, there are four contributions: one color-singlet contribution ($[^3S_1]_1$) three color-octet contributions ($[^1S_0]_8$), ($[^3S_1]_8$), and ($[^3P_0]_8$). For the prompt yield, we add the feed-down contributions from the χ_{cJ}. The contributions of the χ_{cJ} are also shown in Fig. 1. To obtain the feed-down contributions, we multiply the corresponding yields by the corresponding branching fractions,

$$BR(\chi_{c0} \to J/\psi) = 0.0116 \, , BR(\chi_{c1} \to J/\psi) = 0.344 \, , BR(\chi_{c2} \to J/\psi) = 0.21 \, . \quad (7)$$

This gives the p+p baseline contribution for the J/ψ. We have included experimental data on prompt J/ψ production at $\sqrt{S} = 2.76$ TeV [6]. Each of the species will undergo a different modification due to CNM and QGP effects. To obtain the p+A and A+A yields, we calculate the modified yields for each species and combine them again to obtain the net modification.

The right panel of Fig. 1 shows a calculation of the yields of $b\bar{b}$ states. We obtain the results for $\Upsilon(1S)$, $\Upsilon(2S)$ and $\Upsilon(3S)$ separately (but include the relevant χ_b feed-down for each). If only the $\Upsilon(1S)$ yields are measured, then we can obtain that by adding to these yields, the feed-down from $\Upsilon(2S)$ (branching fraction 31%) and $\Upsilon(3S)$ (branching fraction 16.4%). The theoretical calculation gives a good description of the experimental data within the limitations of the LO NRQCD approach. Specifically, the calculated spectra are somewhat harder than the experimental measurements and cannot be pushed down to low p_T. Any difference in normalization cancels out in the calculation of R_{AA}.

3. Quarkonium production in A+A collisions

3.1. Cold Nuclear Matter effects

In heavy ion reactions, the production yields of energetic particles are always affected by cold nuclear matter (CNM) effects. These include nuclear shadowing, initial state energy loss and transverse momentum broadening (also interpreted as the origin of the Cronin effect). In our calculation, the CNM effects are evaluated from the elastic, inelastic and coherent scattering processes of partons in large nuclei, see Refs. [8, 9, 10]. For quarkonia, cold nuclear matter effects are still not well-understood. To our knowledge, Ref. [3] is the first attempt to calculate the Cronin effect for $Q\bar{Q}$ production. We do not try to constrain its magnitude phenomenologically and point out that data from p+A reactions is needed to constrain all CNM effects. We find that including transverse momentum broadening in the same way as is done for light final states reduces the suppression for p_T between 3 and 10 GeV significantly and may actually lead to a small enhancement of the charmonium cross section. It is not clear, given the large error bars, whether the Au+Au and Cu+Cu data at RHIC are better described by the Cronin calculation. In contrast, it is evident that the Cronin effect is not consistent with Pb+Pb data at the LHC, which sees a large attenuation at $p_T \sim 8$ GeV. At present, at high p_T, initial-state energy loss and the Cronin provide an estimate for the uncertainty range of CNM effects.

3.2. Quark-Gluon Plasma effects

In this section we present the details of the dissociation model of quarkonium propagation through the QGP. The essence of the dissociation model for heavy mesons is that they have short formation times and can therefore form in the medium on a time scale $t_{\rm form}$. Interaction with the thermal medium can dissociate the mesons on a time scale $t_{\rm diss}$. The final yield is given by a rate equation which takes into account the formation and dissociation processes. In the next section we first discuss the rate equation abstractly, using $t_{\rm form}$ and $t_{\rm diss}$ as parameters. In the later sections we will estimate $t_{\rm form}$ and calculate $t_{\rm diss}$. For more details on the dissociation model and its application to the phenomenology of open heavy flavor, see [4].

The approach to estimating the formation time of quarkonium states differs considerably from the approach used for open heavy flavor[5, 4] or light particles [11] that come from the fragmentation of a hard parton. For quarkonia, the $Q\bar{Q}$ state is prepared instantly ($\sim 1/\sqrt{p_T^2 + M^2}$) in the hard collision and subsequently expands to the spatial extent determined by the size of the asymptotic wavefunction. In this case all spatial directions are important. The velocity of the heavy quarks in the meson and a typical upper limit of the meson formation time can be evaluated as follows:

$$\beta_Q = \sqrt{\frac{\kappa^2}{\kappa^2 + m_Q^2}}\,, \qquad t^{\rm max}_{\rm rest\ frame} = \frac{a}{\beta_Q}\,, \tag{8}$$

where the typical momenta κ and sizes a are obtained by solving the Schödinger equation for the corresponding quarkonium state. In this paper we are interested in high transverse momentum mesons, in which case there is a boost in the direction of propagation and, consequently, time dilation

$$t^{\rm max}_{\rm form}(p_T, \alpha) = \gamma\, t^{\rm max}_{\rm rest\ frame}(\alpha) = \gamma \frac{a}{\beta_Q}, \qquad \gamma = \frac{\sqrt{p_T^2 + M^2}}{M}\,. \tag{9}$$

Here, γ is the meson boost factor. Since the formation process is non-perturbative and can not be modeled accurately, the values of $t_{\rm form}$ obtained from Eq. 9 should be considered as an estimate of the upper bound. Therefore, in addition to calculating the final yields for $t_{\rm form} = t_{\rm fmax} = \frac{\gamma a}{\beta_Q}$, we also calculate the yields for $t_{\rm form} = t_{\rm fmin} = \frac{\gamma a}{2\beta_Q}$.

The propagation of a $Q\bar{Q}$ state in matter is accompanied by collisional interactions mediated at the partonic level, as long as the momentum exchanges can resolve the partonic structure of the meson. Two effects are related to these interactions: a) a broadening in the distribution of quarkonium states relative to the original direction; b) a modification of the quarkonium wavefunction. The former effect integrates out as long as we consider inclusive production. The latter effect leads to the dissociation of the meson state. Let us define:

$$\chi\mu^2\xi = \int_{t_0}^{t} d\tau \frac{\mu^2(\mathbf{x}(\tau),\tau)}{\lambda_q(\mathbf{x}(\tau),\tau)}\xi\,, \qquad \mathbf{x}(\tau) = \mathbf{x}_0 + \beta(\tau-\tau_0)\,. \tag{10}$$

In Eq. (10) μ^2 is the typical squared transverse momentum transfer given by the Debye screening scale, $\mu = gT$ for a gluon-dominated plasma. λ_q is the mean scattering length of the quark and $\xi \sim$ few is an enhancement factor from the power law tail of the differential scattering cross section. Finally, $\mathbf{x}_0$ is the position of the propagating $Q\bar{Q}$ and β is the velocity of the heavy meson. By evaluation the overlap between the initial and final meson wavefunctions the meson survival probability $P_{\rm surv.}$ can be obtained. The dissociation rate is then given by

$$t_{\rm diss.}(p_T,\alpha) = \frac{dP_{\rm diss.}}{dt} = -\frac{dP_{\rm surv.}}{dt} \tag{11}$$

The uncertainty in $t_{\rm diss}$ arises from the uncertainty in the coupling between the heavy quarks and the medium (described by the strong coupling constant g) and the enhancement that arises from the power law tails of the Moliere multiple scattering in the Gaussian approximation to transverse momentum diffusion [5] (described by ξ).

Let us denote by $N_{Q\bar{Q}}^{\rm hard}(p_T,\alpha)$ the number of perturbatively produced point-like $Q\bar{Q}$ states at transverse momentum p_T. Up to an overall multiplicative Glauber scaling factor T_{AB}, these p_T number distributions are directly proportional to the cross sections discussed in Section 2. The rate of formation of the corresponding hadronic state, with the appropriate quantum numbers $\{\alpha\}$, is given by the inverse formation time $1/t_{\rm form}(p_T,\alpha)$. In the presence of a medium, the meson multiplicity, which we denote by $N_{Q\bar{Q}}^{\rm meson}(p_T,\alpha)$, is reduced by collisional dissociation processes at a rate $1/t_{\rm diss}(p_T,\alpha)$. Finally, the number of dissociated $Q\bar{Q}$ pairs with a net transverse momentum p_T is $N_{Q\bar{Q}}^{\rm diss.}(p_T,\alpha)$.

The dynamics of such a system is governed by the following set of ordinary differential equations:

$$\frac{d\,N_{Q\bar{Q}}^{\rm hard}(t;p_T,\alpha)}{dt} = -\frac{1}{t_{\rm form}(t;p_T,\alpha)} N_{Q\bar{Q}}^{\rm hard}(t;p_T,\alpha)\,, \tag{12}$$

$$\frac{d\,N_{Q\bar{Q}}^{\rm meson}(t;p_T,\alpha)}{dt} = \frac{1}{t_{\rm form}(t;p_T,\alpha)} N_{Q\bar{Q}}^{\rm hard}(t;p_T,\alpha) - \frac{1}{t_{\rm diss.}(t;p_T,\alpha)} N_{Q\bar{Q}}^{\rm meson}(t;p_T,\alpha)\,, \tag{13}$$

$$\frac{d\,N_{Q\bar{Q}}^{\rm diss.}(t;p_T,\alpha)}{dt} = \frac{1}{t_{\rm diss}(t;p_T,\alpha)} N_{Q\bar{Q}}^{\rm meson}(t;p_T,\alpha)\,, \tag{14}$$

subject to the constraint $N_{Q\bar{Q}}^{\rm hard}(t;p_T,\alpha) + N_{Q\bar{Q}}^{\rm meson}(t;p_T,\alpha) + N_{Q\bar{Q}}^{\rm diss.}(t;p_T,\alpha) = N_{Q\bar{Q}}^{\rm hard}(p_T,\alpha)$, and is uniquely determined by the initial conditions

$$N_{Q\bar{Q}}^{\rm hard}(t=0;p_T,\alpha) = N_{Q\bar{Q}}^{\rm hard}(pQCD;p_T,\alpha)\,, \tag{15}$$

$$N_{Q\bar{Q}}^{\rm meson}(t=0;p_T,\alpha) = 0\,, \tag{16}$$

$$N_{Q\bar{Q}}^{\rm diss.}(t=0;p_T,\alpha) = 0\,. \tag{17}$$

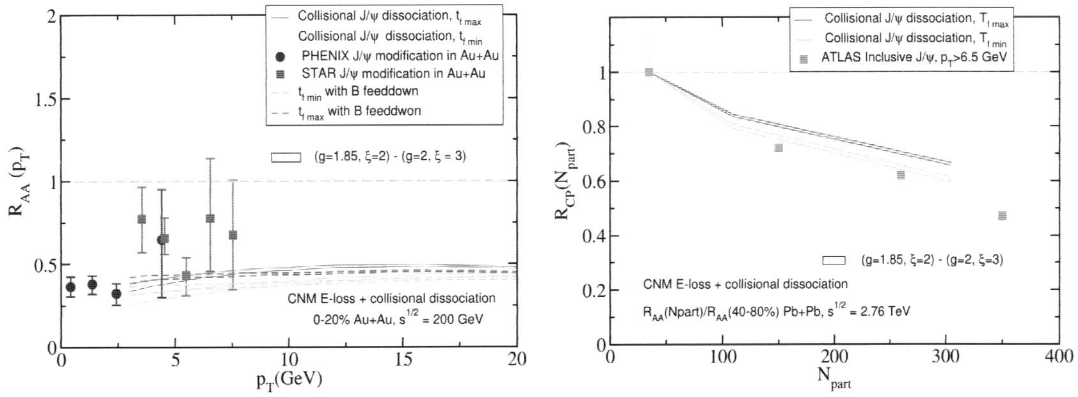

Figure 2. Theoretical model predictions for J/ψ R_{AA} in nucleus-nucleus collisions compared to experimental data [12, 13]. Left panel: RHIC Au+Au collisions at $\sqrt{S} = 0.2$ TeV versus p_T. Right panel is for LHC Pb+Pb collisions at $\sqrt{S} = 2.76$ TeV versus $N_{\rm part}$.

Note that in Eqs. (14) the evolution of the dissociated $\bar{Q}Q$ pair into D- or B- mesons is not shown since it does not couple back to Eqs. (12), (13) (with an appreciable strength).

Realistic simulations include the velocity and $\{\alpha\}$ dependence of the formation rate of all quarkonium states and the time, position and $\{\alpha\}$ dependence of their dissociation rate.

4. Numerical results for the nuclear modification factors

In this section we neglect the Cronin effect but include initial-state cold nuclear matter energy loss and shadowing. Let us first consider the nuclear modification factor for J/ψ mesons. From Fig. 2 we see that the measurement of R_{AA} in RHIC Au+Au collisions [12] shows a suppression factor of about 0.8 at $p_T \sim 6$ GeV. Even including the uncertainty in our model parameters, we obtain a somewhat higher suppression (for $p_T \sim 6$ GeV, $R_{AA} \sim 0.35 - 0.45$ for Au+Au) than the one currently observed at RHIC. In Fig. 2, our results for the prompt yields of J/ψ mesons are marked by upper and lower yellow bands corresponding to the upper ($t_{\rm form}^{\rm max}$) and lower ($t_{\rm form}^{\rm min} = t_{\rm form}^{\rm max}/2$) limits of our formation time estimate respectively. The bands themselves correspond to our estimate of the uncertainly in the sets of parameters that determine the coupling of the heavy quarks with the in-medium partons [$g = 1.85$, $\xi = 2$ (minimum considered coupling gives the upper limit of the yellow band) and $g = 2$, $\xi = 3$ (maximum considered coupling for the lower limit of the yellow band)]. The pronounced effect of the variation of the formation time can be intuitively seen as follows. From Eq. 14, we see that the dissociation mechanism is operative only when $N_{Q\bar{Q}}^{meson}$ is substantial, i.e. after $t_{\rm form}$. Since the upper limit for formation time of quarkonia can be on the order of several fm/c, the density of the medium at $t_{\rm form}^{\rm max}$ is reduced considerably due to Bjorken expansion, giving weaker dissociation and smaller suppression.

In the right panel of Fig. 2 we show the p_T-averaged suppression,

$$R_{AA}(N_{\rm part})\,({\rm or}\ R_{CP}(N_{\rm part})) = \frac{\int_{p_{\rm min.}} dp_T\, R_{AA}(p_T;N_{\rm part})\,({\rm or}\ R_{CP}(p_T;N_{\rm part}))\,\frac{d\sigma}{dydp_T}}{\int_{p_{\rm min.}} dp_T\,\frac{d\sigma}{dydp_T}}\,, \tag{18}$$

of J/ψ mesons versus centrality. We present a comparison to the ATLAS central-to-peripheral data [13]. Our theoretical predictions compare well to the measurement in mid-central and

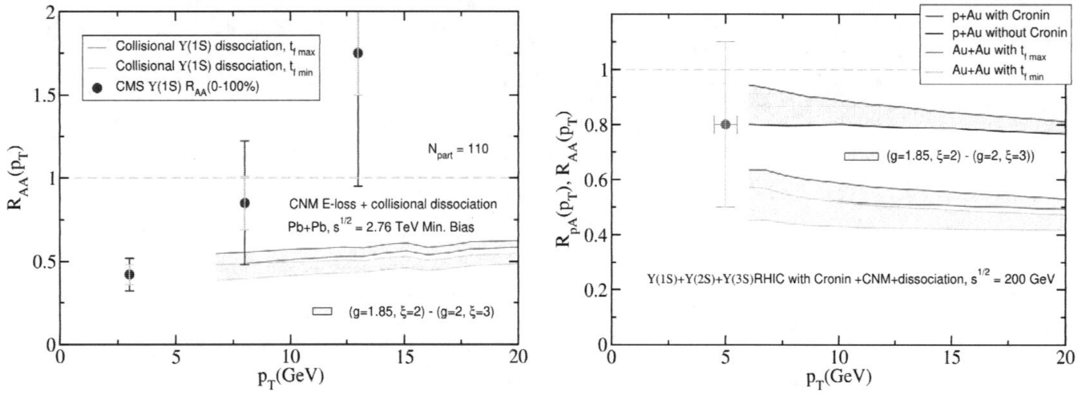

Figure 3. Theoretical model predictions for Υ R_{AA} in nucleus-nucleus collisions. Left panel: data from CMS is for $0-100\%$ centrality [6] compared to the theoretical prediction at the minimum bias $N_{\rm part} \approx 110$. Theoretical model predictions for minimum-bias $R_{pA}(\Upsilon)$ and central 0-20% $R_{AA}(\Upsilon)$. Right panel: RHIC p+Au and Au+Au collisions at $\sqrt{S} = 0.2$ TeV. Data is from STAR [14]

peripheral reactions. The deviation between data and theory is only for $N_{\rm part} > 300$. (In the case of CMS data the deviation between data and theory is for $N_{\rm part} > 200$.)

The CMS experiment at the LHC has also measured R_{AA} for $\Upsilon(nS)$ states in Pb+Pb collisions at $\sqrt{S} = 2.76$ TeV per nucleon pair [6]. We cannot calculate the equivalent ratio of R_{AA}s for inclusive production because our formalism for the production and propagation of Υs is not applicable to $p_T(\Upsilon) \leq 6$ GeV. In our approach the meson should be boosted relative to the medium. Instead, in the left panel of 3 we show comparison to the minimum bias p_T-differential $\Upsilon(1S)$ CMS nuclear modification measurement. In this case, the theoretical calculation is performed for the average number of participants for minimum bias collisions. The combined results for R_{AA} (0-20% central) and R_{pA} (minimum-bias) for bottomonia are given in the right panel of Fig. 3. The effect of transverse momentum broadening is much smaller for bottomonia when compared to the one for charmonia. This can be intuitively understood as follows. The mechanism for Cronin enhancement in this calculation is that initial state scattering increases the typical transverse momentum carried by the incident partons by a few GeV. For quarkonia, there is an additional scale m_H. For bottomonia the mass scale is considerably larger than the transverse momentum broadening scale and few additional GeV do not increase the yield significantly. Preliminary Υ suppression measurements are now available at RHIC [14]. More differential p_T measurements will shed light on the similarities and differences in the CNM effects at RHIC and at the LHC.

5. Conclusions

In summary, we carried out a detailed study of high transverse momentum quarkonium production and modification in heavy ion reactions at RHIC and at the LHC [3]. We used a NRQCD approach to calculate the baseline quarkonium cross sections. We found that for J/ψ mesons the theoretically computed spectrum is slightly harder than the one observed in the experiment. For all Υ states $(1S, 2S, 3S)$ the agreement is within a factor of two when we consider both the TeVatron and the LHC data. In reactions with heavy nuclei, we presented theoretical model predictions for the nuclear modification of quarkonium yields at high p_T in minimum bias p(d)+A and 0-20% central A+A collisions. We focused on the consistent inclusion of both cold (CNM) and hot (QGP) nuclear matter effects in different colliding systems at different center-of-mass energies. We compared our results to published and preliminary experimental data, where applicable.

We found that ignoring the Cronin effect leads to a small overestimate of the suppression of J/ψ mesons in the p_T region between 5 GeV and 10 GeV in central Cu+Cu and Au+Au collisions at $\sqrt{S} = 0.2$ TeV at RHIC. Including initial-state transverse momentum broadening leads to a somewhat better agreement between theory and the current experimental data only for the Cu+Cu reactions. A smaller Cronin enhancement will work better. We demonstrated that CNM effects can be easily constrained in d+A reactions at RHIC. For example, the d+Au calculation that includes power corrections and cold nuclear matter energy loss predicts 20% suppression of the J/ψ cross section. Including transverse momentum broadening may lead to as much as 50% enhancement in the region of the Cronin peak. We also found that the Cronin-like modification of the Υ spectrum is much smaller. Current data on high-p_T quarkonium production at RHIC does not indicate the presence of thermal effects at the level of the quarkonium wavefunction within our theoretical framework.

The conclusions from our theoretical model comparison to the $\sqrt{S} = 2.76$ TeV LHC data are not as clear cut. Our calculations underestimated the suppression for J/ψ production reported by the ATLAS and CMS experiments in the most central Pb+Pb collisions. On the other hand, it works quite well in mid-central and peripheral reactions. We found that the Cronin enhancement at the LHC is smaller than the one at RHIC due to the harder spectra. However, any Cronin enhancement appears incompatible with the experimental results. For Υ mesons, p_T-differential data in A+A collisions is scarce. CMS measurements of minimum bias $\Upsilon(1S)$ indicate that the low p_T suppression may decrease or disappear at high p_T. We plan to address the possibility of thermal effects at the level of the quarkonium wavefunction in a separate publication. We also plan to extend the meson dissociation mechanism to photon-tagged heavy flavor [15, 16] and heavy flavor-tagged jets [17].

References

[1] T. Matsui and H. Satz,, Physics Letters B, **178**, 416-422 (1986).
[2] G. T. Bodwin, E. Braaten and G. P. Lepage, Phys. Rev. D **51**, 1125 (1995) [Erratum-ibid. D **55**, 5853 (1997)] [arXiv:hep-ph/9407339].
[3] R. Sharma and I. Vitev, arXiv:1203.0329 [hep-ph].
[4] R. Sharma, I. Vitev and B. W. Zhang, Phys. Rev. C **80**, 054902 (2009) [arXiv:0904.0032 [hep-ph]].
[5] A. Adil and I. Vitev, Phys. Lett. B **649**, 139 (2007) [hep-ph/0611109].
[6] S. Chatrchyan *et al.* [CMS Collaboration], [arXiv:1201.5069 [nucl-ex]].
[7] V. Khachatryan *et al.* [CMS Collaboration], Phys. Rev. D **83**, 112004 (2011) [arXiv:1012.5545 [hep-ex]].
[8] J. -W. Qiu and I. Vitev, Phys. Lett. B **587**, 52 (2004) [hep-ph/0401062].
[9] R. B. Neufeld, I. Vitev and B. -W. Zhang, Phys. Lett. B **704**, 590 (2011) [arXiv:1010.3708 [hep-ph]].
[10] M. Gyulassy, P. Levai and I. Vitev, Phys. Rev. D **66**, 014005 (2002) [nucl-th/0201078].
[11] C. Markert, R. Bellwied and I. Vitev, Phys. Lett. B **669**, 92 (2008) [arXiv:0807.1509 [nucl-th]].
[12] Z. Tang [STAR Collaboration], J. Phys. G **38**, 124107 (2011) [arXiv:1107.0532 [hep-ex]]; references therein.
[13] G. Aad *et al.* [Atlas Collaboration], Phys. Lett. B **697** (2011) 294 [arXiv:1012.5419 [hep-ex]].
[14] R. Reed [STAR Collaboration], J. Phys. Conf. Ser. **270**, 012026 (2011) [Nucl. Phys. A **855**, 440 (2011)]; Quark Matter 2011.
[15] Z. -B. Kang and I. Vitev, Phys. Rev. D **84**, 014034 (2011) [arXiv:1106.1493 [hep-ph]].
[16] W. Dai, I. Vitev and B. -W. Zhang, arXiv:1207.5177 [hep-ph].
[17] Y. He, I. Vitev and B. -W. Zhang, Phys. Lett. B **713**, 224 (2012) [arXiv:1105.2566 [hep-ph]].

Journal of Physics: Conference Series **389** (2012) 012030 doi:10.1088/1742-6596/389/1/012030

J/ψ measurements in STAR Experiment at RHIC

Olga Hájková for the STAR Collaboration

Czech Technical University in Prague, Faculty of Nuclear Sciences and Physical Engineering, Behov 7, 11519, Prague 1, Czech Republic

E-mail: olga.hajkova@fjfi.cvut.cz

Abstract. During the last decade STAR experiment has studied the hot and dense nuclear matter produced in heavy ion collisions at RHIC. The study of quarkonia, such as J/ψ meson, production provides the information about properties of this nuclear matter. It is predicted that due to the Debye screening the quarkonia production is suppressed when Quark Gluon Plasma (QGP) is presented. In order to understand the properties of hot and dense nuclear matter, it is necessary to study the production of quarkonia in hadron-hadron, hadron-ion, and ion-ion collisions separately to distinguish cold nuclear matter effect, such as gluon shadowing, from the suppression due to the formation of QGP. Moreover, the J/ψ hadron azimuthal correlations allow to extract the fraction of J/ψ that originates from B-mesons. This proceedings presents recent J/ψ measurements at mid-rapidity in p+p, and Au+Au collisions at $\sqrt{s_{NN}} = 200$ GeV at STAR, the fraction of J/ψ originates from B-mesons, and J/ψ eliptic flow v_2 as a function of transverse momentum in Au+Au collisions in $\sqrt{s_{NN}} = 200$ GeV at STAR.

1. Introduction

The strongly interacting matter at sufficiently high density or temperature undergoes a phase transition from the hadronic matter to a new state, so-called quark gluon plasma (QGP), when quarks are no longer confined into color neutral bound states. One of the probes, proposed by Matsui and Satz [1], for searching for QGP and for investigating its properties is a study of the quarkonia production. It is predicted that due to the color screening the quarkonia production is suppressed when QGP is presented at sufficiently high temperature. Before resolving whether QGP was formed or not, it is necessary to study the production of J/ψ in hadron-hadron, hadron-ion and ion-ion collisions separately to distinguish cold nuclear matter effects from the suppression due to the formation of QGP. Other effect which play the important role in J/ψ production is recombination from charm quarks [2] [3]. It is expected that in the high-p_T region the J/ψ production is less influenced by the cold nuclear matter effects and charm quark recombination effect. Therefore study of J/ψ production in high-p_T region provides a cleaner probe to search for evidence of color-screening effect [4] [5] [6]. STAR previous measurement in high-p_T region in Cu+Cu colisions shows no suppression, but the measurement of larger system such as Au+Au is necessary to interpret these results.

In this proceedings the measurements of low-p_T and high-p_T J/ψ spectra in p+p and in Au+Au collisons at $\sqrt{s}$=200 GeV in the years 2009 and 2010 in STAR are presented. We also present the measurement of J/ψ elliptic flow v_2 in semi-central collisions in Au+Au collisions.

2. Analysis

Data reported in this proceedings were collected in p+p collisions at $\sqrt{s}$=200 GeV in the year 2009 and in Au+Au collisions at $\sqrt{s_{NN}}$=200 GeV in the year 2010 with Minimum Bias Trigger for low p_T J/ψ, and with High Tower Trigger for high p_T J/ψ. Main detectors used in presented measurements are the Time Projection Chamber (TPC), which is the main charged particle tracking device in the STAR detector used for particle identification and momentum determination, the Barrel Electromagnetic Calorimeter (BEMC), which is used for deposited energy measurement, Barrel Shower Maximum Detector (BSMD) and also as a trigger detector, and the Time Of Flight (TOF). TOF was completely installed and full functional in year 2010 Au+Au run. During the year 2009 p+p run, 72% of TOF trays was installed. TPC and TOF together are great tool for distinguish electrons and hadrons from low to high p_T region. J/ψ were reconstructed through its dilepton decay channel, J/$\psi \rightarrow e^+ + e^-$ with branching ratio 5.9%.

3. J/ψ spectra in p+p and Au+Au collisions

Figure 1 (left) shows the J/ψ invariant mass distribution in p+p collisions at $\sqrt{s}$=200 GeV in the year 2009. Background was reconstructed via like-sign method, unlike-sign signal (solid circles) and background (gray band). We reconstructed 376 J/ψ in high p_T region ($4 < p_T < 12$ GeV/c), with very suitable signal over background ratio $S/B = 22$. As high S/B ratio is very useful for J/ψ-hadron correlation study. Figure 1 (right) shows the fully corrected J/ψ p_T spectra in p+p collisions at $\sqrt{s}$=200 GeV in the year 2009. STAR and PHENIX results are consistent with each other. The solid magenta line shows the calculation from Color Evaporation Model (CEM) [8].

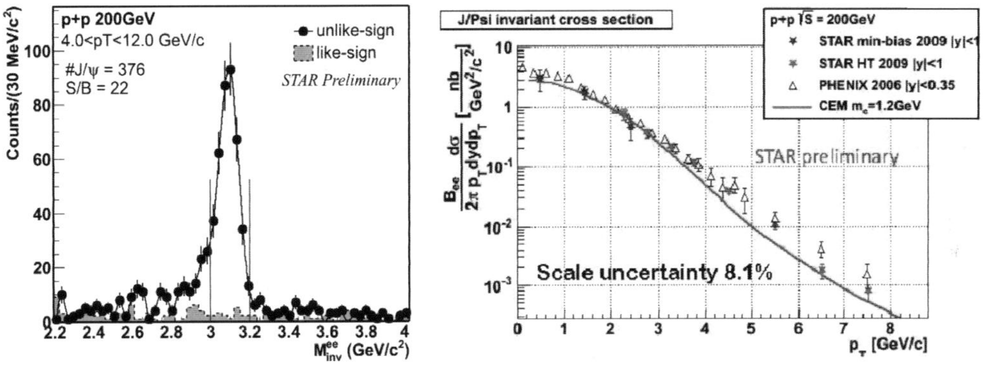

Figure 1. Left: High p_T J/ψ invariant mass distribution in p+p collisions at 200 GeV in the year 2009 reconstructed via dilepton decay channel. Right: Fully correlated J/ψ p_T spectra in p+p collisions in the year 2009. Low p_T spectra reconstructed from Minimum Bias Trigger events (blue stars) and high p_T spectra reconstructed from High Tower Trigger events (red stars). Magenta solid line shows the calculation from Color Evaporation Model (CEM) [7] [8].

Figure 2 (left) shows the high p_T J/ψ invariant mass distribution in Au+Au collisions at $\sqrt{s_{NN}}$=200 GeV in the year 2010. Figure 2 (right) shows fully corrected J/ψ p_T spectra in Au+Au collisions at $\sqrt{s_N N}$=200 GeV with different centralities. STAR and PHENIX results are consistent with each other in their overlapping p_T region, we extend our results p_T region up to 10 GeV/c [3]. The solid and dash-dotted lines present Tsallis Blast-Wave model (TBW) fit to the STAR data points and TBW theoretical predictions which assuming the same radial flow and freeze-out conditions for J/ψ as for light hadrons. Measured spectra do not follow the TBW model predictions from light hadrons in low p_T region [9]. This could indicate significant contribution to the J/ψ spectra from charm quark recombination at low p_T or small J/ψ radial flow.

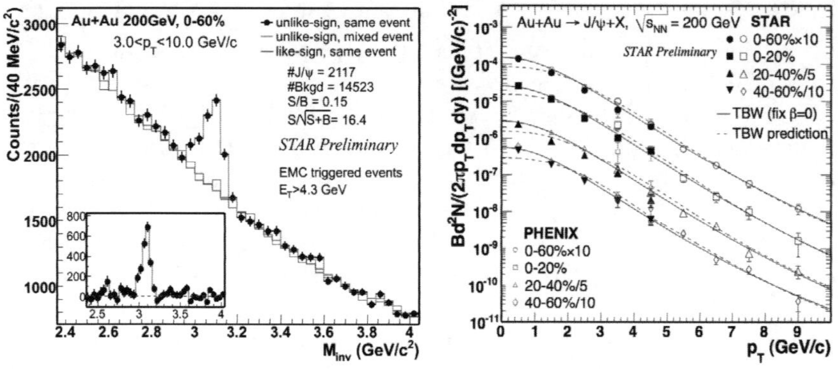

Figure 2. Left: High p_T J/ψ invariant mass distribution in Au+Au collisions at $\sqrt{s}$=200 GeV in the year 2010. Right: Fully correlated J/ψ p_T spectra in Au+Au collisions for different centralities. Solid and dash-dotted lines represent TBW fit to the STAR data points and TBW theoretical predictions respectively [3] [9].

4. J/ψ nuclear modification factor R_{AA}

Figure 3 (left) shows the J/ψ nuclear modification factor R_{AA} in Au+Au collisions at $\sqrt{s_{NN}}$ = 200 GeV as a function of the number of participants for low-p_T and high-p_T J/ψ separately. The result in Cu+Cu collisions at $\sqrt{s_{NN}}$ = 200 GeV for high-p_T is also plotted [10]. Both results for high-p_T J/ψ are consistent with each other, and they show no suppresion in semi-central collisions. In central collisions high-p_T J/ψ is significantly suppresed. The low-p_T J/ψ measurements show suppresion in both, central and semi-central collisions. Low p_T PHENIX results are also plotted and are consistent with low-p_T STAR results [3]. Figure 3 (right) shows the J/ψ nuclear modification factor R_{AA} in Au+Au collisions at $\sqrt{s_{NN}}$ = 200 GeV as a function of p_T for 0-20% central and semicentral (40-60%) collisions separately. The solid and dashed lines represent theoretical calculation include both primordial production or regeneration from charm quarks [11] [12].

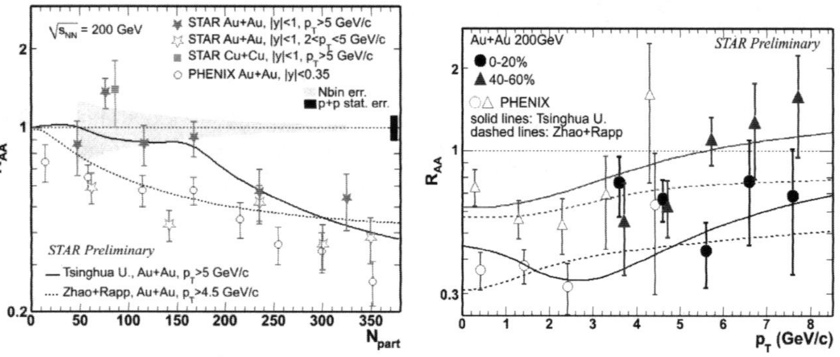

Figure 3. Left: J/ψ nuclear modification factor (R_{AA}) in Au+Au collisions at $\sqrt{s_{NN}}$ = 200 GeV as a function of multiplicity for low p_T (openstars) and high p_T (fullstars). The gray square presents the result in Cu+Cu collisions at $\sqrt{s_{NN}}$ = 200 GeV. The solid and the dashed lines represent theoretical calculations [10] [11] [12]. Right: J/ψ nuclear modification factor (R_{AA}) as a function of p_T in central (0-20%, black circles) and semi-central (40-60%, blue triangles) collisions. The solid and the dashed lines represent theoretical calculations [11] [12].

5. J/ψ elliptic flow v_2

Figure 4 shows J/ψ v_2 as a function of p_T for semi-central Au+Au collisions at $\sqrt{s_{NN}} = 200$ GeV. The data points are plotted as black points, lines are represent different theoretical models. J/ψ v_2 is consistent with zero at high-p_T and disfavors the scenario that coalescence from thermalized charm quarks dominate J/ψ production at $p_T > 2$ GeV/c [13] - [18].

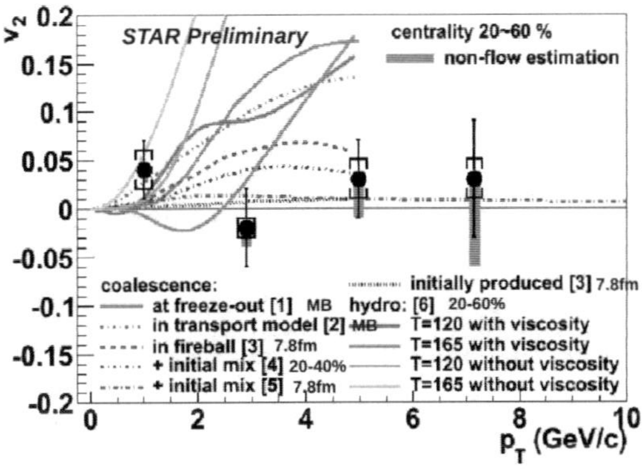

Figure 4. J/ψ v_2 as a function of p_T for semi-central Au+Au collisions (20-60%), lines represent theoretical models [13] - [18]

6. Summary and Outlook

In this proceedings, results of J/ψ measurements in p+p collisions at $\sqrt{s}$=200 GeV and in Au+Au collisions at $\sqrt{s_{NN}}$=200 GeV from STAR are presented. The measurement of nuclear modification factor indicates that suppression in high-p_T region is smaller than in low-p_T region. The preliminary results of J/ψ v_2 show that elliptic flow v_2 is consistent with zero at high-p_T and disfavors the scenario that coalescence from thermalized charm quarks dominate J/ψ production at $p_T > 2$ GeV/c.

This work was supported by grant INGO LA09013 of the Ministry of Education, Youth and Sports of the Czech Republic, and by the Grant Agency of the Czech Technical University in Prague, grant No. SGS10/292/OHK4/3T/14 .

[1] T. Matsui, H. Satz, Phys. Lett. B 178 (1986) 416.
[2] M. C. Abreu et al., Phys. Lett. B 499 (2001) 8596.
[3] A. Adare et al., Phys. Rev. Lett. 98 (2007) 232301.
[4] Y.-p. Liu, Z. Qu, N. Xu, P.-f. Zhuang, Phys. Lett. B 678 (2009) 7276.
[5] X. Zhao, R. Rapp, Phys. Rev. C 82 (2010) 064905.
[6] H. Liu, K. Rajagopal, U.A.Wiedemann, Phys. Rev. Lett. 98 (2007) 182301.
[7] A. Adare et al., Phys. Rev. D 82,(2010) 012001.
[8] M. Bedjidian, et al., hep-ph/0311048.
[9] Z.Tang et al., arXiv:1101.1912, JPG 37, 08194 (2010).
[10] B. I. Abelev et al., Phys. Rev. C 80 (2009) 41902.
[11] Y.-p. Liu, Z. Qu, N. Xu, P.-f. Zhuang, Phys. Lett. B 678 (2009) 7276.
[12] X. Zhao, R. Rapp, Phys. Rev. C 82 (2010) 064905.
[13] V. Greco, C.M. Ko, R. Rapp, Phys. Lett. B 595 (2004) 202.
[14] L. Ravagli, R. Rapp, Phys. Lett. B 655, (2007) 126.
[15] L. Yan, P. Zhuang, N. Xu, Phys. Rev. Lett. 97, (2006) 232301.
[16] X. Zhao, R. Rapp, 24th WWND (2008).
[17] Y. Liu, N. Xu, P. Zhuang, Nucl. Phy. A 834, (2010) 317.
[18] U. Heinz, C. Shen, private communication.

Journal of Physics: Conference Series **389** (2012) 012031 doi:10.1088/1742-6596/389/1/012031

Cold nuclear matter effects on J/ψ production from PHENIX

D McGlinchey (for the PHENIX Collaboration)

Department of Physics, Florida State University, 77 Chieftain Way, Tallahassee FL 32306-4350 USA

E-mail: dmcglinchey@fsu.edu

Abstract. It has become clear in recent studies of d+Au collisions that quantifying the cold nuclear matter effects on J/ψ production is critical to interpreting the results of J/ψ production in heavy ion collisions at RHIC. Recent measurements of the rapidity and impact parameter dependence of J/ψ production in d+Au collisions at $\sqrt{s_{NN}}$ =200 GeV by PHENIX show significant suppression at forward rapidity for collisions with small impact parameters, and require a modification which is stronger than linearly or exponentially dependent on the nuclear thickness to describe the forward rapidity data. New measurements of the transverse momentum dependence of the J/ψ nuclear modification in d+Au collisions show a cold nuclear matter effect which increases with increasing p_T and suggests a cold nuclear matter contribution in heavy ion collisions which is large at high p_T.

1. Introduction

Measurements of quarkonia production in heavy ion (HI) collisions provide a unique probe of the screening length in the Quark Gluon Plasma (QGP) produced during HI collisions. The different binding energies, and therefore radii, of the various quarkonia states, coupled with precision measurements of their production in HI collisions should provide key insights into critial properties of the QGP.

A large suppression of J/ψ production in Au+Au collisions relative to p+p collisions has been measured by PHENIX at both mid and forward rapidities [1; 2]. However, interpreting these results in terms of suppression within a hot nuclear medium is impossible without a detailed understanding of the effects of producing a J/ψ in a nuclear target, often termed Cold Nuclear Matter (CNM) effects. In order to provide the best chance at interpreting the heavy ion results, precision measurements of CNM effects at the same collision energy as the HI results are necessary.

A number of effects are predicted to modify J/ψ production in a nuclear target, including nuclear breakup, nuclear shadowing, initial state energy loss, p_T-broadening, and gluon saturation (for a recent review, see [3]). Disentangling these effects is a daunting task but is facilitated by precision measurements over a broad range in collision centrality, rapidity (y) and transverse momentum (p_T).

Nuclear shadowing, the modification of parton distribution functions inside nuclei (nPDF's), can be constrained by precision measurements of J/ψ modification in p+A collisions as a function of y and p_T. Varying the measured rapidity and p_T changes the sampled space in Feynman-x and energy transfer (Q^2). J/ψ production at RHIC energies can provide crucial constraints

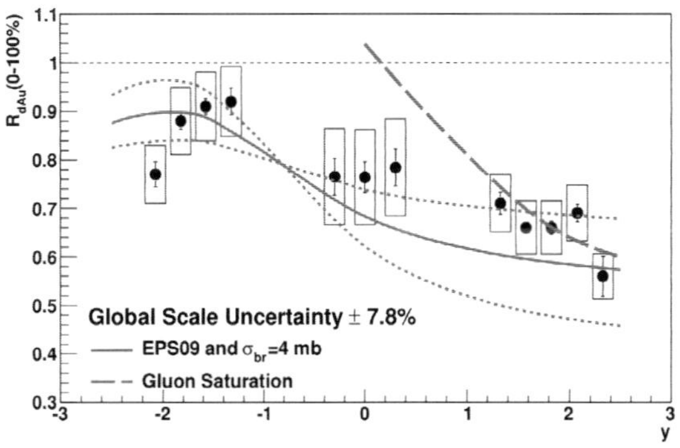

Figure 1. J/ψ modification relative to p+p collisions (R_{dAu}) as a function of rapidity integrated over collision centrality (0-100%) [4]. The curves are described in the text.

on the gluon distributions which are only accessed indirectly by the deep inelastic scattering data which provides the bulk of the information used to derive the nPDF's. Measurements as a function of collision centrality can also provide new and interesting insights into the geometric dependence of these nPDF distributions.

Measurements of the modification of J/ψ production in nuclear targets as a function of p_T provides constraints on p_T-broadening effects, which are not constrained by measurements of the rapidity or collision centrality alone.

Towards this end, PHENIX has made precision measurements of J/ψ production in d+Au collisions at $\sqrt{s_{NN}}$ =200 GeV as a function of rapidity, p_T, and collision centrality using data taken in 2008 [4; 5]. These results show a number of interesting trends leading to new and exciting conclusions.

2. Rapidity dependence of the J/ψ nuclear modification

The nuclear modification of J/ψ production relative to p+p collisions (R_{dAu}) is shown as a function of rapidity integrated over collision centrality and p_T in Figure 1. The data show increasing suppression with increasing rapidity, with up to 40% suppression observed at forward rapidity (deuteron going direction) compared to 10% suppression observed at backward rapidity (Au going direction).

Also shown in Figure 1 are calculations of the nuclear modification from two models. The red curves are calculations which include nuclear shadowing, taken from the EPS09 nPDF set [6] combined with an example nuclear breakup cross section of σ_{br} = 4 mb which is constant across all rapidities. The EPS09 nPDF's provide the nuclear modification of the free proton parton distribution functions as a function of x and Q^2 for a given nucleus, A. Also included in EPS09 are 30 alternate sets which explore the plus and minus 1σ uncertainties in each of the parameters involved in their parameterization. The solid red curve in Figure 1 uses the central, or best fit, set from EPS09, while the two dashed red curves represent the sets which provide the maximum variation in the shape. In [4] the σ_{br} value is chosen by eye to best match the backward rapidity R_{dAu} when combined with the nuclear shadowing from the central EPS09 set, rather than fit to the data. This calculation shows good agreement with the measured R_{dAu} across all rapidities.

The second calculation, shown by the green dashed line in Figure 1, incorporates gluon saturation at small-x [7]. In this model, partons in the projectile nucleon's can interact with multiple partons, possibly from different nucleons, in the target nucleus through coherent gluon scattering. The calculation is only valid in the small-x region, and therefore is only valid at

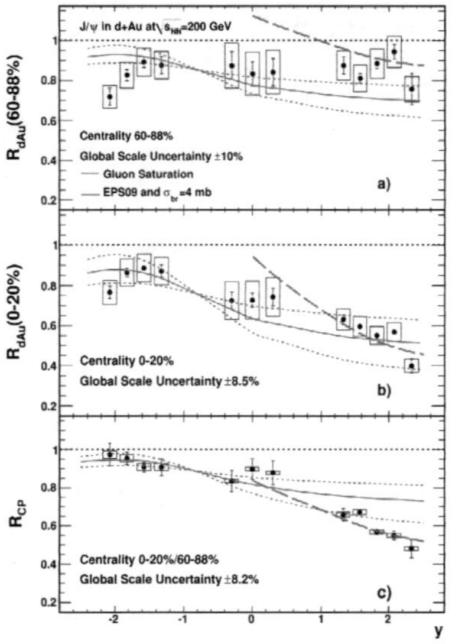

Figure 2. The J/ψ nuclear modification factors R_{dAu} (0-20%)(a), R_{dAu} (60-88%), and R_{CP} as a function of rapidity [4]. The curves are calculations discussed in the text.

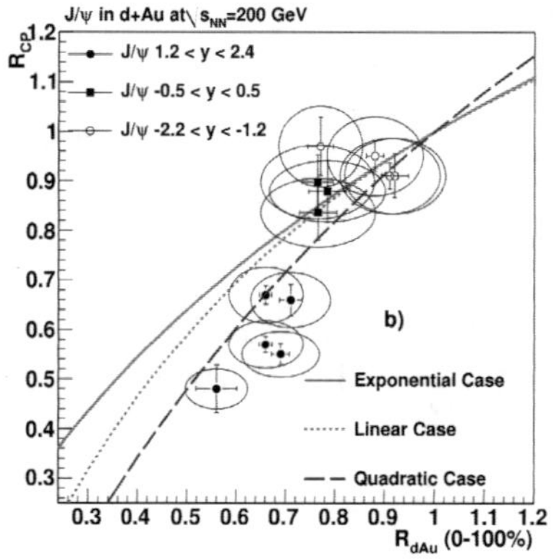

Figure 3. The J/ψ R_{CP} versus the collision centrality integrated J/ψ R_{dAu} [4]. The curves are discussed in the text.

forward rapidity at RHIC energies. The calculation shows good agreement with the forward rapidity data, however it deviates from the data at midrapidity, where the validity of the model is questionable.

The geometric dependence of the nuclear modification is probed by measuring the modification as a function of collision centrality. The J/ψ R_{dAu} for central (0-20%) and peripheral (60-88%) collisions is shown as a function of rapidity in Figure 2. Also shown in Figure 2 is the ratio of the modification in central events to that in peripheral events (R_{CP}). The advantage of the variable R_{CP} is that many of the systematic uncertainties cancel, as the ratio is taken within the same data set. The data show constant suppression with rapidity in peripheral events, with an overall level of suppression consistent with unity within the systematic uncertainties. In contrast, a large suppression is observed at forward rapidity in central events, with a decrease in suppression with decreasing rapidity. This trend is also shown in the J/ψ R_{CP}, which shows that the backward rapidity data is virtually unmodified between central and peripheral events, whereas the forward rapidity shows a large increase in suppression between central and peripheral events.

Calculations of the centrality dependence from the models presented in Figure 1 are also included in Figure 2. The gluon saturation shows good agreement with the data at forward rapidity, as in the centrality integrated case, but diverges from the data at midrapidity. EPS09 (and most other nPDF sets as well) calculate an average modification in the PDF's across the nucleus. Therefore, when calculating the centrality dependence of the nuclear shadowing, some geometric dependence of the modification must be introduced. In [4], a linear dependence of the density weighted longitudinal thickness through the gold nucleus at the location of the struck nucleon, r_T,

$$\Lambda(r_T) = \frac{1}{\rho_0} \int dz \rho(r_T, z) \tag{1}$$

is introduced, where $\rho(r_T, z)$ is taken, on average, to follow a Woods-Saxon distribution. EPS09, with a linear dependence on $\Lambda(r_T)$, combined with $\sigma_{br} = 4$ mb agrees well with the R_{dAu} in central collisions across all rapidity. However, in peripheral collisions, it over predicts the suppression at forward rapidity. This is clearly shown in the difference between the calculation and the measured R_{CP} at forward rapidity.

To further investigate the geometric dependence of the d+Au data, three functional forms for the modification as a function of the longitudinal thickness, each with one free strength parameter a were tested,

$$\text{Linear} : M(r_T) = 1 - a\Lambda(r_T) \quad (2)$$
$$\text{Quadratic} : M(r_T) = 1 - a\Lambda(r_T)^2 \quad (3)$$
$$\text{Exponential} : M(r_T) = e^{-a\Lambda(r_T)}. \quad (4)$$

The nuclear modification factors can then be calculated for a given form of the modification, with a given value for the strength parameter a, with r_T distributions of the PHENIX centrality bins obtained from a Glauber model of d+Au collisions. For a given form of the modification, a single value of a provides a unique relationship between R_{CP} and R_{dAu} (0-100%). Varying a for a given form of the modification creates a locus of points on the R_{CP} vs. R_{dAu} (0-100%) plane, on which any suppression with that geometric dependence must lie. The results for the three functional forms of the modification are shown in Figure 3, along with the data.

Figure 3 shows that at backward and midrapidity, the data is unable to distinguish between the three functional forms, as all three forms agree within uncertainties. At forward rapidity, however, there is no overlap between the data and the linear or exponential dependencies, and indeed, even the quadratic case is in poor agreement with the data. This shows that a longitudinal thickness dependence which is greater than exponential or linear is required by the forward rapidity data.

3. Transverse momentum dependence of the J/ψ nuclear modification

The J/ψ R_{dAu} as a function of p_T for centrality integrated d+Au collisions is shown in Figure 4 for three rapidity ranges, corresponding to the three PHENIX spectrometers [5]. When measuring the p_T dependence of J/ψ production the yield is integrated over the full rapidity of each spectrometer to maximize the p_T reach of the data, rather than using the same binning as shown, for example, in Figure 1. At mid and forward rapidities a suppression is observed at low-p_T which decreases with increasing p_T, becoming consistent with an R_{dAu} of 1 at $p_T \approx 3-4$ GeV/c. A different behavior is observed at backward rapidity, where suppression is still observed at low-p_T but with a much more rapid decrease in suppression with increasing p_T. The R_{dAu} at backward rapidity is consistent with, or greater than, 1 for $p_T > 2$ GeV/c.

Also shown in Figure 4 are two sets of calculations for the p_T dependence of R_{dAu}. The pink dot-dashed curves are calculations by Kopeliovich et al. [8] which include nuclear shadowing (taken from the nDSg nPDF set), nuclear breakup, and the Cronin effect. The black dashed lines represent calculations by Ferreiro *et al.* [9] which include nuclear shadowing (also taken from the nDSg nPDF set) and nuclear breakup with $\sigma_{abs} = 4.2$ mb. In the calculation by Kopeliovich et al. a parameterization of the dipole cross section is used to calculate a nuclear breakup effect which is dependent on the J/ψ kinematics. At mid and forward rapidities, both sets of calculations agree reasonably well with the observed R_{dAu}. At backward rapidity, neither calculation reproduces the rapid decrease in suppression seen at low-p_T. This may indicate a different nPDF description is required at low-x, or an alternate physical effect which modifies the shape of the p_T distribution differently at backward rapidity than mid or forward rapidity.

The centrality dependence of the J/ψ R_{dAu} as a function of p_T is shown in Figures 5, 6, and 7 for backward, mid, and forward rapidity respectively. A strong centrality dependence is seen

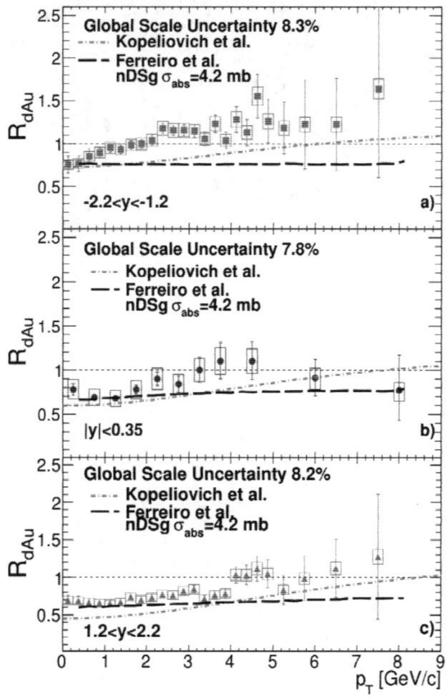

Figure 4. The J/ψ R_{dAu} for 0-100% centrality integrated d+Au collisions as a function of rapidity for backward(a), mid(b), and forward(c) rapidities [5]. The curves are model calculations discussed in the text.

in all three rapidity regions. In central events, suppression is seen at low-p_T with a decrease in suppression with increasing p_T. Transitioning from central to peripheral events produces a flattening of the p_T distribution, with the R_{dAu} for peripheral events being relatively constant with p_T, with an R_{dAu} consistent with 1 within the systematic uncertainties on the data at all three rapidities.

Two sets of calculations by Ferreiro *et al.* are also shown in Figures 5, 6, and 7, where the nuclear shadowing is assumed to be linearly dependent on the local density. The two sets of calculations only differ by the nPDF set used. The solid lines utilize the EKS98 nPDF set, which has a similar modification to EPS09, while the dashed curve utilizes the nDSg nPDF set. At mid and forward rapidities the difference manifests itself in an apparent overall scaling difference, with both sets of calculations agreeing well with the shape of the measured R_{dAu}. At backward rapidity, neither calculation is in agreement with the data. The EKS98 produces a p_T dependence which has increasing suppression with increasing p_T, the opposite trend observed in the data. This difference between the EKS98 and nDSg calculations at backward rapidity is likely due to the difference in the modification in the EMC region of the shadowing curve. At high-x, EKS98 predicts a transition from an enhancement in the gluon distribution to a suppression, whereas nDSg predicts little modification in either direction over the same region of x and Q^2. This could be an indication that the backward rapidity data prefer no modification, or possibly a continued enhancement, of the gluon distribution at high-x over the suppression included in EKS98.

A third set of calculations by Sharma and Vitev [10] is compared with the midrapidity centrality integrated R_{dAu} in Figure 8. This model utilizes non relativistic quantum chromodynamics (NRQCD) to describe J/ψ production. Nuclear shadowing is included through power suppressed coherent final-state scattering. Initial state energy loss and the Cronin effect

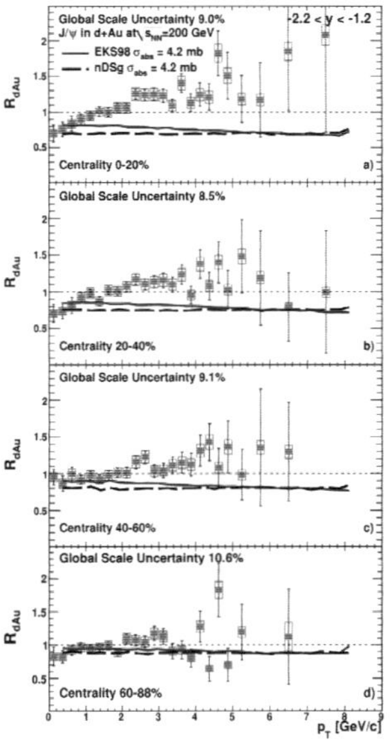

Figure 5. The J/ψ R_{dAu} as a function of p_T for 0-20%(a), 20-40%(b), 40-60%(c) and 60-88% d+Au collisions at backward rapidity [5]. The curves are model calculations discussed in the text.

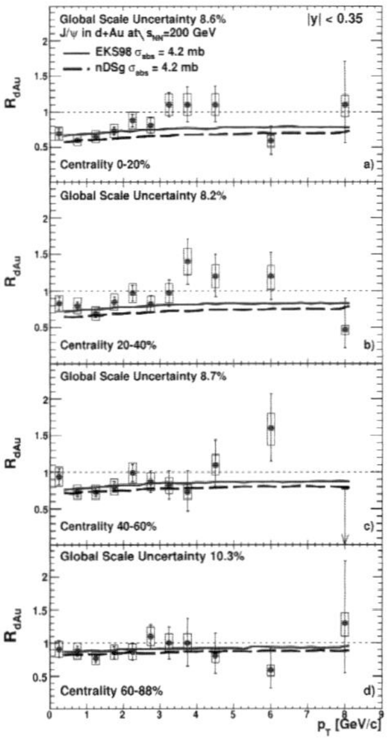

Figure 6. The J/ψ R_{dAu} as a function of p_T for 0-20%(a), 20-40%(b), 40-60%(c) and 60-88% d+Au collisions at mid rapidity [5]. The curves are model calculations discussed in the text.

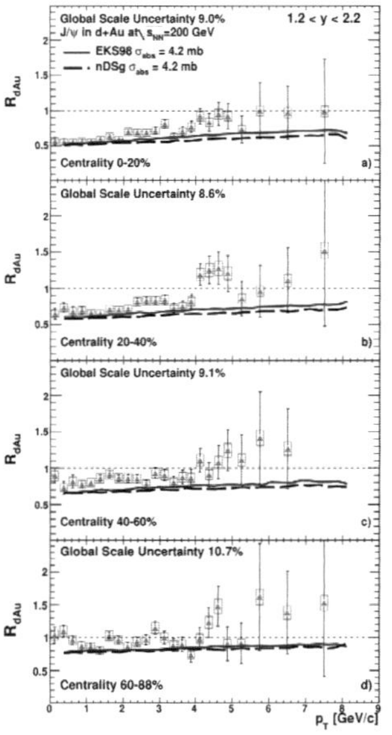

Figure 7. The J/ψ R_{dAu} as a function of p_T for 0-20%(a), 20-40%(b), 40-60%(c) and 60-88% d+Au collisions at forward rapidity [5]. The curves are model calculations discussed in the text.

are also included. The solid red curve in Figure 8 shows the full calculation including the Cronin effect, while the dashed green line shows the calculation without the Cronin effect. The comparison of the two curves gives a direct indication of the strength of the Cronin effect, which is evidently over predicted compared to the data.

4. Implications on the measured suppression in heavy ion collisions

With precision measurements of the CNM effects on J/ψ production in hand a detailed parameterization of the data, which describes well the measured rapidity, p_T and centrality dependence, is required to make accurate qualitative predictions of the CNM effects in HI collisions. However, the data alone give a number of implications about the hot nuclear matter effects on J/ψ production in HI collisions.

Measurements of the J/ψ nuclear modification factor, R_{AA}, in Cu+Cu collisions by PHENIX [11] and STAR [12] as a function of p_T is shown in Figure 9. Based on the high-p_T behavior of R_{dAu} an $R_{AA} \approx 1$ would be expected from CNM effects alone, possibly indicating that the observed R_{AA} in Cu+Cu at high-p_T could indicate a lack of suppression both from CNM and hot nuclear matter effects, rather than an enhancement in the J/ψ production due to recombination effects. A detailed propagation of the CNM effects to HI collisions is required before making an stronger quantitative conclusions.

Similar arguments can give insight into the CNM baseline in Au+Au without a full propagation of the d+Au results. The measured p_T dependence of the J/ψ R_{AA} by PHENIX is shown in Figure 10. Strong suppression is seen for $p_T < 4$ GeV/c in central and mid-

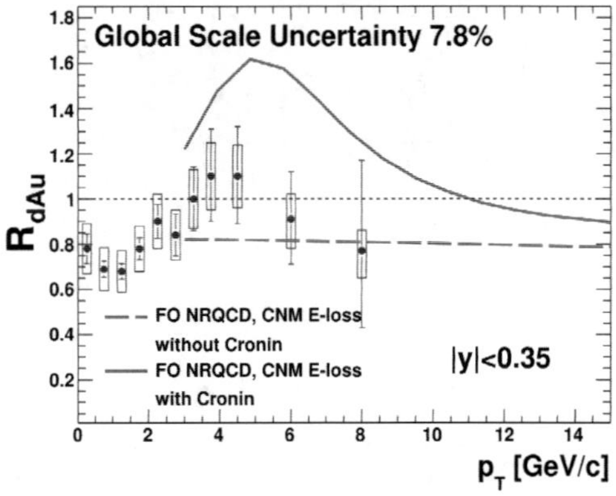

Figure 8. The J/ψ R_{dAu} as a function of p_T for 0-100% d+Au collisions [5]. The curves are calculations from [10] described in the text.

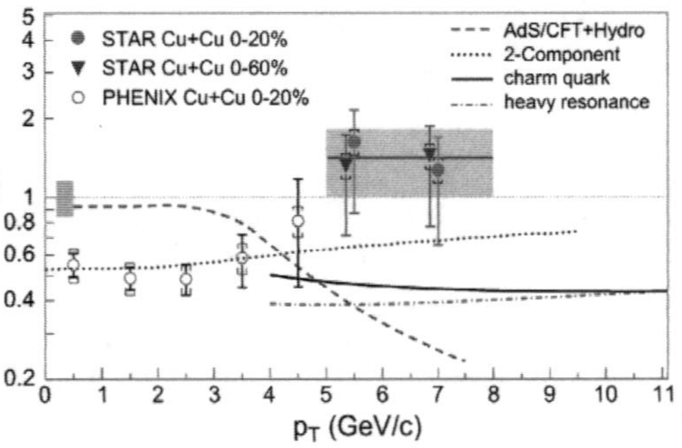

Figure 9. The J/ψ R_{AA} in Cu+Cu collisions as a function of p_T at midrapidity [12].

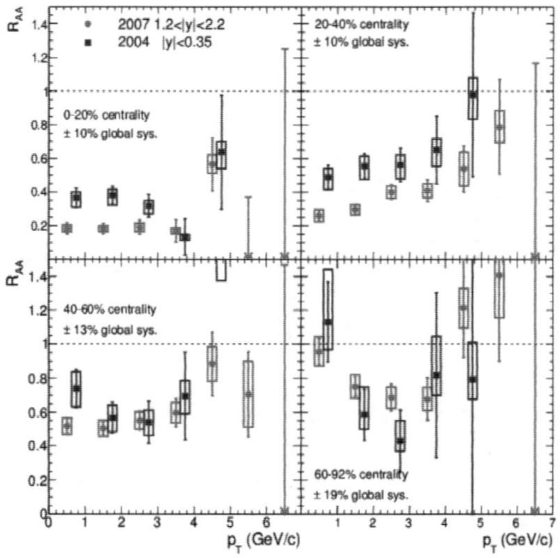

Figure 10. The J/ψ R_{AA} in Au+Au collisions as a function of p_T at mid and forward rapidities [2].

central events, while the suppression pattern is unclear for peripheral events due to the large uncertainties on the data. At midrapidity the cold nuclear matter R_{AA} ($R_{AA}(CNM)$) can be approximated by R^2_{dAu}. It should be stressed that this is only an approximation, and no substitute to a detailed propagation of the d+Au results. At low-p_T the d+Au data suggest an $R_{AA}(CNM) \approx 0.5$, while the observed $R_{AA} \approx 0.4$ in central events, indicating possible suppression beyond CNM effects from this crude approximation. At forward rapidity the $R_{AA}(CNM)$ can be approximated by folding the backward and forward R_{dAu}, giving $R_{AA}(CNM) \approx 0.5$ for low-p_T at forward rapidity compared to the measured $R_{AA} \approx 0.2$ for central collisions. This shows possibly substantial suppression compared to CNM effects. While this is only an approximation, it does show that the CNM effects are likely large in HI collisions, and cannot be neglected.

5. Summary

The PHENIX experiment has made precision measurements of the nuclear modification of J/ψ production in d+Au collisions as a function of rapidity, p_T, and collision centrality, paving the way for a detailed understanding of CNM effects on J/ψ production at RHIC energies. These results include many interesting physical implications. The centrality dependence of the forward rapidity R_{dAu} require a nuclear modification which is stronger than linearly or exponentially dependent on the longitudinal nuclear thickness. The p_T dependence of R_{dAu} show an R_{dAu} which is consistent with 1.0 for $p_T > 4$ GeV/c at mid and forward rapidity, with an $R_{dAu} > 1$ for $p_T > 4$ GeV/c at backward rapidity. The backward rapidity R_{dAu} also show a very different shape of the p_T dependence when compared to mid and forward rapidity, a change which is unexplained by current theoretical models.

A parameterization able to describe the rapidity, p_T, and centrality dependence of the J/ψ R_{dAu} will allow the propagation of the CNM effects to HI collisions, allowing quantitative calculations of the hot nuclear matter effects at RHIC energies. This is the next step towards understanding quarkonia production in heavy ion collisions.

References

[1] Adare A *et al.* (PHENIX Collaboration) 2007 *Phys.Rev.Lett.* **98** 232301
[2] Adare A *et al.* (PHENIX Collaboration) 2011 *Phys.Rev.* C **84** 054912
[3] Brambilla N, Eidelman S, Heltsley B, Vogt R, Bodwin G *et al.* 2011 *Eur. Phys. J. C* **71** 1534
[4] Adare A *et al.* (PHENIX Collaboration) 2011 *Phys.Rev.Lett.* **107** 142301
[5] Adare A *et al.* (PHENIX Collaboration) 2012 (*Preprint* 1204.0777)
[6] Eskola K, Paukkunen H and Salgado C 2009 *JHEP* **0904** 065
[7] Kharzeev D and Tuchin K 2006 *Nucl.Phys.* **A770** 40–56
[8] Kopeliovich B, Potashnikova I and Schmidt I 2011 *Nucl. Phys. A* **864** 203–212
[9] Ferreiro E, Fleuret F, Lansberg J, Matagne N and Rakotozafindrabe A arXiv:1201.5574 (2012).
[10] Sharma R and Vitev I arXiv:1203.0329 (2012).
[11] Adare A *et al.* (PHENIX Collaboration) 2008 *Phys. Rev. Lett.* **101** 122301
[12] Abelev B *et al.* (STAR Collaboration) 2009 *Phys. Rev. c* **80** 041902

Journal of Physics: Conference Series **389** (2012) 012032 doi:10.1088/1742-6596/389/1/012032

Probing the sea contribution to the protons' spin

Richard Hollis, for the PHENIX Collaboration
University of California, Riverside, Riverside, CA 92521
E-mail: rhollis@ucr.edu

Abstract. Longitudinal single-spin asymmetries (A_L) of parity-violating W bosons are measured with PHENIX via the decay daughter muons in $\sqrt{s} = 500\,\text{GeV}$ $p + p$ collisions at RHIC. These proceedings report on the latest $W^{\pm} \rightarrow \mu^{\pm}\nu$ asymmetry results obtained during the 2011 RHIC run using the forward muon-arm detector in PHENIX.

1. Introduction

Over twenty years ago, a naive picture of the protons' spin would have expected that the constituent quark spin would form a large contribution to the total proton spin. Upon the first measurements of the protons' spin this picture was found not to be true and has since sparked much theoretical and experimental interest as to the make-up of this fundamental property of the proton [1].

Several components have been theorized to be necessary in the spin formation with all pieces summing to precisely $\frac{1}{2}\hbar$. The sum rule (Eqn. 1) includes quark components (quark and anti-quark), a gluon component, and an orbital angular momentum component. The relative contributions at different x – the fraction of the momentum carried by the individual quark or gluon –are the quantities we wish to measure. These are expressed as spin-dependent parton distribution functions (see Ref. [2] for theoretical work).

$$\frac{1}{2} = \frac{1}{2}\Delta\Sigma + \Delta G + L_Z \qquad (1)$$

Investigation over the succeeding years using (semi-inclusive) deep inelastic scattering ((SI)DIS) have lead to a precise understanding of the spin-dependence of the parton-distribution functions for quarks, but to a lesser degree for the contribution of the anti-quark [2]. To investigate this at RHIC, W bosons provide an almost ideal probe as they directly probe the sea quark contribution to the proton's spin via their leading order production in hard scattered $u + \bar{d} \rightarrow W^+$ ($\bar{u} + d \rightarrow W^-$) interactions. RHIC collisions of polarized protons at $\sqrt{s} = 500\,\text{GeV}$ are energetic enough for W production. Underlying the parity-violating W production, the weak decay $W^{\pm} \rightarrow l^{\pm} + \nu(\bar{\nu})$ preserves the asymmetries for which the resultant $l^{+(-)}$ which are expected to be large in magnitude at high-rapidity, yet opposite in sign.

In these proceedings, we report on the most recent result on the work to elucidate the sea quark contribution to the protons' spin using the PHENIX detector at RHIC. The necessary upgrades to the forward muon arms, the analysis to reduce the backgrounds and the current preliminary results will be discussed in the context of the published central-arm results and theoretical expectations.

Published under licence by IOP Publishing Ltd

2. The PHENIX Detector towards a W measurement

The W production cross-section at RHIC energies is very small compared to the inelastic or the lepton production cross-section. For the central arms ($W \rightarrow e\nu$, where the ν is not aparent in the aparatus) measurement, this small cross-section is aided by several factors. Firstly, in the decay of W's at mid-rapidity the energy distribution peaks close to half the mass of the W. Such a Jacobian peak has a shape which is distinct from the background continuum of electrons. Secondly, in a related way, the dominating contribution to the energy spectrum at high energy (around $20<E<30$ GeV) is the $W \rightarrow e$ process. The $W \rightarrow e$ contribution can be relatively enhanced by imposing an isolation requirement in a close-proximity region to reduce the background e candidates. At high energy such electron candidates are dominantly from photon conversions and mis-identified hadrons. For a further description of the $W \rightarrow e$ signal extraction see Ref. [3].

By measuring electrons with high precision and the ability to efficiently trigger on them, the central arms provide the basis for the W measurements. However, the amplitude of the asymmetry at mid-rapidity is expected to be small for W bosons, presenting a significant challenge with that measurement. A larger asymmetry for W's is expected to be observed at forward rapidity, where different experimental challenges exist. The underlying Jacobian peak observed in the central region is diminished at forward rapidity, in fact completely washed out with the resolution of the forward muon arms at high momentum. At all momenta, a large background exists from mis-identified low momentum hadrons, which has to be reduced by a factor of 100 or more to be able to extract the true W signal muons. Finally, prior to Run 11, the ability to trigger on forward muons was limited to a low-momentum ($p>1.5$ GeV/c) threshold, which does not provide enough rejection to effectively trigger on the W muons. As the μs from the W decay appear at high momenta, an online selectivity of momenta is highly desirable, in fact necessary if one wishes to record all such muons[1]. Thus, an upgrade to the electronics in the Muon-tracker was made to allow essentially straight-line trajectories to be triggered online. A further upgrade saw a new detector implemented at the furthest z position in the Muon-Tracker/Identifier system. The Resistive plate chambers, RPCs, have sufficient timing to allow for a definite beam-clock association to the candidate μs. An additional iron absorber was also added into the front of the muon tracker to help reduce sources of hadronic background which decay within the muon track volume and could mimic a high-momentum signal muon.

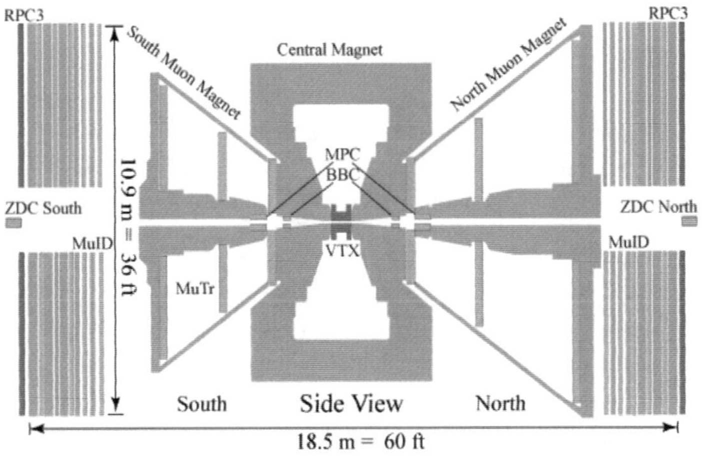

Figure 1. Side-view of the PHENIX detector, highlighting the muon tracker and muon identifier (active regions in green). The outer red boxes depict the RPCs used in this work.

The new triggering system currently matches hit combinations in the muon tracker, which are consistent with straight lines, to a predefined map. To determine the map, straight line

[1] As the rate of interactions is greater than the rate at which PHENIX can record data to tape, minimum bias (or high-frequency type) triggers must be prescaled to prevent the recording rate to slow down – known as dead-time.

Journal of Physics: Conference Series **389** (2012) 012032 doi:10.1088/1742-6596/389/1/012032

trajectories are passed through a geometrical description of the muon tracker and RPC. All possible combinations are stored in a look-up table. To further restrict the trigger to accept only signal muons the time recorded in the RPCs must be consistent with a particle originating from the collision region. This requirement specifically rejects beam-related backgrounds which come from outside the PHENIX detector volume.

3. Background and Signal Extraction

The expected rate for W production is only a few hundred lepton tracks are expected in a given year's running at RHIC are to be measured in the acceptance of the central or muon arms. As such, the number of other tracks in these detectors is much larger in comparison and must be significantly reduced (by several orders of magnitude) in order to extract the desired signal.

As discussed, for the central-arm measurement the Jacobian peak helps to distinguish the signal e's from the background spectrum, see Figure 2. The background spectrum folds in the estimated cross-section of electrons from heavy flavor decays and photon conversion. The uncertainty in the background accounts for photon conversion probablity uncertainties, overall background normalization, and the uncertainty associated with extrapolating to high momenta.

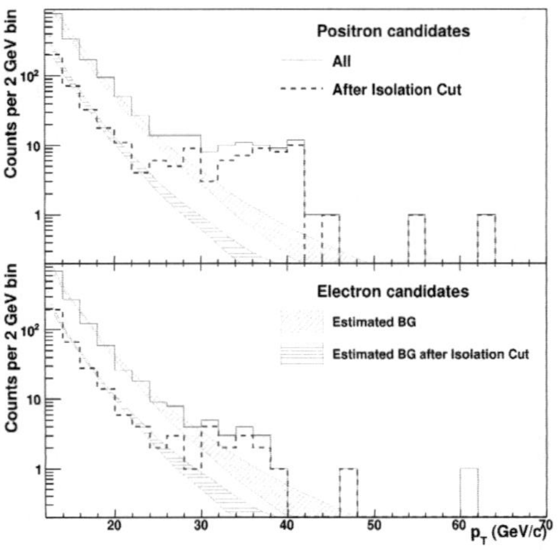

Figure 2. Measured positron (upper) and electron (lower) spectrum from the PHENIX central-arm electromagnetic calorimeter, from Ref.[3]. The red/blue lines represent the data without/with an isolation cut. The estimated background is shown as the bands. The charge-sign is determined from the Drift Chamber bend prior to the calorimeter.

There are two broad types of backgrounds concerning the muon arms: *irreducible* – real muons from charm and bottom decays, and *reducible* – muons which originate from light hadrons. The latter are the largest source of background, but through their decay topology, can be reduced. The former cannot be removed directly, but their contribution can be estimated from the data itself.

To minimize the contamination from the reducible backgrounds, one must first understand how they come about. Hadrons generally lose all their energy in the absorber and do not impinge the muon tracker active elements. However, a small fraction (sub-percent) can punch-through and form a track in the muon tracker. As the time of flight is quite long into the muon arms, charged pions and kaons can decay within the muon-tracker elements into a real muon (and neutrino). Such a decay has two features: a kink in the apparent track of the particle (as seen by the muon tracker) and a "deep" road through the muon-identifier (which is only likely for real muons). Although rare, the first feature can occur such that a low momentum track may appear as a high-momentum track after the decay. This contaminates the high-momentum sample. The deep road means that this "high-momentum" track also has a valid muon-identifier road. This combination does cause a significant problem as the number of hadrons produced in $p+p$ collisions is large compared to the W muons. Fortunately, these new tracks are not perfect and have subtle differences from the true high-momentum tracks. The fit-χ^2 to the muon-tracker portion of the track is typically larger for hadronic backgrounds. The matching between the muon-tracker track and the muon-identifier road is worse owing to the former being reconstituted from two different trajectories. The residual distributions, then, from these elements are used to cut finely on true signal and to reduce the effect from backgrounds. See the right panel of Figure 3 for an example of the matching residual differences between muons and decay-hadrons in the muon tracker.

Billions of Monte-Carlo simulations of single hadrons, as well as PYTHIA events, were used to determine the optimal cut positions on the residual distributions. Several cut levels (to retain between 90% and 99% of signal muons from each cut) were chosen, where the lowest cut values (at 90%) represent the most stringent cuts. As a cross-check, these distributions were formed from samples of real data muons to ensure that the detector behaves the same in both simulation and data. Hundreds of hours of cosmic-ray triggered events were reconstructed to allow for a study of the cut position in that environment. Tracks which pass through both muon arms provide a crucial test of these residual distributions, as well as of the momentum resolution of the high-momentum tracks (discussed below). In a collision environment, muons resultant from J/ψ decays also provide a zeroth-order cross-check of the simulations, albeit at much lower transverse momentum.

For the irreducible backgrounds, removal is not possible as these are true muons, and look just like the W-originating muons, except that the spectral shapes (in p_T) are distinct. For this, samples of PYTHIA events for various sources of real muons were generated, simulated through a GEANT description of the detector setup and reconstructed as though this was a real data sample. The final reconstructed simulation was scaled to the sampled data luminosity and compared to the recorded data at high momentum. The cross-section for these muons are dominantly at a lower momentum than that of the W muons, the only way that these can impact the signal is by being smeared out to higher momenta. As the resolution of the muon tracker degrades rapidly for high-momentum tracks, some of these other muons are pushed to higher momenta and appear as our W signal. The amount of smearing which is present in the data is modeled in the simulation and thus the impact can be studied. To check that the smearing is accurately described, the momentum of the cosmic-muon pairs, which are reconstructed in opposite arms, are compared. As both originate from the same muon (with the same momentum) the smearing in the tracker can be derived from this sample and compared to the simulations. They were found to be in agreement, even at the highest momentum, see the left panel of Figure 3. Finally, to check the overall scale of the backgrounds, the dimuon spectrum in data was compared to that from the PYTHIA simulations, scaled to the measured cross-section. This provides an estimate as to the overall scale in the simulation. The sum of all background muon sources agrees reasonably well with the data, after the application of the residual cuts.

4. Results

The preliminary results on the single-μ cross-section for positive (upper panel) and negative (lower) forward μ are presented in Figure 4. The data were collected at $\sqrt{s} = 500\,\text{GeV}$ $p+p$ collisions during RHIC Run 11. In the figure, the upper green data shows the raw p_T distributions, without any track selectivity. This illustrates the rejection factor ($\mathcal{O}(100)$) which is needed to reduce the backgrounds to the same order of magnitude as the W signal. The red (lower green) show the efficiency corrected final data using the muon-identifier only (muon-tracker) trigger. The statistical reach of the muon-identifier trigger is limited due to the necessary prescale factor applied to allow data to be read out. The filled histograms represent the expected distribution from data. The blue (purple) shades represent the residual irreducible (reducible) backgrounds. The red-filled histogram represents the muons from the signal. The red fill represents the muons from the W signal. Note that this is a logarithmic scale, the true

Journal of Physics: Conference Series **389** (2012) 012032 doi:10.1088/1742-6596/389/1/012032

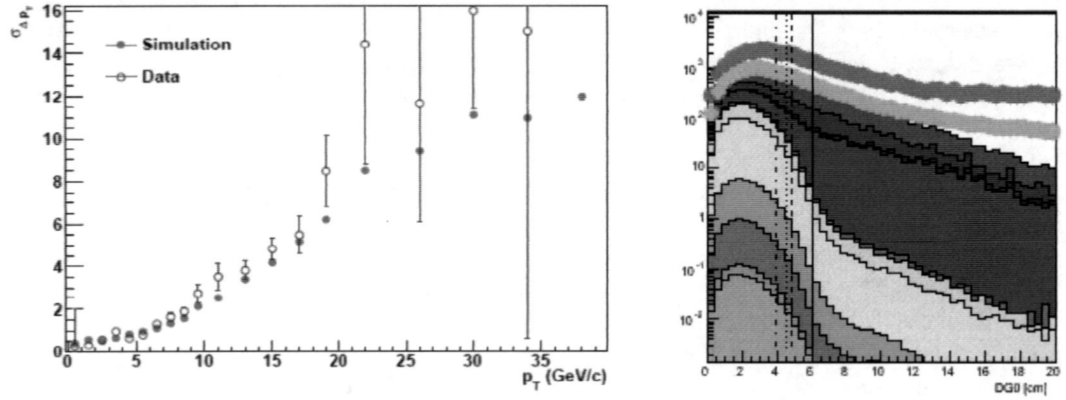

Figure 3. The left panel shows the comparison between the momentum resolution determined from simulation (red) and cosmic data (blue). The right panel illustrates the residual parameter DG0 (spatial difference between extrapolated muon-tracker track and the muon-identifier road). The blue/purple color-filled histograms illustrate the distribution contribution from hadronic sources, green and orange are real muons (not from W's), and red represents the $W \to \mu$ signal. The vertical lines represent cut-levels; the red (green) upper distributions are data with μ-ID (high-p_T) triggers.

signal-to-background is 0.21 to 0.42, dependent on the charge sign and rapidity, integrated over the region $p_T > 20\,\mathrm{GeV}/c$.

The single-spin asymmetries extracted for single muons are shown in Figure 5. The luminosity sampled to form this measurement was $\mathcal{L} = 25\,\mathrm{pb}^{-1}$, with an average polarization of $\mathcal{P} \sim 50\,\%$. The statistical uncertainty is shown as a line, the systematic uncertainty is shown as a box. The largest contribution to the systematic uncertainty is due to the uncertainty in the background sources (both reducible and irreducible). To determine the systematic errors, the backgrounds were varied by a factor of two in calculating the asymmetry. This is only the first measurement of forward W's with a factor of four more data (from RHIC Run 12 onwards) planned to be recorded over the course of the next few years. Data from Run 12, where more $p+p$ data at $\sqrt{s} = 510\,\mathrm{GeV}$ were recorded, includes a new silicon pre-tracker for the muon-tracker which may help to constrain the background contributions in future W measurements.

Published results from the central arm $W \to e\nu$ measurement [3] are shown in Figure 6. The sampled luminosity in this dataset from Run 9 was approximately $8.6\,\mathrm{pb}^{-1}$, with a luminosity of $\sim$40%. Analysis is ongoing toward results from Run 11 and 12 which will have a higher statistical significance.

5. Summary

The PHENIX experiment at RHIC has measured longitudinal single-spin asymmetries from the maximally parity violating $W \to l\nu$ decay at mid- and forward-rapidity. For the forward measurement a series of detector upgrades were necessary to enable an effective online high-momentum selectivity of such candidates. From the Run 9 data, the first W cross-section and asymmetry measurements were made at mid-rapidity. This was followed by forward measurements using data collected in Run 11. The next few years of RHIC running, with fully commissioned PHENIX upgrades will provide a further factor of four in statistics, coupled with better constraints on the background contributions from muons and light hadrons.

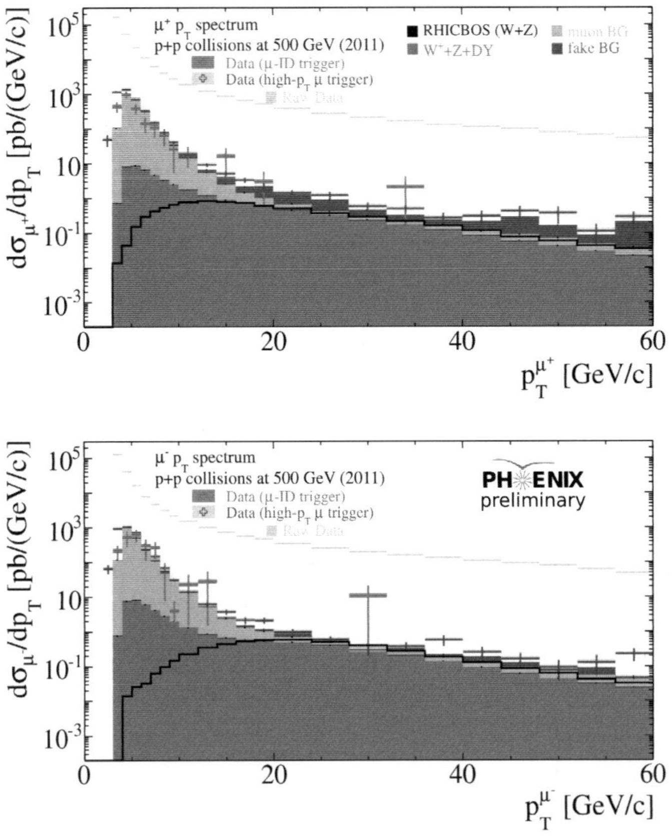

Figure 4. Preliminary positive (upper panel) and negative (lower) forward μ cross-section measured in $\sqrt{s} = 500\,\mathrm{GeV}$ $p+p$ collisions during RHIC Run 11. The upper green histogram shows the raw data, without any track selectivity, the red (lower green) show the efficiency corrected final data using the muon-identifier only (muon-tracker) trigger. The red-filled histogram represents the muons from the signal. The blue (purple) represent the residual irreducible (reducible) backgrounds.

[1] J.Ashman *et al. Nucl. Phys.* **B328** (1989) 1.
[2] D.de Florian *et al. Phys. Rev.* **D80** (2009) 034030.
[3] A.Adare *et al. Phys. Rev. Lett* **106** (2011) 062001.

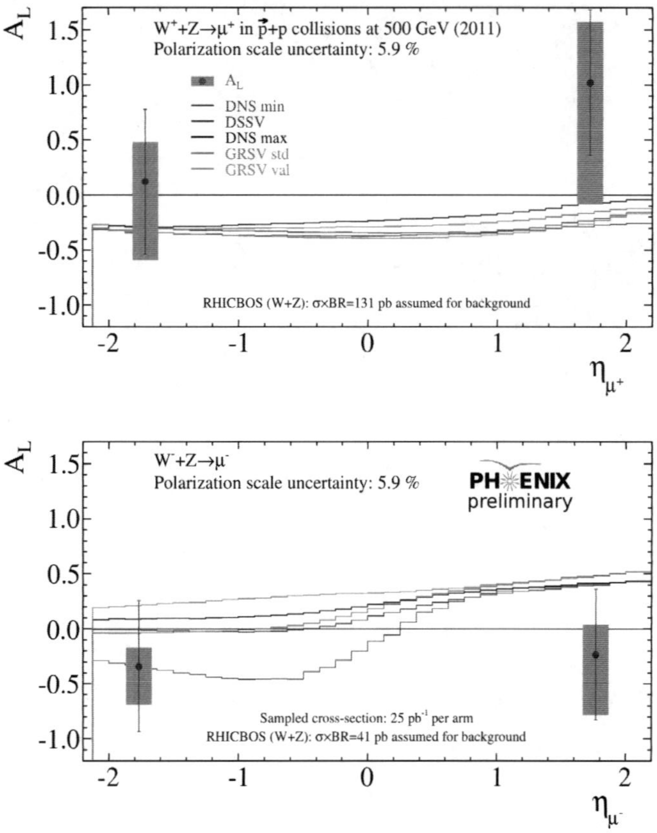

Figure 5. Preliminary positive (upper panel) and negative (lower) forward W single-spin asymmetries measured in $\sqrt{s} = 500\,\mathrm{GeV}$ $p + p$ collisions during RHIC Run 11. The data and associated statistical (line) and systematic (box) errors are measured in the forward (η>0) and backward (η<0) regions. The lines depict different model expectations.

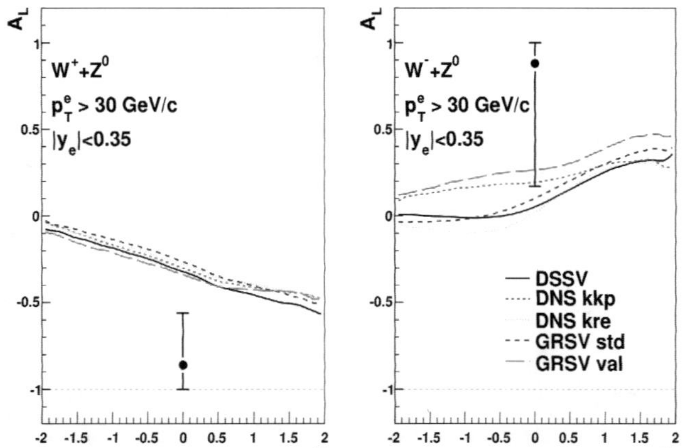

Figure 6. Measured positron (upper) and electron (lower) asymmetries from the PHENIX central-arm electromagnetic calorimeter, from Ref. [3]. The error bars represent the combined statistical and systematic uncertainties, the colored lines show model expectations.

sPHENIX Spin and Forward Physics

K N Barish for the PHENIX collaboration

Department of Physics & Astronomy, University of California, Riverside, 900 University Ave., Riverside CA 92521 USA

E-mail: Kenneth.Barish@ucr.edu

Abstract. The PHENIX experiment is designing a forward upgrade, fsPHENIX, to accompany the central sPHENIX upgrade. fsPHENIX will include the addition of a spectrometer in the forward direction capable of measuring electrons, photons, and hadrons. The motivation for the forward spectrometer includes the measurement and separation of the Sivers and Collins effects via single transverse spin asymmetries in the Drell-Yan process and jet observables with the use of polarized proton collisions, the study of cold nuclear matter effects in proton-nucleus collisions including the calibration of quarkonium, and A+A measurements aimed at giving a 3D "image" of the medium via flow measurements and an understanding of the system expansion via photon measurements.

1. Introduction

The PHENIX experiment is proposing to replace the current PHENIX spectrometer at mid-rapidity with sPHENIX [1], which also frees up space to allow for the addition of a spectrometer, fsPHENIX, in the forward direction capable of measuring electrons, photons, and hadrons. Such a forward detector ($1<|\eta|<4$) is being designed for: 1) the study of forward transverse asymmetries in polarized p+p collisions with the aim of separating the Sivers and Collins effects and to understand factorization and universality of transverse momentum dependent parton distribution functions (TMDs), 2) cold nuclear matter measurements in p+A collisions to provide a calibration of quarkonium (J/ψ and Ψ families) and information about the initial state of heavy ion collisions, and 3) A+A collisions to provide a 3D "image" of the medium and to understand the system expansion via photons.

2. Forward Transverse Asymmetries (polarized p+p collisions)

Over the past 10 years, PHENIX and RHIC spin have extracted the gluon helicity distribution, $\Delta g(x)$, in the proton through the measurement of longitudinal double spin asymmetries in polarized proton collisions. While these measurements have been a great success, they have also revealed that taking only the longitudinal snapshot of the proton does not allow us to fully account for the proton's spin. In particular, the transverse spin structure is needed to access the orbital angular momentum. In the end, the transverse structure of the nucleus will be needed to complete our understanding of the proton.

Transverse single spin asymmetries emerge as left-right asymmetries in scattering processes with one of the colliding particles momentum perpendicular to the polarization vector: $A_N=(d\sigma_\uparrow - d\sigma_\downarrow)/(d\sigma_\uparrow + d\sigma_\downarrow)$. They were originally expected to be very small at RHIC energies, as in collinear QCD A_N should scale like $A_N \sim m_q \alpha_s / p_T$. However they were observed in pion production in various experiments ranging in center-of-mass energies from $\sqrt{s}$=4.9 GeV and 19.6 GeV and have persisted all the way to √s=500 GeV in forward meson production. These observations have led to extensions of the pQCD description which introduce chiral-odd objects into the process, thereby allowing the necessary spin flip to create a transverse asymmetry. A consistent theoretical framework has been developed to describe transverse momentum dependent observables in hard scattering processes. At large transverse momenta the Qiu-Sterman mechanism describes Transverse Momentum Dependent (TMD) observables as higher-twist effects (twist-3 quark-gluon and gluon-gluon correlators) [2]. So far, the predicted fall-off with p_T has not been observed. At smaller transverse momentum, corresponding to the intrinsic transverse momentum of partons in hadrons, the Sivers [3] and Collins [4] effects can generate transverse spin asymmetries. The Sivers mechanism is an initial-state effect due to the correlation between the proton spin and the intrinsic transverse quark momentum. It requires orbital angular momentum of the quarks and provides access to non-collinear PDFs. The Collins effect is a final-state effect due to the correlation between the proton and quark spin (quark transverse spin distribution or transversity) and a spin dependent fragmentation function (the Collins fragmentation function). These mechanisms lead to a consistent picture for single transverse asymmetries in hard processing as the calculations agree in the region of overlap [5]. Phenomenological analyses of both effects exclusively have been shown to reproduce the data in the observed kinematic range. Comparison between experiments (polarized p+p and semi-inclusive DIS measurements) relies on assumed functional forms of the transverse momentum dependent distribution functions. This is one of the topics of active debate in combination with universality of these functions and the formalism of factorization of initial and final state interactions with the hard scattering pQCD process. Collinear factorization for hadron-hadron scattering is well established and universality of the parton distributions is justified. There is less experimental data for polarized case, but the data is supportive and most theoretical foundations are common. This is the foundation for ΔG and $\Delta\bar{q}$ programs. Going beyond the twist-2 collinearly factorized picture is essential to explore QCD dynamics and fully understand the spin structure of the nucleon. Exploring the validity of factorization and universality of transverse momentum dependent (TMD) parton distributions factorization is key.

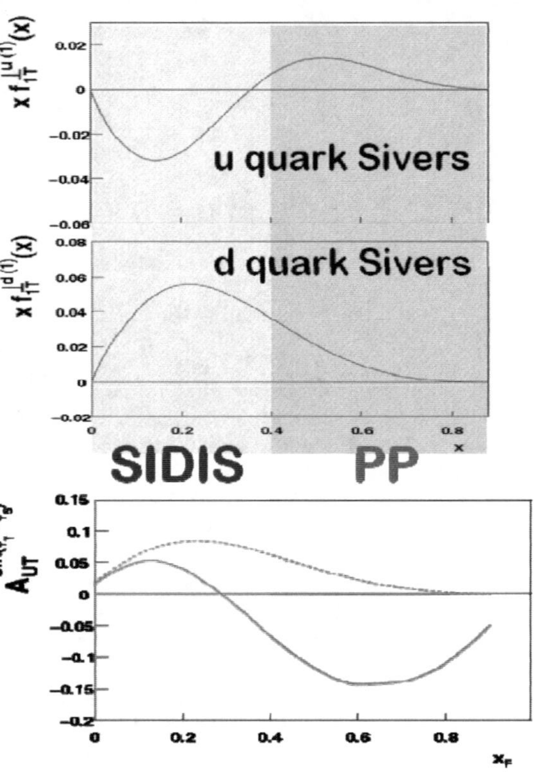

Figure 1. Sivers Drell-Yan asymmetry as extracted by [6] in a "Global" analysis. The world data poorly constrain the function for x_F>0.4. One of them main motivations for forward sPHENIX is to map out the function in this region.

An attempt has been made to extract the Sivers function through a global fit of the A_N data [6]. The results are shown in Figure 1, and it is clear it is effectively unconstrained for x_F>0.4, i.e. the region accessible by PHENIX. In order make progress, it is essential to map out Sivers function over a wide kinematic range. Drell-Yan is a particularly attractive channel as it involves no parton fragmentation and is a direct correlation of intrinsic transverse quark properties and proton spin. Coupled with a solid theoretical understanding of factorization, it provides a fundamental test of our understanding of *QCD*. In √s=500 GeV, we expect the largest asymmetry of about 8% at rapidities around 3 to 4, see Figure 2 [7]. Figure 3 shows the results for the background reduction for different acceptances of the forward spectrometer in the relevant invariant mass range. Our studies so far include only very basic assumptions of detector performances. These initial feasibility studies show the QCD background decreases with increasing rapidity and the measurement becomes possible. Measurement of jet asymmetries in the forward direction is also sensitive to the quark-gluon correlation function, providing an independent measurement of the Sivers function (Figure 4) [8]. The Collins effect manifests itself through an azimuthal anisotropy in the distribution of hadrons in final state jets with respect to the proton spin component normal to the scattering plane. Large asymmetries are expected in the forward direction (Figure 5) [9]. A measurement of the left-right asymmetry of identified particles inside a jet would then be directly sensitive to the Collins fragmentation function.

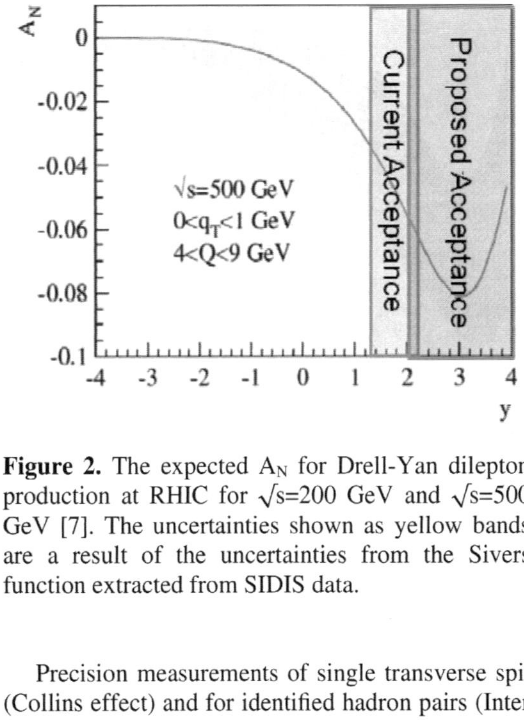

Figure 2. The expected A_N for Drell-Yan dilepton production at RHIC for $\sqrt{s}$=200 GeV and $\sqrt{s}$=500 GeV [7]. The uncertainties shown as yellow bands are a result of the uncertainties from the Sivers function extracted from SIDIS data.

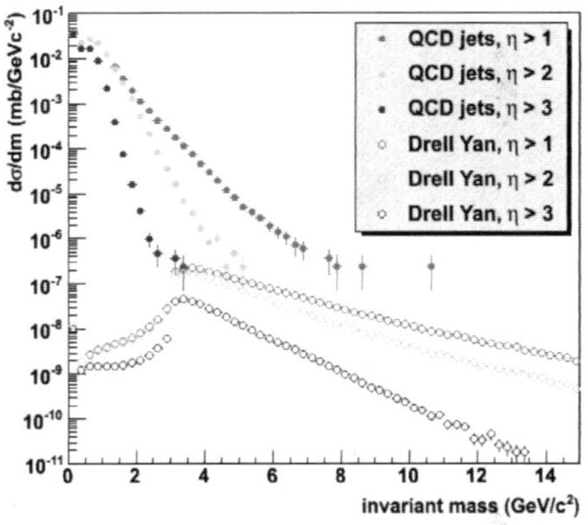

Figure 3. Event generator studies of Drell-Yan versus hard QCD jet background for different detector acceptances in the forward direction as a function of invariant mass. The dip below 3 GeV/c^2 is a result of the settings of the PYTHIA event generator.

Precision measurements of single transverse spin asymmetries for single identified charged pions (Collins effect) and for identified hadron pairs (Interference Fragmentation Function, IFF) will lead to the first measurement of quark transversity distributions at large x > 0.35. A comparison of quark transversity extracted from TMD Collins observables and from collinear IFF asymmetries will provide a powerful test of the TMD framework and evolution.

In order to disentangle these effects and determine their respective contributions, new measurements are needed over a wide kinematic range, preferentially at higher transverse momenta p_T extending to large x_F. Also, it is of high importance to make use of different probes, i.e., measure asymmetries of different particles or correlations of particles. The Drell-Yan measurements require charge sign determination and e/π separation, while the jet measurements require good jet reconstruction (electromagnetic and hadronic calorimetry), particle ID as the Collins effect different for different hadrons (RICH), and a magnetic field and tracker to determine the charge sign of the hadrons.

The forward sPHENIX upgrade, then, is designed to measure jet correlations/structure and Drell-Yan. The measurements will allow the separation of Sivers & Collins and test TMD parton distribution factorization and universality. The ultimate goal is that the measurements lead to a complete understanding of the theoretical framework so that the theory can make progress in providing a quantitative link between the Sivers function and orbital angular momentum.

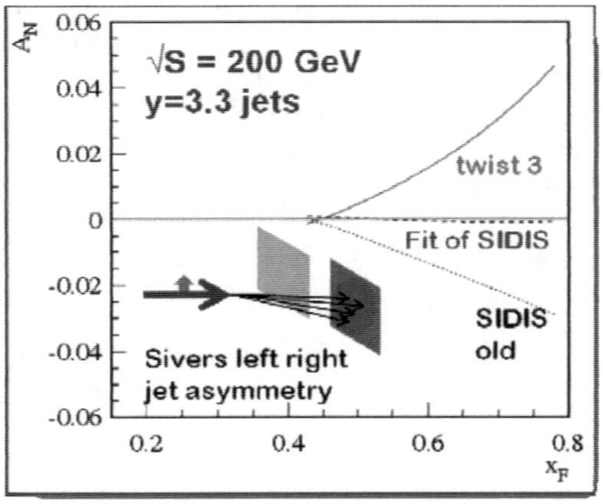

Figure 4. The expected Sivers A_N for jets [8].

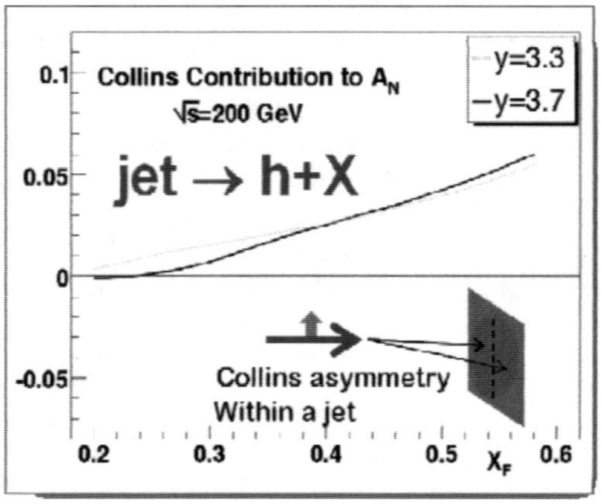

Figure 5. Expected Collins A_N effect in jets [9].

3. Cold Nuclear Matter (p +A Collisions)

It is a central goal of the sPHENIX forward upgrade to systematically survey the cold nuclear matter effects in high energy nuclear collisions. The physics motivations include the calibration of quarkonium (J/ψ and ϒ families) and precise knowledge of the initial state of heavy-ion collisions. RHIC can make unique contributions with its √s dependence, the possibility of using different species, and access to low Q^2. The design is driven by the need for large data samples, access to 1>η>4, and detectors sensitive to electrons, photons, charged hadrons, and jets.

Quarkonimum measurements in heavy-ion collisions are directly related to the temperature of the system. However, currently we are limited by our understanding of quarkonium. In order to fully calibrate quarkonium, there are a number of topics that must be quantitatively understood including parton initial energy loss, quark PDFs, gluon PDFs, and the absorption or breakup cross-sections. Measurements of Drell Yan vs √s, γ-jet, open heavy flavor, and quarkonium in p+A collisions is sensitive to the various effects in different ways (Figure 6). Using a redundant set of measurements will allow the isolation of the necessary components. The sPHENIX forward upgrade will significantly extend the kinematic reach and channels of measurement.

A critical part of gaining a quantitative insight into the properties of the sQGP is developing a complete understanding of the initial state - the distribution of partons, primarily gluons, in the colliding Au nuclei. Compelling theoretical arguments have been made that the distribution of gluons in a heavy nucleus "saturates" at small x, the so-called "Color Glass Condensate" (CGC), suppressing very low-x gluons and pushing them to higher-x. This is a calculable regime of gluons at high density but weak coupling. The effect is enhanced by large nuclei which provide an $A^{1/3}$ amplification (relative to p+p) since there are simply more partons available. For the same reason the effect is also sensitive to the centrality of the collision. Figure 7 shows a schematic illustration of the region of x and Q where this saturation should take place. The exact location of the line in Figure 7 between pQCD dominated physics and CGC physics is not well known. It is known, however, that within the framework of the CGC, this line will move depending on the centrality of the collision as indicated by the arrow. The saturation assumed by these authors is present in the initial state before the nuclei collide. Hence, the study of proton-nucleus collisions will be important in distinguishing these effects from final state effects such as the formation of a quark-gluon plasma. Moving forward in rapidity, to regions covered by forward sPHENIX, therefore, may provide measurements within the CGC region. The CGC predicts a suppression of the away side jet at low-x for forward η, and the effect should become larger as the collisions become more central.

	Parton initial energy loss	Quark pdf	Gluon pdf	Absorbtion or breakup cross section	recombination
DY vs √s	✓ (quark)	✓	✓		
γ-jet	✓ (quark)	d+A	✓		
open heavy	✓ (gluon)		✓	✓	
Quarkonia	✓ (gluon)		✓	✓	
A+A quarkonia centrality					✓
A+A open heavy					✓

Figure 6. Sensitivity of channels to make quarkonium calibration.

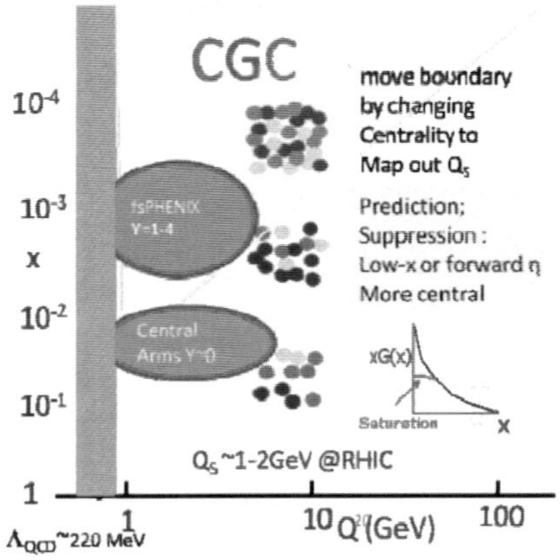

Figure 7. A schematic diagram of regions in the nucleus, showing the CGC region bordered by a line representing Q_S.

There may be an intriguing connection between the CGC and TMDs. TMD factorization is violated for dijet production in hadron hadron collisions. However, it is possible to get back effective TMD factorization in the case of small x partons at high density (i.e, the "CGC regime"), which is probed by quarks or photons. While TMD parton distributions are not universal, they can be constructed from building blocks which are. In particular, they can be constructed by the unintegrated Gluon PDFs $G^{(1)}(x,q_\perp)$ and $G^{(2)}(x,q_\perp)$. Interestingly, the quantities derived via the CGC and TMD approaches are identical. There is, therefore, an equivalence between the TMD and CGC approaches in the CGC regime [10]. In order to explore this connection, we need to measure photon-jet and dijets at low-x in p or p+A collisions.

4. Heavy-ions (A+A Collisions)

The forward spectrometer give the opportunity to take a 3D "image" of the medium. At mid-rapidity we only see the evolution/final state of the "slice" of the fluid which is initially at rest longitudinally. Flow measurements in the forward direction can help complete the picture (Figure 8) [11].

Measurement of photons in the forward direction also provides information on the system expansion/early evolution and gives access to the high baryon density region (Figure 9) [12].

Journal of Physics: Conference Series **389** (2012) 012033 doi:10.1088/1742-6596/389/1/012033

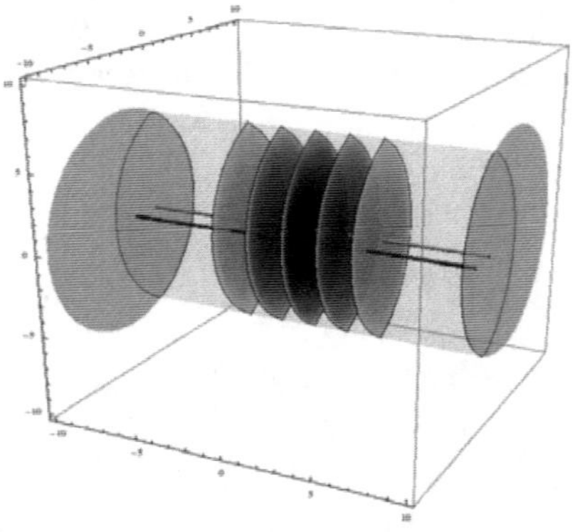

Figure 8. 3D "image of the medium [11],

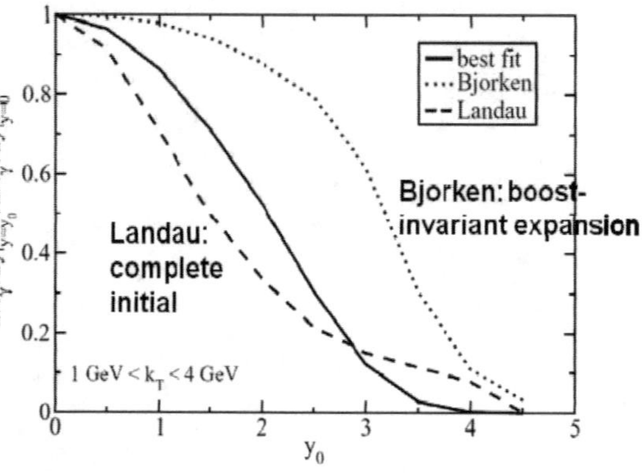

Figure 9. Forward photon measurements is sensitive to system evolution [12].

5. Detector Considerations

Precision Drell-Yan measurements require excellent dilepton identification and the ability to reduce backgrounds from correlated charm and beauty decays. The forward sPHENIX Drell-Yan measurements will be via dielectrons, and will require an electromagnetic calorimeter and charged particle tracking as well as heavy flavor tagging for background rejection.

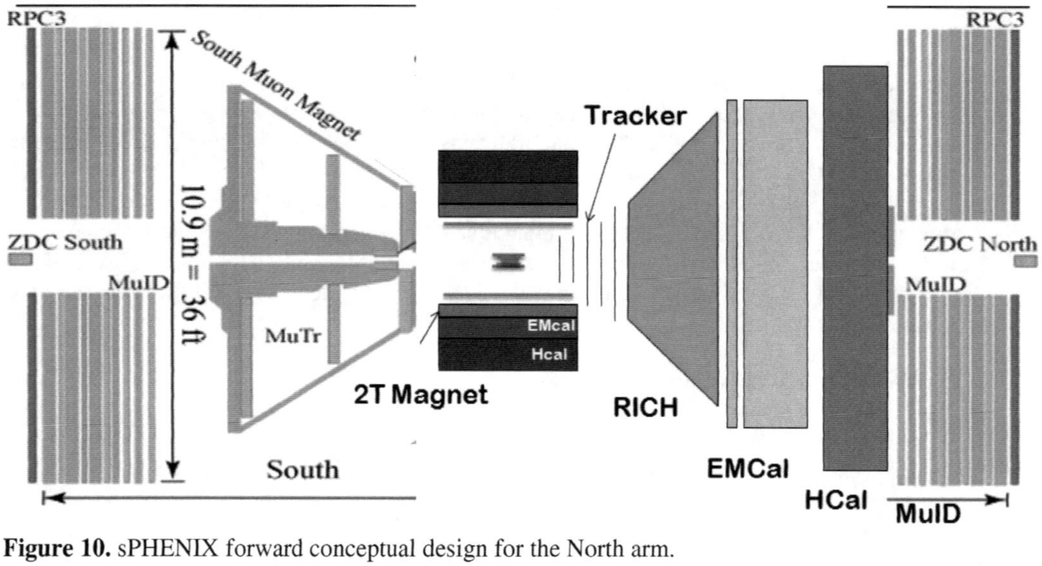

Figure 10. sPHENIX forward conceptual design for the North arm.

The forward sPHENIX detector will greatly extend our ability to measure the Collins and Sivers asymmetries at forward rapidities by performing full jet reconstruction. Experimentally, the measurement of transverse spin effects within jets will require electromagnetic and hadronic calorimetry for jet reconstruction, particle tracking to determine the fractional momentum of hadrons or hadron pairs in the jet and particle ID to avoid cancelling of transverse spin effects in the fragmentation of different hadrons. Precise studies of the Sivers and Collins effects in different channels with the sPHENIX forward upgrade will make it possible to decompose the large transverse single spin asymmetries that historically have been observed in polarized proton-proton collisions and to identify and quantify the contributions from different spin effects.

The fsPHENIX upgrade is currently in a conceptual phase. A straw-man conceptual design is shown in Figure 10. The exact detector performance design requirements for these physics channels are being determined. The currently envisioned sPHENIX forward detector will have an acceptance from a pseudorapidity of $1.2 < \eta < 4$. Particle identification is based on a Ring Imaging Cherenkov Detector. It believed that a hadronic calorimeter with a modest energy resolution will energy smearing for Collins measurements in jets within acceptable limits. The forward electromagnetic calorimeter may consist of a re-stack of the current PHENIX electromagnetic calorimeters (EMCal) and the MPC-EX towers. Early simulation results indicate that the performance of the EMCal will be sufficient.

6. ePHENIX

A subsequent upgrade adding an electron detector in the opposite direction would further evolve sPHENIX into a detector for inclusive, semi-inclusive and exclusive processes in deep inelastic electron-proton and electron-nucleus scattering, referred to as ePHENIX, utilizing a future high intensity electron beam at RHIC.

7. Outlook

sPHENIX Forward will significantly extend physics capabilities. The additional of the new forward detector will enable us to understand large SSA, separating contributions from the Sivers effect and Collins function. The forward detector will also provide calibration for quarkonium measurements and probe CNM effects in d+A (connections with CGC and TMDs?). Potential also exists to explore 3D image of medium in Au+Au collisions.

A staged implementation approach is planned. The first stage will include only the EMCal, charged particle ID, and charge sign. This is sufficient for the Drell-Yan and Quarkonia measurements. Then add jet followed by identified hadron capabilities. The sPHENIX forward spectrometer would also be well matched with ePHENIX.

References

[1] Hanks A "sHENIX to Study Quark Gluon Plasma Properties" these proceedings.; Aidala C *et al.* "sPHENIX: An Upgrade Concept from the PHENIX Collaboration" arXiv:1207.6378
[2] Qiu J and Sterman G, 1999 *Phys.Rev. D* **59** 014004
[3] D. Sivers, 1990 *Phys.Rev.* D **41** 83
[4] Collins J C, Nucl. 1993 *Nucl.Phys. B* **396** 161
[5] Ji X, Qiu J-W, Vogelsang W, Yuan F, 2006 *Phys.Rev.Lett.* **97** 082002
[6] Prokudin A, Kang Z-B, 2012 *Phys Rev D* **85** 074008
[7] Kang Z and Qiu J. 2010 *Phys.Rev. D* **81** 054020
[8] Kang Z et al. 2011 *Phys.Rev. D* **83** 094001
[9] Yuan F, 2008 *Phys.Rev. B* **666** 44
[10] Dominques F, et al. 2011 *Phys.Rev.* D **83** 105005
[11] Stankus P, "Recent Flow Results from PHENIX", these proceedings.
[12] Renk T, 2005 *Phys.Rev. C* **71**, 064905

Journal of Physics: Conference Series **389** (2012) 012034 doi:10.1088/1742-6596/389/1/012034

Single transverse-spin asymmetry of very forward neutron production in polarized $p + p$ collisions at PHENIX

Yuji Goto (for the PHENIX collaboration)

RIKEN Nishina Center for Accelerator-Based Science, Wako, Saitama 351-0198, Japan
RIKEN BNL Research Center, Brookhaven National Laboratory, Upton, New York 11973-5000, USA

E-mail: goto@bnl.gov

Abstract. The cross section and x_F dependence of A_N of very forward neutron production in polarized $p + p$ collisions at $\sqrt{s} = 200$ GeV were measured in the PHENIX experiment at RHIC. The measured cross sections were consistent with the x_F scaling claimed by the ISR experiment which measured neutron production in $p + p$ collisions at $\sqrt{s}$ from 30.6 to 62.7 GeV. These cross sections for large x_F neutron production, as well as those in $e + p$ collisions at HERA, are mainly described by one-pion exchange. At PHENIX, significant negative A_N was observed in the forward region, and no significant backward A_N was observed. We also measured $\sqrt{s}$ and p_T dependence of A_N of very forward neutron production at $\sqrt{s}$ from 62.4 to 500 GeV. The observed large asymmetry of neutron production is considered to come from the interference between the spin-flip amplitude of the pion exchange and spin-non-flip amplitude of all Reggeon exchange. Therefore, the neutron asymmetry has a sensitivity to other Reggeon exchange amplitude even if it is a small amplitude.

1. Introduction

The origin of the nucleon spin 1/2 has been investigated with polarized proton collisions in the PHENIX experiment at RHIC. Polarized DIS experiments found that the origin of the nucleon spin was not explained by the quark spin. First goal of the RHIC spin project is a measurement of the gluon spin contribution to the nucleon spin. The helicity structure of the nucleon is measured by the double-helicity asymmetries (A_{LL}) in longitudinally-polarized proton collisions.

For A_{LL} measurements, a good local polarimeter at the interaction point (IP) is necessary for systematic understanding of the beam polarization. At RHIC, polarized protons are stored with transverse polarization, and it is monitored by the proton-carbon CNI (Coulomb-nuclear interference) polarimeter [1] and polarized atomic hydrogen gas-jet polarimeter [2] located around the IP12 area. For the longitudinally-polarized collisions in the collider experiments at PHENIX (IP8) and STAR (IP6), spin-rotator magnets rotate the proton polarization into the longitudinal direction. The longitudinal polarization is monitored at each IP by the local polarimeter by using physics processes with single transverse-spin asymmetry (A_N).

Before beginning the longitudinal-spin polarized proton collisions, we needed to find a good physics process to monitor the polarization at PHENIX. One candidate process was a forward π^0 asymmetry which was found at Fermilab-E704 experiment. At PHENIX, there was a space in the very forward area (zero-degree area) to locate a polarimeter. Since the π^0 asymmetry in the very forward region had not been measured at Fermilab-E704, the very forward π^0 asymmetry was measured at RHIC. We installed an EM calorimeter in the very forward region on one side of the IP12 in 2001-2002 and measured the asymmetry of photons mainly from π^0 decay. In this experiment, we found that the photon asymmetry, or π^0 asymmetry in the very forward region was not large enough for the polarimeter, but a large asymmetry of very forward neutron production [3], $A_N = -0.090\pm0.006(stat)\pm0.009(syst)$, was discovered. The neutron asymmetry was confirmed with a hadron calorimeter which was installed on the other side of the IP12.

The discovery of the large asymmetry of neutron production is also a new important information to understand the production mechanism of very forward neutron production. Cross section of very forward neutron production had been measured at ISR and Fermilab [4, 5]. They measured a forward peak in the x_F distribution around $x_F = 0.8$ and found only a small $\sqrt{s}$ dependence. One-Pion Exchange (OPE) model gave a reasonable description on the measurements. In OPE models [6, 7, 8, 9, 10, 11, 12], the incoming proton emits a pion which scatters on the other proton as described in Figure 1.

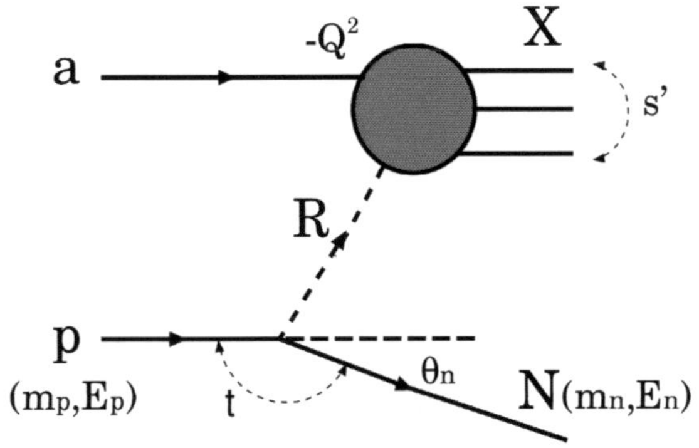

Figure 1. A schematic diagram of the neutron production, $p + a \to n + X$, on the Reggeon exchange model shown with Lorentz invariant variables s', Q^2 and t. "a" is a proton or electron/positron for $p + p$ or $e + p$ reactions. R indicates a Regge trajectory with isospin odd such as π, ρ, a_2 and Pomeron-π in the Regge theory. In case of the pion exchange, R means π.

Very forward neutron cross section was measured at HERA in $e + p$ collisions with a high resolution p_T distribution [13, 14]. They found a suppression of the forward peak. CERN-NA49 also measured very forward neutron cross section in $p + p$ reaction [15]. Their cross section was twice large as those measured at ISR and Fermilab, and consistent with that at HERA. In order to understand the production mechanism of the forward neutron, more data are necessary, and the asymmetry measurement gives a new information.

2. Experimental setup

At PHENIX, we measured the cross section and A_N of very forward neutron production in polarized $p + p$ collisions with the local polarimeter. The PHENIX local polarimeter was made with an existing ZDC (Zero-Degree Calorimeter) by adding a position-sensitive SMD (Shower Maximum Detector) [16]. A plan view of the experimental setup for very forward neutron measurement at PHENIX is shown in Figure 2.

The ZDC is composed of Cu-W alloy absorbers with PMMA-based communication grade optical fibers and it corresponds to 1.7 nuclear interaction length for one module. A

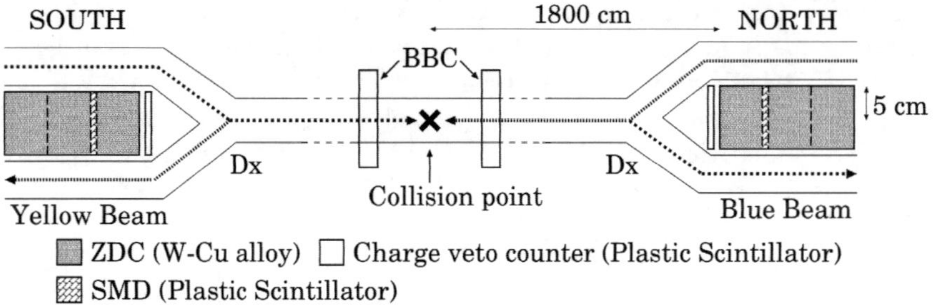

Figure 2. A plan view of the experimental setup at PHENIX (not to scale). Shown are the principle components for the neutron physics.

photomultiplier collects Cherenkov lights via the optical fibers in one module. Three ZDCs are located in series at ±1800 cm away from the collision point within the small acceptance, covering 10 cm in the transverse plane. It has 5.1 nuclear interaction length and 149 radiation length by combining three ZDC modules, and has energy resolution about 20% for 100 GeV neutron.

The SMD consists of x-y scintillator strip hodoscopes and they are inserted between the first and second ZDC modules at the position of maximum hadronic shower approximately. The x-coordinate (horizontal) is sampled by 7 scintillator strips of 15 mm width, while the y-coordinate (vertical) is sampled by 8 strips of 20 mm width, tilted by 45 degree. It gives a position by calculating the center of gravity of shower generated in the first ZDC module. Its position resolution is estimated to be about 1 cm for 50 GeV neutron in the simulation study.

Detectors are located downstream of the RHIC-DX dipole magnet so that charged particles from collisions are expected to be swept out. A forward scintillation counter, covering 10 cm × 12 cm, has been installed in front of the ZDC to remove charged-particle backgrounds from other sources.

As a beam luminosity monitor, Beam Beam Counters (BBCs) are used. The BBC consists of 64 sets of PMTs and 3 cm thick quartz Cherenkov radiator. BBCs are mounted around beam pipe located ±144 cm away from the collision point. They cover ±(3.0–3.9) and 2π in η and ϕ spaces, respectively.

The data was collected by two sets of triggers for the neutron measurement. One was the ZDC trigger for neutron inclusive measurements by requiring energy deposit in either side of the ZDC (the north side or the south side) above 5 GeV. The other was the ZDC⊗BBC trigger, a coincidence trigger of the ZDC trigger with BBC hits which were defined as one or more charged particles in both sides of BBCs.

3. Results

3.1. Measurements at $\sqrt{s}$ = 200 GeV

The cross section in the integrated p_T region, $0 < p_T < 0.11 \times x_F$ GeV/c was measured by assuming p_T distribution from the ISR result. The result is shown in Figure 3. We couldn't measure p_T distribution due to the limited acceptance and position resolution. Our cross section result showed consistency with the ISR data. x_F scaling was satisfied in higher center of mass energy.

The azimuthal modulation of very forward neutron production is shown in Figure 4. The single transverse-spin asymmetry with the ZDC trigger was $A_N = -0.066 \pm 0.006(stat)$, and that with the ZDC⊗BBC trigger was $A_N = -0.083 \pm 0.004(stat)$. The result was consistent with the IP12 measurement with BBC hits in the trigger requirement within the errors.

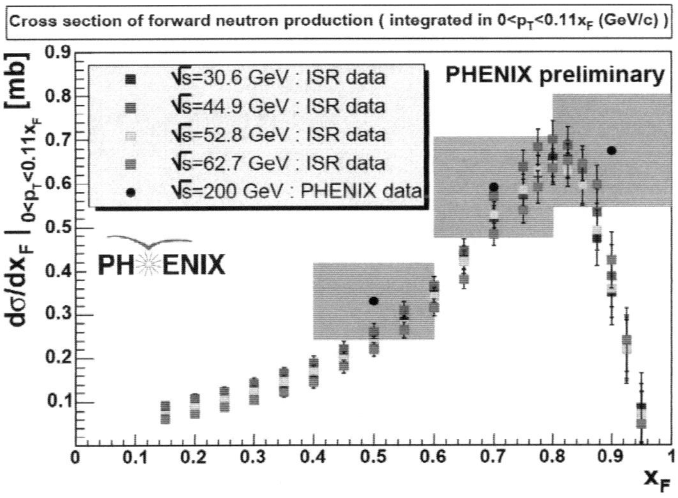

Figure 3. Cross section of very forward neutron production from $p+p$ collisions at $\sqrt{s} = 200$ GeV (circles). Squares show the ISR results of various center of mass energies.

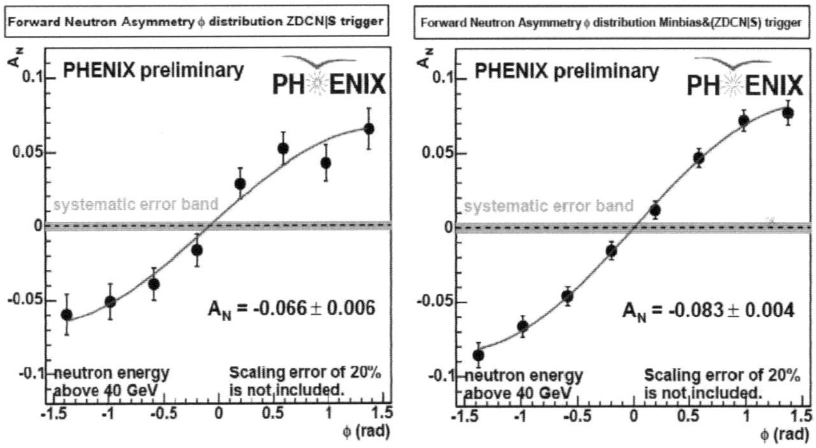

Figure 4. The azimuthal modulation of very forward neutron production with the ZDC trigger (left) and that with the ZDC⊗BBC trigger (right).

The x_F dependence of A_N of very forward neutron production is shown in Figure 5. A significant negative A_N was seen in the positive x_F region and there was no energy dependence within the errors in both trigger sets. No significant backward neutron asymmetry was observed.

3.2. $\sqrt{s}$ *dependence*

We also studied $\sqrt{s}$ dependence of A_N from three different collision energies, 62.4 GeV, 200 GeV, and 500 GeV. The result is shown in Figure 6. At higher collision energy, the absolute value of the asymmetry was higher. The hit position dependence on the detector, or the polar angle dependence of the neutron direction was measured at each energy, although it was largely smeared by the position resolution. The result was converted to the p_T dependence. It showed

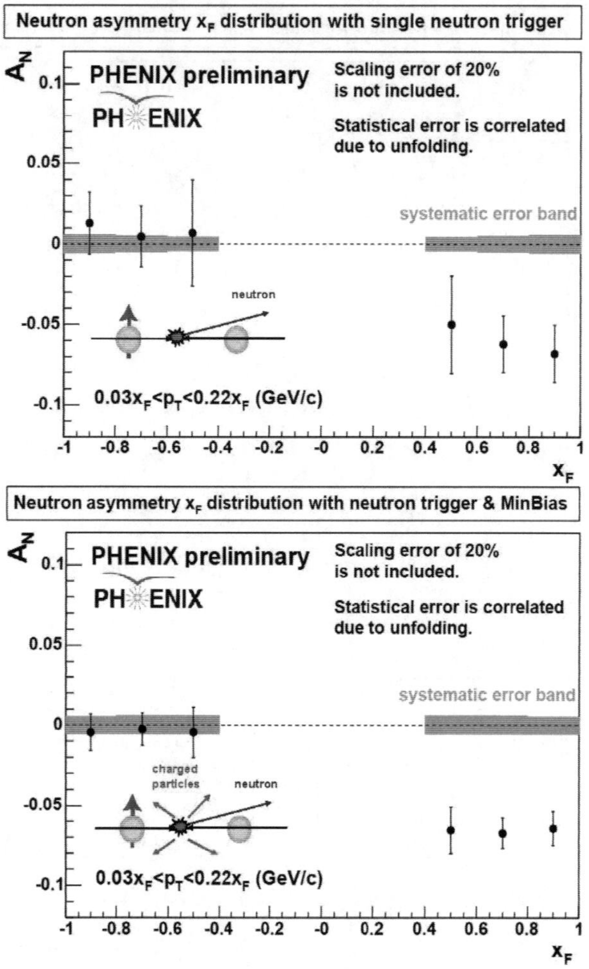

Figure 5. The x_F dependence of very forward neutron production at $\sqrt{s} = 200$ GeV with the ZDC trigger (top) and that with the ZDC⊗BBC trigger (bottom).

a hint of p_T scaling property of A_N of the very forward neutron production.

4. Discussion

The asymmetry is caused by interference between spin-flip and non-flip amplitudes with a relative phase. Kopeliovich et al. [17] studied the asymmetry from only pion exchange. Because spin-flip amplitude and non-flip amplitude have the same phase, absorption correction was considered for a relative phase. The absorption correction was also important for cross section calculation, but the gained shift between spin-flip and non-flip amplitude was too small to explain the large asymmetry. Next Kopeliovich et al. [18] studied interference with other Reggeons. They considered interference of pion and a_1, or pion and ρ in the 1^+S state. As a result, the data agreed well with independence of energy. The asymmetry has a sensitivity to the presence of different mechanisms, e.g. Reggeon exchange with spin-non-flip amplitude even if they are small amplitudes.

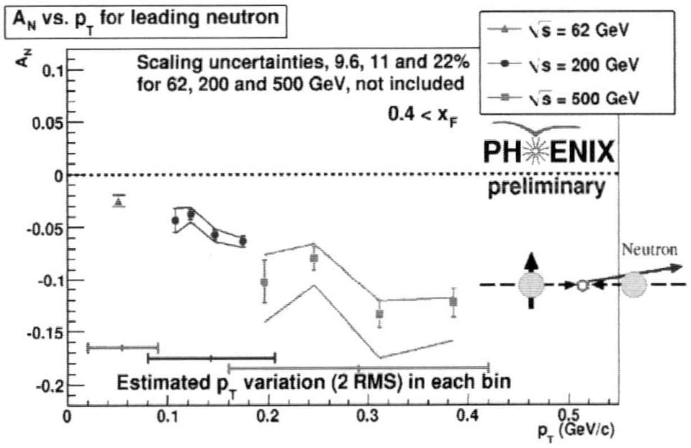

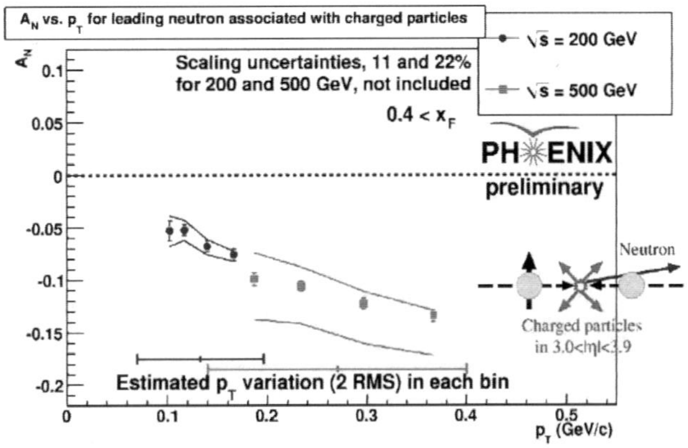

Figure 6. The measured asymmetries of very forward neutron production as functions of p_T with the ZDC trigger (top) and that with the ZDC$\otimes$BBC trigger (bottom).

5. Summary

We measured the x_F dependence of A_N and cross section of neutron production in very forward kinematics with the PHENIX detector at $\sqrt{s}$=200 GeV. Significant negative A_N was observed in the forward region, and no significant backward A_N was observed. The observed asymmetries were consistent with the previous result at RHIC-IP12 [3] within the error. Asymmetry showed almost no x_F dependence. The measured cross section at $\sqrt{s}$=200 GeV has no violation of x_F scaling with the results from the ISR experiment. These cross sections for large x_F neutron production, as well as those in $e + p$ collisions at HERA, are mainly reproduced in one pion exchange model.

We also measured $\sqrt{s}$ and p_T dependence of A_N of very forward neutron production at $\sqrt{s}$ from 62.4 to 500 GeV. The observed large asymmetry of neutron production is considered to come from the interference between the spin-flip amplitude of the pion exchange and spin-non-flip amplitude of all Reggeon exchange. Therefore, the neutron asymmetry has a sensitivity to other Reggeon exchange amplitude even if it is a small amplitude.

Journal of Physics: Conference Series **389** (2012) 012034 doi:10.1088/1742-6596/389/1/012034

References

[1] O. Jinnouchi *et al.*, RHIC/CAD Accelerator Physics Note 171, 2004.
[2] H. Okada *et al.*, Phys. Lett. B **638**, 450 (2006).
[3] Y. Fukao *et al.*, Phys. Lett. **B650**, 325-330 (2007). [hep-ex/0610030].
[4] J. Engler *et al.*, Nucl. Phys. **B84**, 70 (1975).
[5] W. Flauger, F. Monnig, Nucl. Phys. **B109**, 347 (1976).
[6] A. Capella, J. Tran Thanh Van, J. Kaplan, Nucl. Phys. **B97**, 493 (1975).
[7] B. Kopeliovich, B. Povh, I. Potashnikova, Z. Phys. **C73**, 125-131 (1996). [hep-ph/9601291].
[8] N. N. Nikolaev, J. Speth, B. G. Zakharov, [hep-ph/9708290].
[9] N. N. Nikolaev, W. Schafer, A. Szczurek, J. Speth, Phys. Rev. **D60**, 014004 (1999). [hep-ph/9812266].
[10] U. D'Alesio, H. J. Pirner, Eur. Phys. J. **A7**, 109-119 (2000). [hep-ph/9806321].
[11] A. B. Kaidalov, V. A. Khoze, A. D. Martin, M. G. Ryskin, Eur. Phys. J. **C47**, 385-393 (2006). [hep-ph/0602215].
[12] A. Bunyatyan, B. Povh, Eur. Phys. J. **A27**, 359-364 (2006). [hep-ph/0603235].
[13] S. Chekanov *et al.* [ZEUS Collaboration], Nucl. Phys. **B776**, 1-37 (2007). [hep-ex/0702028].
[14] F. D. Aaron *et al.* [H1 Collaboration], Eur. Phys. J. **C68**, 381-399 (2010). [arXiv:1001.0532 [hep-ex]].
[15] T. Anticic *et al.* [NA49 Collaboration], Eur. Phys. J. **C65**, 9-63 (2010). [arXiv:0904.2708 [hep-ex]].
[16] C. Adler, A. Denisov, E. Garcia, M. J. Murray, H. Strobele, S. N. White, Nucl. Instrum. Meth. **A470**, 488-499 (2001). [nucl-ex/0008005].
[17] B. Z. Kopeliovich, I. K. Potashnikova, I. Schmidt and J. Soffer, AIP Conf. Proc. **1056**, 199 (2008) [arXiv:0807.1449 [hep-ph]].
[18] B. Z. Kopeliovich, I. K. Potashnikova, I. Schmidt, J. Soffer, [arXiv:1109.2500 [hep-ph]].

Journal of Physics: Conference Series **389** (2012) 012035 doi:10.1088/1742-6596/389/1/012035

Charge asymmetry dependency of π^+/π^- elliptic flow in Au + Au collisions at $\sqrt{s_{NN}} = 200$ GeV

Hongwei Ke[1,2] **(for the STAR Collaboration)**

1. Institute of Particle Physics, Central China Normal University, Wuhan 430079, China
2. Brookhaven National Laboratory, Upton, NY 11973, USA

E-mail: `kehw@rcf.rhic.bnl.gov`

Abstract. In this proceedings, we present STAR's measurement of v_2 difference between positively charged pions and negatively charged pions at low transverse momentum for Au + Au collisions at $\sqrt{s_{NN}} = 200$ GeV. The integrated v_2 of π^+ (π^-), $0.15 < p_T < 0.5$ GeV/c, decreases (increases) linearly with the increasing charge asymmetry $A_\pm$. The v_2 difference between π^+ and π^- is proportional to $A_\pm$ and the slope parameters have their order of magnitude of about 0.01. All of these observed features are consistent with the predication based on Chiral Magnetic Wave. The centrality dependence of the v_2 difference is different than that was predicted.

1. Introduction

In relativistic heavy ion collisions, a significant fraction of incident energy is deposited in the collision region. Due to the existence of the large energy density, a phase of extremely hot and dense matter consisting of quarks and gluons, the Quark Gluon Plasma (QGP) [1, 2, 3, 4], is expected to be created. On the other hand, in non-central collisions, spectators keep moving along the incident directions. Since spectators are charged objects, moving spectators will generate electric current which induces a magnetic field perpendicular to the reaction plane (defined by the line connecting the colliding nuclei and the incident direction). It has been estimated that the magnetic field created in heavy ion collisions can be as strong as up to $\sim 10^{15}$ Tesla at the beginning of collisions while it decreases rapidly with time [5]. With QGP under a such strong magnetic field, it is pointed out that the chiral magnetic effect (CME) [5, 6, 7, 8, 9] and the chiral separation effect (CSE) [10, 11] are expected to exist. The CME in Quantum Chromodynamics (QCD) is that the external magnetic field induces a vector current with a finite axial chemical potential μ_A, which describes the asymmetry between the densities of left- and right-handed quarks.

$$\boldsymbol{j}_V = \frac{N_c e}{2\pi^2}\mu_A \boldsymbol{B}. \qquad (1)$$

If there exist meta-stable P and CP odd domains, space-time regions occupied by a classical field with a nonzero topological charge, under CME it would lead to the asymmetry in the emission of positively and negatively charged particles perpendicular to the reaction plane in non-central collisions. Such effect has been intensively studied by several experiments [12, 13, 14, 15]. The other effect in the discussion, CSE, is the separation of chiral charge along the axis of the external magnetic field at finite density of vector charge. The resulting axial current is given by

$$\boldsymbol{j}_A = \frac{N_c e}{2\pi^2}\mu_V \boldsymbol{B}. \tag{2}$$

The local fluctuation of axial current from Eq. (2) will induce a local fluctuation of the axial chemical potential, and thus according to Eq. (1) a fluctuation of electric current. The resulting fluctuation of electric charge density will then in turn, again, induce an axial current according to Eq. (2). With alternating CME and CSE this cycle continues, and it gives rise to a gapless excitation regarded as Chiral Magnetic Wave (CMW). Please note that the CMW does not require the local parity violation although one of its ingredients, namely the CME, is responsible for making the local parity violation signal visible in relativistic heavy ion collisions.

The CMW has a consequence that also could be observed in experiments. As pointed out in a recent theoretical work [16], a CMW in QGP will form a electrical quadrupole moment which can lead to more positive charge near the poles of the created fireball and more negative charge near the equator. That means, this configuration will cause a difference in elliptic flow, v_2 [17], between positively and negatively charged particles. Taking into account the quadrupole moment, the azimuthal angle distribution of positive and negative particles is given by Eq. (3),

$$\begin{aligned}\frac{dN_\pm}{d\phi} &= N_\pm\left[1 + 2v_2\cos(2\phi)\right] \\ &\approx \bar{N}_\pm\left[1 + 2v_2\cos(2\phi) \mp A_\pm r\cos(2\phi)\right]\end{aligned} \tag{3}$$

Here $\bar{N}_+$ ($\bar{N}_-$) is number of positive (negative) particles, $A_\pm \equiv (\bar{N}_+ - \bar{N}_-)/(\bar{N}_+ + \bar{N}_-)$ is the net charge asymmetry and $r \equiv 2q_e/\bar{\rho}_e$, while q_e represents the electrical quadrupole and $\bar{\rho}_e$ is charge density. Authors in [16] also pointed out this difference in elliptic flow should be see via π^+ and π^- due to their similar absorption cross sections in hadronic matter at finite baryon density. The elliptic flow difference between π^+ and π^- is expected to be proportional to the net charge asymmetry as described in Eq. (4).

$$\Delta v_2^{\rm CMW} \equiv v_2(\pi^-) - v_2(\pi^+) \approx rA_\pm \tag{4}$$

There are other models that also predicted different elliptic flow between particles and anti-particles. Quark transport model predicted v_2 order of particles and anti-particles according to their quark components [18]. This model is base on quark coalesce and assumes transported quarks have lager v_2 than produced quarks. AMPT model with hadronic potential also predicts different v_2 of particles and anti-particles [19]. However, as currently implemented the two models mentioned above did not provide a net charge asymmetry dependency of the v_2 difference.

In STAR experiment, the difference in integrated v_2 between particles and anti-particles has been observed. In this paper, we aim to study such difference differentially as a function of net charge asymmetry, and compare that to features predicted by the CMW.

2. Data Analysis and Results

STAR detector complex [20] has large acceptance and full azimuthal coverage, which is ideal for studying elliptic flow. In this analysis, we used $\sim$ 238M minimum-bias Au + Au events at $\sqrt{s_{NN}}$ = 200 GeV taken by STAR during year 2010. STAR Time Projection Chamber (TPC) [21] is used to reconstruct tracks and identify particles. All particles used in this analysis are required to be in the middle rapidity range $|\eta| < 1.0$. For flow analysis, pions are identified by ionization energy loss in TPC and required to have $0.15 < p_T < 0.5$ GeV/c. All charged particles within $0.15 < p_T < 12$ GeV/c are used to calculate net charge asymmetry except low p_T protons, $p_T < 0.4$ GeV/c. Those protons are dominated by protons produced from knockout/nuclear interactions of pions with inner detector material, and thus rejected. Antiprotons within the same p_T range are also excluded from the net charge asymmetry calculation in order to balance the cut on protons.

To estimate the elliptic flow of π^+ and π^-, we use Q-cumulants method, which is also called direct cumulants method [22]. Q-cumulants method estimates different order of harmonics via multi-particle correlations. This relatively new method needs only on pass over tracks and has comprehensive detector inefficiency corrections. In practice of this analysis, two-particle correlation is used to estimate $v_2\{2\}$ of pions and an η-gap of 0.3 pseudo-rapidity unit on each side is applied to suppress the short-range correlation.

With in a certain centrality bin, we calculate the net charge asymmetry $A_\pm$ event by event and divide the sample into five sub-groups so that each sub-group has roughly the same number of events. Considering finite detector acceptance and tracking inefficiency, $A_\pm$ obtained directly from final state hadrons is different than the true $A_\pm$, thus it needs to be corrected. In discussion below, we call the $A_\pm$ before the correction the observed $A_\pm$, and otherwise, the true $A_\pm$. The correction procedure on the observed $A_\pm$ will be discussed later. Fig. 1 shows the observed $A_\pm$ distribution of 30-40% most central Au + Au events at $\sqrt{s_{NN}} = 200$ GeV, in which red lines divide the statistics equally into five bins.

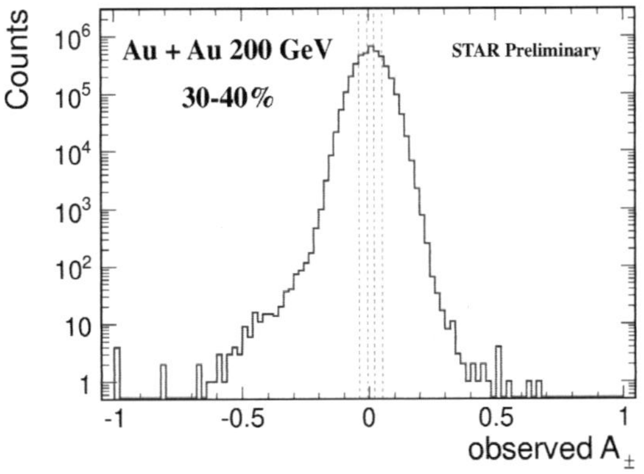

Figure 1. Observed net charge asymmetry $A_\pm$ distribution of 30-40% most central Au + Au events at $\sqrt{s_{NN}} = 200$ GeV. The vertical dashed red lines show the sub-groups of events.

We measured the integrated $v_2\{2\}$ of π^+ and π^- with $0.15 < p_T < 0.5$ GeV/c and present it as a function of observed $A_\pm$. In Fig. 2(a), v_2 of π^+ and π^- both show a linear relationship with respect to the observed $A_\pm$. With the increasing observed $A_\pm$, v_2 of π^- increases and v_2 of π^+ decreases. As a consequence, the v_2 difference between π^+ and π^-, $\Delta v_2 = v_2(\pi^-) - v_2(\pi^+)$, increases linearly with the increase of the observed $A_\pm$. All of the features mentioned here are consistent with the prediction made in [16].

The slope of the linear relationship between Δv_2 and $A_\pm$ is of special interest because it has a connection to the electrical quadrupole induced by a CMW in QGP. However, one needs to correct the observed $A_\pm$ before a meaningful slope can be extracted. Applying the correction on observed $A_\pm$ is conducted in two steps. First, we measure the tracking efficiency of charged pions in embedding data and use it as the tracking efficiency of all charged hadrons. This efficiency reflects the combined effect of detector and track reconstruction inefficiency. Secondly, we generated some Monte-Carlo events by using the event generator Hijing [23]. Net charge asymmetry in Hijing events is measured before and after applying the tracking efficiency and all

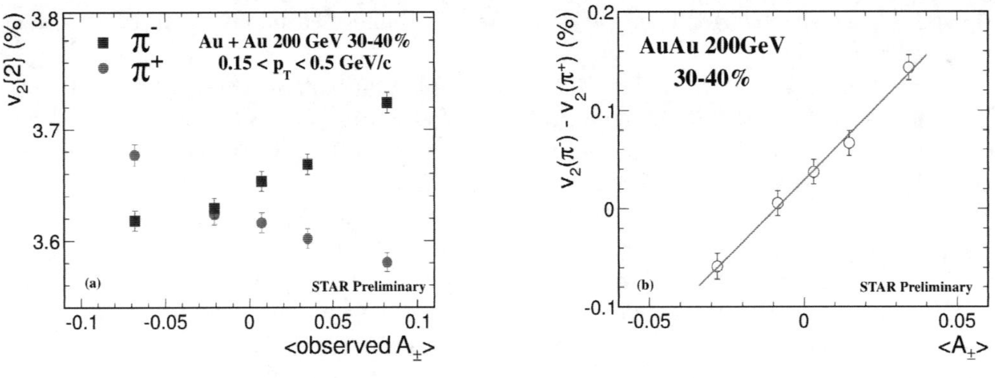

Figure 2. (a) Integrated $v_2\{2\}$ of π^+ and π^-, $0.15 < p_T < 0.5$ GeV/c, shows as a function of observed $A_\pm$, 30-40% most central Au + Au events at $\sqrt{s_{NN}} = 200$ GeV/c. (b) v_2 difference with respect to corrected net charge asymmetry $A_\pm$ fits to a straight line.

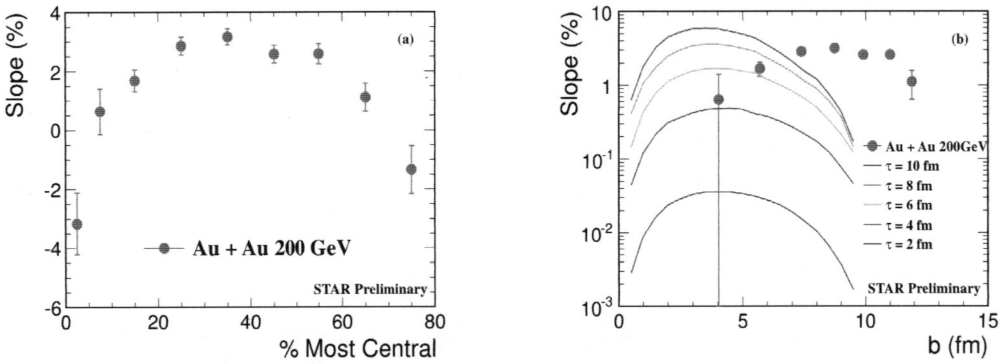

Figure 3. (a) Slope extracted from the relationship of Δv_2 with respect to $A_\pm$ for all centralities. Statistical errors only. (b) Slope parameters compare to the theoretical prediction made in [16], which is shown by the solid lines.

of track selection cuts as applied on real data. The relationship between the two measurements of net charge asymmetry is used to map the observed $A_\pm$ to the true $A_\pm$. Roughly, the corrected charge asymmetry $A_\pm$ is a half of its corresponding observed $A_\pm$.

In Fig. 2(b), the Δv_2 is plotted a function of $A_\pm$. A clearly linear relationship is observed. Fitting the data points to a straight line, the slope parameter can be extracted. This procedure is repeated for all centrality classes.

The slope parameters measured for events in the nine centrality bins are shown in Fig. 3(a). Here only statistical uncertainties are shown. It can be seen that slopes have the order of magnitude of about 0.01, which is consistent with the theoretical prediction based on CMW. The slopes are relatively large in mid-centrality and relatively small in peripheral and central collisions. In order to quantitatively compare to the prediction made in [16], we map out our centrality definition to impact parameter b according to a previous STAR study based on Monte-Carlo simulation [24]. The comparison is shown in Fig. 3(b). It can be seen that, although the magnitude of slopes in general agree with the prediction, the centrality dependency is a little different - the data is peaked at a larger b than the prediction. As the strong correlation exist between b and the magnetic field, this observation calls for a more advanced calculation on the magnetic field that is used to calculate the Δv_2.

3. Summary

We measured the v_2 of π^+ and π^- of Au + Au collisions at $\sqrt{s_{NN}} = 200$ GeV in different centrality and different net charge asymmetry $A_\pm$ bins. The integrated v_2 of π^+ (π^-), $0.15 < p_T < 0.5$ GeV/c, decreases (increases) linearly with the increasing $A_\pm$. The v_2 difference between π^+ and π^- is proportional to $A_\pm$ and the slope parameters have their order of magnitude of about 0.01. All of these observed features are consistent with the predication based on CMW in [16]. However, the predicted trend of the slope with respect to impact parameter is different than the measurement.

Acknowledgments

This work is supported in part by National Natural Science Foundation of China under Grants 11075060, 11135011 and 11105060.

References

[1] Arsene I and et al (BRAHMS Collaboration) 2005 *Nuclear Physics A* **757** 1 – 27
[2] Back B and et al (PHOBOS Collaboration) 2005 *Nuclear Physics A* **757** 28 – 101
[3] Adams J and et al (STAR Collaboration) 2005 *Nuclear Physics A* **757** 102 – 183
[4] Adcox K and et al (PHENIX Collaboration) 2005 *Nuclear Physics A* **757** 184 – 283
[5] Kharzeev D E, McLerran L D and Warringa H J 2008 *Nuclear Physics A* **803** 227 – 253
[6] Dmitri and Kharzeev 2006 *Physics Letters B* **633** 260 – 264
[7] Kharzeev D and Zhitnitsky A 2007 *Nuclear Physics A* **797** 67 – 79
[8] Fukushima K, Kharzeev D E and Warringa H J 2008 *Phys. Rev. D* **78**(7) 074033
[9] Kharzeev D E 2010 *Annals of Physics* **325** 205 – 218
[10] Son D T and Zhitnitsky A R 2004 *Phys. Rev. D* **70**(7) 074018
[11] Metlitski M A and Zhitnitsky A R 2005 *Phys. Rev. D* **72**(4) 045011
[12] Abelev B I and et al (STAR Collaboration) 2009 *Phys. Rev. Lett.* **103**(25) 251601
[13] Abelev B I and et al (STAR Collaboration) 2010 *Phys. Rev. C* **81**(5) 054908
[14] Ajitanand N N, Esumi S and Lacey R A (PHENIX Collaboration) 2010 In Proc. of the RBRC Workshops (Brookhaven National Laboratory, Upton, NY, 2010), Vol. 96.
[15] Selyuzhenkov I and Collaboration A 2012 (*Preprint* `arXiv:1203.5230v1`)
[16] Burnier Y, Kharzeev D E, Liao J and Yee H U 2011 *Phys. Rev. Lett.* **107** 052303
[17] Poskanzer A M and Voloshin S A 1998 *Phys. Rev. C* **58** 1671–1678
[18] Dunlop J C, Lisa M A and Sorensen P 2011 *Phys. Rev. C* **84**(4) 044914
[19] Xu J, Chen L W, Ko C M and Lin Z W 2012 *Phys. Rev. C* **85**(4) 041901
[20] Ackermann K H and et al (STAR Collaboration) 2003 *Nucl. Instrum. Methods A* **499** 624 – 632
[21] Anderson M and et al (STAR Collaboration) 2003 *Nucl. Instrum. Methods A* **499** 659 – 678
[22] Bilandzic A, Snellings R and Voloshin S 2011 *Phys. Rev. C* **83** 044913
[23] Gyulassy M and Wang X N 1994 *Computer Physics Communications* **83** 307 – 331
[24] Abelev B I and et al (STAR Collaboration) 2009 *Phys. Rev. C* **79** 034909

Journal of Physics: Conference Series **389** (2012) 012036 doi:10.1088/1742-6596/389/1/012036

Correlations and fluctuations studied with ALICE

Michael Weber for the ALICE collaboration
University of Houston
E-mail: m.weber@cern.ch

Abstract.
The measurement of particle correlations and event–by–event fluctuations of physical observables allows to study a large variety of properties of the matter produced in ultra relativistic heavy–ion collisions. We will present results for two–particle correlations, mean transverse momentum fluctuations, and net charge fluctuations in Pb–Pb collisions at $\sqrt{s_{\mathrm{NN}}} =$ 2.76 TeV.

1. Motivation

In ultra relativistic heavy–ion collisions the created nuclear matter reaches high temperatures, with a sufficient energy density to produce a deconfined state of quarks and gluons. The study of this so-called quark gluon plasma (QGP) as predicted by quantum chromodynamics (QCD) [1] is the aim of the ALICE experiment at the Large Hadron Collider (LHC) at CERN [2, 3]. At RHIC (Relativistic Heavy–Ion Collider) energies evidence for the existence of this new state of matter has already been published [4]. Entering a new energy regime with the LHC heavy–ion program, the existence of the QGP has further been established by first experimental results [5, 6]. The questions that now naturally emerge in this context can be summarized as:

- What are the properties of the QGP, e.g. η/s, medium transport coefficients?
- What is the evolution from QGP formation to final state hadrons which are measured in the experiments, e.g. thermalization, phase transition, hadronization?

The study of correlations among the produced particles and the event–by–event fluctuations of quantities such as the net charge and the mean transverse momentum is proposed to provide some of the answers. We will present with the help of two examples the strength of these methods to learn about the properties and evolution of the QGP. The analyses are performed on data taken in Pb–Pb collisions at $\sqrt{s_{\mathrm{NN}}} = 2.76$ TeV making use of the excellent tracking capabilities of the ALICE Time Projection Chamber (TPC). The primary vertex is provided by both the TPC and the Silicon Pixel Detector (SPD) and the centrality of the collisions is determined by scintillation detectors (VZERO-A and VZERO-C) as described in [7].

2. Correlations

2.1. Particle yield modification in jet–like azimuthal dihadron correlations

One way to learn about the properties of the QGP is to study the in–medium energy loss of high transverse momentum partons, which are produced in initial hard scattering processes [8] and are measurable in the suppression of high–p_T hadrons in the nuclear modification factor R_{AA} [6]. Further constraints on the quenching mechanisms can be retrieved from the per trigger yield of associated particles [9], which is defined for azimuthal dihadron correlations as

$$C(\Delta\varphi) = \frac{1}{N_{trig}} \frac{dN_{assoc}}{d\Delta\varphi}. \tag{1}$$

After background subtraction the yield Y is integrated for momentum intervals, where inter– and intra–jet correlations are the dominating source of particle correlations (8 GeV/$c < p_{\mathrm{T,trig}} <$ 15 GeV/c and 3 GeV/$c < p_{\mathrm{T,assoc}} < p_{\mathrm{T,trig}}$) in two regions: the near–side peak ($|\Delta\varphi| < 0.7\ rad$) and the away–side peak ($|\Delta\varphi - \pi| < 0.7\ rad$). The ratio of yields in central Pb–Pb to pp collisions I_{AA} is shown in Fig. 1 in comparison to different shower evolution models [10]. In the away side a strong suppression ($I_{AA} \simeq 0.6$) of associated particles due to quenching is observed. At the near–side a moderate enhancement ($I_{AA} \simeq 1.2$) is observed, which can be understood as (i) a change of the fragmentation function, (ii) a possible change of the quark/gluon jet ratio in the final state due to the different coupling to the medium, and (iii) a bias on the parton p_T spectrum after energy loss due to the trigger particle selection. Only one model (ASW) is able to reproduce both, the near– and the away–side I_{AA}, showing the strong constraints that can be obtained from this measurement.

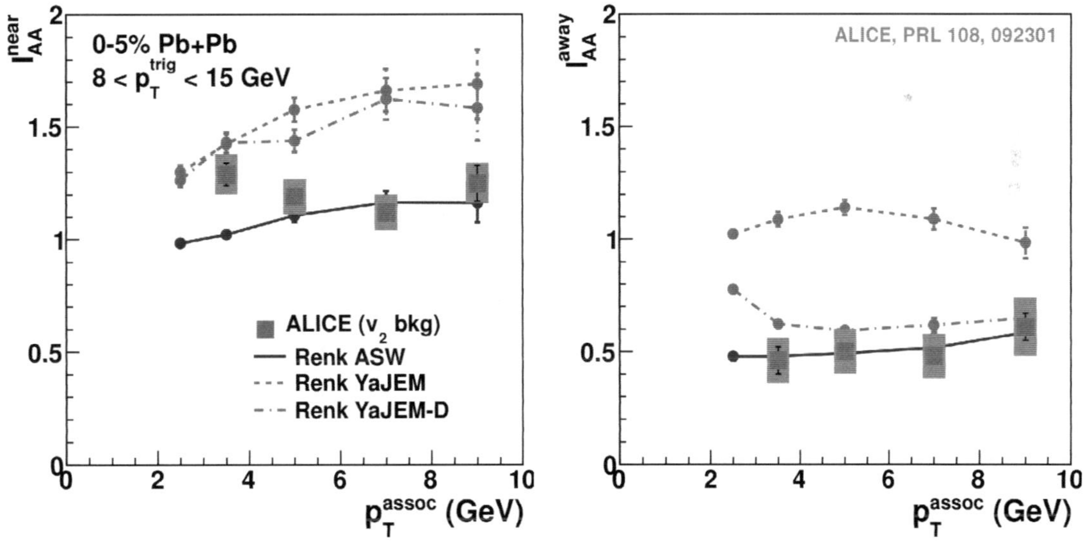

Figure 1. Charged hadron I_{AA} for trigger particle $p_T = 8 - 15$ GeV/c [9] compared to different shower evolution models [10].

2.2. Harmonic decomposition

Moving from the jet–like particle correlations to the soft regime, where collective processes are the dominant source of particle correlations, the question of initial conditions in the heavy–ion collision can be tackled. To study the shape evolution of triggered particle pair distributions in $\Delta\varphi$ quantitatively [11], the correlation function

$$C(\Delta\eta, \Delta\varphi) = \frac{N_{mixed}}{N_{same}} \frac{N_{same}(\Delta\eta, \Delta\varphi)}{N_{mixed}(\Delta\eta, \Delta\varphi)} \tag{2}$$

was calculated from the distributions of particle pairs formed from the same event $N_{same}(\Delta\eta, \Delta\varphi)$ and from different events $N_{mixed}(\Delta\eta, \Delta\varphi)$ for different trigger and associated particle transverse momentum ranges. The ratio of mixed-event N_{mixed} to same-event N_{same} pair counts is included as a normalization factor. In the bulk dominated regime ($p_T < 3-4$ GeV/c) in addition to the near– and away–side peak at small pseudorapidity intervals between trigger and associate particle a near-side ridge and a peaked structure are observed for $0.8 < \Delta\eta < 1.8$ (see left panel of Fig. 2). The decomposition in Fourier components, both even and odd:

$$V_{n\Delta} = \langle \cos n\Delta\varphi \rangle = \frac{\int d\Delta\varphi C(\Delta\varphi) cos n\Delta\varphi}{\int d\Delta\varphi C(\Delta\varphi)} \tag{3}$$

are shown in the right panel of Fig. 2. It is sufficient to use components up to $n = 5$ to describe the correlation distributions. Furthermore, it is found that for $p_{\mathrm{T,assoc}} < 4$ GeV/c the two–particle Fourier coefficients $V_{n\Delta}$ are consistent with the single-particle harmonic coefficients obtained from a global fit, except for $n = 1$. In Fig. 3 this is shown for $n = 2$ as an example. In addition, the v_n values obtained from the global fit agree with two–particle cumulant measurements for $n = 2-5$ [12]. This decomposition can be understood as the collective response of produced particles to anisotropic initial conditions. However, this decomposition is not necessarily unique to hydrodynamic flow. Factorization may be observed whenever both particles are correlated to one another through their correlation to a common symmetry plane having large longitudinal extent.

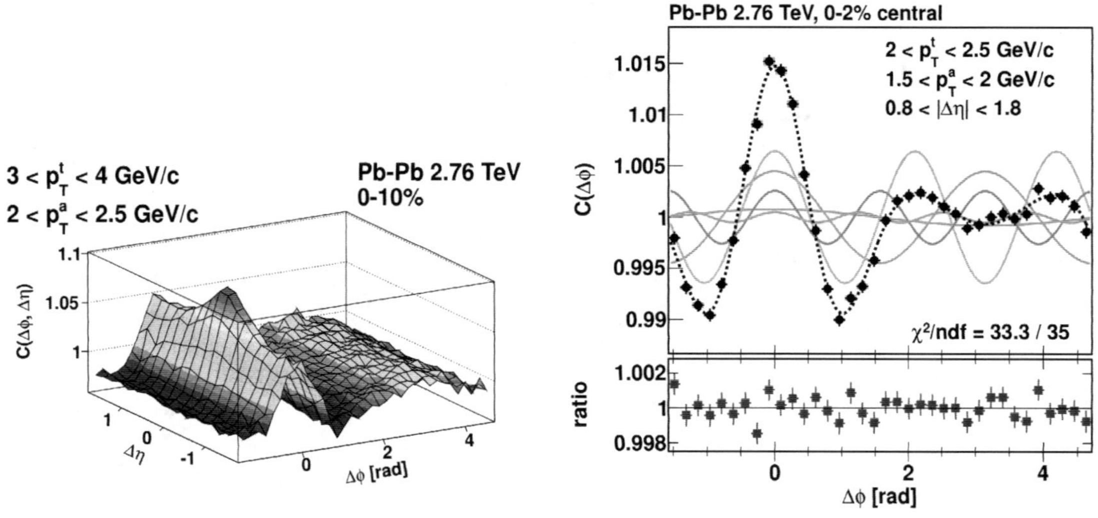

Figure 2. Left panel: Example of two–particle correlation functions for central Pb–Pb collisions at low to intermediate transverse momentum. Right panel: $C(\Delta\varphi)$ for particle pairs at $|\Delta\eta| > 0.8$. The Fourier harmonics for $V_{1\Delta}$ to $V_{5\Delta}$ are superimposed in color. Their sum is shown as the dashed curve. The ratio of data to the $n \leq 5$ sum is shown in the lower panel. Figure from [11].

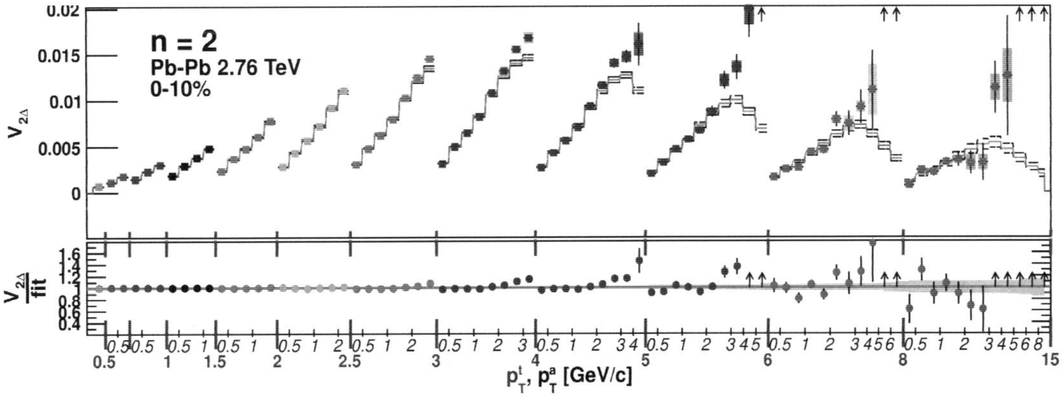

Figure 3. Global fit examples in 0–10% central events for n = 2. The measured $V_{2\Delta}$ coefficients are plotted on an interleaved $p_{\mathrm{T,trig}}$, $p_{\mathrm{T,assoc}}$ axis in the upper panels, and the global fit function is shown as the red curves. The global fit systematic uncertainty is represented by dashed lines. The lower section shows the ratio of the data to the fit, and the shaded bands represent the systematic uncertainty propagated to the ratio. In all cases, off–scale points are indicated with arrows. Figure from [11].

3. Fluctuations

3.1. Mean-p_T fluctuations

It was proposed in [13, 14] to use event–by–event fluctuations of the mean transverse momentum in heavy–ion collisions as an indicator of the critical behaviour of the system in the vicinity of a phase transition. Furthermore the influence of early–time dynamics like geometric fluctuations of the colliding system can leave their imprint on these fluctuations. As a reference pp collisions are used, where the influence of known physics, like resonance decays, HBT effects, (mini–) jets can be studied. The two–particle correlator C_m for mean–p_T fluctuations is defined by the covariances of all pairs of particles i and j in the same event with respect to the inclusive mean transverse momentum $\langle p_{\mathrm{T}}\rangle_m$ in a multiplicity class m:

$$\sigma^2_{dyn} \equiv C_m = \langle \Delta p_{\mathrm{T,i}} \Delta p_{\mathrm{T,j}} \rangle = \frac{1}{\sum_{k=1}^{N_{ev}} N_k^{pairs}} \cdot \sum_{k=1}^{N_{events}} \sum_{i=1}^{N_k} \sum_{j=i+1}^{N_k} (p_{\mathrm{T,i}} - \langle p_{\mathrm{T}}\rangle_m)(p_{\mathrm{T,j}} - \langle p_{\mathrm{T}}\rangle_m) \quad (4)$$

where N_k is the number of particles and $N_k^{pairs} = 0.5 \cdot N_k(N_k - 1)$ is the number of particle pairs in event k. It is a measure of the dynamical fluctuations and C_m vanishes in the presence of only statistical fluctuations.

In Fig. 4 the relative fluctuations, normalized to the inclusive mean transverse momentum $\langle p_{\mathrm{T}}\rangle_m$, are shown as a function of the uncorrected event multiplicity N_{acc}, the number of tracks accepted by the analysis cuts. In both pp and Pb–Pb collisions significant non–statistical fluctuations are observed, diluting with increasing multiplicity. A common scaling in pp and peripheral Pb–Pb can be described by a power law fit $\sqrt{C_m}/\langle p_{\mathrm{T}}\rangle_m(N_{acc}) = A \cdot N_{acc}^b$ to the pp data with $b = -0.385 \pm 0.003$, whereas for higher multiplicities in Pb–Pb collisions an additional reduction of mean transverse momentum fluctuations is evident. This can be understood in the onset of local thermal equilibrium at higher centralities [14]. The results from the HIJING event generator [15], on the contrary, can be described by a single power law fit parameter, differing from the one extracted from the data, which is in agreement with a scaling by $1/\sqrt{N_{acc}}$.

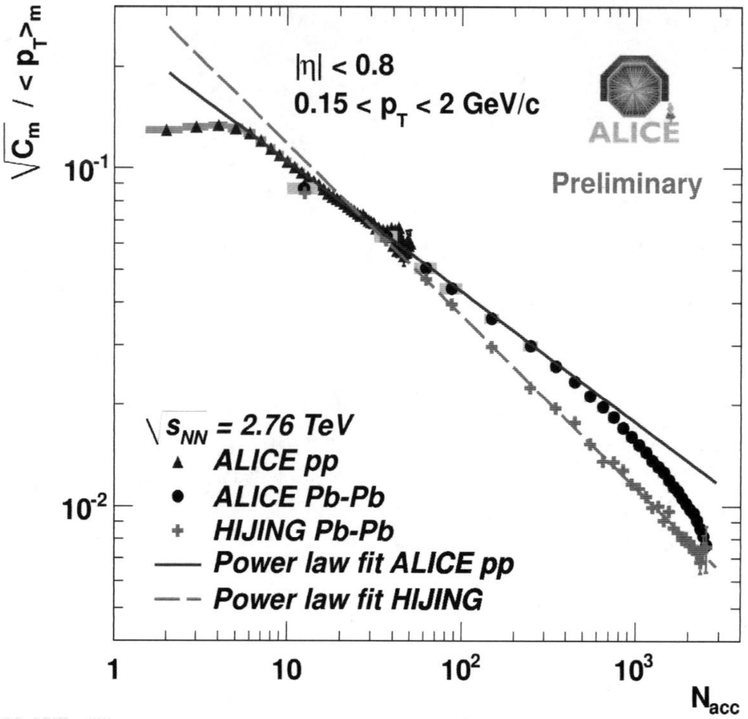

Figure 4. Comparison of relative mean transverse momentum fluctuations $\sqrt{C_m}/\langle p_{\mathrm{T}}\rangle_m$ in pp and Pb–Pb collisions together with a HIJING simulation and power law fits both for pp data and HIJING simulation [15].

3.2. Net charge fluctuations

Deeper insight on the phase transition to a QGP phase can be gained by determining the relevant charge carriers of the system produced in a heavy–ion collision [16]. In a hadron (resonance) gas these are mesons and baryons, in the QGP phase quarks are carrying the electric charge. Using the number of positive N_+ and negative particles N_- in

$$\nu_{+-,dyn} = \frac{\langle N_+(N_+-1)\rangle}{\langle N_+\rangle^2} + \frac{\langle N_-(N_--1)\rangle}{\langle N_-\rangle^2} - 2\frac{\langle N_-N_+\rangle}{\langle N_-\rangle\langle N_+\rangle} \tag{5}$$

as a measure for the event-by-event net charge fluctuations, a significant difference between the two phases is expected. The magnitude of the fluctuations can be expressed by the quantity D, defined as [17]

$$D = 4\frac{\langle \delta Q^2\rangle}{N_{ch}} \approx \nu_{+-,dyn}\langle N_{ch}\rangle + 4 \tag{6}$$

with $\langle \delta Q^2\rangle$ the variance of the net charge Q with $Q = N_+ + N_-$ and $N_{ch} = N_+ + N_-$. In Fig. 5 the values of $\nu^{corr}_{+-,dyn}\langle N_{ch}\rangle$, corrected for global charge conservation, are shown for pp and Pb–Pb collisions as a function of number of participating nucleons N_{part} in a pseudorapidity window $\Delta\eta = 1$ and $\Delta\eta = 1.6$. The shaded bands indicate the predictions for a hadron resonance gas (HG) and the QGP [18]. The data points show a monotonic decreasing centrality dependence and lie clearly below the HG, but above the QGP expectation. On the other hand, the results from the HIJING event generator for both pseudorapidity windows do not show any dependence on N_{part} and are found in the vicinity of the HG band. The observed pseudorapidity dependence of $\nu^{corr}_{+-,dyn}\langle N_{ch}\rangle$ may hint to a dilution of the primordial fluctuations during the evolution of the system from hadronization to kinetic freeze–out because of the diffusion of charged hadrons in rapidity [19, 20].

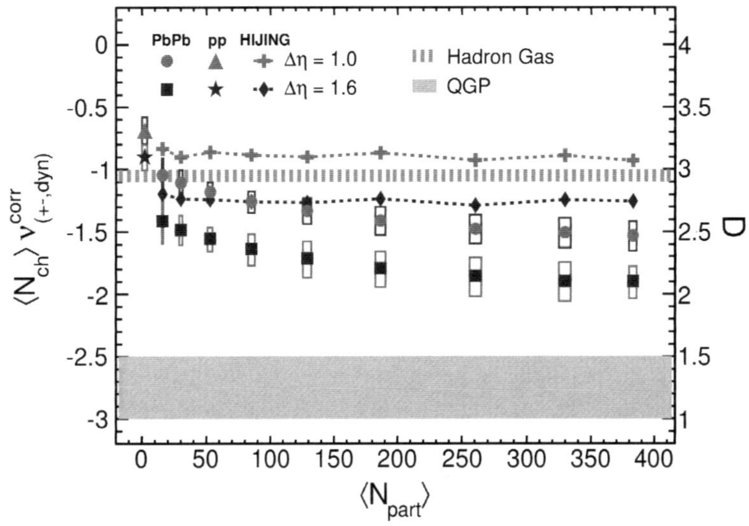

Figure 5. $\nu^{corr}_{+-,dyn}\langle N_{ch}\rangle$ (left axis) and D (right axis) as a function of the number of participants for $\Delta\eta = 1$ and $\Delta\eta = 1.6$ in Pb–Pb collisions at $\sqrt{s_{\rm NN}} = 2.76$ TeV. Also shown are the results from the HIJING event generator for both the $\Delta\eta$ windows and in the shaded bands the expectations for a hadron resonance gas and the QGP [18]. Figure from [16].

4. Summary and outlook

In summary, it was shown that the ALICE experiment offers a large variety of tools to study the properties and evolution of the QGP via particle correlations and event–by–event fluctuations. Energy–loss models can be constrained, initial conditions, composition, and evolution of the matter produced in ultra relativistic heavy–ion collisions can be probed. To further improve our understanding many more observables or extensions of the presented analyses are needed. Among them, we are studying three–particle correlations, harmonic decomposition on an event–by–event basis, $\Delta p_T \Delta p_T$ correlations, and the charge correlations using the balance function.

References

[1] H. Satz, Rep. Prog. Phys. **63**, (2000) 1511;
S.A. Bass, M. Gyulassy, H. Stöcker, W. Greiner, J. Phys. **G25**, (1999) R1;
E.V. Shuryak, Phys. Rep. **115**, (1984) 151;
J. Cleymans, R.V. Gavai, E. Suhonen, Phys. Rep. **130**, (1986) 217.
[2] K. Aamodt *et al.* [ALICE Collaboration], J. Phys. **G30**, (2004) 1517;
K. Aamodt *et al.* [ALICE Collaboration], J. Phys. **G32**, (2006) 1295.
[3] K. Aamodt *et al.* [ALICE Collaboration], JINST **3**, (2008) S08002.
[4] I. Arsene *et al.* [BRAHMS Collaboration], Nucl. Phys. **A757**, (2005) 1.
K. Adcox *et al.* [PHENIX Collaboration], Nucl. Phys. **A757**, (2005) 184.
B. B. Back *et al.* [PHOBOS Collaboration], Nucl. Phys. **A757**, (2005) 28.
J. Adams *et al.* [STAR Collaboration], Nucl. Phys. **A757**, (2005) 102.
[5] K. Aamodt *et al.* [ALICE Collaboration], Phys. Rev. Lett. **105**, (2010) 252302.
K. Aamodt *et al.* [ALICE Collaboration], Phys. Rev. Lett. **107**, (2011) 032301.

Journal of Physics: Conference Series **389** (2012) 012036 doi:10.1088/1742-6596/389/1/012036

[6] G. Aad *et al.* [ATLAS Collaboration], Phys. Rev. Lett. **105**, (2010) 252303.
K. Aamodt *et al.* [ALICE Collaboration], Phys. Lett. **B696**, (2011) 30.
[7] K. Aamodt *et al.*, Phys. Rev. Lett. **106**, (2011) 032301.
[8] J. D. Bjorken, FERMILAB-PUB-82-059-THY (1982).
M. Gyulassy and M. Plumer, Phys. Lett. B 243, (1990) 432.
X.-N. Wang and M. Gyulassy, Phys. Rev. Lett. 68, (1992) 1480.
[9] K. Aamodt *et al.* [ALICE Collaboration], Phys. Rev. Lett. **108**, (2012) 092301.
[10] T. Renk, K. Eskola, *arXiv:*1106.1740.
[11] K. Aamodt *et al.* [ALICE Collaboration], Phys. Lett. **B708**, (2012) 249.
[12] K. Aamodt *et al.* [ALICE Collaboration], Phys. Rev. Lett. **107**, (2011) 032301.
[13] M.A. Stephanov, K. Rajagopal and E.V. Shuryak, Phys.Rev.Lett. **81**, (1998) 4816-9.
M.A. Stephanov, K. Rajagopal and E.V. Shuryak, Phys.Rev. **D60**, (1999) 114028.
[14] S. Gavin, Phys. Rev. Lett. **92**, (2004) 162301.
[15] M. Gyulassy and X. N. Wang, Comput. Phys. Commun. **83**, 307 (1994).
X. N. Wang and M. Gyulassy, Phys. Rev. **D44**, 3501 (1991).
[16] B. Abelev *et al.*, *arXiv:*1207:6068.
[17] S. Jeon, V. Koch, Phys. Rev. Lett. 85 (2000), 2076.
[18] S. Jeon and V. Koch, In QuarkGluon Plasma 3, Ed. R.C. Hwa and X.N. Wang, 430 (2004); arXiv:hep-ph/0304012v1.
[19] E. V. Shuryak, M. A. Stephanov, Phys. Rev. C **63**, (2001) 064903.
[20] M. A. Aziz, S. Gavin, Phys. Rev. C **70**, (2004) 034905.

Journal of Physics: Conference Series **389** (2012) 012037 doi:10.1088/1742-6596/389/1/012037

Clustering of Color sources and the Equation of State in Heavy Ion Collisions at RHIC and LHC Energies

R. P. Scharenberg

Department of Physics, Purdue University, West Lafayette, Indiana-47907, USA

E-mail: schrnbrg@purdue.edu

Abstract. The initial temperature T_i, energy density ε_i, and formation time τ_i of the initial state of the QGP formed in the heavy ion collisions at RHIC and LHC energies are determined using the data driven Color String Percolation Model (CSPM). Multiparticle production by interacting strings stretched between projectile and target form a spanning cluster at the percolation threshold. The relativistic kinetic theory relation for η/s is evaluated as a function of T and the mean free path (λ_{mfp}) using data and CSPM. $\eta/s(T_i, \lambda_{mfp})$ describes the transition from a strongly interacting QGP at $T/T_c \sim 1$ to a weakly coupled QGP at $T/T_c \geq 6$. We find that the reciprocal of η/s is equal to the trace anomaly $\Delta = \varepsilon - 3P/T^4$ which also describes the transition. We couple this initial state of the QGP to a 1D Bjorken expansion to determine the sound velocity c_s^2 of the QGP for $0.85 \leq T/T_c \leq 3$. The bulk thermodynamic quantities and the equation of state are in excellent agreement with LQCD results.

1. Clustering of Color Sources

Multiparticle production is currently described in terms of color strings stretched between the projectile and the target, which decay into new strings and subsequently hadronize to produce observed hadrons. Color strings may be viewed as small areas in the transverse plane filled with color field created by colliding partons. With growing energy and size of the colliding system, the number of strings grows, and they start to overlap, forming clusters, in the transverse plane very much similar to disks in two dimensional percolation theory. At a certain critical density a macroscopic cluster appears that marks the percolation phase transition. This is the Color String Percolation Model (CSPM) [1, 2]. The interaction between strings occurs when they overlap and the general result, due to the SU(3) random summation of charges, is a reduction in multiplicity and an increase in the string tension hence increase in the average transverse momentum squared, $\langle p_t^2 \rangle$. We assume that a cluster of n strings that occupies an area of S_n behaves as a single color source with a higher color field $\vec{Q_n}$ corresponding to the vectorial sum of the color charges of each individual string $\vec{Q_1}$. The resulting color field covers the area of the cluster. As $\vec{Q_n} = \sum_1^n \vec{Q_1}$, and the individual string colors may be oriented in an arbitrary manner respective to each other , the average $\vec{Q_{1i}}\vec{Q_{1j}}$ is zero, and $\vec{Q_n^2} = n\vec{Q_1^2}$.

Knowing the color charge $\vec{Q_n}$ one can obtain the multiplicity μ and the mean transverse momentum squared $\langle p_t^2 \rangle$ of the particles produced by a cluster of n strings [2]

$$\mu_n = \sqrt{\frac{nS_n}{S_1}}\mu_0; \quad \langle p_t^2 \rangle = \sqrt{\frac{nS_1}{S_n}}\langle p_t^2 \rangle_1 \tag{1}$$

where μ_0 and $\langle p_t^2\rangle_1$ are the mean multiplicity and $\langle p_t^2\rangle$ of particles produced from a single string with a transverse area $S_1 = \pi r_0^2$. This implies a simple relation between the multiplicity and transverse momentum $\mu_n\langle p_t^2\rangle_n = n\mu_0\langle p_t^2\rangle_1$, which means conservation of the total transverse momentum produced.

In the thermodynamic limit, one obtains an analytic expression [1, 2]

$$\langle\frac{nS_1}{S_n}\rangle = \frac{\xi}{1-e^{-\xi}} \equiv \frac{1}{F(\xi)^2} \tag{2}$$

where $F\xi)$ is the color suppression factor and $\xi = \frac{N_s S_1}{S_N}$ is the percolation density parameter. Eq.(1) can be written as $\mu_n = nF(\xi)\mu_0$ and $\langle p_t^2\rangle_n = \langle p_t^2\rangle_1/F(\xi)$. The critical cluster which spans S_N, appears for $\xi_c \geq 1.2$ [3]. It is worth noting that CSPM is a saturation model similar to the Color Glass Condensate (CGC), where $\langle p_t^2\rangle_1/F(\xi)$ plays the same role as the saturation momentum scale Q_s^2 in the CGC model [4, 5].

2. Experimental Determination of the Color Suppression Factor $F(\xi)$

The suppression factor is determined by comparing the *pp* and A+A transverse momentum spectra. To evaluate the value of ξ from data for Au+Au collisions, a parameterization of *pp* events at 200 GeV is used to compute the p_t distribution

$$dN_c/dp_t^2 = a/(p_0+p_t)^\alpha \tag{3}$$

where a is the normalization factor. p_0 and α are parameters used to fit the data. This parameterization also can be used for nucleus-nucleus collisions to take into account the interactions of the strings [2]

$$dN_c/dp_t^2 = \frac{a'}{(p_0\sqrt{F(\xi_{pp})/F(\xi)}+p_t)^\alpha} \tag{4}$$

The color suppression factor $F(\xi)$ is related to the percolation density parameter ξ.

$$F(\xi) = \sqrt{\frac{1-e^{-\xi}}{\xi}} \tag{5}$$

3. Energy Density ε

The initial energy density ε_i above T_c is given by [6]

$$\varepsilon_i = \frac{3}{2}\frac{\frac{dN_c}{dy}\langle m_t\rangle}{S_n\tau_{pro}} \tag{6}$$

To evaluate ε_i we use the charged pion multiplicity dN_c/dy at midrapidity and S_n values from STAR for 0-10% central Au-Au collisions with $\sqrt{s_{NN}}$ =200 GeV [7]. We can calculate $\langle p_t\rangle$ using the CSPM thermal distribution Eqs.(9) and (10). For $0.2 < p_t < 1.5$, $\langle p_t\rangle = 0.394 \pm 0.003$GeV, adding the extra energy required for the rest mass of pions at hadronization $\langle m_t\rangle = 0.42 \pm 0.003$GeV. The error on $\langle p_t\rangle$ is due to the error on T_i [8].

The dynamics of massless particle production has been studied in QE2 quantum electrodynamics. QE2 can be scaled from electrodynamics to quantum chromodynamics using the ratio of the coupling constants [9]. The production time τ_{pro} for a boson (gluon) is [10]

$$\tau_{pro} = \frac{2.405\hbar}{\langle m_t\rangle} \tag{7}$$

For Au-Au collisions at $\sqrt{s_{NN}}$ =200 GeV Eqs. (6) and (7) gives $\varepsilon_i = 2.27\pm0.16$ GeV$/fm^3$ at ξ=2.88. In CSPM the energy density ε is proportional to ξ. From the measured value of ξ and ε it is found that ε is proportional to ξ for the range $1.2 < \xi < 10.56$, $\varepsilon_i = 0.788\ \xi$ GeV$/fm^3$ [7, 11] as shown in Fig.1.

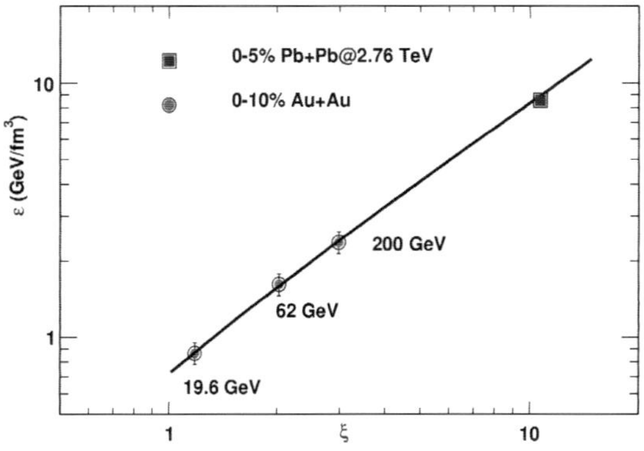

Figure 1. Energy density ϵ as a function of the percolation density parameter ξ. The value for LHC energy is shown as blue square [7, 11].

4. Determination of the Temperature

The connection between the measured ξ and the temperature $T(\xi)$ involves the Schwinger mechanism (SM) for particle production. The Schwinger distribution for massless particles is expressed in terms of p_t^2 [10, 9]

$$dn/dp_t^2 \sim e^{-\pi p_t^2/x^2} \tag{8}$$

where the average value of the string tension is $\langle x^2 \rangle$. The tension of the macroscopic cluster fluctuates around its mean value because the chromo-electric field is not constant. The origin of the string fluctuation is related to the stochastic picture of the QCD vacuum. Since the average value of the color field strength must vanish, it can not be constant but changes randomly from point to point [12]. Such fluctuations lead to a Gaussian distribution of the string tension for the cluster, which transforms SM into the thermal distribution [12]

$$dn/dp_t^2 \sim e^{(-p_t\sqrt{\frac{2\pi}{\langle x^2 \rangle}})} \tag{9}$$

with $\langle x^2 \rangle = \pi \langle p_t^2 \rangle_1 / F(\xi)$. The temperature is expressed as [8, 13]

$$T(\xi) = \sqrt{\frac{\langle p_t^2 \rangle_1}{2F(\xi)}} \tag{10}$$

Recently, it has been suggested that fast thermalization in heavy ion collisions can occur through the existence of an event horizon caused by a rapid de-acceleration of the colliding nuclei [14]. The thermalization in this case is due to the Hawking-Unruh effect [15, 16]. In CSPM the strong color field inside the large cluster produces de-acceleration of the primary $q\bar{q}$ pair which can be seen as a thermal temperature by means of the Hawking-Unruh effect. The string percolation density parameter ξ which characterizes the percolation clusters measures the initial temperature of the system. Since this cluster covers most of the interaction area, this temperature becomes a global temperature determined by the string density. In this way at $\xi_c = 1.2$ the connectivity percolation transition at $T(\xi_c)$ models the thermal deconfinement transition.

Journal of Physics: Conference Series **389** (2012) 012037 doi:10.1088/1742-6596/389/1/012037

We adopt the point of view that the experimentally determined universal chemical freeze-out temperature (T_f) is a good measure of the phase transition temperature, T_c. $\langle p_t^2 \rangle_1$ is evaluated using Eq.(10) at $\xi_c = 1.2$ with $T_f = 167.7 \pm 2.6$ MeV [17]. This gives $\sqrt{\langle p_t^2 \rangle_1} = 207.2 \pm 3.3$ MeV. This calibrates the CSPM temperature scale.

Figure 2 shows a plot of ε/T^4 as a function of T/T_c. The lattice QCD results are from HotQCD Collaboration [18]. It is observed that at LHC energy the CSPM results are in excellent agreement with the lattice QCD results. The lattice and CSPM results are available for T/$T_c < 2$. Table I gives the CSPM values ξ, T, ε and η/s at $T/T_c = 0.88$, 1, 1.16 and 1.57.

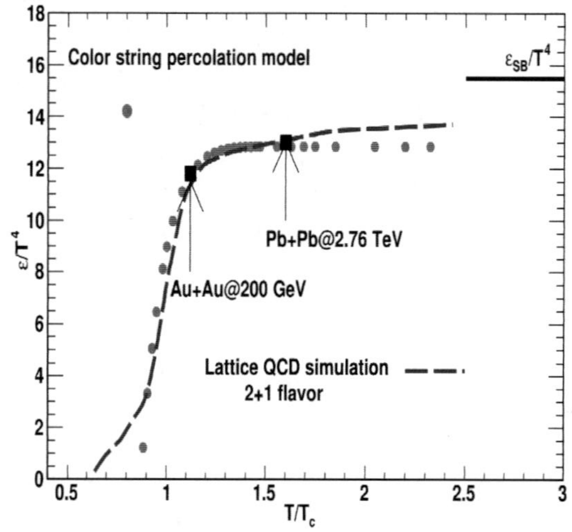

Figure 2. The energy density from CSPM versus T/T_c^{CSPM} (red circles) and Lattice QCD energy density vs T/T_c^{LQCD} (blue dash line) for 2+1 flavor and p4 action [18].

5. The Shear Viscosity to Entropy Density ratio η/s

The relativistic kinetic theory relation for the shear viscosity over entropy density ratio, η/s is given by [19, 20]

$$\frac{\eta}{s} \simeq \frac{T\lambda_{mfp}}{5} \tag{11}$$

where T is the temperature and λ_{mfp} is the mean free path given by

$$\lambda_{mfp} \sim \frac{1}{(n\sigma_{tr})} \tag{12}$$

n is the number density of an ideal gas of quarks and gluons and σ_{tr} the transport cross section for these constituents.

After the cluster is formed it behaves like a free gas of constituents. Eq. (10) can be applied to obtain the shear viscosity. In CSPM the number density is given by the effective number of sources per unit volume [21]

$$n = \frac{N_{sources}}{S_N L} \tag{13}$$

L is the longitudinal extension of the source, L = 1 fm [13]. The area occupied by the strings is related to ξ through the relation $(1 - e^{-\xi})S_N$. Thus the effective no. of sources is given by the

Table 1. The measured percolation density parameter ξ, temperature T, energy density ε and η/s for the meson gas [22]; the hadron to QGP transition; Au+Au at 200 GeV and Pb+Pb at 2.76 TeV . Au+Au is for 0-10% and Pb+Pb is for 0-5% central events.

System	ξ	T (MeV)	$\varepsilon(GeV/fm^3)$	η/s
Meson Gas	0.22	150.0	-	0.76
Hadron to QGP	1.2	167.7 ±2.6	0.94±0.07	0.240±0.012
Au+Au	2.88±0.09	193.6 ±3.0	2.27±0.16	0.204±0.020
Pb+Pb	10.56±1.05	262.2±13.0	8.32±0.83	0.260±0.026

total area occupied by the strings divided by the effective area of the string $S_1F(\xi)$.

$$N_{sources} = \frac{(1-e^{-\xi})S_N}{S_1F(\xi)} \tag{14}$$

In general $N_{sources}$ is smaller than the number of single strings. The number density of sources from Eqs. (12) and (13) becomes

$$n = \frac{(1-e^{-\xi})}{S_1F(\xi)L} \tag{15}$$

In CSPM the transport cross section σ_{tr} is the transverse area of the effective string $S_1F(\xi)$. Thus σ_{tr} is directly proportional to $F(\xi)$ and hence to $\frac{1}{T^2}$. The mean free path is given by

$$\lambda_{mfp} = \frac{L}{(1-e^{-\xi})} \tag{16}$$

For a large value of ξ the λ_{mfp} reaches a constant value. η/s is obtained from ξ and the temperature

$$\frac{\eta}{s} = \frac{TL}{5(1-e^{-\xi})} \tag{17}$$

The behavior of η/s is dominated by the fractional area covered by strings for $\xi < \xi_c$. This is not surprising because η/s is the ability to transport momenta at large distances and that has to do with the density of voids in the matter.

Figure 3 shows a plot of η/s as a function of T/T_c. The estimated value of η/s for Pb+Pb is shown in Fig. 3 at T/T_c = 1.57. The lower bound shown in Fig. 3 is given by AdS/CFT [23]. These results from STAR and ALICE data show that the η/s value is 2.5 and 3.3 times the KSS bound [23].

The theoretical estimates of η/s has been obtained as a function of T/T_c for both the weakly (wQGP) and strongly (sQGP) coupled QCD plasma are shown in Fig. 3 [19]. It is seen that at the RHIC top energy η/s is close to the sQGP. Even at the LHC energy it follows the trend of the sQGP. By extrapolating the η/s CSPM values to higher temperatures it is clear that η/s could approach the weak coupling limit near $T/T_c \sim 6$. The CSPM η/s value for the hadron gas is in agreement with the calculated value using measured elastic cross sections for a gas of pions and kaons [22].

6. The trace Anomaly Δ and the reciprocal of η/s

The Δ and η/s both describe the transition from a strongly coupled QGP to a weakly coupled QGP. We find that the reciprocal of η/s is in quantitative agreement with $\varepsilon - 3P/T^4$ the trace anomaly over this wide temperature range. This result is shown in Fig.4. The minimum in η/s =0.20 at T/T_c 1.15 determines the peak of the interaction measure $\sim$ 5 in agreement with the recent HotQCD values[24].

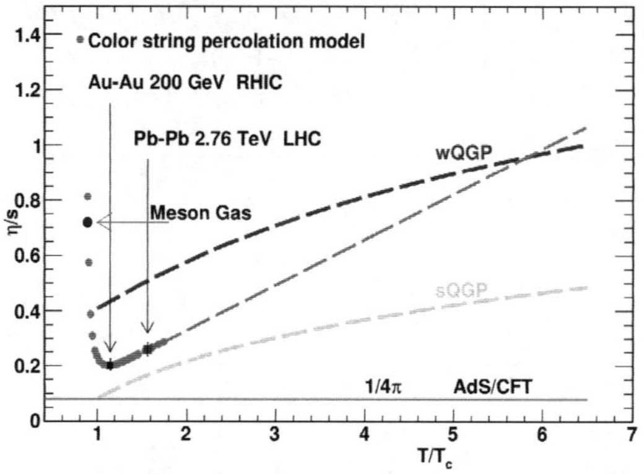

Figure 3. η/s as a function of T/T_c. Au+Au at 200 GeV for 0-10% centrality is shown as solid black square. wQGP and sQGP values are shown as dotted blue and green lines respectively [19]. The red dotted line represents the extrapolation to higher temperatures from the CSPM. The hadron gas value for $\eta/s \sim 0.7$ is shown as solid black circle at $T/T_c \sim 0.88$ [22].

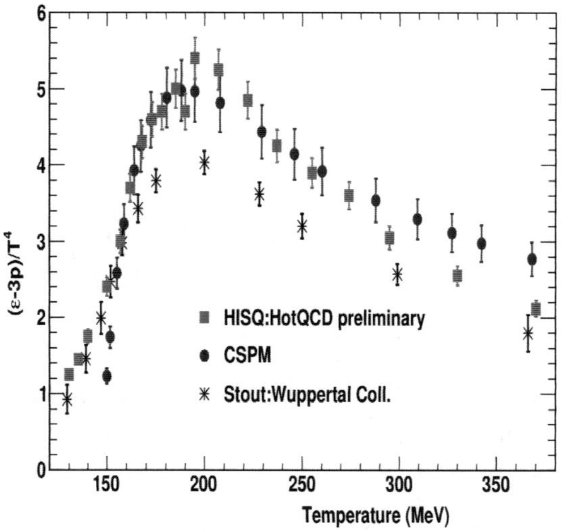

Figure 4. The trace anomaly $\Delta = \varepsilon - 3P/T^4$ vs temperature.

7. The Equation of State

We use CSPM coupled to a 1D Bjorken expansion. The input parameters are the initial temperature T, the initial energy density ε, and the trace anomaly Δ are determined by data. The Bjorken 1D expansion gives the sound velocity

$$\frac{1}{T}\frac{dT}{d\tau} = -C_s^2/\tau \tag{18}$$

$$\frac{dT}{d\tau} = \frac{dT}{d\varepsilon}\frac{d\varepsilon}{d\tau} \tag{19}$$

$$\frac{d\varepsilon}{d\tau} = -Ts/\tau \tag{20}$$

where ε is the energy density, s the entropy density, τ the proper time, and C_s the sound velocity. Since $s = \varepsilon + P/T$ and $P = (\varepsilon - \Delta T^4)/3$ one gets

$$\frac{dT}{d\varepsilon} s = C_s^2 \tag{21}$$

From above equations C_s^2 can be expressed in terms of ξ

$$C_s^2 = (-0.33)\left(\frac{\xi e^{-\xi}}{1-e^{-\xi}} - 1\right) + 0.0191(\Delta/3)\left(\frac{\xi e^{-\xi}}{(1-e^{-\xi})^2} - \frac{1}{1-e^{-\xi}}\right) \tag{22}$$

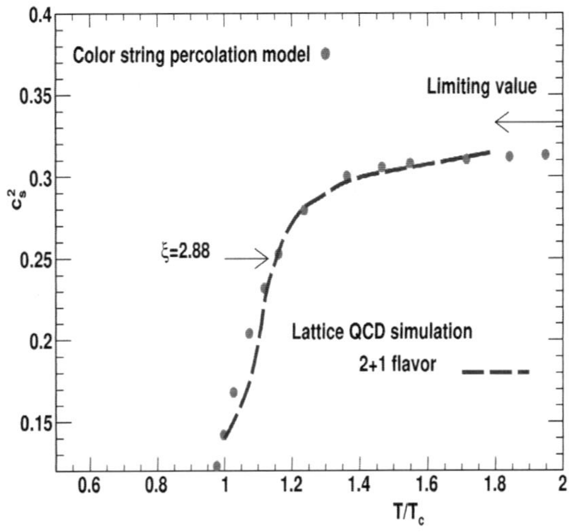

Figure 5. The speed of sound from CSPM versus T/T_c^{CSPM}(red circles) and Lattice QCD-p4 speed of sound versus T/T_c^{LQCD}(blue dash line)[18].

8. Conclusions and open questions

Two central objectives in the experimental characterization of the QGP are the Equation Of State (EOS) and the shear viscosity to entropy ratio η/s. We have found that s/η is equal to the trace anomaly $\Delta = \varepsilon - 3P/T^4$. We determine the bulk thermodynamics, ε/T^4, s/T^3 and the EOS c_s^2 of the QGP as a function of T/T_c(CSPM). The results are in excellent agreement with LQCD numerical simulations as a function of T/T_c(LQCD). CSPM predicts that the QGP will be formed in high multiplicity p-p collisions at LHC. The CSPM η/s ratio can be used to test the quasi-particle hypothesis where the jet transport parameter is $\hat{q} = \frac{3}{2}T^3\frac{s}{\eta}$, which can be compared with $\hat{q}$ measured in jet quenching using the the ALICE EMCAL detector at LHC.

9. Acknowledgments

This research was supported by the Office of Nuclear Physics within the U.S. Department of Energy Office of Science under Grant No. DE-FG02-88ER40412.

References

[1] M. A. Braun and C. Pajares 2000 *Eur. Phys. J.* C **16** 349.
[2] M. A. Braun, F. del Moral and C. Pajares 2002 *Phys. Rev.* C **65** 024907.

[3] H. Satz 2000 *Rep. Prog. Phys.* **63** 1511.
[4] L. McLerran and R. Venugopalan 1994 *Phys. Rev.* D **49** 2233.
[5] J. Dias de Deus and C. Pajares 2011 *Phys. Lett.* B **695** 455.
[6] J. D. Bjorken 1983 *Phys. Rev.* D **27** 140.
[7] B. I. Abelev et al. (STAR Collaboration) 2009 *Phys. Rev.* C **79** 34909.
[8] R. P. Scharenberg, B. K. Srivastava and A. S. Hirsch 2011 *Eur. Phys. J.* C **71** 1510.
[9] C. Y. Wong, Introduction to high energy heavy ion collisions (World Scientific,1994).
[10] J. Schwinger 1962 *Phys. Rev.* **128** 2425.
[11] B. K. Srivastava, R. P. Scharenberg, T. Tarnowsky, (STAR Collaboration) 2006 *Nukleonika* **51**, s109.
[12] A. Bialas 1999 *Phys. Lett.* B **466** 301.
[13] J. Dias de Deus and C. Pajares 2006 *Phys. Lett.* B **642** 455.
[14] D. Kharzeev, E. Levin and K. Tuchin 2007 *Phys. Rev.* C **75** 044903.
[15] S. W. Hawking, Commun. 1975 *Math. Phys.* **43** 199.
[16] W. G. Unruh 1976 *Phys. Rev.* D **14** 870.
[17] F. Becattini, P. Castorina, A. Milov and H. Satz 2010 *Eur. Phys. J.* C **66** 377.
[18] A. Bazavov et al. 2009 *Phys. Rev.* D **80**, 014504.
[19] T. Hirano and M. Gyulassy 2006 *Nucl. Phys.* A **769** 71.
[20] P. Danielewicz and M. Gyulassy 1985 *Phys. Rev.* D **31** 53.
[21] I. Bautista, J. Dias de Deus and C. Pajares arXiv:hep-ph/1102:3837.
[22] M. Prakash et al.1993 *Phys. Rept.* **227** 321.
[23] P. K. Kovtun, D. T. Son and A. O. Starinets 2005 *Phys. Rev. Lett.* C **94** 111601.
[24] P. Petreczky, Lattice 2012, Cairns, Australia, 24 - 30 June 2012.

Fluctuation and flow probes of early-time correlations in relativistic heavy ion collisions

Sean Gavin[1] **and George Moschelli**[2]

[1] Department of Physics and Astronomy, Wayne State University, 666 W Hancock, Detroit, MI, 48202, USA

[2] Frankfurt Institute for Advanced Studies, Johann Wolfgang Goethe University, Ruth-Moufang-Str. 1, 60438 Frankfurt am Main, Germany

E-mail: [1] `sean@physics.wayne.edu` [2] `moschelli@fias.uni-frankfurt.de`

Abstract. Fluctuation and correlation observables are often measured using multi-particle correlation methods and therefore mutually probe the origins of genuine correlations present in multi-particle distribution functions. We investigate the common influence of correlations arising from the spatially inhomogeneous initial state on multiplicity and momentum fluctuations as well as flow fluctuations. Although these observables reflect different aspects of the initial state, taken together, they can constrain a correlation scale set at the earliest moments of the collision. We calculate both the correlation scale in an initial stage Glasma flux tube picture and the modification to these correlations from later stage hydrodynamic flow and find quantitative agreement with experimental measurements over a range of collision systems and energies.

1. Introduction

Fluctuating initial conditions of relativistic heavy ion collisions influence not only flow and but also fluctuation measurements. The random distribution of transverse parton density "lumps" results in arbitrary event shapes. Harmonic flow is generated largely by the global spatial anisotropy of the initial shape and induces a global correlation in the final momenta of measured hadrons. Conversely, the mere existence of parton density lumps results in local correlations; partons emerging from the same source are more likely to share the same phase space throughout the collision lifetime. We argue that the same local correlations that give rise to multiplicity and p_t fluctuations drive fluctuations of the harmonic flow coefficients [1, 2].

To begin, in Sec. 2 we describe multiplicity, p_t, and v_n fluctuations as well as the ridge in terms of two-particle correlations. Using a cumulant expansion of the pair distribution reminiscent of those in Refs. [3, 4], we show that all of the observables are linked by a common correlation function. The normalization of this correlation function sets a common scale that governs most of the centrality and center of mass collision energy dependence.

In Sec. 3 we investigate contributions to the correlation function from transversely local longitudinally long range correlations. We build on an approach started in Refs. [5, 6] in which long range correlations result from the fragmentation of Glasma flux tubes. In this formulation local spatial correlations emerge from fluctuations in the number and distribution of flux tubes. These spatial correlations are then modified by transverse expansion, giving rise to azimuthal correlations. A similar physical picture motivates studies using a wide range of different techniques [7, 8, 9, 10, 11, 12, 13, 14].

Journal of Physics: Conference Series **389** (2012) 012038 doi:10.1088/1742-6596/389/1/012038

Finally in Sec. 4 we discuss how fluctuation measurements calculated in a unified framework can be used to constrain theories of the initial state of heavy ion collisions.

2. Correlations, Fluctuations, and Flow

Correlation measurements largely rely on the distribution of hadron pairs $\rho_2(\mathbf{p}_1, \mathbf{p}_2) = dN/d\mathbf{p}_1 d\mathbf{p}_2$. Using the cumulant expansion method the pair distribution can be written as

$$\rho_2(\mathbf{p}_1, \mathbf{p}_2) = \rho_1(\mathbf{p}_1)\rho_1(\mathbf{p}_2) + r(\mathbf{p}_1, \mathbf{p}_2), \tag{1}$$

where $\rho_1(\mathbf{p}) = dN/d\mathbf{p}$ is the singles distribution, and $r(\mathbf{p}_1, \mathbf{p}_2)$ represents the genuine two-particle correlation function [1, 2, 3, 4].

In this section we outline the role of correlations, $r(\mathbf{p}_1, \mathbf{p}_2)$, in determining fluctuation and flow observables and leave our model description to Sec. 3. In the absence of correlations, the pair distribution factorizes into the square of the singles distribution, $\rho_2(\mathbf{p}_1, \mathbf{p}_2) \to \rho_1(\mathbf{p}_1)\rho_1(\mathbf{p}_2)$. This assumption is often cited as a signature of harmonic flow, given that the event geometry is the dominant contributor to v_n measurements. It is informative, however, to discuss the implications of relaxing this assumption when considering flow fluctuations. To address this we study other fluctuation observables such as multiplicity and momentum fluctuations which do not rely on event geometry but still reflect the existence of genuine correlations. We then find that although harmonic flow coefficients do depend on the global correlations induced by geometry, flow fluctuations emerge largely from the the same correlations as other fluctuation observables.

2.1. Multiplicity Fluctuations

The existence of genuine correlations in the pair distribution signifies non-Poissonian particle production; all particles are not emitted from random independent sources [15]. Rearranging (1) as

$$r(\mathbf{p}_1, \mathbf{p}_2) = \rho_2(\mathbf{p}_1, \mathbf{p}_2) - \rho_1(\mathbf{p}_1)\rho_1(\mathbf{p}_2), \tag{2}$$

one can define the "robust variance",

$$\mathcal{R} = \frac{1}{\langle N \rangle^2} \int r(\mathbf{p}_1, \mathbf{p}_2) d\mathbf{p}_1 d\mathbf{p}_2 = \frac{\langle N(N-1) \rangle - \langle N \rangle^2}{\langle N \rangle^2}. \tag{3}$$

We can rewrite the right most term of (3) as $\mathcal{R} = Var(N) - \langle N \rangle / \langle N \rangle^2$. Possonian particle production requires $Var(N) = \langle N \rangle$, and consequently, non-zero values of $\mathcal{R}$ indicate a correlation in the production of hadrons. We will therefore refer to $\mathcal{R}$ as the correlation strength in this work. Significantly, the PHENIX collaboration measures multiplicity fluctuations using the width of the negative binomial distribution of produced hadrons [16]. The inverse of the width parameter is equivalent to the correlation strength, $\mathcal{R} = k_{NBD}^{-1}$ [17]. The results are shown in the bottom panel of Fig. 1. This quantity sets the scale of all fluctuation measurements using two-particle correlations.

In Sec. 3 we will argue that these correlations originate from the fluctuating initial parton density. Pairs produced together in the hot spot or density lump are more likely to be correlated. We stress here that the shape of the event due to the distribution of hot spots does not contribute to multiplicity fluctuations. Integration of (3) over all momenta, including azimuthal angles, sums all correlated particle pairs, regardless of direction.

2.2. Transverse Momentum Fluctuations

Comparably to multiplicity fluctuations, many describe the dynamic fluctuations of transverse momentum using the covariance

$$\langle \delta p_{t1} \delta p_{t2} \rangle = \frac{\langle \sum_{i \neq j} \delta p_{ti} \delta p_{tj} \rangle}{\langle N(N-1) \rangle}, \tag{4}$$

where $\delta p_{ti} = p_{ti} - \langle p_t \rangle$ and the average transverse momentum is $\langle p_t \rangle = \langle P_t \rangle / \langle N \rangle$ for $P_t = \sum_i p_{ti}$ the total momentum in an event [18, 19, 20, 21]. This quantity vanishes when particles i and j are uncorrelated. Equation (4) sums pairs of differences in particle momenta from the average and can be written in terms of the pair correlation function (2) as

$$\langle \delta p_{t1} \delta p_{t2} \rangle = \int d\mathbf{p}_1 d\mathbf{p}_2 \frac{r(\mathbf{p}_1, \mathbf{p}_2)}{\langle N(N-1) \rangle} \delta p_{t1} \delta p_{t2}. \tag{5}$$

Pairs in which both particles have higher than average momentum add to $\langle \delta p_{t1} \delta p_{t2} \rangle$. Lower-than-average pairs also add to the covariance, while high/low pairs subtract from it. In global equilibrium $\langle \delta p_{t1} \delta p_{t2} \rangle \equiv 0$. The presence of hot spots makes $\langle \delta p_{t1} \delta p_{t2} \rangle > 0$ (as would cold spots) [22]. Motion of the sources would further enhance this quantity [7]. It follows that both jets and flow add to the p_t covariance.

It is important to notice that momentum fluctuations depend on the same correlations as multiplicity fluctuations, the only difference being momentum weights. Like (3) the integration in (5) counts pairs independent of their direction and therefore does not depend on the asymmetry of the system. Because of the momentum weights, (5) is sensitive to the average expansion and temperature and reflects the degree of thermalization of the system.

2.3. Flow Fluctuations

Unlike multiplicity and momentum fluctuations, harmonic flow correlation and fluctuation measurements depend on the azimuthal anisotropy of the collision system. The anisotropy of the collision can be characterized by the moments

$$\langle v_n \rangle = \langle \cos n(\phi - \psi_{RP}) \rangle, \tag{6}$$

where $\langle ... \rangle$ represents an average over particles and events and ψ_{RP} is the reaction plane angle defined by the plane spanned by the impact parameter $\mathbf{b}$ and the beam direction. Since the reaction plane cannot be measured, experiments turn to the two-particle correlation measurement

$$v_n\{2\}^2 = \frac{\langle \sum_{i \neq j} \cos n(\phi_i - \phi_j) \rangle}{\langle N(N-1) \rangle} \tag{7}$$

[3, 4, 23, 24, 25]. Being a two-particle correlation, we can cast (7) in terms of the pair distribution

$$v_n\{2\}^2 = \int \frac{\rho_2(\mathbf{p}_1, \mathbf{p}_2)}{\langle N(N-1) \rangle} \cos n(\phi_1 - \phi_2)\, d\mathbf{p}_1 d\mathbf{p}_2. \tag{8}$$

Using (1), (8) can be written as

$$v_n\{2\}^2 = \langle v_n \rangle^2 + 2\sigma_n^2, \tag{9}$$

where the flow coefficient relative to the reaction plane is $\langle v_n \rangle = \int \rho_1(\mathbf{p}) \cos n(\phi - \Psi_{RP}) d\mathbf{p}$ comes from the single particle distribution and the factor of two in the fluctuation term σ is conventional [26]. The effects of geometry primarily enter from the anisotropy of the single particle distribution characterized by (6). Genuine correlations contributing to (3) and (5) contribute to flow fluctuations

$$\sigma_n^2 = \int d\mathbf{p}_1 d\mathbf{p}_2 \, \frac{r(\mathbf{p}_1, \mathbf{p}_2)}{2\langle N(N-1)\rangle} \cos n\Delta\phi. \tag{10}$$

Ulike (3) and (5), (10) sums the relative azimuthal angles $\Delta\phi = \phi_1 - \phi_2$ of correlated pairs. Anisotropic flow would modify correlated pairs emerging from common density sources since those pairs would experience a different push depending on their location in the transverse plane.

Thus far we have only mentioned correlations among flowing particles. The correlation function $r(\mathbf{p}_1, \mathbf{p}_2)$ also contains correlations emerging from jets, resonance decays, HBT effect, etc.. These correlations may or may not be modified by flow but contribute to all mentioned observables. To eliminate these effects we can consider long range correlations and measurements of observables with rapidity separations.

2.4. The Ridge

Ridge measurements study two-particle azimuthal correlations at small and large (pseudo)rapidity separations $\Delta\eta$. Correlations at rapidity separations larger than $\sim 1-2$ units cannot come from jets, resonance decays or other short range phenomenon and causality dictates that such long range correlations originate in the early stages of the collision [27, 5].

The soft ridge observable developed by STAR can be shown to relate directly to flow coefficients as well as flow fluctuations by

$$\frac{\Delta\rho(\Delta\phi, \Delta\eta)}{\sqrt{\rho_{\mathrm{ref}}}} = \frac{1}{2\pi}\frac{dN}{d\eta}\left(2\sum_{n=1}^{\infty}\langle v_n\rangle^2 \cos n\Delta\phi + \frac{r(\Delta\phi, \Delta\eta)}{\rho_{\mathrm{ref}}}\right). \tag{11}$$

[2]. Here the $\langle v_n\rangle$ terms follow from (6), and ρ_{ref} is the experimental mixed-event background. We can isolate the contribution of flow fluctuations using

$$2\frac{dN}{dy}\sigma_n^2 \approx \int \left.\frac{\Delta\rho(\Delta\phi, \Delta\eta)}{\sqrt{\rho_{\mathrm{ref}}}}\right|_{FS} \cos(n\Delta\phi)\; d\Delta\phi d\Delta\eta, \tag{12}$$

where $(\Delta\rho(\Delta\phi, \Delta\eta)/\sqrt{\rho_{\mathrm{ref}}})|_{FS}$ is the flow-subtracted ridge amplitude on the near side, centered at $\Delta\phi = \Delta\eta = 0$ as reported by STAR [28, 29, 30].

Since ridge measurements rely on both flow and fluctuations, they can be compared to independent flow and flow fluctuation measurements to determine the relative contributions of long and short range correlations. Moreover, calculating all of the observables mentioned in Sec. 2 within the same framework can constrain the common correlation function (2) present in all measurements.

3. Glasma Correlations and Results

We argue that multiplicity, momentum, and flow fluctuations originate from initial state spatial correlations. These correlations emerge from fluctuating parton density distributions. Density lumps or hot spots emerge reflecting the positions of partons sources such as wounded nucleons or, in our case, Color Glass Condensate - Glasma flux tubes. We take correlated particles as those originating from the same source.

The spatial correlation function

$$c(\mathbf{x}_1, \mathbf{x}_2) = n_2(\mathbf{x}_1, \mathbf{x}_2) - n_1(\mathbf{x}_1)n_1(\mathbf{x}_2) \tag{13}$$

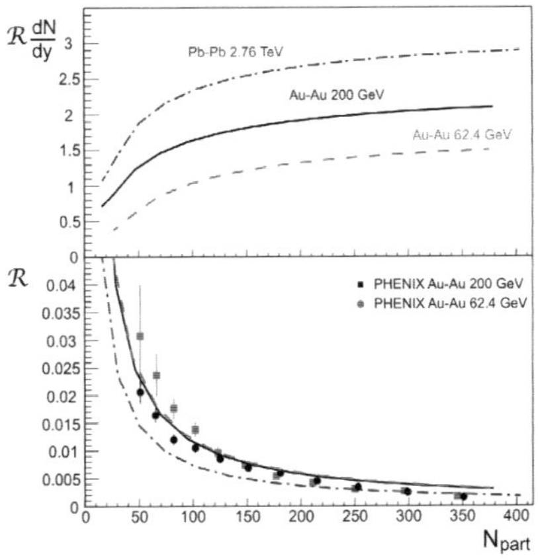

Figure 1. Prediction for $\mathcal{R}dN/dy$ as a function on the number of participants N_{part} at three beam energies (top). Calculated $\mathcal{R}$ compared to PHENIX data from [16] (bottom).

identifies the initial distribution of correlated pairs in the transverse space. If the pair distribution, $n_2(\mathbf{x}_1, \mathbf{x}_2)$, contains no correlations, it factorizes into the square of the singles distribution, $n_1(\mathbf{x}_i)$ and (13) vanishes. We write (13) in a Glasma flux tube picture as

$$c(\mathbf{x}_1, \mathbf{x}_2) = \langle N\rangle^2 \mathcal{R}\, \delta(\mathbf{r}_t)\rho_{FT}(\mathbf{R}_t). \tag{14}$$

The delta function in relative transverse elliptical coordinates $\mathbf{r}_t = \mathbf{r}_{t1} - \mathbf{r}_{t2}$ enforces the condition that correlated partons originate from the same flux tube presuming the flux tube transverse size is small compared to that of the collision area. The flux tube probability distribution, written in average coordinates $\mathbf{R}_t = (\mathbf{r}_{t1} + \mathbf{r}_{t2})/2$, is

$$\rho_{FT}(\mathbf{R}_t) \approx \frac{2}{\pi R_A^2}\left(1 - \frac{R_t^2}{R_A^2}\right). \tag{15}$$

Here the shape of (15) resembles the nuclear thickness function and πR_A^2 is the area of the overlap region. The distribution (15) represents an average over all possible shapes, which we have taken to be a simple ellipse.

To compute the correlation strength $\mathcal{R}$, we imagine each event produces K flux tubes with transverse size $\sim Q_s^{-2}$. In the saturation regime K is proportional to the transverse area R_A^2 divided by the area per flux tube, Q_s^{-2} [31]. Allowing K to fluctuate from event to event with average $\langle K\rangle$, we calculate in Ref. [5] that

$$\mathcal{R} = \frac{\langle N^2\rangle - \langle N\rangle^2 - \langle N\rangle}{\langle N\rangle^2} \propto \langle K\rangle^{-1}. \tag{16}$$

Each Glasma flux tube yields an average multiplicity of $\sim \alpha_s^{-1}(Q_s)$ gluons and as in Ref. [31], the number of gluons in a rapidity interval Δy is then

$$\langle N\rangle = (dN/dy)\Delta y \sim \alpha_s{}^{-1}(Q_s)\langle K\rangle. \tag{17}$$

Finally, the Glasma correlation scale,

$$\mathcal{R}dN/dy = \kappa\alpha_s^{-1}(Q_s^2), \tag{18}$$

follows from the multiplication of (3) and (17). Notice (18) is dimensionless and depends only on the saturation scale, Q_s^2, which can be calculated from first principles. Measurements of the

Journal of Physics: Conference Series **389** (2012) 012038 doi:10.1088/1742-6596/389/1/012038

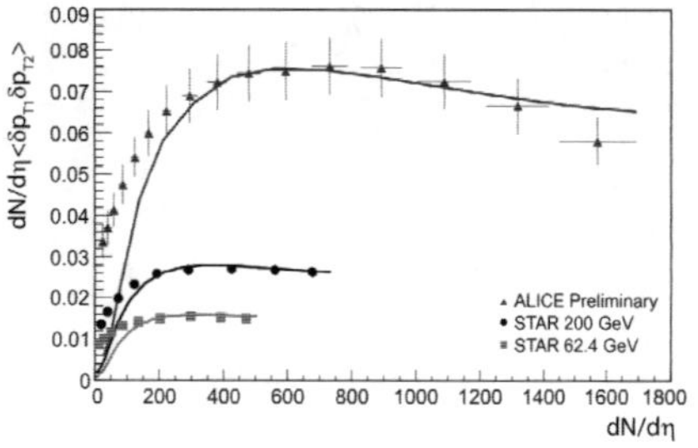

Figure 2. Transverse momentum fluctuations $\langle \delta p_{t1} \delta p_{t2} \rangle dN/d\eta$ as a function on the number of participants N_{part} at three beam energies. Data is from [21, 32].

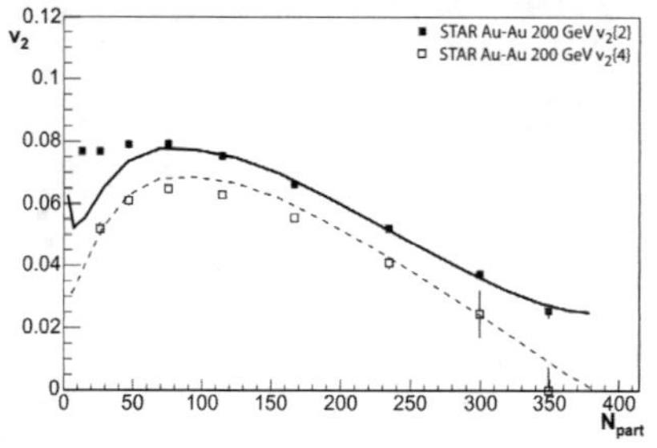

Figure 3. Measured $v_2\{2\}$ and $v_2\{4\}$ from STAR [33] compared to calculations. The difference between the calculated curves is due to flow fluctuations.

ridge at various beam energies, target masses, and centralities fix the dimensionless coefficient κ and are in excellent accord with the leading-order dependence [5, 6].

Intrinsically correlated partons originate from the same transverse position and experience, on average, the same transverse momentum modulation from flow. We represent the genuine two-particle momentum space correlation as

$$r(\mathbf{p}_1, \mathbf{p}_2) = \int c(\mathbf{x}_1, \mathbf{x}_2) f(\mathbf{x}_1, \mathbf{p}_1) f(\mathbf{x}_2, \mathbf{p}_2) d\Gamma_1 d\Gamma_2 \tag{19}$$

We evaluate (19) using a blast-wave description of the Boltzmann phase-space density $f(\mathbf{x}, \mathbf{p})$ on a Cooper-Frye freeze out surface $d\Gamma$ [1, 2]. Notice that full integration of (19) yields $\langle N \rangle^2 \mathcal{R}$ as does (13), tying azimuthal correlations to multiplicity and transverse momentum fluctuations [1].

Using (19) we calculate (5) and find the results shown in Fig. 2. We find good agreement in central collisions where the assumption of gluon saturation necessary for the production of Glasma flux tubes is most valid. Moreover, the onset of thermalization in peripheral collisions should modify the correlation function. In particular, this effect has been shown to modify $\langle \delta p_{t1} \delta p_{t2} \rangle$ at low numbers of participants [22]. Partial thermalization describes peripheral RHIC data very well and, moreover, allows one to describe *pp* and *AA* collisions in the same model [34].

Just as with (5) we can use the same correlation function (19) to study flow fluctuations (10) and therefore examine the effects of event anisotropy. If flow fluctuations arise only from changes in the shape of the collision area, then their governing correlation function would predict $\langle \delta p_{t1} \delta p_{t2} \rangle = 0$ and have almost no dependence on collision energy. To start we calculate (10) for 200 GeV Au+Au collisions. Using this along with the experimental definition

$$\sigma_n^2 = \frac{v_n\{2\}^2 - v_n\{4\}^2}{2}. \tag{20}$$

and taking $v_n\{4\} \approx \langle v_n \rangle$ we calculate $v_2\{2\}$ as the solid line in Fig. 3. The dashed line is (6) calculated from blast wave and flow fluctuations (10) determine the difference between the two curves. Notice that in central collisions the circular event geometry drives $\langle v_2 \rangle$ to zero. Here flow fluctuations are largest and completely determine $v_2\{2\}$.

As with multiplicity fluctuations and p_t fluctuations, the correlation strength $\mathcal{R}$ sets the scale for flow fluctuations. In our model, the increase in growth of flow fluctuations (10) is coupled

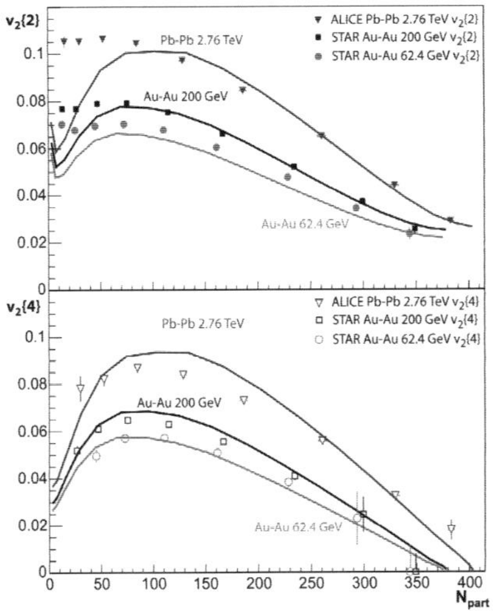

Figure 4. Measured $v_2\{2\}$ (top) from STAR [33] and ALICE [35, 36] compared to calculations using (9) and (10). Same for $v_2\{4\}$ (bottom) computed from (6) in the blast wave.

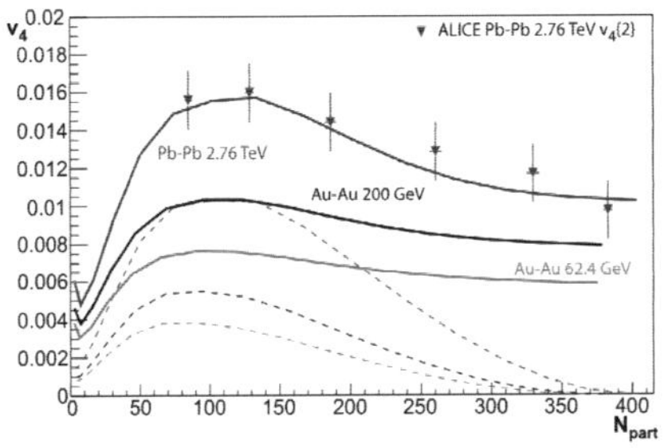

Figure 5. Same calculation as in Fig. 4 but for the fourth order flow coefficients $v_4\{2\}$ (solid) and $v_4\{4\}$ (dashed). ALICE data [35].

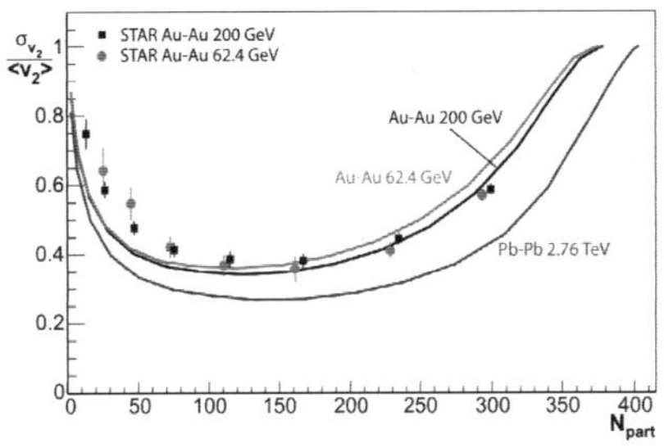

Figure 6. Flow fluctuations in terms of the coefficient of variation for elliptic flow (21) compared to STAR data [33].

to the growth in the saturation scale Q_s^2 via (18). Comparisons of $v_2\{2\}$ and $v_2\{4\}$ to different collision energy measurements are shown in Fig. 4. It is also informative to make comparisons to v_4 as shown in Fig. 5. The effect of the change in blast wave parameters with collision energy is apparent from the separation of the solid lines in the bottom panel of Fig. 4 and the dashed lines in Fig. 5. The importance of the Glasma contribution is similarly evident by the change is separation between $v_n\{4\}$ and $v_n\{2\}$ with collision energy.

STAR has also measured v_2 fluctuations in the form of

$$\frac{\sigma_{v_n}}{\langle v_n \rangle} = \sqrt{\frac{v_n\{2\}^2 - v_n\{4\}^2}{v_n\{2\}^2 + v_n\{4\}^2}} \tag{21}$$

resembling the so-called "coefficient of variation" defined as the standard deviation divided by the mean. Care should be taken here since the definition of σ_n^2, Eq.(20), is not strictly the variance. In Fig. 6 we compare our calculation of (21) to measurement. Calculations at RHIC energies seem to agree reasonably well and we include the calculation for Pb+Pb 2.76 GeV as a prediction.

In Fig. 7 we show v_3 fluctuations from Glasma. We further emphasize that the energy dependence in Fig. 7 is in good accord with data, supporting the Glasma scaling with Q_s^2. The

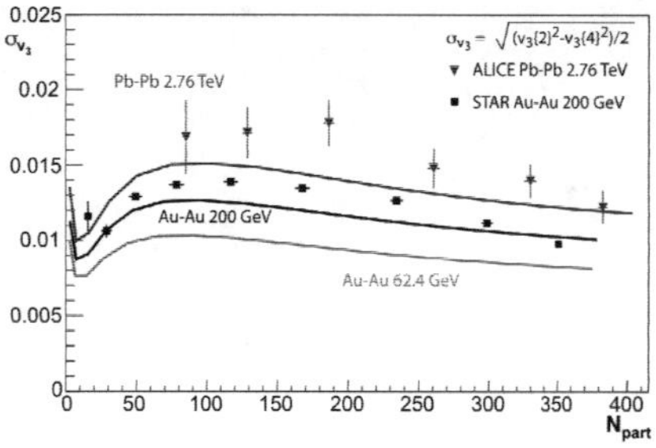

Figure 7. Triangular flow fluctuations, σ_{v_3}, Eq. (10). Data points are computed from $v_3\{2\}$ and $v_3\{4\}$ measurements [37, 35].

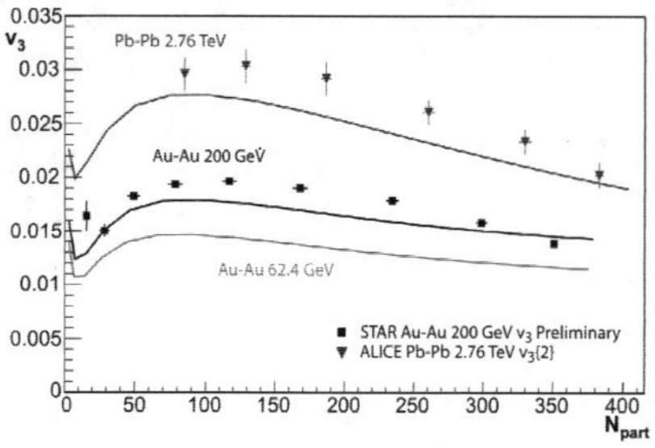

Figure 8. The triangular flow coefficient, $v_3\{2\}$ compared to STAR and ALICE data [37, 35].

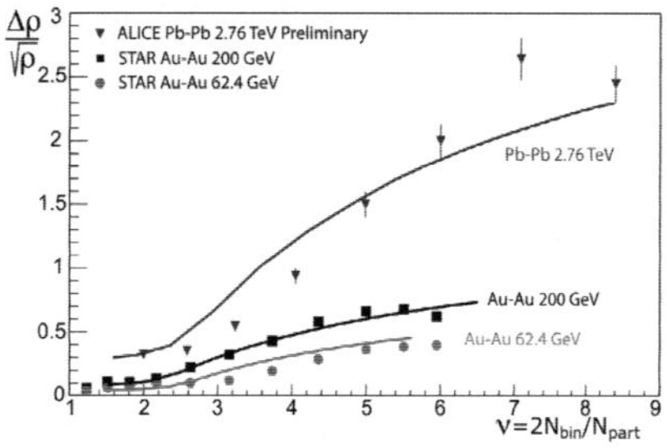

Figure 9. Near side ridge amplitude calculation from Glasma source correlations. Experimental data is from (STAR) [28, 29, 30] and (ALICE) [38].

shape with centrality reflects contributions not only from Glasma, but also our parameterizations of the average reaction plane eccentricity ε and our choice the flux tube distribution ρ_{FT}.

To use (10) and (20) to calculate $v_3\{2\}$, we must come to grips with the fact that our blast wave parametrization assumes $v_3\{4\} = 0$. While this seems to be the case for the STAR measurements, ALICE has measured a non-zero $v_3\{4\}$ for Pb+Pb collisions at 2.76 TeV. To correct for this possible discrepancy, we can provide an ad-hoc parameterization of $v_3\{4\}$ and use (10) and (20) to calculate $v_3\{2\}$. Agreement shown in Fig. 8 is reasonable. As a preferable alternative, we compare our calculated σ_n to fluctuations extracted from ALICE and STAR measurements in Fig. 7.

The characterization of the ridge in terms of Fourier flow coefficients can prove a valuable check for models describing flow and its fluctuations with long range correlations. Relative contributions of flow and flow fluctuations to the ridge decoupled using Eq.(12). Though it would be more informative to extract σ_n directly from ridge measurements, we can compare the flow-subtracted near side ridge amplitude to integration of (2), suitably normalized to the mixed event background, see Refs.[2, 5, 6] for details.

Our calculation of the flow-subtracted near side ridge amplitude can be seen in Fig. 9. The dimensionless coefficient κ in (18) is set by the change in amplitude of the 200 GeV Au+Au measurement and remains fixed for all other calculations. Like the other observables, the change in collision energy is determined by $\mathcal{R}$ and therefore Q_s^2. The agreement with data is quite reasonable. Commonality of the energy dependence between the ridge and flow fluctuations should not come as a surprise. Measurements of $v_2\{2\}$ at ALICE using particle pairs with and without a rapidity separation only show noticeable difference in peripheral collisions [36, 39]. The remaining difference between $v_2\{2\}$ and $v_2\{4\}$ suggests (20) results from a long range correlation. However, long range correlation explanation of the energy dependence of multiplicity and momentum fluctuations does come as a surprise. Jets and resonance decays, for example, should have significant effects on these observables. Further investigation into the relative contributions of long and short range correlations can indicate if indeed long range correlations dominate fluctuation observables at these energies.

4. Discussion

In this work we study the connection between correlations and fluctuations of multiplicity, momentum, and harmonic flow coefficients. We first argue in Sec. 2 that given the existence of genuine two-particle correlations in the hadron pair distribution, Eq. (1), multiplicity fluctuations, momentum fluctuations, flow fluctuations, and the ridge are all linked by a common scale $\mathcal{R}$. We then consider contributions to the genuine correlations due to fluctuating or lumpy initial parton distributions; correlated partons emerge from the same source/lump. We take the origin of initial state parton density lumps from Glasma flux tubes and present our model calculations in Sec. 3.

We focus on two important features of the data: the agreement with the change in collision energy and the link between the correlation and fluctuation observables. Long range correlations such as those measured in the ridge indicate that the correlations originate at early times in the collision. The connection between the ridge and flow fluctuations is evidence that the difference between $v_n\{4\}$ and $v_n\{2\}$ is also a long range phenomenon. Given that flow fluctuations result from the existence of parton density lumps in the initial state, one can compare the behavior of flow fluctuations to that of other consequences of the lumpy initial state such as multiplicity and momentum fluctuations. The correlation strength $\mathcal{R}$ sets a common scale for all of these observables and controls much of the centrality and collision energy dependence. Theories of the initial state can therefore be constrained with measurements of $\mathcal{R}$. In this work we have chosen a CGC-Glasma based description of $\mathcal{R}$ and found reasonable agreement in the centrality and collision energy changes in all of the mentioned observables. This suggests that saturation based descriptions of the initial state are consistent with at least mid-central to central collisions.

Acknowledgments

This work was supported in part by the U.S. NSF grant PHY-0855369 (SG) and The Alliance Program of the Helmholtz Association (HA216/EMMI) (GM)

References

[1] Gavin S and Moschelli G 2012 *Phys.Rev.* **C85** 014905 (*Preprint* 1107.3317)
[2] Gavin S and Moschelli G 2012 (*Preprint* 1205.1218)
[3] Borghini N, Dinh P M and Ollitrault J Y 2001 *Phys. Rev.* **C63** 054906 (*Preprint* nucl-th/0007063)
[4] Borghini N, Dinh P M and Ollitrault J Y 2001 *Phys.Rev.* **C64** 054901 (*Preprint* nucl-th/0105040)
[5] Gavin S, McLerran L and Moschelli G 2009 *Phys.Rev.* **C79** 051902 (*Preprint* 0806.4718)
[6] Moschelli G and Gavin S 2010 *Nucl.Phys.* **A836** 43–58 (*Preprint* 0910.3590)
[7] Voloshin S A 2006 *Phys.Lett.* **B632** 490–494 (*Preprint* nucl-th/0312065)
[8] Pruneau C A, Gavin S and Voloshin S A 2008 *Nucl. Phys.* **A802** 107–121 (*Preprint* 0711.1991)
[9] Lindenbaum S J and Longacre R S 2007 *Eur. Phys. J.* **C49** 767–782
[10] Sorensen P 2008 (*Preprint* 0811.2959)
[11] Peitzmann T 2009 (*Preprint* 0903.5281)
[12] Takahashi J *et al.* 2009 *Phys. Rev. Lett.* **103** 242301 (*Preprint* 0902.4870)
[13] Andrade R P G, Grassi F, Hama Y and Qian W L 2010 (*Preprint* 1008.4612)
[14] Werner K, Karpenko I, Pierog T, Bleicher M and Mikhailov K 2010 (*Preprint* 1004.0805)
[15] Pruneau C, Gavin S and Voloshin S 2002 *Phys. Rev.* **C66** 044904 (*Preprint* nucl-ex/0204011)
[16] Adare A *et al.* (PHENIX) 2008 *Phys. Rev.* **C78** 044902 (*Preprint* 0805.1521)
[17] Gelis F, Lappi T and McLerran L 2009 *Nucl. Phys.* **A828** 149–160 (*Preprint* 0905.3234)
[18] Voloshin S, Koch V and Ritter H 1999 *Phys.Rev.* **C60** 024901 (*Preprint* nucl-th/9903060)
[19] Voloshin S A (STAR) 2001 591–596 (*Preprint* nucl-ex/0109006)
[20] Adamova D *et al.* (CERES) 2003 *Nucl.Phys.* **A727** 97–119 (*Preprint* nucl-ex/0305002)
[21] Adams J *et al.* (STAR) 2005 *PHRVA,C72,044902.2005* **C72** 044902 (*Preprint* nucl-ex/0504031)

[22] Gavin S 2004 *Phys. Rev. Lett.* **92** 162301 (*Preprint* nucl-th/0308067)
[23] Bilandzic A, Snellings R and Voloshin S 2011 *Phys.Rev.* **C83** 044913 (*Preprint* 1010.0233)
[24] Voloshin S A, Poskanzer A M and Snellings R 2008 (*Preprint* 0809.2949)
[25] Sorensen P 2009 (*Preprint* 0905.0174)
[26] Voloshin S A, Poskanzer A M, Tang A and Wang G 2008 *Phys.Lett.* **B659** 537–541 (*Preprint* 0708.0800)
[27] Dumitru A, Gelis F, McLerran L and Venugopalan R 2008 *Nucl. Phys.* **A810** 91 (*Preprint* 0804.3858)
[28] Daugherity M (STAR) 2006 *PoS* **CFRNC2006** 005 (*Preprint* nucl-ex/0611032)
[29] Daugherity M S 2008 *Two-particle correlations in ultra relativistic heavy ion collisions* Ph.D. thesis
[30] Daugherity M (STAR) 2008 *J. Phys.* **G35** 104090 (*Preprint* 0806.2121)
[31] Kharzeev D and Nardi M 2001 *Phys. Lett.* **B507** 121–128 (*Preprint* nucl-th/0012025)
[32] Heckel S (ALICE) 2011 (*Preprint* 1107.4327)
[33] Pruthi N K (The STAR Collaboration) 2011 (*Preprint* 1111.5637)
[34] Gavin S 2004 *J. Phys.* **G30** S1385–S1388 (*Preprint* nucl-th/0404048)
[35] Aamodt K *et al.* (ALICE Collaboration) 2011 *Phys.Rev.Lett.* **107** 032301 (*Preprint* 1105.3865)
[36] Bilandzic A 2012 *Anisotropic flow measurements in ALICE at the large hadron collider* Ph.D. thesis Utrecht University
[37] Sorensen P (STAR Collaboration) 2011 *J.Phys.G* **G38** 124029 8 pages. Conference proceedings for Quark Matter 2011 (*Preprint* 1110.0737)
[38] Timmins A R (ALICE) 2011 (*Preprint* 1106.6057)
[39] Snellings R 2011 *J. Phys.* **G38** 124013 (*Preprint* 1106.6284)

$\Delta\eta - \Delta\phi$ correlations and the ridge structure in STAR

L. C. De Silva for the STAR collaboration

The Department of Physics, University of Houston, 617 SR Building 1, Houston, Texas 77204-500

E-mail: desilva@uh.edu

Abstract. Triggered di-hadron correlation studies using central Au+Au collisions at $\sqrt{S_{NN}}$ = 200 GeV in STAR revealed a novel ridge-like structure in two dimensions ($\Delta\eta$, $\Delta\phi$) for high p_T particles. A similar structure is also present in an inclusive un-triggered di-hadron correlation analysis. We study the $\langle p_T \rangle$ evolution of di-hadron correlations by increasing the lower p_T acceptance of both charged particles. A smooth evolution of data is observed and our results reproduce the triggered analysis structure near $\langle p_T \rangle$ = 2.7 GeV/c. We quantify the correlation structure evolution by fitting a model function. The model function emphasizes possible initial state fluctuation contributions via the use of higher harmonic model components v_n (n=1,2,3,4,5) and a remainder which is modeled via an asymmetric 2d Gaussian. The extracted parameters are compared to model predictions and p+p data at $\sqrt{S_{NN}}$ = 200 GeV and possible origins of the nearside structure are discussed.

1. Introduction

Two dimensional, high p_T di-hadron correlation studies in STAR revealed a novel correlation structure elongated in pseudo-rapidity ($\Delta\eta$) [1]. This structure which is known as the "ridge" was observed by applying an asymmetric transverse momentum cut on trigger (4 GeV/c > p_T > 3 GeV/c) and associate ($p_T > 2GeV/c$) particles [2]. Furthermore, it was revealed that the structure persists up to trigger particle p_T > 6 GeV/c, possibly suggesting a jet-related origin. An independent analysis [3] also observed the ridge correlation for charged particles with p_T > 0.15 GeV/c for both trigger and associate particles. This analysis differs from the above mentioned one in three key aspects. It uses all possible correlated particle pairs, it applies a common lower p_T threshold to all particles and the correlation function is normalized in a way that yields correlated particles per final state charged particle. The measured structure is not attributed to expected physical processes and in the past several attempts have been made to explain the long-range correlations through the medium modification of elementary processes such as jets or flux tube formation [4,5,6].

For our study we adopt the un-triggered analysis method and focus on the correlation structure in the 0-10% centrality bin where we have the highest probability of forming a deconfined medium of quarks and gluons. We then investigate the evolution of the correlation structure as a function of transverse momentum. The structure is studied by increasing the lower transverse momentum acceptance for both particles. The resulting correlation is modeled via an empirical model fit function. We discuss our findings in the context of models that describe medium properties and modified jet phenomena in deconfined QCD matter.

Published under licence by IOP Publishing Ltd

Journal of Physics: Conference Series **389** (2012) 012039 doi:10.1088/1742-6596/389/1/012039

2. Data and analysis

The data used in this analysis were collected during two runs, in 2004 and 2009, using the STAR detector at the Relativistic Heavy Ion Collider (RHIC) Brookhaven National Lab (BNL), Long Island, New York. For the results presented here we have analyzed 11 million Au+Au central trigger events and 264 million p+p minimum bias events at $\sqrt{S_{NN}} = 200$ GeV. Only events with a primary vertex within |x,y| < 3cm and |z| < 25cm were taken into account. Charged particle tracks were reconstructed by the STAR TPC only and constrained to the primary vertex. The correlation function was calculated only for tracks within the kinematical cuts of $|\eta| < 1.0$, full 2π acceptance in azimuth ϕ and $p_T = 0.15$-15 GeV/c. The correlation measure for this analysis is defined as follows:

$$\frac{\Delta\rho}{\sqrt{\rho_{ref}}} = \frac{\rho_{sib} - \rho_{ref}}{\sqrt{\rho_{ref}}} \tag{1}$$

The symbol ρ_{sib} refers to the total correlated and uncorrelated pair density as a function of $\Delta\eta$ and $\Delta\phi$, for hadrons from a single event. ρ_{ref} refers to the number of uncorrelated pairs as a function of $\Delta\eta$ and $\Delta\phi$, for hadrons from different events. $\Delta\rho$ is therefore the number of correlated pairs in a single event. The denominator is the square root of the number of background pairs, which corresponds to the number of particles. The correlation function therefore measures the number of correlated pairs per particle. In order to correct for acceptance effects, we sub-bin the z vertex into ten 5cm bins and allow a multiplicity window of 50 for event mixing. The conversion electron/positron background is reduced by using a 1.5σ dE/dx cut on electrons in the momentum ranges, $0.2 < p_T < 0.45$ GeV/c and $0.7 < p_T < 0.8$ GeV/c. In order to analyze the transverse momentum evolution of di-hadron correlations in the 0-10% centrality bin we steadily increase the lower particle p_T to gain further insight into the origin of the long range correlations. The correlation function evolution for eight selected bins is shown in Fig. 1. We model structures via the use of an empirical function and infer potential underlying physics mechanisms of the near side long range correlation. In the following section we discuss the model fit function.

3. Model function study

The main part of the model function is based on recent theory developments related to initial energy density fluctuation contributions to the two particle correlation structure [7]. It has been pointed out that the fluctuation contributions are not negligible even in the raw data spectrum and should be observable in very central events using high p_T charged particles, as shown in Fig. 2. The remaining structure after subtracting out the higher order Fourier harmonics, which we term the remainder, is modeled via an asymmetric 2d Gaussian, and thus can be attributed to modified jet phenomena. The remainder properties are quantified by comparing to p+p data.

3.1. Evidence for initial density fluctuations in data

Following the theory prediction [7], we focus our attention to the 0-1% centrality bin. Fig. 2 shows the projected 2d correlation structure on to $\Delta\phi$ as a function of p_T revealing a double hump structure in the $2 < p_T < 5$ GeV/c interval for the away side correlations. This observation is consistent with the kinematic ranges predicted by theory and most likely due to triangular flow [8].

3.2. Model function

We initially attempt to describe the data using a maximum number of Fourier coefficients, which for our data requires coefficients up to n = 5 in the $\cos(n\Delta\phi)$ series. However the remaining

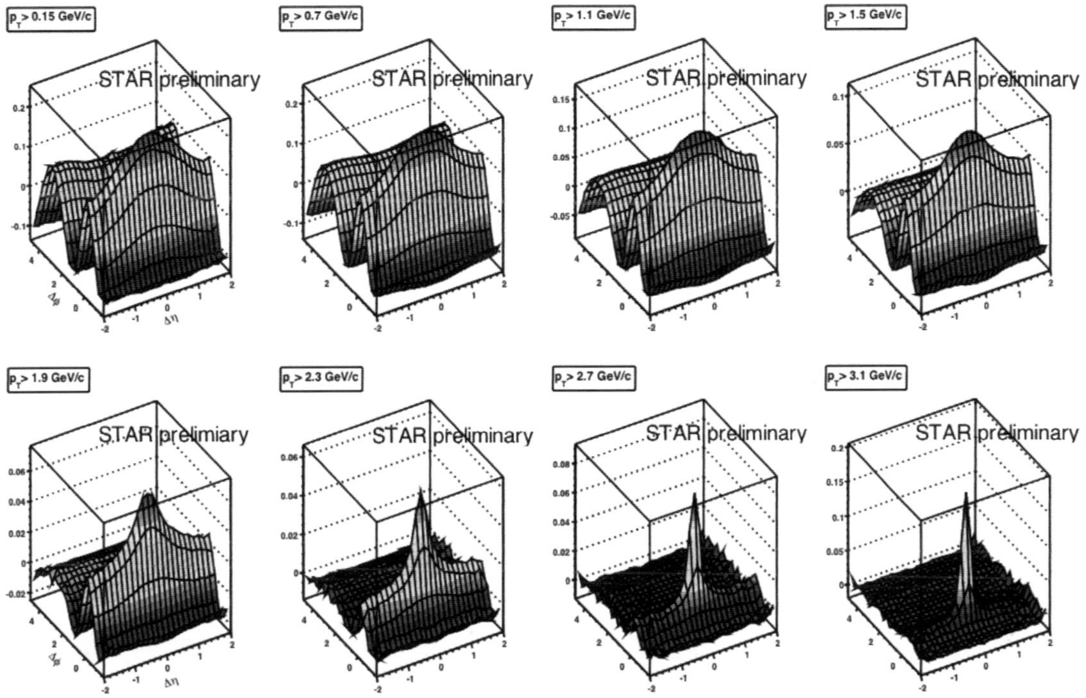

Figure 1. Evolution of the raw normalized correlation structure for selected cuts on p_T of the particles (increasing $\langle p_T \rangle$ from left to right) at 0−10% centrality bin.

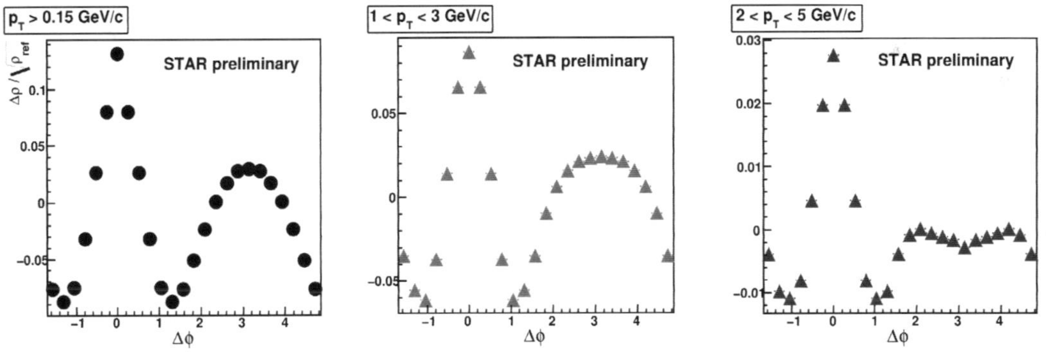

Figure 2. Angular correlations as a function of p_T cuts on the pairs for 0-1% central data.

structure reveals a significant residual, thus requiring an additional asymmetric 2d Gaussian term. The complete function can be broken down as follows.

$$\begin{aligned}
F &= c_0 && \propto \textit{Offset } (f_1) \\
&\quad + c_1 * cos(1\Delta\phi) && \propto \textit{Away-side momentum conservation } (f_2) \\
&\quad + c_2 * cos(2\Delta\phi) && \propto \textit{Second order Fourier harmonic } (f_3) \\
&\quad + c_3 * cos(3\Delta\phi) && \propto \textit{Third order Fourier harmonic } (f_4) \\
&\quad + c_4 * cos(4\Delta\phi) && \propto \textit{Fourth order Fourier harmonic } (f_5) \\
&\quad + c_5 * cos(5\Delta\phi) && \propto \textit{Fifth order Fourier harmonic } (f_6) \\
&\quad + c_6 * exp(-0.5 * ((\Delta\phi/c_7)^2 + (\Delta\eta/c_8)^2)) && \propto \textit{Modified jet fragmentation } (f_7)
\end{aligned}$$

$$F = f_1 + f_2 + f_3 + f_4 + f_5 + f_6 + f_7$$

In the following section we discuss and interpret the model parameters we extract.

4. Results and Discussion

4.1. Results

In this paper we mainly focus on the evolution of the model components shown in Fig. 3. Fig. 3(a) shows the evolution of the remainder component after subtracting out the harmonics (n=1-5) contributions to the correlation structure. Fig. 3(b) shows the sum of all harmonics (n=2-5) contributions. The remainder parameters might shed some light on potential modification to jets in the medium whereas the harmonics parameters can be used to test predictions from hydrodynamical models.

The $\Delta\eta$-$\Delta\phi$ asymmetry of the remainder widths are shown in Fig. 4. We observe an asymmetry below $\langle p_T \rangle \approx 2.5$ GeV/c. The largest asymmetry is observed at $p_T \approx 1.3$ GeV/c where the $\Delta\eta$ width is a factor of two greater than the $\Delta\phi$ width.

We compare the extracted parameters to the same parameters from our analysis of p+p data in Fig. 5. In the region above 2.5 GeV/c where the $\Delta\eta$-$\Delta\phi$ widths are symmetric in Au+Au, the widths are about a factor of two larger than in p+p. Below that $\langle p_T \rangle$ value, we observe modification in both $\Delta\eta$ and $\Delta\phi$ remainder withs compared to corresponding p+p widths. Furthermore, the maximum relative width modulation (AA/pp) is a factor of two greater in $\Delta\eta$ width compared to $\Delta\phi$ width with the maximum modulations occuring at two different $\langle p_T \rangle$ values.

Fig. 6 shows the comparison between the remainder and the sum of all harmonics amplitudes. The amplitudes are comparable in the $0.9 < p_T < 2.1$ GeV/c range. The summed harmonics amplitude, which extracts the near-side long range correlation in our model, saturates at a small value at high p_T. The correlation strength of the added harmonic structure peaks at a $\langle p_T \rangle \approx 0.9$ GeV/c which is higher than the $\langle p_T \rangle$ for bulk particles thus indicating that long-range correlations might carry particles coming from jet fragmentation (hard processes). The asymmetric remainder amplitude drops with $\langle p_T \rangle$ and the symmetric remainder amplitude rises. The behavior of the symmetric remainder amplitude at $p_T > 2.5$ GeV/c can be understood based on more correlated jet fragment pairs in a jet like cone.

Finally we show scaling relations using higher order harmonics in order to compare our findings to certain hydro predictions [9]. Plotted in fig. 7 are the ratios $v_4^{1/4}/v_2^{1/2}$ and $v_3^{1/3}/v_2^{1/2}$. Theory predicts those ratios should be independent of $\langle p_T \rangle$ in the bulk p_T regime. Even though we observe the predicted trends, in order to extract more information about the medium properties from the scaling ratios, we need to compare our results to theory applied to similar kinematic constraints (centrality and p_T) than the data.

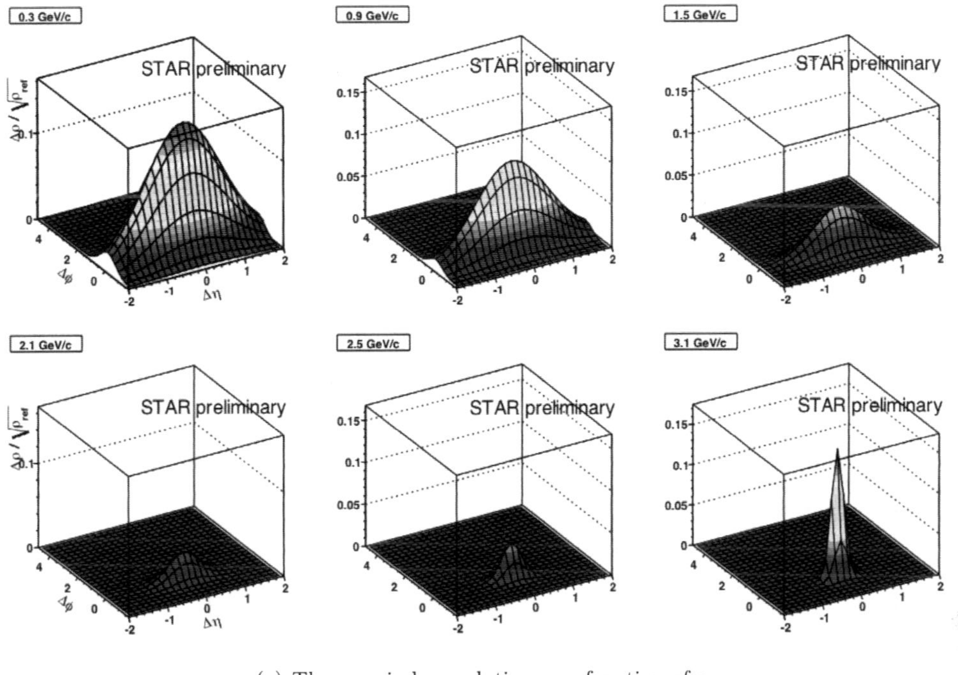

(a) The remainder evolution as a function of p_T

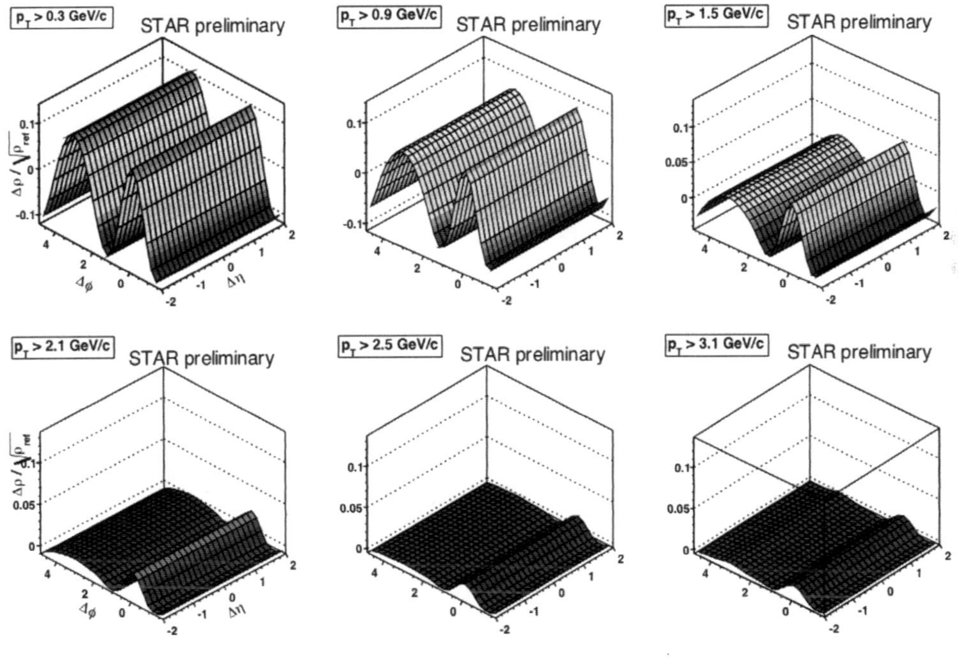

(b) $v_2 + v_3 + v_4 + v_5$ evolution as a function of p_T

Figure 3. The evolution of the "remainder" (a) and the summed v_n structure (n = 2 - 5) (b), as a function of the lower cut on p_T of the particles.

4.2. Discussion

We studied the evolution of di-hadron correlations in angular variables $\Delta\eta$, $\Delta\phi$ using Au+Au collisions at $\sqrt{S_{NN}} = 200$ GeV. Our main focus was to study the long-range correlation observed

Journal of Physics: Conference Series **389** (2012) 012039 doi:10.1088/1742-6596/389/1/012039

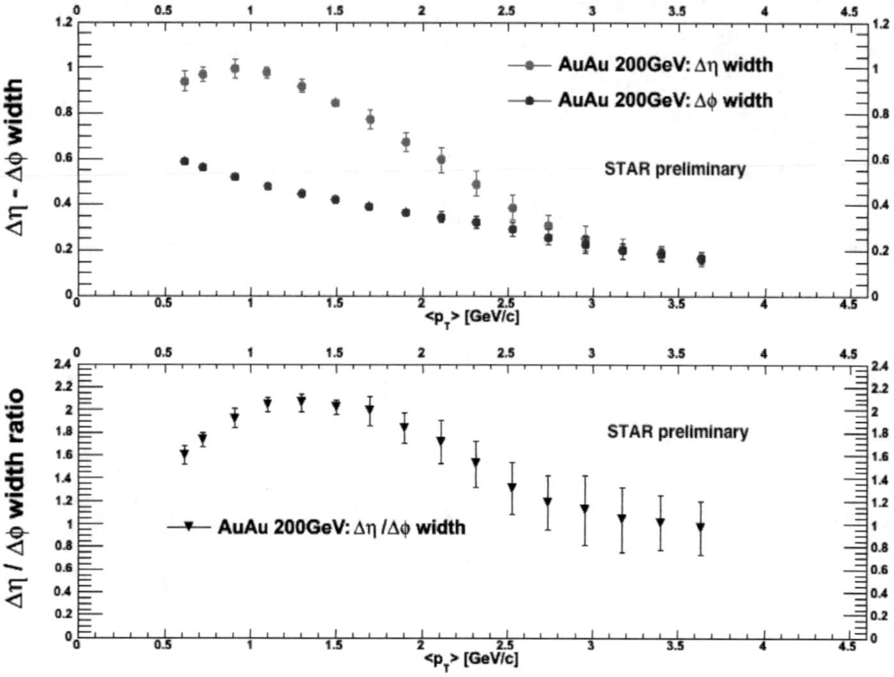

Figure 4. The $\langle p_T \rangle$ evolution of $\Delta\eta$, $\Delta\phi$ widths of the remainder.

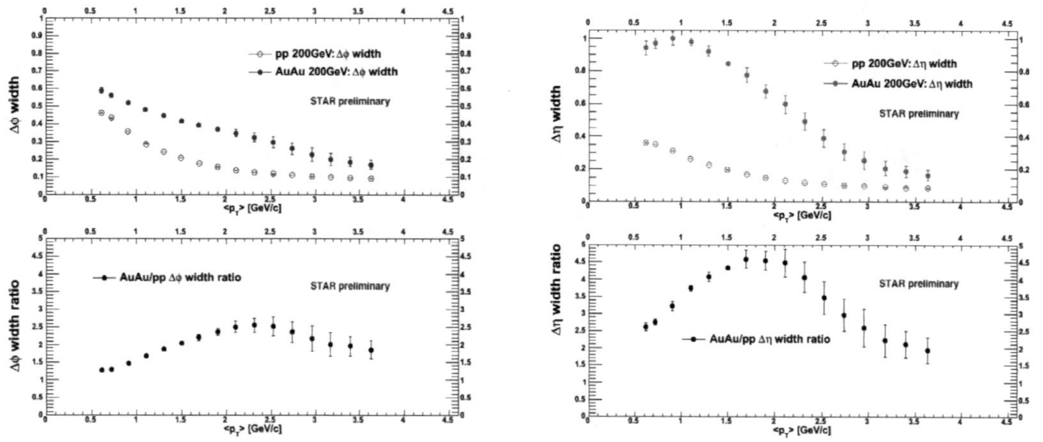

(a) $\Delta\phi$ width comparison

(b) $\Delta\eta$ width comparison

Figure 5. Comparison of the remainder $\Delta\eta$, $\Delta\phi$ widths to corresponding p+p widths.

in the central events bin by increasing the lower p_T acceptance of both charged particles. By focusing on 0-10% centrality bin, we study the most violent heavy ion collisions where a formation of a quark gluon plasma is expected. Therefore our study of the long-range correlation could reveal further insight into the properties of the deconfined medium.

The measured structures evolve smoothly as a function of $\langle p_T \rangle$. The near-side correlation evolves from an elongated 2d Gaussian-like structure towards a structure which has a flat pedestal and a jet like peak on top of it. The away side structure broadens and flattens at higher p_T. Both the

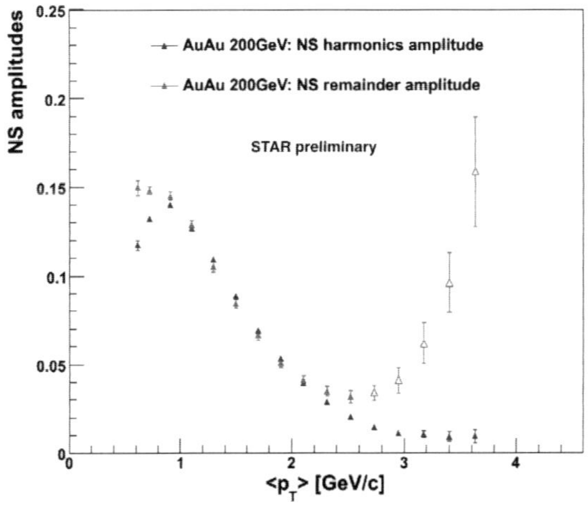

Figure 6. Remainder amplitude variation compared to summed harmonic ($v_2 + v_3 + v_4 + v_5$) amplitude.

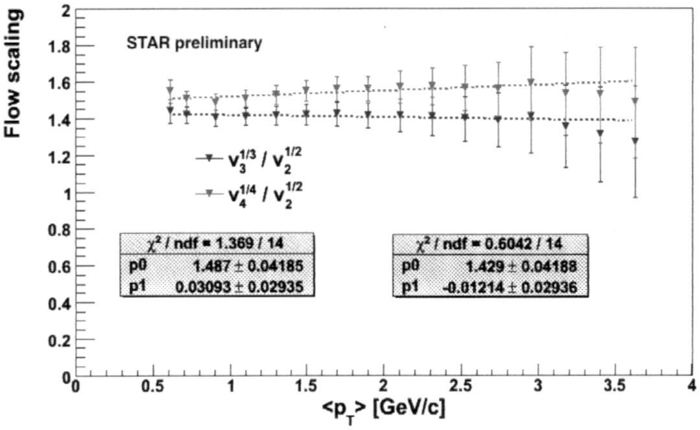

Figure 7. Higher order harmonic scaling relations using, v_2, v_3 and v_4.

near side long-range correlation and away side amplitude reduces with increasing p_T. However, we observe a jet like peak strength growth (see fig. 1) after $\langle p_T \rangle \approx 2.5$ GeV/c possibly due to an increase of correlated pairs within a fragmenting high p_T parton even though the total yield in the correlation function (including uncorrelated background) drops, as expected.

The data evolution was modeled using an empirical model function which has nine free parameters. We extract the contribution to the correlations due to initial energy density fluctuations via the use of higher order Fourier harmonics $\cos(n\Delta\phi)$, n = 1,2,3,4,5. In addition, the data require an asymmetric 2d Gaussian model component in order to fit the residual structure after subtracting the harmonics contribution.

The evidence for initial energy density fluctuations in the correlation data (see Fig. 2) implies that we do create a medium that behaves collectively. We further test the collectivity of the medium by testing our predictions against hydrodynamical scaling relations (see Fig. 7) and observe the predicted trends. Theory states that any deviation of the scaling ratios from ideal

hydro predictions could be due to partial thermalization effects [9]. However our present scaling data could not be directly compared to those predictions due to discrepancies in kinematic cuts. Nonetheless, our extracted v_n parameters are useful in studying medium properties(e.g. viscosity and diffusion coefficients) and initial conditions (e.g. CGC vs. Glauber initial distributions) of the colliding heavy ions.
The remainder component after subtracting harmonics, can be viewed in two parts. Below $\langle p_T \rangle \approx 2.5$ GeV/c the remainder widths are asymmetric and the amplitude drops as a function of $\langle p_T \rangle$. In the same $\langle p_T \rangle$ region, the widths are modified compared to p+p data. The relative asymmetries as well as width modulation with respect to p+p data occur at different $\langle p_T \rangle$ values. Nonetheless, the remainder below $\langle p_T \rangle \approx 2.5$ GeV/c is elongated in $\Delta\eta$ and more modified than the $\Delta\phi$ width in comparison to p+p parameters. Therefore the cause of the asymmetric remainder component is open to interpretation and the measured modulation factors could be useful input for jet shape modification models. On the other hand, the symmetric remainder component above $\langle p_T \rangle \approx 2.5$ GeV/c shows a width modification of a factor two compared to p+p collisions. Theorists could use this information to constrain partonic energy loss models and calculation of energy transport coefficients in the deconfined medium.
In summary, we discussed that the near side long range correlation structure can be understood as arising from initial energy density fluctuations and possible modified jet phenomena. The modified asymmetric 2d Gaussian needs to be studied further in the context of jet fragmentation. Charge dependent and particle identified measurements are two suggested approaches for further studies.

References
[1] J. Adams et al. (STAR Collaboration), Phys. Rev. Lett. 95, 152301 (2005)
[2] J.Putschke for the STAR collaboration, J. Phys. G34, S679 (2007)
[3] G. Agakishiev et al. (STAR Collaboration), arXiv:1109.4380
[4] J. Porter and T. Trainor, J. Phys.: Conf. Ser. 27, 98 (2005)
[5] S. Voloshin, Phys. Lett. B632, 490 (2006)
[6] S. Gavin, G. Moschelli, L. McLerran, Phys. Rev. C79, 051902 (2009)
[7] Xin-Nian Wang et al., Phys. Rev. Lett. 106, 162301(2011)
[8] B. Alver et al., Phys. Rev. C 81, 054905 (2010)
[9] C. Gombeaud et al., Phys. Rev. C 81, 014901 (2010)

Journal of Physics: Conference Series **389** (2012) 012040 doi:10.1088/1742-6596/389/1/012040

Separating jets from bulk matter in heavy ion collisions at the LHC

K. Werner[(a)], Iu. Karpenko[(b)], M. Bleicher[(c)], T. Pierog[(d)], S. Porteboeuf-Houssais[(e)]

[(a)] SUBATECH, University of Nantes – IN2P3/CNRS– EMN, Nantes, France
[(b)] Bogolyubov Institute for Theoretical Physics, Kiev 143, 03680, Ukraine
[(c)] FIAS, Johann Wolfgang Goethe Universitaet, Frankfurt am Main, Germany
[(d)]Karlsruhe Inst. of Technology, KIT, Campus North, Inst. f. Kernphysik, Germany
[(e)]LPC, Universit Blaise Pascal, CNRS-IN2P3, Clermont-Ferrand, France

E-mail: werner@subatech.in2p3.fr

Abstract. We discuss a theoretical scheme that accounts for bulk matter, jets, and the interaction between the two. The physical picture of our approach is the following: Initial hard scatterings result in mainly longitudinal flux tubes, with transversely moving pieces carrying the p_t of the partons from hard scatterings. These flux tubes constitute eventually both bulk matter (which thermalizes, flows, and finally hadronizes) and jets, according to some criteria based on partonic energy loss. High energy flux tube segments will leave the fluid, providing jet hadrons via the usual Schwinger mechanism of flux-tube breaking caused by quark-antiquark production. But the jets may also be produced at the freeze-out surface. Here we assume that the quark-antiquark needed for the flux tube breaking is provided by the fluid, with properties (momentum, flavor) determined by the fluid rather than the Schwinger mechanism. Considering transverse fluid velocities up to 0.7c, and thermal parton momentum distributions, one may get a "push" of a couple of GeV to be added to the transverse momentum of the string segment. This will be a crucial effect for intermediate p_t jet hadrons.

The starting point is a multiple scattering approach corresponding to a marriage of Gribov-Regge theory and perturbative QCD (pQCD). An elementary scattering corresponds to a parton ladder, containing a hard scattering calculable based on pQCD, including initial and final state radiation (for details see [1]). These ladders are identified with flux tubes, which are mainly longitudinal objects, with transversely moving parts, carrying the transverse momenta of the hard scatterings. These objects are also referred to as kinky strings. One should note that here multiple scattering does not mean just a rescattering of hard partons, it rather means a multiple exchange of complete parton ladders, leading to many flux tubes. In this case, the energy sharing between the different scatterings will be very important, to be discussed later. The consistent quantum mechanical treatment of the multiple scattering is quite involved, it is based on cutting rule techniques to obtain partial cross sections, which are then simulated with the help of Markov chain techniques [2].

As said before, the final state partonic system corresponding to elementary parton ladders are identified with flux tubes. The relativistic string picture [3, 4, 5] is very attractive, because its dynamics is essentially derived from general principles as covariance and gauge invariance. The simplest possible string is a surface $X(\alpha, \beta)$ in 3+1 dimensional space-time, with piecewise constant initial velocities $\partial X/\partial\beta$. These velocities are identified with parton velocities, which provides a one to one mapping from partons to strings. For details see [2, 1]. The high transverse momentum (p_t) partons will show up as transversely moving string pieces. Despite the fact that in the TeV energy range most processes are hard, and despite the theoretical importance of very high p_t partons, it should not be forgotten that the latter processes are rare, most kinks carry only few GeV of transverse momentum, and the energy is nevertheless essentially longitudinal. In case of elementary reactions, the strings will break via the production of quark-antiquark pairs according to the so-called area law [6, 7, 2, 1]. The string segments are identified with final hadrons and resonances. This picture has been very successful to describe particle production in electron-positron annihilation or in proton-proton scattering at very high energies.

In heavy ion collisions and also in high multiplicity events in proton-proton scattering at very high energies, the density of strings will be so high that the strings cannot decay independently as described above. Here we have to modify the procedure as discussed in the following (see ref. [8]). The starting point are still the flux tubes (kinky strings) originating from elementary collisions. These flux tubes will constitute both, bulk matter which thermalizes and expands collectively, and jets. The criterion which decides whether a string piece ends up as bulk or jet, will be based on energy loss. In the following we consider a flux tube in matter, where "matter" first means the presence of a high density of other flux tubes, which then thermalize. A more quantitative discussion will follow.

Three possibilities should occur, referred to as A, B, C, see Fig. 1(a,b) :

A String segments far from the surface and/or being slow will simply constitute matter, they loose their character as individual strings. This matter will evolve hydrodynamically and finally hadronize ("soft hadrons").

B Some string pieces (like those close to transversely moving kinks) will be formed outside the matter, they will escape and constitute jets ("jet hadrons").

C There are finally also string pieces produced inside matter or at the surface, but having enough energy to escape and show up as jets ("jet hadrons"). They are affected by the flowing matter ("fluid-jet interaction").

Let us discuss how the above ideas are realized. In principle the formation and expansion of matter and the interaction of partons with matter is a dynamical process. However, the initial distribution of energy density and the knowledge of the initial momenta of partons (or string segments) allows already an estimate about the fate of the string segments. By "initial time" we mean some early proper time τ_0 which is a parameter of the model. Strictly speaking, energy loss concerns partons, modifying eventually the kink momenta in our picture, and the momenta of the string segments after breaking will be reduced. We will therefore base our discussion on energy loss on string segments.

We estimate the energy loss ΔE of string segments along their trajectory to be

$$\Delta E = k_{\mathrm{Eloss}}\, E_0 \int (\rho V_0)^{3/8} \max(1, \sqrt{E/E_0}\,)\, dL/L_0, \tag{1}$$

inspired by [9], where ρ is the density of string segments at initial proper time τ_0, V_0 is an elementary volume cell size (technical parameter, taken to be 0.147 fm^3), L_0 is a (technical) length scale (taken to be 1 fm), E the energy of the segment in the "Bjorken frame" moving with a rapidity y equal to the space-time rapidity η_s, dL is a length element, and k_{Eloss} and E_0 are parameters. We introduce an energy cutoff E_0 to have sufficient energy loss for slowly moving segments.

A string segment will contribute to the bulk (type A segment), when its energy loss is bigger than its energy, i.e.

$$\Delta E \geq E. \tag{2}$$

Journal of Physics: Conference Series **389** (2012) 012040 doi:10.1088/1742-6596/389/1/012040

All the other segments are allowed to leave the bulk (type B or C segments). Only the bulk segments are used to determine the initial conditions for hydrodynamics, following the same procedure as explained in [1, 8]. Starting from this initial condition, the bulk matter will evolve according to the equations of ideal hydrodynamics till "hadronization", which occurs at some "hadronization temperature" T_H [1]. Hadronization means that we change from matter description to particle description, but hadrons still interact among each other, realized via a hadronic cascade procedure [10], already discussed in [1].

After having performed the hydrodynamic expansion, we have to come back to the string segments which escape the bulk because their energy is bigger than the energy loss. We employ a formation time: the string segments are formed at times t distributed as $\exp(-t/\gamma\tau_{\text{form}})$, with some parameter τ_{form} which is taken to be 1fm/c. If the formation time is such that the segment is produced outside the "hadronization surface" defined by T_H, the segment will escape as it is (type B segment).

Most interesting are the segments which are formed inside but still escape, because they have $E > \Delta E$. These are type C segments. They escape, but their properties change. Actually such a segment leaves "matter" at the hadronization surface at a particular space-time point x, which is characterized by some collective flow velocity $\vec{v}(x)$. We assume that the string breaking in this case is modified such that the quark and antiquark (or diquark) necessary for the string breaking are taken from the flowing fluid rather than being produced via the Schwinger mechanism. So the new string segment is composed of a quark and antiquark (diquark) carrying the flow velocity, and the string piece in between, which has not been changed. This string piece may or may not carry large momentum, depending on whether it is close to a kink or not, the former possibility shown in Fig. 1(c).

In any case, due to the fluid-jet interaction, the properties of this segment change drastically compared to the normal fragmentation:

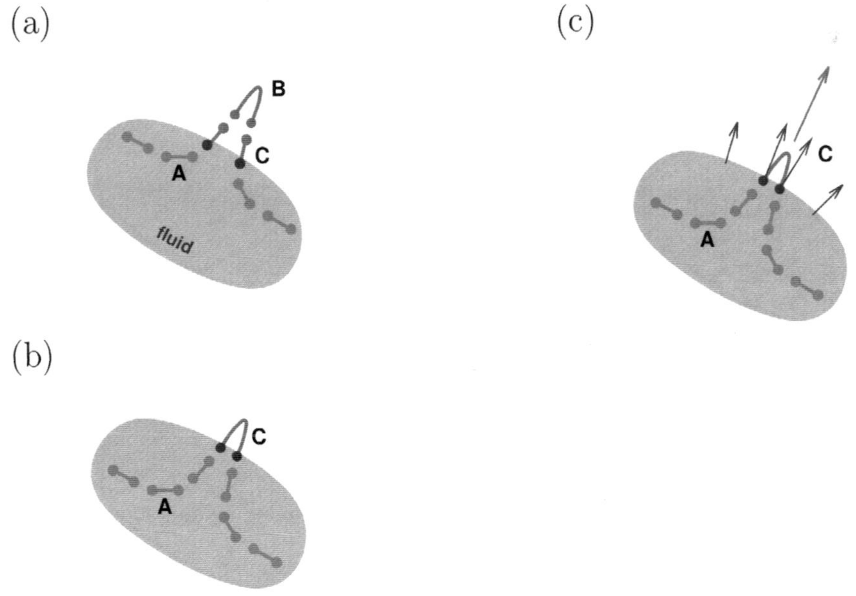

Figure 1. (Color online) Flux tube in matter (from other flux tubes, blue colored area). One distinguishes three types of behavior for string segments, noted as A, B, C (see text). The highest p_t string segment may be of type B (a) or of type C (b). A type C segment picks up quark and antiquark from the fluid, carrying momenta and flavor according to the fluid properties (c).

Journal of Physics: Conference Series **389** (2012) 012040 doi:10.1088/1742-6596/389/1/012040

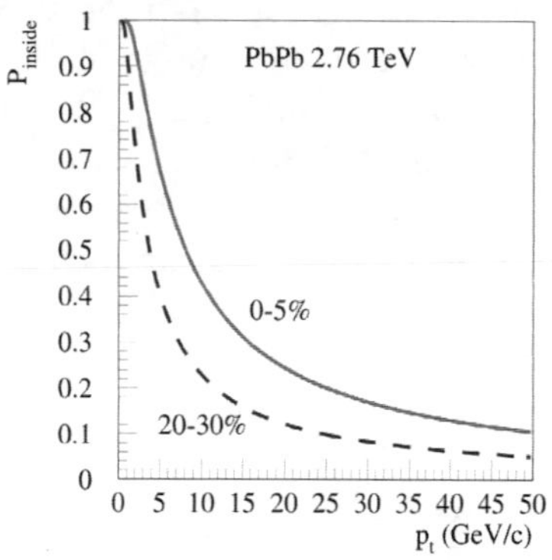

Figure 2. (Color online) The estimate P_{inside} to form (pre)hadrons inside the fluid, as a function of p_t, for Pb-Pb collisions at 2.76 TeV. We show the curves for the 0-5% and the 20-30% most central events.

- The quark and antiquark (or diquark) from the fluid provide a push in the direction of the moving fluid.
- The quark (antiquark) flavors are determined from Bose-Einstein statistics, with more strangeness production compared to the Schwinger mechanism.
- The probability p_{diq} to have a diquark rather than an antiquark will be bigger compared to a highly suppressed diquark-antidiquark breakup in the Schwinger picture (p_{diq} is a parameter).

Our procedure has 4 parameters: k_{Eloss} (=0.042), E_0 (=6 GeV), τ_{form}(=1 fm/c), p_{diq}(=0.22). It allows to cover in a single scheme the production of jets, of bulk, and the interaction between the two.

A crucial ingredient to the mechanism of fluid-jet interaction is the formation time of jet hadrons (the hadrons which leave the fluid). An estimate of the probability distribution of the formation times t of jet hadrons with gamma factors γ is given as [8]

$$P_{\text{inside}} = 1 - \exp\left(-\frac{(r_{\text{Pb}} - b/2)\, m}{p_t\, \tau_{\text{form}}}\right). \tag{3}$$

In Fig. 2, we show the result for the 0-5% and the 20-30% most central events in Pb-Pb collisions at 2.76 TeV, using $c\tau_{\text{form}} = 1\,\text{fm}$, $mc^2 = 1\,\text{GeV}$, $r_{\text{Pb}} = 6.5\,\text{fm}$, and for the average impact parameters $b = 1.8\,\text{fm}$ (0-5%) and $b = 7.8\,\text{fm}$ (20-30%).

By construction, the probability $P_{\text{fluid-jet}}$ of having a fluid-jet interaction is equal to the probability of forming (pre)hadrons inside the fluid, so its estimate is given by P_{inside}. From Fig. 2, we see that the probability is quite large for intermediate values of p_t, but even large values (50 GeV/c) are significantly affected. Whether the effect of the interaction can be seen in some observable is a different question and will be discussed later.

The bulk matter extracted as described above provides the initial condition for a hydrodynamic evolution. As explained in [1], we compute the energy momentum tensor and the flavor flow vector at some position x (at $\tau = \tau_0$) from the four-momenta of the bulk string segments. The time $\tau = \tau_0$ is as well taken to be the initial time for the hydrodynamic evolution. We employ three-dimensional ideal hydrodynamics as described in [8]. We try to mimic viscous effects by taking artificially large values of the flux tube radii (we take 1 fm), in order to get smoother initial conditions. This has the effect of reducing the elliptical flow by 20-30%, as needed

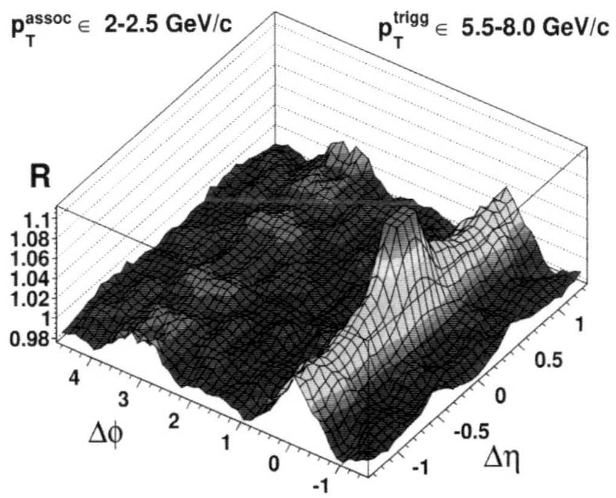

Figure 3. (Color online) Calculated dihadron correlation function for p_t^{trigg} in the interval 5.5-8.0 GeV/c and p_t^{assoc} in the range 2-2.5 GeV/c in the 0–10% most central Pb-Pb collisions at 2.76 TeV .

Our prescription for bulk-jet separation and interaction should strongly affect dihadron correlations, which provide much more information than simple spectra. With all parameters (k_{Eloss}, E_0, τ_{seg}, p_{diq}) being fixed from the considerations in the last section, we now compute dihadron correlation functions defined as

$$R(\Delta\eta, \Delta\phi) = \frac{M}{S} \times \frac{S(\Delta\eta, \Delta\phi)}{M(\Delta\eta, \Delta\phi)}, \tag{4}$$

where S is the number of pairs in real events, and M the number of pairs for mixed events.

In Fig. 3, we show a correlation function for p_t^{trigg} in the interval 5.5-8.0 GeV/c and p_t^{assoc} in the range 2-2.5 GeV/c, in the 0–10% most central Pb-Pb collisions at 2.76 TeV. Although the trigger p_t is too large to originate from freeze-out (from the flowing fluid), one observes a ridge structure, which is due to the fluid-jet interaction. Let us consider the situation of an initial azimuthal anisotropy in the energy density which is transported into a corresponding anisotropy in the flow. We sketch in Fig. 4 the (somewhat exaggerated) situation of a triangular transverse flow pattern with maximal flow around $\phi = 0^o$, 120°, and 240° (with respect to the y-axis). The flow maxima are indicated by blue arrows. Again it is very important that this flow pattern is (not necessarily in magnitude, but in shape) very similar at different longitudinal positions – in the figure indicated by the two transverse planes **P** and **P**′, corresponding to two different space-time rapidities η_s and η'_s. A soft hadron (**S**) produced at η_s at the fluid surface close to the position of maximal flow (for example at $\phi = 0^o$), will be boosted by the latter one and therefore carry information about this flow. A jet hadron (**J**) produced at η'_s at the same angle ($\phi = 0^o$) close to the surface, will pick up a quark and an antiquark, both carrying flow, which adds the corresponding transverse momentum to the p_t of the string segment (red element in the figure). It is the same flow which affects the jet hadron at η'_s and the soft hadron at η_s, which creates the dihadron correlation at $\Delta\phi = 0$, the "ridge". The correlation remains visible, even when the flow contribution to the jet hadron is only 10%, this is why the correlation is still present even for trigger transverse momenta beyond 10 GeV/c.

In semi-peripheral Pb-Pb collisions at 2.76 GeV/c, one observes for example for p_t^{trigg} in the interval 5.5-8.0 GeV/c and p_t^{assoc} in the range 2-2.5 GeV/c a clear elliptical flow structure, again due to the mechanism of fig. 4.

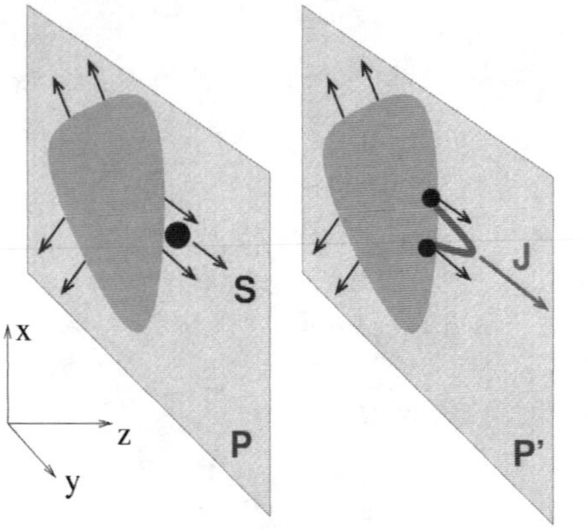

Figure 4. (Color online) Sketch of two cuts of the fluid volume corresponding to the space-time rapidities η_s and η'_s, the two corresponding transverse planes being $\mathbf{P}$ and $\mathbf{P}'$. We show the example of a triangular flow pattern – the same at η_s and η'_s.

The correlation functions are essentially flat as a function of $\Delta\eta$, for large $\Delta\eta$. One therefore gets complete information about the long range correlations by integrating over $\Delta\eta$,

$$R(\Delta\phi) = \frac{1}{2(B-A)} \int_{A<|\Delta\eta|<B} R(\Delta\eta, \Delta\phi)\, d\Delta\eta, \tag{5}$$

where we use $A = 0.8$ and the maximum $B = 2$. This function agrees perfectly with its Fourier decomposition,

$$R(\Delta\phi) = 1 + \sum_{n=1}^{5} 2V_{n\Delta} \cos(n\Delta\phi), \tag{6}$$

using the first five terms. This is very convenient, because it allows to discuss the features of the correlation functions for different options for p_t^{trigg} and p_t^{assoc} by simply considering the Fourier coefficients.

In fig. 5, we plot some coefficients $V_{n\Delta}$ as a function of p_t^{trigg} . We compare our simulation (stars) with the results from ALICE [11] (circles). We see clearly the dominance of elliptical flow: the $n = 2$ coefficients are by far the largest. Nevertheless, also the higher harmonics contribute. We see in all cases an increase of the coefficients with p_t^{assoc} and with p_t^{trigg} up to values of around 2-3 GeV/c. At the latter values the hydrodynamic flow contributes the most to the correlation between soft hadrons from the fluid. For higher transverse momenta, the coefficients get smaller, because the correlation between soft particles dies out. But $V_{2\Delta}$does not at all drop to zero at high p_t because here the correlations between soft and jet particles come into play – the jet particles which suffered a push by the fluid, as discussed earlier (fluid-jet interaction). The fluid transfers at maximum few GeV/c of transverse momentum to the jet, but this is easily visible in the correlation (even at 20 GeV/c).

Whereas dihadron correlations provide the most complete information about particle production – in particular concerning the role of the "flowing" fluid, one may get the essential information by considering the elliptical flow coefficient v_2 of single particle production. Here we consider a definition where the reference plane is given by the event plane angle ϕ^{backward} (ϕ^{forward}), obtained from counting all particles in the opposite hemisphere [12]. In Fig. 6, we plot v_2 as a function of the transverse momentum for different centralities in Pb-Pb collisions at 2.76 TeV. The magnitude of the elliptical flow coefficients increase at low p_t to reach a maximum around 2-3.5 GeV/c and then drop slowly at large p_t.

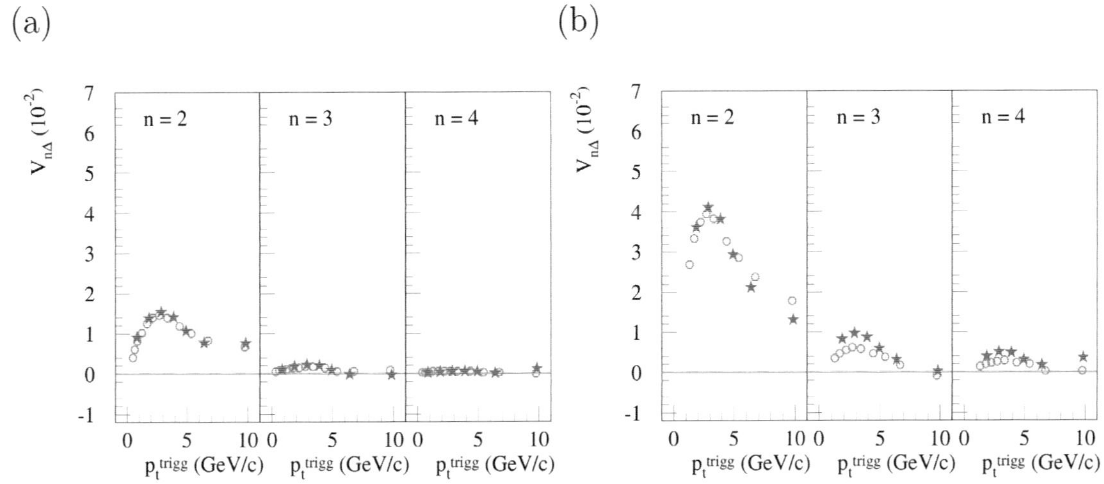

Figure 5. (Color online) The Fourier coefficients $V_{n\Delta}$ as a function of p_t^{trigg} for p_t^{assoc} within 0.25-0.5 GeV/c (a), 1-1.5 GeV/c (b), in the 40-50% most central Pb-Pb collisions at 2.76 TeV. We compare the ALICE data [11] (circles) with calculations (red stars) .

The behavior at high p_t is the most interesting aspect: even at 10 GeV/c, there is a significant amount of elliptical flow, due to the fluid-jet interaction, which pushes jet particles in the direction of the collective flow at the freeze-out surface (and this effect will continue up to even higher p_t, but we are simply running out of statistics). The high p_t behavior is closely related to the formation time discussion we had earlier. The non-vanishing v_2 at high p_t is mainly due to fluid-jet interactions, and the values follow indeed the estimated probability P_{inside} to form the jet hadron inside the fluid, see ref. [8].

To summarize: We presented a theoretical scheme which accounts for bulk matter, jets, and the interaction between the two. The criterion for bulk-jet separation is based on parton energy loss. But in addition to the latter mechanism, there are very important new phenomena which have not been discussed so far: The interaction between jet hadrons and soft ones (from fluid freeze-out), and the interaction between the fluid and jet hadrons at the moment of the creation of the latter ones. Particle production between zero and (at least) 20 GeV/c is affected. We understand quantitatively azimuthal anisotropies in single particle production and dihadron (long range) correlations at large values of p_t.

ACKNOWLEDGMENTS: This research has been carried out within the scope of the ERG (GDRE) "Heavy ions at ultra-relativistic energies", a European Research Group comprising IN2P3/CNRS, Ecole des Mines de Nantes, Universite de Nantes, Warsaw University of Technology, JINR Dubna, ITEP Moscow, and Bogolyubov Institute for Theoretical Physics NAS of Ukraine. Iu.K. acknowledges partial support by the State Fund for Fundamental Researches of Ukraine (Agreement of 2011) and National Academy of Sciences of Ukraine (Agreement of 2011). K.W. would like to thank Ludmila Malinina and Konstantin Mikhailov for useful discussions. M.B. thanks the Hessian LOEWE initiative for financial support.

[1] K. Werner, Iu. Karpenko, T. Pierog, M. Bleicher, K. Mikhailov, Phys. Rev. C 82, 044904 (2010)
[2] H. J. Drescher, M. Hladik, S. Ostapchenko, T. Pierog and K. Werner, Phys. Rept. 350, 93, 2001
[3] Y. Nambu, Proc. Intl. Conf. on Symmetries and Quark Models, Wayne State Univ., 1969
[4] J. Scherk, Rev. Mod. Phys. 47, 123 (1975)
[5] C. Rebbi, Phys. Rep. 12 (1974) 1
[6] X. Artru, Phys. Rep. 97, 147 (1983)
[7] D. A. Morris, Nucl. Phys. B288, 717, 1987

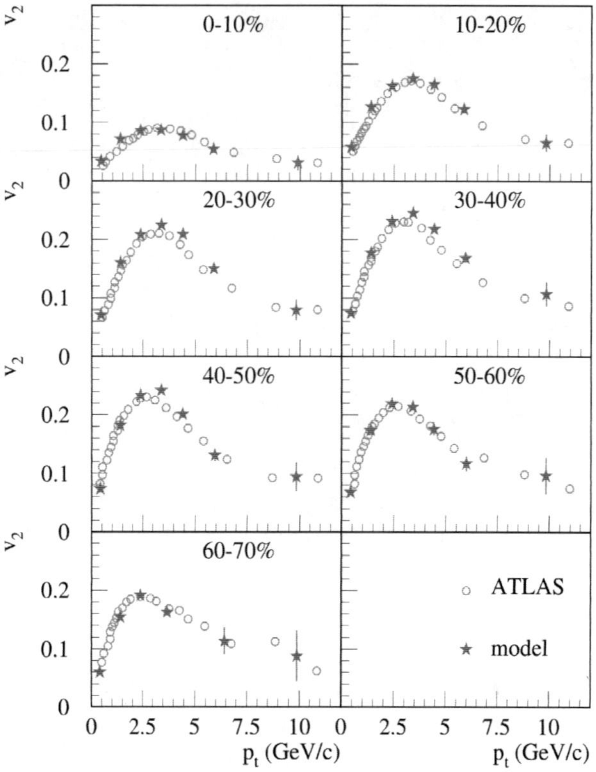

Figure 6. (Color online) p_t dependence of elliptical flow (defined with respect to the opposite hemisphere sub-event plane) for different centralities in Pb-Pb collisions at 2.76 TeV. We compare the ATLAS data [12] (circles) with calculations (red lines).

[8] K.Werner, Iu.Karpenko, M. Bleicher, T.Pierog, S. Porteboeuf-Houssais, arXiv:1203.5704, Phys. Rev. C 85, 064907 (2012)
[9] R. Baier et al., Nucl.Phys.B483:291-320,1997; S. Peign, AIP Conf.Proc.1038:139-148,2008
[10] M. Bleicher et al., J. Phys. G25 (1999) 1859; H. Petersen, J. Steinheimer, G. Burau, M. Bleicher and H. Stocker, Phys. Rev. C78 (2008) 044901
[11] ALICE collaboration, K. Aamod et al., Phys.Lett. B708 (2012) 249-264, arXiv:1109.2501
[12] ATLAS collaboration, G. Aad et al., Phys.Lett. B707 (2012) 330-348, arXiv:1108.6018

Photons from spectators

Edwin Norbeck and Yasar Onel

Department of Physics and Astronomy, University of Iowa, Iowa City, IA 52242 USA

E-mail: edwin-norbeck@uiowa.edu and yasar-onel@uiowa.edu

Abstract. For Pb-Pb collisions at the LHC, the spectators, those parts of the Pb nuclei that are not involved in nucleon-nucleon collisions, continue on in the original beam direction, but they are not inert objects. If they are not too large a fraction of the original nucleus they are completely disintegrated into neutrons and protons. The spectators are subject to huge electric fields, fields causing accelerations so large that the quarks in the nucleons will radiate almost all of the energy they acquire from the field. Because of the extreme Lorentz transform most of the photons can be found in the 10 cm between the ingoing and outgoing beam pipes 140 m from the collision point. Spectator radiation may be the source of cosmic-ray photons seen in emulsion stacks at 100,000 ft.

1. Introduction

In a reaction between two highly relativistic heavy ions the overlap region where nucleon-nucleon collisions occur becomes quark-gluon plasma. Those parts of the original ions that are outside of the overlap region are known as spectators, figure 1. There have been extensive studies of many aspects of the quark-gluon plasma, but the physics of the spectators has been mostly ignored. The spectators are pieces of nuclear matter subject to extreme and unusual conditions. The spectators from LHC Pb-Pb collisions are subject to huge electromagnetic fields, which cause the spectators to emit EM radiation. Because of the large velocity of the spectators most of the radiation is confined to small angles, $\theta < 1.0$ mrad. When the spectators are small, less than half of the original nucleus, they are completely disintegrated into protons and neutrons. The disintegration process almost certainly produces photons. When the overlap region is small, and the spectators large, some of the spectator matter is in the form of a small nucleus, which is usually in an excited state that emits photons. The photons may be of moderate energy, in the MeV range, but the Lorentz transformation boosts them into to the GeV range.

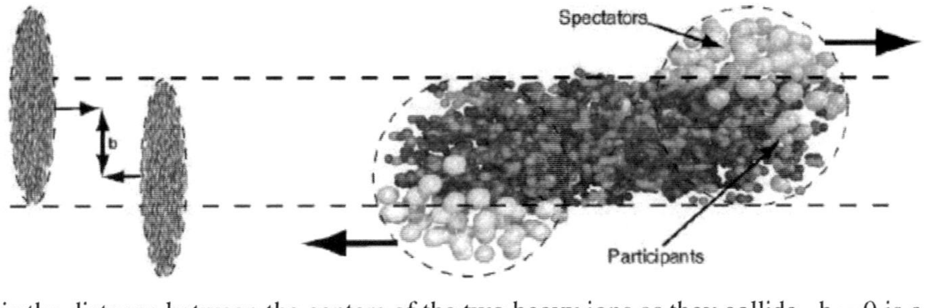

Figure 1. b is the distance between the centers of the two heavy ions as they collide. b = 0 is a central collision. b is often called "impact parameter."

Journal of Physics: Conference Series **389** (2012) 012041 doi:10.1088/1742-6596/389/1/012041

Little is known about these photons, but there are many interesting possibilities, some of which are discussed below.

2. Photons in the CMS ZDC

The part of the CMS [1] experiment at the LHC [2] that could observe the spectator photons is the Zero Degree Calorimeter, ZDC [3]. The ZDC is about 9 cm wide and is located between the incoming and outgoing beam pipes at 140 m from the center of CMS (the closest point for which the two beams are in separate pipes). There are two ZDCs, one on each side of CMS.

The Lorentz transformation for a photon at 90° in the spectator frame is easily understood. The transverse momentum is not affected by the transformation. The photon can be regarded as a particle of a given energy; this energy (momentum) is multiplied by the relativistic γ. The transverse momentum is unchanged, and the longitudinal momentum ends up being the transverse momentum multiplied by γ. So the tangent of the angle is $1/\gamma$. With an isotropic distribution for photons in the spectator frame, half of the photons are within the angle $1/\gamma$.

For the 2010 and 2011 Pb-Pb runs the Pb beams were at 1.375 TeV/nucleon with $\gamma = 1463$. Half of the photons are in a cone with a diameter of 19 cm at the ZDC. After the forthcoming 2-times energy upgrade the cone will have a diameter of only 9.6 cm so that almost half of the photons will hit the ZDC, which is a square with sides of almost 9 cm. The ZDC also detects the spectator neutrons; the protons and light nuclei are swept away by magnets along the 140 m path.

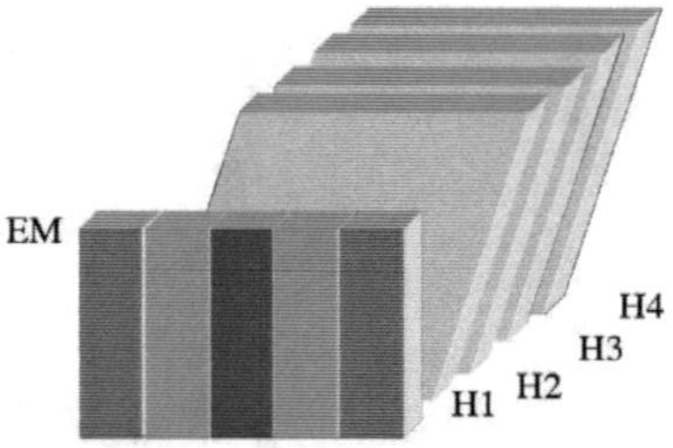

Figure 2. The parts of the ZDC.

Figure 2 is a schematic of the ZDC. The EM, electromagnetic, part is 19 X_0 thick and is divided horizontally into five sections. It is followed by four thicker sections of 1.4 nuclear interaction lengths each that together stop most of the neutrons. The absorbing material is W metal. Hadrons start their showers in the EM part 50% of the time. There is a prediction by the ALICE group that to the 1 sigma level the neutrons should be in a spot only 6 mm across at 116 m (7 mm at 140 m) [4]. If this were correct the neutrons would be in one or two of the central sections and the photons would be in all of the sections. With this assumption, the data show that the proton/neutron ratio increases as the reactions become more central. It appears that ALICE assumed that the transverse momentum is entirely due to a traditional Fermi momentum distribution in the Pb nucleus. More recent calculations by ALICE [5] show a broadening of the neutron distributions resulting from the transverse beam divergence and the beam diameter at the interaction point along with a directed flow of the spectator neutrons [6].

Another broadening effect results from short-range correlations [7]. The strong short range repulsion and intermediate range attraction between nearby nucleons provides a deep potential well such that the relative motion of the nucleon pair has a momentum substantially larger than the Fermi momentum characteristic of the nucleus. If during the collision one of the pair is swept away into the participant region, the one remaining in the spectator is left with the large momentum. It can escape into the direction of the open hemisphere, figure 3, with considerable energy or go into the spectator and raise its temperature. This process should lead to an asymmetric distribution of neutrons at the

ZDC. A measurement of this distribution could provide a measure of the reaction plane and also the direction of the impact parameter vector.

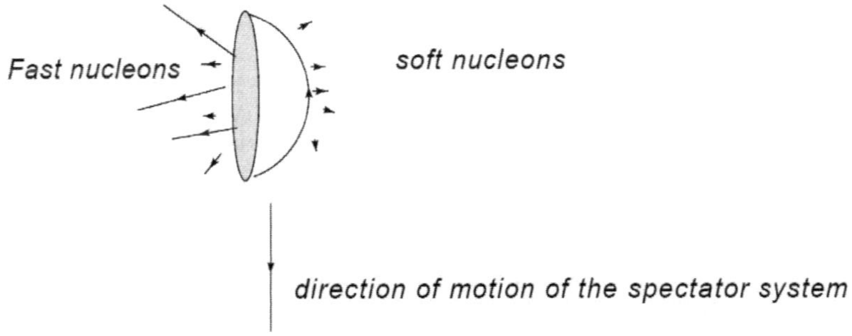

Figure 3. Asymmetric distribution of spectator nucleons

The fraction of the neutrons that are part of a broken pair becomes larger as the spectator becomes smaller with decreasing impact parameter. At this time it is not clear whether the change in the five point EM distributions is caused by a change in the photon/neutron ratio or simply an increase in the transverse energy of the neutrons as the spectators become smaller.

Another possible contribution to the flattening of the five point EM distributions as reactions become more central is the production of Λ^0 baryons along the boundary between the spectator and the quark-gluon plasma. The signal from a Λ^0 going into the ZDC is the same as for a neutron, but if it decays in flight to $n + \pi^0$ some of the photons from the π^0 decay will be seen in the EM section. For the 2011 runs 70% of the Λ^0 decay along the 140 m path. The decay into $p + \pi^-$ will be seen by the ZDC if it occurs in the last half of the 140 m, which is free of magnets.

Detailed studies of the photon and neutron distributions would require improvements to the EM section. Presently it consists of 2 mm thick W-metal plates interleaved with ribbons of quartz fibers that are divided horizontally into five groups. Čerenkov light from each of the five fiber bundles is viewed with a photomultiplier tube. For an improved version the fiber ribbons could be replaced with pixilated silicon detectors. For better measurements of the neutron distributions the last layer would be preceded by a 1 cm thick W-metal plate to block the remnant of the photon showers and to increase the detected fraction of the neutrons to 55%. The improved EM section would be used only with heavy ions, for which the luminosity is orders of magnitude smaller than for protons. Even with heavy ions, radiation damage must be considered. It may be necessary to use the more radiation resistant diamond detectors. This is a small detector with transverse dimensions of only 9 cm by 15 cm. The electronics would be similar to that in the CMS silicon tracker.

3. Cosmic Ray Data

Unexplained jets of pure photons have been observed with balloon flights using nuclear emulsion stacks at 100,000 ft [8]. There were a dozen or so GeV photons in a small region less than 50 μm across. No charged particles of any kind were found to accompany the incident photons. Many single photons were observed, as might be expected from spectators derived from heavy ions, such as iron, striking nitrogen and oxygen at 140,000 ft. Most of the photons from such spectators would be well separated from the nucleon component. The photons could easily reach the detectors; there is only 12 g/cm^2 of air above 100,000 ft, and $X_0 = 36.5$ g/ cm^2 for air.

It is worth considering the possibility that a highly directed beam of photons might sometimes be emitted from a spectator. Assuming a $\gamma = 10{,}000$, the Lorentz contraction makes 40,000 ft of travel only 4 ft. A photon beam that expands to 50 μm in one or two meters seems within the realm of possibility.

At the LHC with $\gamma = 2925$ the 140 m to the detector is Lorentz contracted to only 4.8 cm with a corresponding reduction in the spreading of the photon beam by one or two orders of magnitude. The only way to differentiate between a dozen photons at essentially the same point and one photon of a higher energy is by the depth penetration of the shower.

4. Extreme Electric Fields

The magnitudes of the various features for Pb-Pb at the full LHC design energy of 5.5 TeV/nucleon-pair are spectacular. The spectators are subjected to a brief but huge electric field directed radially outward from the beam axis. The electric field at a distance of 10 fm from the center of a Pb nucleus at rest is 1.18×10^{21} V/m. The value of the relativistic gamma of one Pb seen by the other is 1.71×10^{7}. Multiplying these gives $E = 2 \times 10^{28}$ V/m. This should be compared with $E_{crit} = 1.6 \times 10^{18}$ V/m, which will produce electron-positron pairs out of the vacuum by the Schwinger mechanism [9]. E_{crit} is also the field that will cause an electron to accelerate at a rate such that the reaction to the radiated power given by the Larmor formula is a major contributor to the motion of the electron. There are currently several projects that are trying to reach E_{crit} with high powered lasers. With electric fields that are millions of times larger the effect of the field on u and d quarks would be to produce radiation but virtually no change in the kinetic energy of the quarks.

The time that the force is applied must be divided by gamma so that the total momentum transferred, $F \times t$, is independent of gamma. If there were no radiation an electron would acquire about 28 MeV. From another point of view, the uncertainty principle sets an upper limit of 28 MeV on the transverse momentum of a photon emitted by a moving Pb ion, see p. 4 of [10].

5. Superradiance and Solitons

The hyper critical acceleration issue raises fascinating questions and possibilities. What actually happens when a force attempts to accelerate a charged particle by much more than the critical value? With the LHC the electric field seen by the spectator is 2×10^{28} V/m, but for only 3×10^{-30} seconds. By the uncertainty principle the energy available during the interaction could be over 200 TeV. The initial radiation is most likely not a photon but a particle-antiparticle pair. High energy photons interact with matter by producing pairs. Compton scattering is only effective for photon energies round m_ec^2. It would be the annihilation of the pairs that produces the EM radiation. The final electromagnetic energy radiated per quark, in the spectator system, is sufficiently low that the wave length of the radiation is larger than the spectator dimensions. This suggests the possibility of superradiance effects. A superradiant medium can be considered as a mirrorless laser pumped in a percussional way at time $t = 0$. In the current literature the medium has the energy stored as excited states of atoms. In our case the energy would be in particle-antiparticle pairs, electrons excited out of the Fermi sea. A d quark would have less energy associated with it than a u quark because of its smaller charge and larger mass. Perhaps most of the EM energy would be emitted as two highly directed bunches.

A wave form that propagates without changing shape is called a soliton. They were first observed in water canals in which the normal effects of dispersion are canceled by nonlinearities. Such one dimensional solitons have proved useful with light signals travelling in specially tailored fibers.

A light beam in free space will normally expand because of diffraction, however a light beam with amplitude cross section given by a J_0 Bessel function is diffraction free [11]. Such beams are sometimes called 2-D solitons.

There have been many unsuccessful attempts to explain elementary particles as 3-D solitons, where the shape of the wave packet remains constant along all three axes. There does exist at least one 3-D electromagnetic soliton [12]. It requires a radial energy distribution of the form of a J_0 Bessel function and a time distribution of an Airy function, figure 4. Experiments [12] have shown that a reasonable approximation to such a shape degrades only slowly as it propagates. The balloon data require only that the transverse dimension expand slowly; the temporal features of a 3-D soliton are not needed.

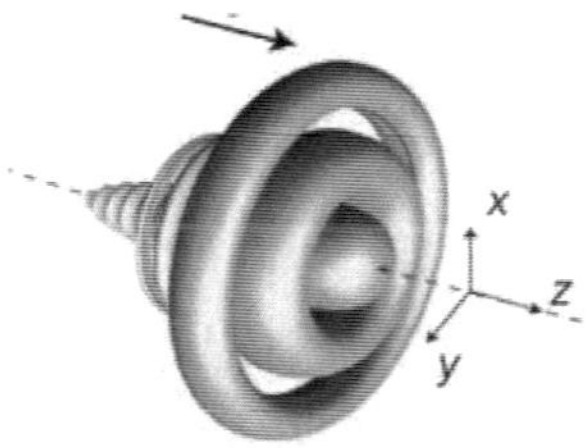

Figure 4. Central part of intensity distribution of a 3-D EM soliton.

6. Conclusions

There have been no definitive measurements of photons from spectators from highly relativistic heavy ion collisions, although such photons may have been seen in cosmic rays. At LHC Pb-Pb collisions energies of 5.5 TeV/nucleon-pair the Lorentz transformation puts all photons from the forward hemisphere of the spectator into a cone that is only 9.5 cm across at 140 m from the interaction point. Better theoretical predictions are needed along with improved detectors for measuring the distributions of photons and neutrons. Designing such detectors will be challenging but not expensive, because the area to be covered is only 135 cm^2.

References

[1] CMS Collaboration 2008 *JINST* **3** S08004
[2] Evans L and Bryant P 2008 *JINST* **3** S08001
[3] Grachov O A, Murray M J, Ayan A S, Debbins P, Norbeck E, Onel Y and d'Enterria D 2006 *AIP Conf. Proc.* 867:258–265; *e-Print* nucl-ex/0608052
[4] 1999 *Zero-Degree Calorimeter Technical Design Report* CERN/LHCC 99-5
[5] ALICE Collaboration 2006 *J. Phys.* G **32** 1295-2040
[6] Poskanzer A M and Voloshin S A 1998 *Phys. Rev.* C **58** 1671
[7] Alvioli M and Strikman M 2011 *Phys. Rev.* C **83** 044905; White S and Strikman M 2009 ArXiv:0910.3205v2 [nucl-ex]
[8] Collins G B, Ficenec J R, Stevens D M, Trower W P and Fischer J 1973 *Phys. Rev.* D **8** 982
[9] Hebenstreit F, Alkofer R and Gies H 2008 *Phys. Rev.* D **78** 061701(R)
[10] Baltz A J *et al* 2008 *Phys. Rep.* **458** 1–171
[11] Durnin J, Miceli J J and Eberly J H 1987 *Phys. Rev. Lett.* **58** 1499
[12] Chong A, Renninger W H, Christodoulides D N and Wise F W Feb. 2010 *Nature Phot.* **4** 103

Rho Meson Diffraction off Au Nuclei

R. Debbe for the STAR Collaboration
Brookhaven National Laboratory Upton NY 11973, USA
E-mail: debbe@bnl.gov

Abstract. The STAR Ultra Peripheral Collisions program has collected a substantial sample of ρ mesons and for the first time at RHIC energies it has been able to extract the distribution of momentum transfert t from diffractive elastic scattering off the Au ion. The resulting difraction pattern is consistent with coherent scattering off a nuclear object the size of the Au nuclei. Measurements of this nature can offer insights and guidance to the ongoing preparations for the new electron ion programs.

1. Introduction

Our perception of nuclei has evolved drastically since the early days of Rutherford's experiments which made them appears as almost point like and massive objects. Higher energy electron scattering refined that knowledge providing the spatial distribution of charge in nuclei. The actual extent of the strong force was investigated with beams of pions and protons but that program suffered from the difficulties in disentangling the interference between Coulomb and nuclear interactions. At that point nuclei were well described by an almost spherical distribution refered to as the Wood-Saxon distribution. With the access to much higher energies, electron scattering opened a whole new way to study nucleons as it made accessible the study of their composite partonic nature. In particular, the HERA program identified the surprising fact that up to 20% of the interactions had the target nucleon left intact. This fact can be understod as the result of these high energy interactions being able to resolve partons which carry even smaller fractions of the nucleon momentum and are subject to non-linear effects. The ideal probe to study the nuclear component of nuclei would be one that does not carry charge. Such probe can be found in the photon fluctuations; a photon of enough energy can be itself or it may fluctuate into neutral vector mesons, quark pairs or lepton pairs with corresponding decreasing probability. Such beams of photons and its fluctuations are available at the RHIC and LHC colliders. High fluxes of photons described with the Weizsäker-Williams formalism [1] are present at RHIC with energies that range up to 24 GeV in the photon-nucleon center of mass.

We report the measurement of the transverse momentum of the ρ^0 meson after it undergoes an elastic scattering on the corresponding Au ion. For the first time, a diffraction pattern with several secondary peaks is clearly visible. Previous similar measurements were performed at low energy at SLAC [2] and at DESY with bremsstrahlung photons [3], in both cases the distributions of transverse momentum of the ρ mesons only shows the main diffraction peak. This measurements come as preparations are taking place for a new Electron Ion Collider (EIC) that will revolutionize, one more time, our ability to study nucleons and nuclei. This new program will, in addition to existing variables used to characterize the state of partons, include information about their distribution in the transverse plane. That information will be extracted from distributions similar to the ones reported in this work as the transverse momentum of the recoil target is a conjugate of the impact parameter of the collision.

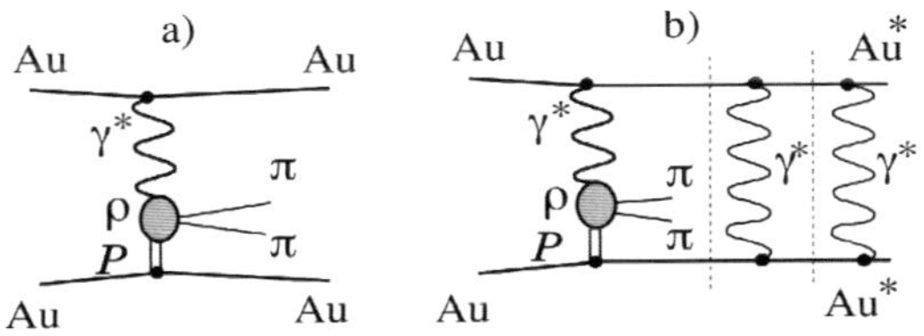

Figure 1. Schematic diagrams for UPC events. Panel a shows the most unbiased events. Panel b shows UPC events with further photon exchanges that exite both nuclei.

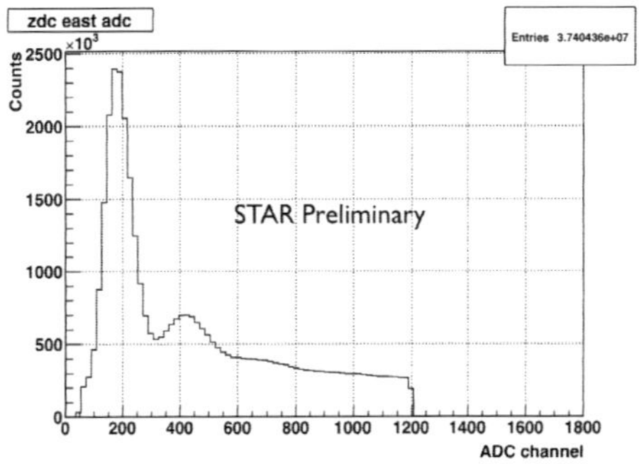

Figure 2. The distribution of charge collected in one of the ZDC detectors, a clear one neutron peak is visible followed by a smaller one related to two neutrons.

2. The STAR Experimental setup

The data used for this analysis were collected with the STAR detector at RHIC from Au+Au collisions with center of mass energy per nucleon $\sqrt{s_{\rm NN}}$=200 GeV. The momentum of charged particles were detected with the STAR Time Projection Chamber (TPC) in two units of pseudo-rapidity centered around 0 ($|\eta| < 1$) and full azimuthal coverage. The TPC records up to 45 samples of the ionization left in the detector gas which allows for a good resolution particle identification based on energy loss. The charge left by particles inside the TPC drifts along its axis and is read out on the East and West sides which are divided into six sectors each. The TPC is also used to identify the vertex of the collisions and together with the bending power of the 0.5 Tesla magnetic field, it provides a momentum resolution equal to $\Delta p_T/p_T = 0.005+0.004p_T$. More details about the TPC can be found in [4]. The cylindrical TPC is completely surrounded by the Time Of Flight (TOF) detector consisting of 23040 Resistive Parallel Plate gas detectors arranged in cells, groups of which form modules installed in two sets of trays along the East and West sides of the TPC, 10 TOF trays overlap the azimuth coverage of one TPC readout sector. The TOF detector was used to trigger the UPC events and provides good time-of-flight measurements, although this analysis doesn't make use of that information. STAR has two Zero Degrees Calorimeters (ZDC) installed at ±18 meters away from the nominal interaction point. These calorimeters are optimized for the detection of beam energy neutrons. These detector are instrumental in the definition of the trigger used for this analysis, more details about them can be found in [5].

2.1. Triggers

At the time the data used for the present analysis was collected, two triggers were defined to collect events described by the diagrams of Fig. 1. The event shown in panel a of that figure are collected with a trigger named UPC_Topo and events similar to those depicted in panel b were collected with the so called UPC_Main trigger. For both triggers the TOF detector is used to set limits in the number of tracks detected in the event. The ZDC calorimeters are used to identify events where the nuclei undergo a Coulomb exitation (a Giant Dipole Resonance GDR) decaying into one or more neutrons emmitted along the beam. The Beam Beam Counter (BBC)

Journal of Physics: Conference Series **389** (2012) 012042 doi:10.1088/1742-6596/389/1/012042

is used as a veto of events with any activity at high rapidity. The UPC_Topo trigger is the most inclusive of the triggers and is designed to tag the least biased UPC events; the ones without constraints on the number of neutrons detected in the ZDC calorimeters. This trigger is based on the TOF detector ability to find a pair of back-to-back particles by selecting events with TOF hits in two azimuthally opposite sectors (as mentioned above, a sector is formed with 10 trays). To avoid contamination from cosmic rays impinging vertically on the TPC the top (0) and bottom (3) sector are used as vetoes.

The cross section for ρ photoproduction tagged by UPC_Topo trigger is $\sim$ 10 times higher than the one for the UPC_Main events [6], but the less restrictive nature of the first trigger makes it inefficient as it collect many spurious events, in contrast the UPC_Main trigger is cleaner and more efficienct because of the conditions imposed on the number of neutrons detected in the ZDC detector.

2.2. Data sets

This report is based on the analysis of 22.9 million UPC_Main events written to tape by the STAR data aquisition system. This represents two thirds of the total number of events collected that year. (Events from an earlier part of that run period may be included in future analyses once further tests are conducted on the appropriate setup of the trigger.) Once the data is fully reconstructed a first selection of events is made. During run 10 the luminosity of RHIC was high enough such that the TPC being a slow detector had recorded several collisions (it takes $\sim 36\ \mu$ sec for the ionization left by a track close to the middle of the detector to drift to the readout planes). The events that are read in this first pass can contain several vertices. The analysis of each events starts by looping over the vertices, pairs of tracks originating from the same vertex are formed whenever the vertex was constructed out of at most 12 tracks. These pairs are then written to root tree files. Each track in the pair carries all the information extracted from the TPC as well as the corresponding TOF relevant variables whenever available. Information about energy deposited in the EM calorimeter cells is also written for each track. The root tree files are then processed separately to identify the events that where identified ρ mesons were produced exclusively in events that recorded a single neutron in the ZDC calorimeters.

3. The ρ^0 meson identification

During the third and final pass on the data, events are first selected to have no more that one neutron in each one of the ZDCs, this selects the 1n1n exitation of the ions. Figure 2 shows the pedestal subtracted signal from the East ZDC, The cut for a single neutron is placed at ADC channel 300. Tracks that are part of the triggered events are identified through their vertex and the one associated to the TOF hits present in the data. Pairs that do not have that connection to the TOF hits that triggered the events are rejected. Figure 3 shows one event in the YZ view of the TPC, tracks originating from the same vertex are presented as lines of the same color. The tracks shown with black lines are the ones connected to at least one TOF hit.

The identity of each track in the selected pairs is obtained from the amount of ionization energy deposited along their trajectory inside the TPC volume. Tracks that have less than 14 samples (hits) are eliminated from the set because they are found to be produced by low momentum particles spiraling along the solenoidal magnetic field. The resolution of the dE/dx measurement in the TPC is very good and allows a clean identification of pions up to $\sim 1\ GeV/c$ covering the range of momenta for pions originating from ρ mesons produced practically at rest. Figure 4 shows the correlation between dE/dx and the momentum of the particles. Pions are clearly visible as the mostly horizontal band of "minimum ionization". With this identification it is now possible to select events that contain pion pairs. Pairs with different charge will then be considered as candidates decays from a ρ^0 meson. Pairs of tracks with equal charge $++$ or $--$ are also used to build our best estimate so far of the background for the production of ρ^0

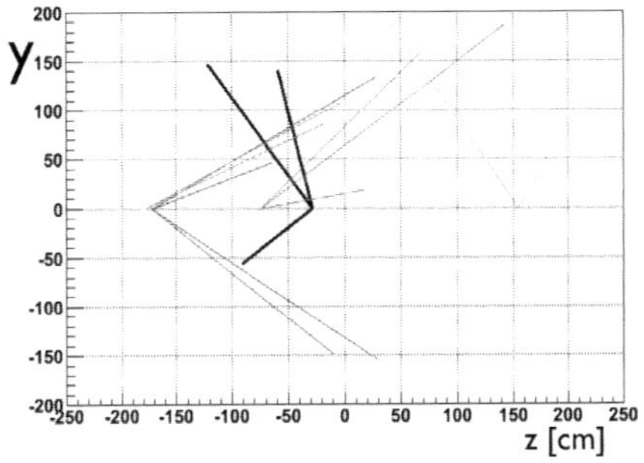

Figure 3. The projection of tracks on the YZ view of the STAR TPC is shown with black lines for a UPC triggered event. Other events recorded in the TPC are shown with different colors.

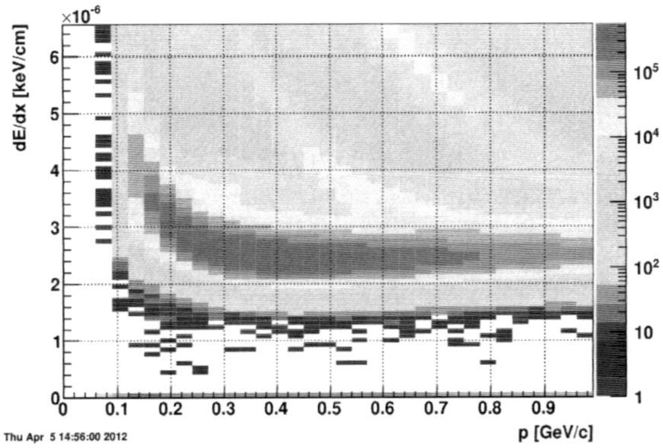

Figure 4. The average ioniation energy loss (in keV/cm) recorded for tracks detected in the TPC is shown as function of their measured momentum (in GeV/c). The pion band is dominant in this figure but both kaon and proton bands are also visible.

mesons in UPC events. The invariant mass of both types of pairs is constructed and is shown in Fig. 5 where the black histogram shows the mass distribution for pairs of opposite charge, and the red one shows the mass distribution from equal charge pairs, which from now on will be considered as the background to this ρ^0 meson study. The wide mass distribution centered at 770 MeV is dominant in the black histogram, as well as a very narrow peak at the kaon mass, most likely produced by the decay of K^0_S into two pions.

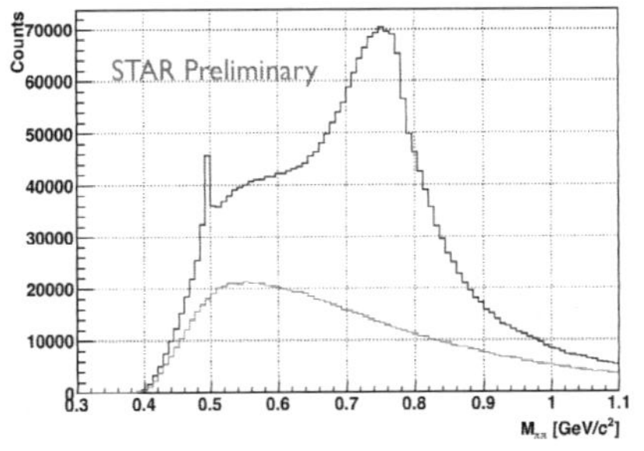

Figure 5. Invariant mass distribution for pion pairs of opposite charge shown with a black histograms, together with the similar distribution filled with pion pairs of equal charge.

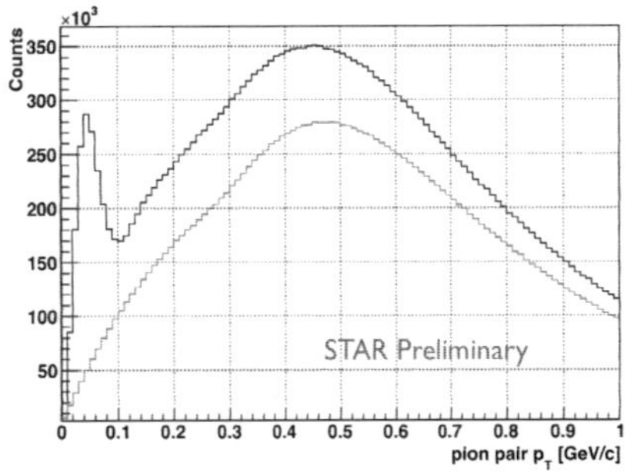

Figure 6. Transverse momentum distribution of pairs of opposite charge are shown with a black histogram. The red histogram shows similar distribution but this time for pion pairs of equal charge.

Figure 6 shows the distribution of the transverse momentum of the pion pair for opposite charge pions in the black histograms and the same charge pion pairs shown in the red histogram. The distribution shows a prominent peak below 100 MeV/c and a much broader distribution at higher value of p_T. The low transverse momentum peak is related to ρ mesons scattering coherently off the Au ion. In other words, the uncertainty principle connects the narrow low p_T peak with a broad spatial structure; the Au nucleus. An earlier STAR publication [7] dealing with UPC ρ^0 meson coherent production imposed a cut on the p_T of the pion pair at 150 MeV/c to select coherent scattering, in this analysis no such cut is applied and a subtraction of the incoherent component of the ρ scattering is used to enhance the coherent events. The assumption made is that the coherent component has a rapidly falling tail that extends to higher values of p_T and is completely hidden in Fig. 6 by the less steeply falling incoherent component as well as by the increasing contribution of background contamination. The best estimate of the pion background in our data sample is the corresponding shape and magnitude of the equal sign pairs in every studied distribution, as done in the STAR publication [8].

The last condition applied to the pairs on this third pass requires that the vertex of the selected event is formed with only two tracks. This condition, together with the veto imposed on the BBC detectors should then make our sample of ρ^0 mesons as exclusively produced in UPC events.

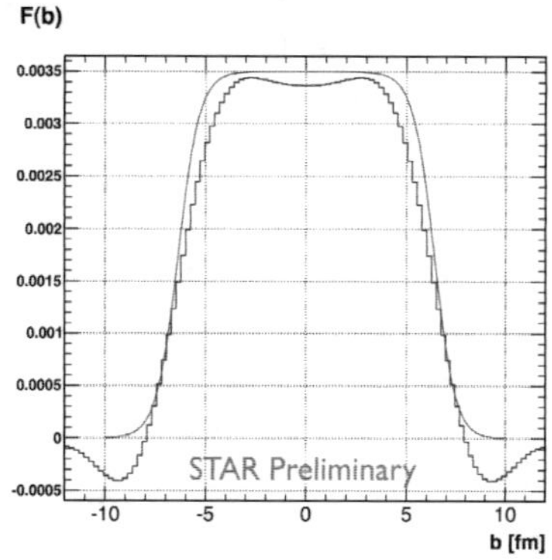

Figure 7. One dimensional Fourier transformation of the measured diffraction pattern shown a a black histogram. The Wood Saxon functional form with R=6.38 fm and $a_0 = 0.535\ fm$ is shown with a blue curve.

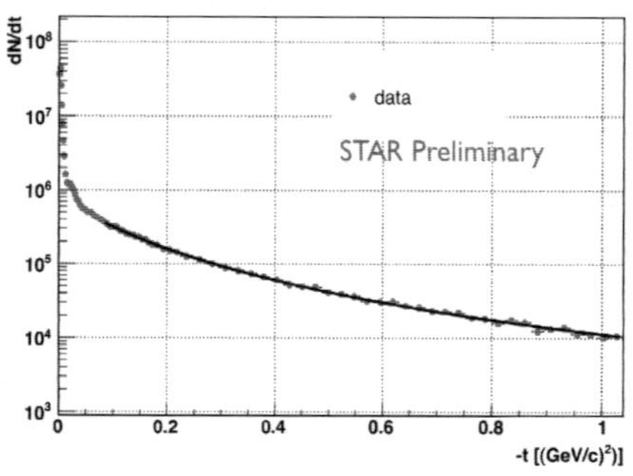

Figure 8. Background subtracted distribution of ρ mesons as function of -t shown with red markers. A fit to a power law function is shown with a black curve. No acceptance correction is applied.

The ρ meson is finally defined with cuts applied to the invariant mass distribution of the pion pairs that satified the conditions listed above. The lowest value cut on mass is set at 500 MeV and the top value of the same cut at 1 GeV.

4. The diffraction pattern

We make the important assumption that the t of the ρ^0 meson completely balances the one of the target Au ion and what we are producing is the t distribution of the recoil ion. The appropriateness of such assumption will be discussed again when we compare the measurements to model calculations.

The t Mandelstam variable for the ρ defined as $t = (P_1 - P_3)^2$ where the four-vectors P_1 and P_3 are related to the ρ meson before and after the elastic interaction with the Au ion. In the lab. frame this quantity is written as $t = t_{min} - p_T^2$ where $t_{min} = -M^2_{\pi^+\pi^-}/(2\gamma_L e^{|2y|})$ with γ_L being the Lorentz boost of the ion beam (equal to 108 at top RHIC ion energy) and y the rapidity of the ρ mesons. The value of t_{min} is small and was included for completeness.

Two histograms with variable bin size are prepared to be filled with the value of t for the exclusively produced ρ mesons detected in events with GDR excitations in both Au ions recording a single neutron in each ZDC. The bin width ranges from 0.0015 $(GeV/c)^2$ at $t \sim 0$ to 0.024 $(GeV/c)^2$ at $t = 0.3(GeV/c)^2$. The bin size was chosen to be always wider than the corresponding t resolution ($\Delta t/t = 0.01 + 0.008\sqrt{t}$). One histogram ($dN_{\rho}^{candidates}/dt$) is used to record the -t distribution of opposite sign pions pairs. The other histogram ($dN_{background}/dt$) is filled with the value of -t for pion pairs of equal charge. As mentioned above, this second histogram is considered to carry all the information about the backgroung to ρ photo-production in UPC events. Studies to characterize the validity of this choice are still under way. At every bin, the contents of both histograms are divided by the corresponding bin width to produce the actual distributions. The $dN_{background}/dt$ distribution is then subtracted bin by bin from $dN_{\rho}^{candidates}/dt$ and the statistical errors are recalculated. The resulting distribution (dN_{ρ}^{signal}/dt) is shown with red dots in Fig. 8 where the first diffractive peak is clearly visible at $-t \sim 0$, a clear hint of the second diffractive peak is also present at this point in the analysis. A long slowly falling tail is present at higher values of -t. At this point the assumption is made that the coherent component of the ρ meson scattering off the Au nuclei is negligeable and the totality of this tail is populated by incoherent interactions with individual nucleon. A fit to a "power law" shape ($A/(1 + |t|/p_0)^n$) is performed for $-t > 0.2$ $(GeV/c)^2$ This functional form reproduces well the incoherent tail ($\chi/NDF = 0.97$ $A = 7.609 \times 10^4, p_0 = 0.23 \pm 0.1$ and $n = 2.8 \pm 0.4$). The separation between coherent and incoherent contributions to the elastic scattering of the ρ meson on the Au nuclei is done statistically by subtracting the value of the fitted function at each bin center of the background subtracted distribution (dN_{ρ}^{signal}/dt). The resulting distribution is shown in Fig. 9 where all normalization and other corrections were applied.

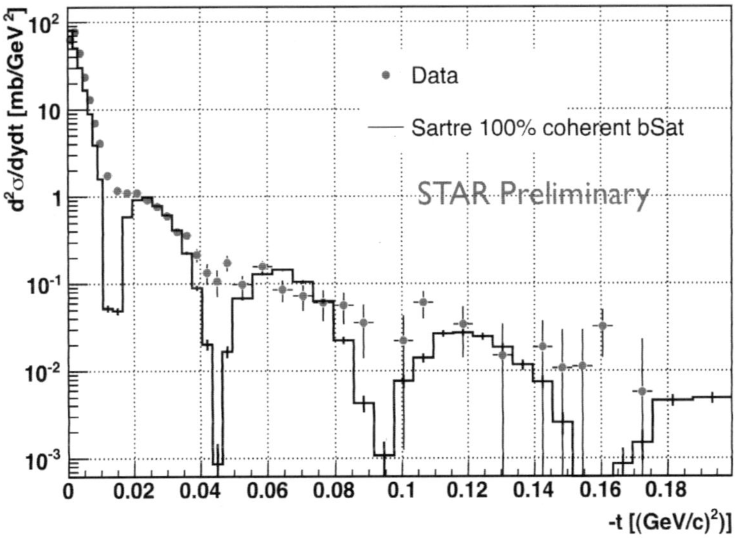

Figure 9. The diffraction pattern extracted from elastic scattering from the Au nuclei shown with red markers. The black histogram is the result of a calculation performed with Sartre.

The ρ meson reconstruction efficiency and the magnitude of the TPC acceptance are extracted making use of of STAR embedding procedures; pion pairs from the decay of one million ρ^0

mesons generated in UPC events with the StarLight [6] event generator are simulated and their trajectories and energy deposited along them are produced with a complete and up to date GEANT simulation of the STAR detector setup as it was at the time of RHIC run 10. The hits recorded in this simulation are then merged with information recorded during actual data taking in run 10 with the most unbiased trigger (zerobias). Embedding into these events produces the best description of the backgroung in UPC events; a mix of pile-up, beam gas events as well as low multiplicity hadronic interactions. The complete STAR reconstruction chain is then used to process these data and the result is presented in several formats. To extract a first version of the reconstruction efficiency and the acceptance, a summary output was used where the information about the input generated particles as well as the parameters of the reconstructed tracks that match to a generated particle. The matching between reconstructed tracks and generated particles is based on the number of common hits. Ratios between 2D histograms counting matched and generated particles were calculated for different combinations of the ρ meson variables: transverse momentum and rapidity, azimuthal angle and rapidity and -t and rapidity. The value of efficiencies$\times$acceptance as function of -t was extracted. For the study reported here the rapidity dependance is averaged and the resulting value of the correction is equal to 44% and is found to be flat in -t.

The integrated luminosity for the runs included in this analysis is equal to $679(\mu b)^{-1}$ and is based on one of the STAR minimum bias triggers (*minbias_monitor*) which is estimated to cover up to 6 barns of the total Au+Au cross section.

4.1. Comparison to model calculations

The histogram shown in Fig. 9 was produced with the Sartre event generator [9] which in turn is based on the impact parameter dependent saturation model bSat [10]. Ten million ρ mesons were generated with the model set to mimic the Au+Au Ultraperipheral collisions at RHIC. The geometry of the target Au ion is simulated in a Glauber based Monte Carlo with individual nucleons distributed according to a Wood Saxon distribution with R=6.38 fm and $a_0 = 0.535\ fm$. The cross section of the quark dipole is connected to the gluon distribution in the target nuclei and in particular, the black histogram in Fig. 9 includes the effects of saturation. The agreement between the data and the simulation is remarkable. The slopes of both first peaks are close to each other, the location and magnitude of the other diffraction peaks formed in the measurement are also well reproduced by the calculation. To first order, this agreement is strong indication that the methods used to isolate the coherent component of the ρ elastic scattering in the present analysis were correct, even though a quantification of such agreement remains to be done. A change in the di-quark cross section in the Sartre event generator which this time does not include the effects of saturation produces strong changes in the position of the deeps in the calculated diffraction pattern as is shown with the magenta histogram in Fig. 10

The Fourier transform of the measured diffraction pattern has been calculated making use of the azimuthal symmetry of the system : $f(b) = \int_0^{2\sqrt{t_{max}}} \sqrt{F(x^2)} J_0(xb/\hbar c) \frac{x}{2\pi} dx$ The result of the transformation is shown as a histogram in Fig. 7, together with the outline of the Wood-Saxon distribution drawn with a blue curve. No attempt has been made yet at assigning a systematic uncertainty to this calculation. The high b sides of the transform are related to $-t \sim 0$ in the diffraction pattern and the fact that they take negative values may be an artifact of bin size or the presence of destructive interference already measured in the same system [11]. The region close to b=0 is not well defined because the measurement has poor statistics at high -t. The falling edges of the distribution will turn out to be the best suited to extract some physics out of this measurement. Figure 11 shows the comparison between the diffraction pattern extracted from data (red markers) and the one caculated with the Starlight event generator (blue markers). This comparison shows a remarkable agreement on the depths of the diffraction deeps which

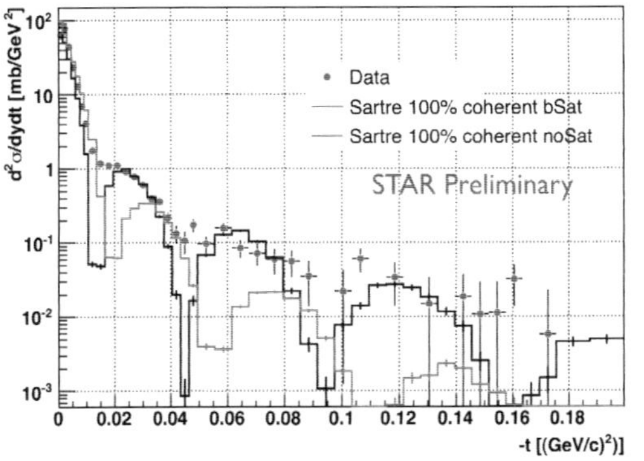

Figure 10. Diffraction patter shown with two Sartre calculations, one with the presence of saturation in the gluon distribution of the ion target shown as a black histogram, and the other were such non linear effect have been turned off, is shown with a magenta histogram.

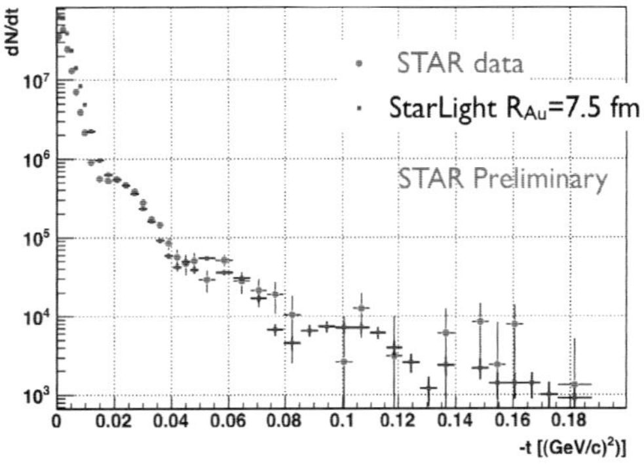

Figure 11. The unnormalized distribution for the -t value of ρ mesons generated with Starlight is shown with blue markers to compare them with the same distribution extracted from data. The comparisons highlights the good match of peaks and valleys in both cases.

within the Starlight calculation are the result of non zero transverse momentum component in the Weizsäcker-Williams photon exchanged between the ions [12]. A similar calculation performed for UPC events at 130 GeV produced shallow diffraction deeps [13]. Such facts make our original assumption of purelly longitudinal photons too simplistic but it does not preclude the possible unfolding of the photons transverse momentum contribution from our present results.

5. Summary

The STAR UPC program has measured coherent elastic scattering off Au nuclei with beams of exclusively produced ρ^0 mesons in Ultra Peripheral Collisions of Au ions at $\sqrt{s_{\rm NN}}= 200$ GeV. The diffraction pattern extracted using the p_T of the scattered meson may not be an exact match to the one measured with the recoiling target ion, but such goal appears attainable with the help of a QED calculation to unfold the transverse contribution from the Weizsäcker-Williams photons. This measurement opens up a possible program to study the gluon form factor of nuclei, albeit with a restricted reach in Q^2.

References

[1] Bertulani C *et al.* 2005 *Ann. Rev. Nucl. Part. Phys.* **55** 271
[2] Bulos F *et al.* 1969 *Phys. Rev. Lett.* **22** 490
[3] Alvensleben H *et al.* 1970 *Phys. Rev. Lett.* **24** 786
[4] Anderson M *et al.* 2003 *Nuclear Instrum. Methods* **A499** 659
[5] Adler C, Strobele H, Denisov A, Garcia E, Murray M *et al.* 2001 *Nucl.Instrum.Meth.* **A461** 337–340
[6] Baltz A Klein S and Nystrand J 2002 *Phys. Rev. Lett.* **89** 012301
[7] Adler C *et al.* 2002 *Phys. Rev. Lett.* **89** 272302
[8] Abelev B I *et al.* (STAR) 2008 *Phys. Rev.* **C77** 034910 (*Preprint* 0712.3320)
[9] Toll T and Ulrich T 2011 vol INT-PUB-11-034 p 385 (*Preprint* nucl-th/1108.1713)
[10] Kowalski H and Teaney D 2003 *Phys.Rev.* **D68** 114005 (*Preprint* hep-ph/0304189)
[11] Abelev B *et al.* 2009 *Phys. Rev. Lett.* **102** 112301
[12] Klein S and Nystrand J 2000 *Phys. Rev. Lett.* **84** 2330
[13] Frankfurt L, Strikman M and Zhalov M 2003 *Phys. Rev.* **C67** 034901

AUTHOR INDEX